International
REVIEW OF
Neurobiology
Volume 77

International
REVIEW OF
Neurobiology
Volume 77

The Pharmacology of Neurogenesis and Neuroenhancement

EDITED BY

DONARD S. DWYER
Louisiana State University, Shreveport
Louisiana, USA

ELSEVIER

AMSTERDAM • BOSTON • HEIDELBERG • LONDON
NEW YORK • OXFORD • PARIS • SAN DIEGO
SAN FRANCISCO • SINGAPORE • SYDNEY • TOKYO
Academic Press is an imprint of Elsevier

Academic Press is an imprint of Elsevier
525 B Street, Suite 1900, San Diego, California 92101-4495, USA
84 Theobald's Road, London WC1X 8RR, UK

This book is printed on acid-free paper. ♾

For information on all Academic Press publications
visit our Web site at www.books.elsevier.com

ISBN-13: 978-0-12-373678-9
ISBN-10: 0-12-373678-1

Printed and bound in the United Kingdom

Transferred to Digital Printing, 2011

CONTENTS

Contributors ix
Preface xi

Regenerating the Brain

David A. Greenberg and Kunlin Jin

I. Introduction 1
II. Neurodegenerative Diseases: Therapeutic Targets for Neurogenesis 2
III. Building Brains: Evolution and Development 3
IV. Endogenous and Exogenous Sources of New Neurons 4
V. Endogenous Adult Neurogenesis 5
VI. Injury-Induced Neurogenesis 6
VII. Neurogenesis in Stroke 8
VIII. Neurogenesis in AD 14
IX. Neurogenesis in HD 15
X. Neurogenesis in PD 16
XI. Neurogenesis in Motor Neuron Disease 19
XII. Convergent and Divergent Features of Injury-Induced Neurogenesis 19
XIII. Unanswered Questions 20
References 23

Serotonin and Brain: Evolution, Neuroplasticity, and Homeostasis

Efrain C. Azmitia

I. Introduction 32
II. Evolution: From Unicellular to Humans 33
III. Holistic Brain Function Starts at Development 38
IV. Homeostasis of Brain 42
V. Clinical Implications of Loss of Homeostasis 47
References 49

Therapeutic Approaches to Promoting Axonal Regeneration in the Adult Mammalian Spinal Cord

Sari S. Hannila, Mustafa M. Siddiq, and Marie T. Filbin

I. Introduction 58
II. Growth-Inhibiting Properties of the Glial Scar 60
III. Inhibition of Axonal Regeneration by CNS Myelin 61
IV. Cell-Based Strategies to Promote Regeneration 73
V. Future Directions 90
References 91

Evidence for Neuroprotective Effects of Antipsychotic Drugs: Implications for the Pathophysiology and Treatment of Schizophrenia

Xin-Min Li and Haiyun Xu

I. Introduction 108
II. Altered Levels of NTs and Their Receptors by Antipsychotic Drugs 109
III. The Regulation of Neurogenesis by Antipsychotic Drugs 114
IV. Neuroprotective Effects of Antipsychotic Drugs 118
References 130

Neurogenesis and Neuroenhancement in the Pathophysiology and Treatment of Bipolar Disorder

Robert J. Schloesser, Guang Chen, and Husseini K. Manji

I. Introduction 144
II. Clinical Evidence for a Role of Neuroplasticity in BPD 145
III. Preclinical *In Vivo* and *In Vitro* Studies 150
IV. Conclusions 164
References 165

Neuroreplacement, Growth Factor, and Small Molecule Neurotrophic Approaches for Treating Parkinson's Disease

Michael J. O'Neill, Marcus J. Messenger, Viktor Lakics, Tracey K. Murray, Eric H. Karran, Philip G. Szekeres, Eric S. Nisenbaum, and Kalpana M. Merchant

I. Introduction 180
II. Etiology of PD 182
III. Current Treatment for PD 184
IV. Rodent Models of PD Used to Evaluate Putative Neuroprotective Therapies 186

V. Approaches to Brain Repair in PD 187
VI. Summary and Conclusions 203
References 204

Using *Caenorhabditis elegans* Models of Neurodegenerative Disease to Identify Neuroprotective Strategies

BRIAN KRAEMER AND GERARD D. SCHELLENBERG

I. Introduction 220
II. *C. elegans* Models of Neurodegenerative Disorders 220
III. Identifying Candidate Modifier Genes of Neurodegeneration 238
IV. Molecular Inhibitors of Neurodegeneration 239
V. Neuroprotective Strategies 240
VI. Utility of HTS for Drugs in Worms 241
VII. Conclusions 242
References 242

Neuroprotection and Enhancement of Neurite Outgrowth with Small Molecular Weight Compounds from Screens of Chemical Libraries

DONARD S. DWYER AND ADDIE DICKSON

I. Introduction 248
II. Brain Deficit Disorders: The Challenges 249
III. Treatment Objectives 250
IV. Application to Neurodevelopmental Disorders: Schizophrenia 254
V. Screening and Identification of Novel Compounds for Drug Discovery 272
VI. Neurodevelopmental Disorders and Neurodegenerative Diseases 274
VII. Future Directions 277
References 278

INDEX 291
CONTENTS OF RECENT VOLUMES 311

CONTRIBUTORS

Numbers in parentheses indicate the pages on which the authors' contributions begin.

Efrain C. Azmitia (31), Department of Biology and Psychiatry, Center for Neural Science, New York University, New York 10003, USA

Guang Chen (143), Laboratory of Molecular Pathophysiology, Mood and Anxiety Disorders Research Program, National Institute of Mental Health, Bethesda, Maryland 20892, USA

Addie Dickson (247), Department of Psychiatry, Louisiana State University Health Sciences Center, Shreveport, Louisiana 71130, USA

Donard S. Dwyer (247), Department of Psychiatry, Louisiana State University Health Sciences Center, Shreveport, Louisiana 71130, USA

Marie T. Filbin (57), Department of Biological Sciences, Hunter College, City University of New York, New York 10021, USA

David A. Greenberg (1), Buck Institute for Age Research, Novato, California 94945, USA

Sari S. Hannila (57), Department of Biological Sciences, Hunter College, City University of New York, New York 10021, USA

Kunlin Jin (1), Buck Institute for Age Research, Novato, California 94945, USA

Eric H. Karran (179), Eli Lilly and Co. Ltd., Lilly Research Centre, Erl Wood Manor, Windlesham, Surrey GU20 6PH, United Kingdom

Brian Kraemer (219), Geriatric Research Education and Clinical Center, Veterans Affairs Puget Sound Health Care System, Seattle Division, Seattle, Washington 98108, USA; Department of Medicine, Division of Gerontology and Geriatric Medicine, University of Washington, Seattle, Washington 98195, USA

Viktor Lakics (179), Eli Lilly and Co. Ltd., Lilly Research Centre, Erl Wood Manor, Windlesham, Surrey GU20 6PH, United Kingdom

Xin-Min Li (107), Neuropsychiatry Research Unit, Department of Psychiatry, University of Saskatchewan, Saskatoon, SK, S7N 5E4, Canada

Husseini K. Manji (143), Laboratory of Molecular Pathophysiology, Mood and Anxiety Disorders Research Program, National Institute of Mental Health, Bethesda, Maryland 20892, USA

Kalpana M. Merchant (179), Eli Lilly and Co. Ltd., Lilly Corporate Center, Indianapolis, Indiana 46285, USA

Marcus J. Messenger (179), Eli Lilly and Co. Ltd., Lilly Research Centre, Erl Wood Manor, Windlesham, Surrey GU20 6PH, United Kingdom

Tracey K. Murray (179), Eli Lilly and Co. Ltd., Lilly Research Centre, Erl Wood Manor, Windlesham, Surrey GU20 6PH, United Kingdom

Eric S. Nisenbaum (179), Eli Lilly and Co. Ltd., Lilly Corporate Center, Indianapolis, Indiana 46285, USA

Michael J. O'Neill (179), Eli Lilly and Co. Ltd., Lilly Research Centre, Erl Wood Manor, Windlesham, Surrey GU20 6PH, United Kingdom

Gerard D. Schellenberg (219), Geriatric Research Education and Clinical Center, Veterans Affairs Puget Sound Health Care System, Seattle Division, Seattle, Washington 98108, USA; Department of Medicine, Division of Gerontology and Geriatric Medicine, University of Washington, Seattle, Washington 98195, USA; Department of Neurology, University of Washington, Seattle Washington 98195, USA; Department of Pharmacology, University of Washington, Seattle Washington 98195, USA

Robert J. Schloesser (143), Laboratory of Molecular Pathophysiology, Mood and Anxiety Disorders Research Program, National Institute of Mental Health, Bethesda, Maryland 20892, USA

Mustafa M. Siddiq (57), Department of Biological Sciences, Hunter College, City University of New York, New York 10021, USA

Philip G. Szekeres (179), Eli Lilly and Co. Ltd., Lilly Research Centre, Erl Wood Manor, Windlesham, Surrey GU20 6PH, United Kingdom

Haiyun Xu (107), Neuropsychiatry Research Unit, Department of Psychiatry, University of Saskatchewan, Saskatoon, SK, S7N 5E4, Canada

PREFACE

Recent advances in psychiatry and neurology have led to the realization that many brain disorders result from neurodevelopmental insults or neurodegenerative processes at the regional and cellular levels. Nevertheless, the majority of drugs used to treat these conditions primarily alter brain neurochemistry rather than the underlying deficits in cells and circuitry. In addition, there is now greater appreciation for the fact that plasticity in the adult brain includes not only the remodeling of synaptic connections and dendritic morphology, but also the birth and integration of new neurons—a process known as neurogenesis. The important role of neurogenesis during development is long established. The possibility that continuous generation of new neurons in adults contributes to normal brain function may offer novel therapeutic strategies. In fact, emerging evidence suggests that some of the clinical improvement observed in patients treated with antidepressant and other psychotropic drugs is due to the induction of neurogenesis by these drugs. A less dramatic therapeutic goal would be to optimize neuronal energy metabolism, promote cell survival, and enhance neuronal function (e.g., neurite outgrowth and synaptic activity). This overall strategy has previously been referred to as neuroenhancement. The idea is to maximize functional activity in defective neurons, and to prevent the loss of cells and/or neurites and dendrites in degenerative diseases or following brain injury.

The aim of this volume is to review the cutting-edge research related to pharmacological aspects of neurogenesis and neuroenhancement. The first two chapters are more general in nature and set the stage for the subsequent reviews. Greenberg and Jin begin with a comprehensive overview of neurogenesis, including the role of this process in evolution and development as well as neurodegenerative diseases and stroke. The article concludes with a series of thought-provoking questions and issues that must be resolved by future studies. The chapter by Azmitia introduces the important relationship between neuroplasticity and homeostasis and the role of serotonin in maintaining dynamic interactions in the brain. The trophic actions of serotonin in the brain offer encouragement that small-molecule approaches to neuroenhancement should be possible. The next three chapters deal with the use of small molecules (e.g., cAMP) or drugs to promote regeneration (Hannila *et al.*), survival (Li and Xu), and plasticity and neurogenesis (Schloesser *et al.*). More specifically, Hannila *et al.* discuss various factors that control the regeneration of axons in the CNS, the effects of cAMP and the antidepressant, rolipram, on axon regrowth, and the latest findings concerning cell-based strategies for regeneration. The chapter by Li and Xu is an excellent summary of the effects of antipsychotic drugs on neuroprotection and neurogenesis. This includes insight into mechanisms of action and implications for the treatment of psychotic disorders. Schloesser *et al.* review recent studies showing that drugs used to treat bipolar disorder stimulate neurogenesis and enhance neuronal function, which may contribute to their therapeutic benefits. The chapter includes an illuminating discussion of possible targets and cellular/molecular mechanisms involved in the response to mood stabilizers. The final three chapters examine present and future efforts to discover new drugs for neurodegenerative and neurodevelopmental disorders based on neuroprotection and neuroenhancement. O'Neill *et al.* report exciting progress in the discovery of new drugs for the treatment of Parkinson's disease. Their chapter describes the potential of trophic factors and small molecules that modulate trophic factor activity (e.g., AMPA receptor potentiators) to provide protection in animal models of Parkinson's disease. Kraemer and Schellenberg review promising efforts to establish the model organism, *Caenorhabditis elegans*, as a tool for drug discovery in neurodegenerative disease. A goal of this research is to use mutant strains undergoing neurodegeneration to identify neuroprotective drugs in large-scale screening of compound libraries. Finally, Dwyer and Dickson describe the application of structure-based drug design to the discovery of novel drugs that produce neuroenhancement. Drugs from this effort may be useful in the treatment of neurodevelopmental disorders, including schizophrenia, Rett syndrome, and autism.

I wish to thank Elsevier Press for their support and encouragement for this project and, in particular, Cindy Minor and Rogue Schindler for all their fine efforts. I would also like to express my gratitude to the various authors and coauthors of the chapters. Their hard work, creativity, and steadfast dedication to research that challenges prevailing concepts and dogma made this volume possible. Collectively, we acknowledge other researchers in the field whose work was cited here and who have made significant contributions to this area of investigation. We all wish to thank the readers of this volume for their interest in our work and desire to learn more about the field. It is our hope that the ideas presented here will stimulate further discussion and future investigation of the therapeutic potential of neurogenesis and neuroenhancement.

REGENERATING THE BRAIN

David A. Greenberg and Kunlin Jin

Buck Institute for Age Research, Novato, California 94945, USA

I. Introduction
II. Neurodegenerative Diseases: Therapeutic Targets for Neurogenesis
III. Building Brains: Evolution and Development
IV. Endogenous and Exogenous Sources of New Neurons
V. Endogenous Adult Neurogenesis
VI. Injury-Induced Neurogenesis
VII. Neurogenesis in Stroke
VIII. Neurogenesis in AD
IX. Neurogenesis in HD
X. Neurogenesis in PD
XI. Neurogenesis in Motor Neuron Disease
XII. Convergent and Divergent Features of Injury-Induced Neurogenesis
XIII. Unanswered Questions
References

Recent discoveries related to adult neurogenesis suggest that the long-sought goal of regenerating injured adult brain tissue may be achievable in principle. A variety of human neurological diseases and rodent models of these diseases—including stroke, Alzheimer's disease, Huntington's disease, Parkinson's disease, and amyotrophic lateral sclerosis—are associated with an increase in the brain's neuroproliferative capacity. In some cases, this capacity can be enhanced further by growth factors or drugs. Therefore, therapeutic manipulation of endogenous neurogenesis may be feasible. In addition to its potential clinical utility, the study of injury-induced adult neurogenesis is helping to reveal mechanisms of neuronal proliferation, migration, and differentiation that may also operate during normal development.

I. Introduction

The regeneration of human tissue has been a theme since antiquity. Osiris, the Egyptian god of earth and vegetation, was murdered by his brother Seth, and Osiris' body was thrown into the Nile. When Osiris' sister Isis recovered the body, Seth chopped it into 14 parts, which he scattered. However, Isis searched for and found these parts, which she reassembled to resurrect Osiris. In Greek mythology, the titan

INTERNATIONAL REVIEW OF
NEUROBIOLOGY, VOL. 77
DOI: 10.1016/S0074-7742(06)77001-5

0074-7742/07 $35.00

Prometheus stole fire from the gods on Mount Olympus and gave it to man. For this and other offenses, Zeus ordered that Prometheus be chained to a rock, where the eagle Ethon would eat out his liver; the liver grew back each day, only to be eaten again.

Tissue regeneration is a well-known phenomenon to the zoologist—as exemplified by the ability of creatures like sponges, hydra, flatworms, annelids, sea stars, newts, and salamanders to regrow body parts. It is also recognized in clinical medicine, where it occurs in the setting of wound healing. In some organ systems of vertebrates, however, the conventional wisdom has been that little or no regeneration can take place, and this is perhaps nowhere more commonly held to be the case than in the central nervous system (CNS).

In the 1960s and 1970s, Joseph Altman and colleagues at Purdue University mapped the development of the rodent brain, using [^{3}H]thymidine to birthdate neurons. They found that in some brain regions, the birth of new neurons could be observed well into adulthood. However, this insight seems to have receded from attention until relatively recently, when it provided the basis for a new field of investigation—that of adult neurogenesis.

What makes adult neurogenesis more than just a biological oddity akin to the regrowth of a sea star's arms is the possibility it suggests for therapeutic application. The CNS is notoriously difficult to repair, and injuries or diseases that affect it often dramatically change a patient's quality of life. One has only to see a quadriplegic victim of spinal cord injury, an individual with profound amnesia due to alcoholic Korsakoff's syndrome, or a survivor of stroke who can no longer comprehend or produce language, to be impressed with the extent to which brain damage can depreciate life.

A substantial body of evidence indicates that cerebral injury of several types can stimulate neurogenesis in the adult brain. The most compelling outstanding issue regarding the potential clinical importance of this phenomenon is to determine if it can generate functional neurons that contribute to enhanced recovery.

II. Neurodegenerative Diseases: Therapeutic Targets for Neurogenesis

Acute and chronic neurodegenerative diseases are common, disabling, and poorly responsive to current treatment. Stroke, the most frequent cause of acute neurodegeneration, has a prevalence of ~4.8 million and an incidence of ~700,000 individuals per year in the United States, where it is the third leading cause of death (American Heart Association, 2004). Even among those who survive stroke, disability due to hemiparesis, gait disorders, aphasia, and other syndromes is common, and ~20% of these patients require institutional care at 6-months poststroke. This long-term disability contributes to the average lifetime cost for stroke care of ~\$140,000 and an annual national cost of ~\$54 billion. The only major recent advance in treatment, the use of thrombolytic agents to dissolve clots in the acute aftermath of stroke, has had limited impact because it appears to be effective only within about the first 3 hours after onset of symptoms (Brott and Bogousslavsky, 2000).

Chronic neurodegenerations include Alzheimer's disease (AD), Parkinson's disease (PD), hereditary polyglutamine disorders like Huntington's disease (HD), and amyotrophic lateral sclerosis (ALS). These diseases affect different (albeit overlapping) regions of the CNS and vary in prevalence, from ~4.5 million cases in AD to ~1.5 million cases in PD, and ~30,000 cases in HD and ALS in the United States. However, all typically culminate in an extended period of functional disability preceding death. Except for PD, in which drugs and surgery are available to reduce symptoms, at least temporarily, even symptomatic treatment for chronic neurodegenerations is extremely limited at present. In AD, acetylcholinesterase inhibitors and the *N*-methyl-*D*-aspartate (NMDA)-type glutamate receptor antagonist memantine exert modest behavioral effects in some patients (Cummings, 2004). In ALS, another NMDA antagonist, riluzole, has provided minimal benefit (Rowland and Shneider, 2001). Perhaps most notably, no treatment exists for any of these diseases that can restore lost function.

Impaired brain function in acute and chronic neurodegenerative diseases results primarily from cell (especially neuronal) loss. One reason for the limited responsiveness of neurodegenerative diseases to treatment may be that it is more difficult to overcome the loss of cells than the impairment of selected cellular functions. As an example, among neurological disorders, the greatest therapeutic successes have come in conditions where cell loss is not a major feature, such as epilepsy and migraine. Even in PD, where cell loss is relatively circumscribed, pharmacological restoration of a key cellular function like dopaminergic neurotransmission, without the temporal, spatial, and stimulus-coupled regulation that a cellular context provides, has been an imperfect stratagem.

Based on this experience, it is reasonable to conclude that cell-replacement therapy, technically challenging though it may be, is worth pursuing. In addition to the prospect of more completely restoring brain function, cell-replacement therapy has the further advantage that it might be effective at later stages of a disease. This is an important consideration not only in disorders like stroke, which often evolve too quickly for acute treatment to be instituted, but also in chronic neurodegenerations, where cell loss is already extensive before the onset of symptoms.

III. Building Brains: Evolution and Development

Evidence for the feasibility of cell replacement in the brain and principles to guide cell-replacement research come from several sources, including evolution and development. The challenge of cell replacement for neurodegenerative diseases is, in simple terms, to (re)build the brain. This is a task that is faced in one form or another: (1) in evolution, as brain size increases and (2) in ontogeny, as the brain develops from the neural tube.

As larger brains evolved, they appear to have done so primarily through an increase in neuron number, rather than, for example, neuron size or proportional connectivity (Streidter, 2005). This suggests that supplying new cells might also be the principal requirement for brain rebuilding. The evolutionary principle of epigenetic population matching suggests that trophic influences of surviving brain cells may help direct new neurons to reestablish appropriate connections. A related concept, the parcellation hypothesis, predicts a mechanism for pruning of exuberant axonal connections to help restore normal patterns of circuitry. Finally, the phenomenon of connectional invasion presages a capacity for restoring connections over an altered neuronal landscape and, perhaps, forming alternative, compensatory circuitry.

Development is the most extensively employed archetype for studying adult neurogenesis, providing voluminous information about mechanisms and patterns of neuronal proliferation, migration, differentiation, and settling (Bayer and Altman, 1995). For example, molecular mechanisms of trophic factor stimulation, cell cycle regulation, programmed cell death, and neurodifferentiation, as well as pathways for the migration of newborn neurons, appear to be highly conserved between ontogeny and adult neurogenesis. These observations do not imply that principles guiding evolution or development are necessarily transferable to regeneration, only that they offer biological precedents that may be useful starting points for investigation.

IV. Endogenous and Exogenous Sources of New Neurons

At least two sources of cells for neuronal replacement after neurodegeneration can be envisioned: (1) cells mobilized from within the affected individual and (2) cells obtained from an exogenous source, or donor, and transplanted into a recipient. In either case, the stage of differentiation of the cells employed could vary as well, from pluripotent, self-renewing stems cells to more developmentally restricted progenitor or precursor cells. The possibility that endogenous cells might be available to achieve cell replacement is based on the now well-established principle that new cells continue to be produced throughout life, not only in organs (such as bone marrow, skin, and intestines) where cell turnover is rapid, but also in tissues like the brain, where cell turnover is more limited.

Endogenous and exogenous replacement of brain cells both have theoretical advantages and disadvantages. Endogenous cell replacement is inherently less invasive, circumvents immunologic compatibility problems, and makes maximal use of endogenous mechanisms that direct cell proliferation, survival, differentiation, migration, settling, and functional integration. With exogenous replacement strategies, on the other hand, larger numbers of cells can be obtained, many sources of cells can be used (e.g., embryonic vs adult stem cells), and the state of precursor cell differentiation can be optimized *ex vivo* prior to transplantation into the recipient.

V. Endogenous Adult Neurogenesis

The need to produce new cells continues beyond the primary period of development, as cells succumb to use or injury, and must be replaced. This is accomplished through the activity of adult stem cells, which preferentially reconstitute cells of the tissues in which they reside. The best-known examples are found in organs, such as bone marrow, skin, and intestine, where cell turnover is rapid and continues throughout life. However, new cells are also produced throughout life in tissues like the brain, where cell turnover is more limited.

Proliferating neuronal precursors can be identified by labeling with [^{3}H] thymidine or bromodeoxyuridine (BrdU), transfection with viral vectors, or immunoreactivity for proliferation markers such as proliferating cell nuclear antigen (PCNA). As newborn neurons mature, they express successive waves of developmentally regulated proteins, including polysialylated (embryonic) nerve cell adhesion molecule (ENCAM), neuronal differentiation antigen NeuroD, βIII-tubulin, and Hu. As they migrate to their ultimate destinations, they can be identified by antibodies for doublecortin (DCX), and eventually NeuN and MAP-2. In rodents, adult neurogenesis occurs primarily in two brain regions—the subventricular zone (SVZ), especially it is in most rostral extent along the anterior horns of the lateral ventricles (Kirschenbaum and Goldman, 1995; Lois and Alvarez-Buylla, 1993; Luskin, 1993), and the subgranular zone (SGZ) of the hippocampal dentate gyrus (DG) (Altman, 1963). Neurons arising in the SVZ migrate primarily along the rostral migratory stream (RMS) to the olfactory bulb (OB), where they replace granule and periglomerular cells, although the human SVZ may not have this capacity (Sanai *et al.*, 2004). Alternative pathways for migration from the adult SVZ, such as the lateral cortical stream (Nguyen-Ba-Charvet *et al.*, 2004) and ventral migratory mass (De Marchis *et al.*, 2004), have also been described. Neurons arising in the SGZ migrate into the adjacent DG granule cell layer (GCL). Although its physiological role is incompletely understood, adult neurogenesis has the capacity to generate functional neurons (Song *et al.*, 2002; van Praag *et al.*, 2002), which may help to replace cells lost to physiological cell death. Some reports suggest that additional brain regions may also generate new neurons in the adult brain (Gould *et al.*, 1999; Lie *et al.*, 2002;

Markakis *et al.*, 2004), but the extent to which this occurs under physiological conditions in primates is controversial (Kornack and Rakic, 2001).

Neurogenesis undergoes physiological regulation by glucocorticoids, sex hormones, growth factors, excitatory neurotransmission, learning, and stress (Cameron *et al.*, 1995, 1998; Horner and Gage, 2000), and can be stimulated by drugs, including lithium, antidepressants, antipsychotics, NMDA antagonists, phosphodiesterase inhibitors, anti-inflammatories, and statins. Neuronal precursor cells can be cultured *in vitro* and several growth factors stimulate neurogenesis in such systems, including epidermal growth factor (EGF) (Reynolds and Weiss, 1992), basic fibroblast growth factor-2 (FGF-2) (Ray *et al.*, 1993), brain-derived neurotrophic factor (BDNF) (Kirschenbaum and Goldman, 1995), erythropoietin (EPO) (Shingo *et al.*, 2001), stem-cell factor (SCF) (Jin *et al.*, 2002a), heparin-binding EGF-like growth factor (HB-EGF) (Jin *et al.*, 2002b), and vascular endothelial growth factor (VEGF) (Jin *et al.*, 2002c; Sun *et al.*, 2003). In addition, some studies have shown that cultured progenitor cells (Kilpatrick and Bartlett, 1993; Taupin *et al.*, 2000; Temple, 1989) or tissue explants containing axons that project to neuroproliferative zones (Dehay *et al.*, 2001) release factors into conditioned medium that can regulate neurogenesis. Administration or overexpression of EGF (Kuhn *et al.*, 1997), FGF-2 (Kuhn *et al.*, 1997; Wagner *et al.*, 1999), insulin-like growth factor-I (IGF-I) (O'Kusky *et al.*, 2000), BDNF (Benraiss *et al.*, 2001; Pencea *et al.*, 2001; Zigova *et al.*, 1998), SCF (Jin *et al.*, 2002a), HB-EGF (Jin *et al.*, 2002b), or VEGF (Jin *et al.*, 2002c; Sun *et al.*, 2003) has also been shown to enhance neurogenesis in neuroproliferative zones of the adult brain *in vivo*.

VI. Injury-Induced Neurogenesis

Pathological processes can also stimulate neurogenesis in the brain (Gould and Tanapat, 1997; Snyder *et al.*, 1997), and appear in some cases to redirect the migration of nascent neurons from normal routes like the RMS and toward the site of pathology. For example, apoptotic degeneration of corticothalamic neurons in mice is followed by the restoration of corticothalamic connections, and appears to involve neurogenesis, because the cells that make these connections can be labeled with the cell-proliferation marker BrdU and express immature neuronal markers such as DCX and Hu (Magavi *et al.*, 2000). Similarly, seizure-induced brain injury resulting from status epilepticus in the rat both stimulates neurogenesis and diverts neuronal precursors from the RMS and into the affected forebrain (Parent *et al.*, 2002a). Injury-induced neurogenesis, which has been observed in excitotoxic damage (Gould and Tanapat, 1997; Yoshimura *et al.*, 2001), seizures (Parent *et al.*, 1997), oxidative stress-induced apoptosis

(Magavi *et al.*, 2000), and global (Liu *et al.*, 1998; Takagi *et al.*, 1999) or focal (Gu *et al.*, 2000; Jin *et al.*, 2001) cerebral ischemia, may contribute to CNS recovery and repair.

One of the most striking features of injury-induced neurogenesis is its ability to redirect migrating neurons away from their normal transit (e.g., from SVZ to OB) and into the region of injury. This is observed in epilepsy (Parent *et al.*, 2002a) as well as in cerebral ischemia (Arvidsson *et al.*, 2002; Jin *et al.*, 2003; Nakatomi *et al.*, 2002; Parent *et al.*, 2002b). How migration is redirected by injury is unknown, but the altered migration of SVZ precursors into the ischemic cerebral cortex via the lateral cortical stream is reminiscent of the partial redirection that occurs in *Slit1* knockout mice, which lack normal Slit/Robo chemorepulsive signaling (Nguyen-Ba-Charvet *et al.*, 2004).

Whether different forms of cerebral injury trigger neurogenesis through the same or different mediators is unknown. Even the role of cell death in stimulating neurogenesis is uncertain—on one hand, if neurons proliferate to replace cells that are lost through injury or disease, cell loss might be a prerequisite for neurogenesis. On the other hand, neurogenesis can be increased by seizures, which do not necessarily kill cells, by sublethal forebrain ischemia, and by physiological stimuli such as exercise, so cell death may not be a *sine qua non* for injury-induced neurogenesis. Neurogenesis induced by injury in the absence of cell death cannot *replace* cells, but it might have other functions. For example, these new neurons could provide surviving but damaged neurons with trophic factors, or set up parallel neuronal connections to bypass or supplement malfunctioning circuits.

In addition to general attributes shared by injury-induced neurogenesis from diverse causes, unique features of neurogenesis are also likely to arise in different diseases. This makes finding a universally applicable approach to therapeutic neurogenesis unlikely, but may facilitate tailoring therapy to specific disease contexts.

As illustrated in Table I, neurodegenerative diseases exhibit differences in time course, tissue distribution, cellular vulnerability, and etiology. Understanding how injury stimulates neurogenesis in each of these models will provide information about the temporal, spatial, cytopathological, and pathophysiological requirements for eliciting neurogenesis in response to brain injury.

The intrinsic differences among the diseases in question also dictate that the approach to cell-replacement therapy is likely to require modification depending on the disease being targeted. For example, the new cells produced will need to be directed toward different phenotypic fates and regional destinations if they are to have a functional impact.

An important question that has received little attention is whether endogenous neural precursor cells that might be mobilized for brain repair are themselves affected by the diseases against which they are to be targeted. For example,

TABLE I
DISTINGUISHING FEATURES OF SELECTED ACUTE AND CHRONIC NEURODEGENERATIONS

Disease	Time course	Distribution	Cells affected	Etiology
Stroke	Acute	Focal, unilateral; vascular territory	Neurons, glia, endothelium	Ischemic
Alzheimer's	Chronic	Diffuse; cortical	Neurons	Sporadic > genetic
Parkinson's	Chronic	Focal, bilateral; nigrostriatal tract	Dopaminergic nigrostriatal neurons	Sporadic > genetic
Huntington's	Chronic	Focal, bilateral; striatum and cortex	Medium spiny neurons	Genetic
Motor neuron disease	Chronic	Focal, bilateral; upper, and lower motor neurons	Anterior horn cells	Sporadic > genetic

such cells might be functionally impaired *ab initio* or destined for an early death. If this were the case, it might suggest that in these diseases, approaches employing exogenous sources of replacement cells will be preferable.

VII. Neurogenesis in Stroke

Stroke results from interruption of blood flow to a region of the brain (cerebral ischemia), usually due to thrombotic or embolic arterial occlusion, producing localized cell death and corresponding loss of brain function. Stroke is the third leading cause of death in the United States, with an incidence of ~500,000 events per year and a prevalence of about 4 million individuals. Brain pathology in stroke is characterized by focal necrosis, with loss of neurons, glia, and endothelium. The pathophysiology of stroke is incompletely understood, but excitotoxicity, intracellular calcium overload, and loss of energy and ionic homeostasis have been invoked, leading to a combination of necrotic and programmed cell death. Current treatment options are limited, consisting for the most part of preventive measures, such as antiplatelet agents and anticoagulants, and acute thrombolysis, for which only a small percentage of patients are candidates because the time window of therapeutic opportunity is only about 3 hours. Animal models of stroke, such as intravascular suture occlusion of the middle cerebral artery, can recapitulate the pathophysiological and histopathological features of stroke, although they have been less successful in predicting the success of experimental therapeutics.

In addition to intracellular adaptations that promote the survival of ischemic cells, such as increased expression of anti-apoptotic proteins, the brain may also

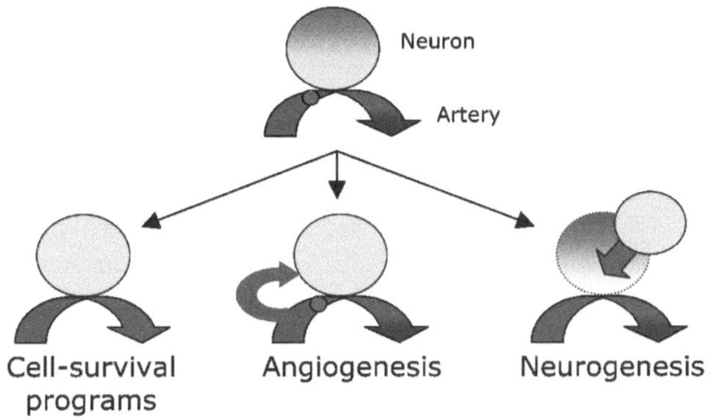

FIG. 1. Mechanisms that may be employed to survive or recover from cerebral ischemia include mobilization of cell-survival (e.g., anti-apoptotic) programs, recruitment of a new vascular supply via angiogenesis, and replacement of dead neurons through neurogenesis. (See Color Insert.)

respond to ischemia with increased production of cells that can restore blood supply (angiogenesis) or replace dead or damaged neurons (neurogenesis) (Fig. 1), or glia (gliogenesis). Both of these cell types are lost after focal cerebral ischemia of sufficient severity to produce infarction, and both also continue to be produced in the adult mammalian brain throughout life.

Early studies on injury-induced neurogenesis showed that excitotoxic or mechanical injury to the DG GCL in the rat increased proliferation of granule neuron precursors in the adjacent SGZ, as shown by enhanced incorporation of [^{3}H]thymidine and BrdU, induction of PCNA, and coexpression of mature neuronal markers in [^{3}H]thymidine-labeled cells (Gould and Tanapat, 1997). Seizures were also shown to trigger neurogenesis in the SGZ, with BrdU-labeled cells expressing the neuronal markers turned on after division, 64 kDa (TOAD-64), βIII-tubulin, and microtubule-associated protein 2 (MAP-2) (Parent *et al.*, 1997). Moreover, apoptosis induced by oxidative stress in mouse corticothalamic neurons increased BrdU labeling in cells that went on to express the neuronal markers DCX, Hu, and NeuN (Magavi *et al.*, 2000). These observations suggested that other forms of injury, including ischemia, might similarly elicit increased production of brain neurons.

Several studies were undertaken to test this possibility. In global cerebral ischemia in the gerbil, neurogenesis was increased in the SGZ (Liu *et al.*, 1998), with enhanced BrdU labeling of cells coexpressing the neuronal markers NeuN, MAP-2, and calbindin. These cells migrated into the GCL, where they took on phenotypic attributes of mature neurons. Global ischemia also enhanced proliferation of BrdU-labeled neuronal precursors in mouse DG (Takagi *et al.*, 1999). One study of neurogenesis in focal cerebral ischemia used a photothrombotic model of stroke in the rat, and showed enhanced incorporation of BrdU in periinfarction cortex, with some labeled cells coexpressing MAP-2 or NeuN (Gu *et al.*, 2000).

In studies that employed the middle cerebral artery occlusion (MCAO) model to investigate ischemia-induced neurogenesis, MCAO increased the proliferation of neuronal precursors, identified by BrdU labeling and the expression of immature neuronal (DCX) and cell-proliferation (PCNA) markers, in both DG and SVZ (Jin *et al.*, 2001). This study revealed two potentially important features of ischemia-induced neurogenesis. First, neurogenesis occurred in areas that were not themselves affected by the injury, suggesting that a mechanism must exist to couple injury to neurogenesis over a distance. Second, unilateral injury produced a bilateral neurogenesis response, which is likely to have implications for the manner in which the existence of an injurious lesion is conveyed to the brain's neuroproliferative zones.

Other studies have demonstrated increased numbers of BrdU/Musashi1-immunopositive cells in DG (Takasawa *et al.*, 2002), or of BrdU or ENCAM-immunoreactive cells in SVZ and cerebral cortex ipsilateral to MCAO (Zhang *et al.*, 2001), or at longer (1–2 months) postischemic intervals, an increase in BrdU-labeled cerebral cortical cells that also expressed the immature neuronal marker βIII-tubulin or the mature neuronal marker MAP-2 or NeuN (Jiang *et al.*, 2001; Takasawa *et al.*, 2002).

Additional interesting observations regarding ischemia-induced neurogenesis include the finding that a large percentage of newborn neurons produced in the setting of focal cerebral ischemia may undergo programmed cell death (Takasawa *et al.*, 2002), which is also the case during basal neurogenesis (Ekdahl *et al.*, 2001; Pompeiano *et al.*, 2000), and that (global) ischemia-induced neurogenesis occurs not only in rodents, but in nonhuman primates as well (Tonchev *et al.*, 2003). Others have reported that neural stem cells found in the medial SVZ are preferentially resistant to hypoxic-ischemic death, compared to more differentiated ENCAM-positive neuronal progenitors in the lateral SVZ (Romanko *et al.*, 2004), which could have implications for the ability of different populations of neuronal precursor cells to survive as they migrate into an ischemic microenvironment. Finally, regarding the mechanisms involved in ischemia-induced neurogenesis, MCAO has been shown to shift the orientation of mitotic cleavage in SVZ neuronal progenitors to favor symmetric over asymmetric cell division (Zhang *et al.*, 2004b).

Several studies have explored possible mediators of ischemia-induced neurogenesis (Fig. 2). Hypoxia stimulated proliferation and enhanced survival and neuronal differentiation of E12 rat mesencephalic cells in culture (Studer *et al.*, 2000), and this was associated with increased expression of EPO, which was shown subsequently to promote neurogenesis *in vivo* (Shingo *et al.*, 2001). MCAO-induced neurogenesis, measured by BrdU incorporation and staining

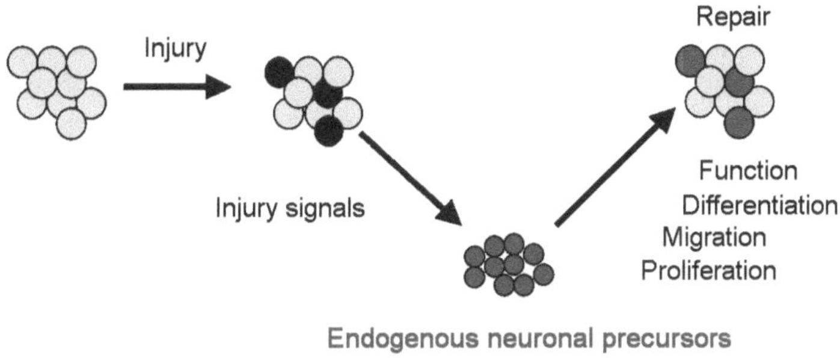

FIG. 2. Elements of ischemia-induced neurogenesis. The factors that mediate injury signaling, precursor cell proliferation, migration and differentiation of newborn neurons, and the assumption of mature neuronal functions are being investigated. (See Color Insert.)

for NeuN, was reduced markedly in FGF-2-knockout mice, and restored by herpes simplex virus, type 1 (HSV1) amplicon-based FGF-2 administration (Yoshimura *et al.*, 2003). Intraventricular infusion of a tyrosine kinase receptor B-Fc fragment of immunoglobulin G (TrkB-Fc) fusion protein, which complexes BDNF and thereby decreases free BDNF levels, had no effect on the total number of BrdU-positive cells, but increased BrdU/Hu- and BrdU/NeuN-positive cells in DG 2–6 weeks after global ischemia, which was interpreted as evidence for inhibition of ischemia-induced neuronal differentiation by BDNF (Gustafsson *et al.*, 2003), but which is at odds with data from other groups supporting a positive effect of BDNF on basal neurogenesis (Benraiss *et al.*, 2001; Louissaint *et al.*, 2002; Pencea *et al.*, 2001).

In another study, signaling factors that might be involved in ischemia-induced neurogenesis were sought by preparing cerebral cortical cultures from embryonic mouse brains and depriving them of oxygen to simulate ischemia *in vitro* (Jin *et al.*, 2002a). Hypoxia increased BrdU incorporation into cells that expressed cell-proliferation and immature neuronal markers, but which did not show DNA damage or caspase-3 activation. Hypoxia-conditioned medium (HCM) also stimulated BrdU incorporation into normoxic cultures, suggesting the presence in this medium of one or more neurogenesis-promoting factors. Candidate factors present in HCM at levels above those in control medium included SCF (Jin *et al.*, 2002a) and HB-EGF (Jin *et al.*, 1998, 2004d), both of which reproduced the neurogenesis-promoting effect of hypoxia *in vitro* and of ischemia *in vivo*. VEGF, another growth factor that is induced by hypoxia or ischemia, also stimulated neurogenesis (Jin *et al.*, 2002c; Sun *et al.*, 2003).

One of the most striking features of ischemia-induced neurogenesis is that after they proliferate, newborn neurons migrate into ischemic brain areas. In a study of transient global forebrain ischemia in rats, BrdU/NeuN-immunopositive cells that appeared to arise from nearby periventricular precursors appeared to migrate to the hippocampus and replace CA1 pyramidal neurons killed by ischemia (Nakatomi *et al.*, 2002). Unilateral focal ischemia produced by transient MCAO in the rat was associated with migration of BrdU-labeled cells that coexpressed DCX, and later NeuN, from the SVZ into the ischemic striatum, but not into the ischemic cerebral cortex (Arvidsson *et al.*, 2002). In another rat study, SVZ neurogenesis, measured using BrdU and immunostaining for cell type-specific markers, was greatly increased 10–21 days after transient MCAO, and chains of newborn neurons appeared to migrate from the SVZ to the ischemic striatum. Some of these cells expressed markers characteristic of medium spiny neurons, which are preferentially vulnerable to ischemia, consistent with differentiation toward a phenotype appropriate for cell replacement (Parent *et al.*, 2002b).

The migration of cells labeled by cell-proliferation markers and antibodies against neuronal proteins was mapped for up to 2 weeks after a 90-min episode of focal cerebral ischemia caused by occlusion of the middle cerebral artery (Jin *et al.*, 2003). DCX-immunoreactive cells in the rostral SVZ migrated into the ischemic penumbra of the striatum and cerebral cortex. Thus, these cells were diverted from their normal transit from the SVZ, via the RMS, to the OB. How does this occur? First, some newborn neurons appear to migrate *en masse* directly into the adjacent striatum. Others enter the RMS, but then branch off in chains to enter the striatum. Neurons destined for the ischemic cerebral cortex also depart the RMS to reach their destination; in so doing, they migrate rostrocaudally, just dorsal to the corpus callosum in a trajectory that resembles developmental migration in the lateral cortical stream, and then toward the cortical surface. Thus, one feature of cell migration in injury-induced adult neurogenesis appears to be the reuse of largely vestigial migratory pathways. Interestingly, although we observed increased neuroproliferation in DG after focal cerebral ischemia, new neurons arising there did not appear to migrate into the ischemic brain regions.

Studies of neuromigration in ischemia have also been conducted using transplanted cells. In one study, rat embryonic stem (ES) cells expressing green fluorescent protein (GFP) were labeled with a magnetic resonance imaging (MRI) contrast agent and grafted into the cortico-callosal border, contralateral to an ischemic lesion produced by transient MCAO 2 weeks earlier (Hoehn *et al.*, 2002). These cells migrated along the corpus callosum and into the opposite hemisphere, where they differentiated primarily into NeuN-positive cells, and accumulated in the ischemic striatum. In another MRI study, SVZ cells isolated from young adult rats were labeled with ferromagnetic particles and injected into the cisterna magna of adult rats, 48 hours after embolic MCAO (Zhang *et al.*, 2003). These cells, which showed phenotypic neuronal features by light and electron microscopy, migrated, at ~65 μm/hour, into the striatal and cortical ischemic penumbra, in preference to the same regions of the nonischemic hemisphere. Human CNS stem cells transplanted into the ischemic rat cerebral cortex after MCAO migrated transcortically as much as 1.2 mm, into the ischemic penumbra, expressing DCX and βIII-tubulin in sequence (Kelly *et al.*, 2004). These and the previously described studies on neuromigration in the ischemic brain suggest that endogenous mechanisms exist to guide such cells toward ischemic brain lesions.

Numerous factors are likely to be involved in the migration of newborn neurons in response to ischemia. For example, neuronal migration under normal circumstances depends on the assembly and disassembly of microtubules, which are thought to be involved in nuclear translocation within the migrating neuron. Microtubule-associated proteins regulate microtubule structure and function by promoting polymerization or depolymerization of microtubules, and we have demonstrated roles for the microtubule-*stabilizing* protein DCX and the microtubule-*destabilizing* protein stathmin in neuromigration in the nonischemic adult rat brain (Jin *et al.*, 2004b). Reelin (Reln), an extracellular matrix (ECM) serine protease involved in the migration of newborn neurons during development, appears to have a continuing role in neuromigration in the adult brain, including the ischemic brain (Won *et al.*, 2006).

Ischemia-induced neuroproliferation and neuromigration are both likely to involve responses to factors released from injured brain tissue. In the case of proliferation, additional information that regulates, for example, the lineage along which cells will differentiate, may also be encoded in the identities and the relative amounts of different factors that are released. Neuromigration in response to ischemia may be more complex, requiring extracellular signposts provided by the ECM and by other cell types to direct migrating cells in transit. Studies on ischemia-induced neuromigration point to some general principles of ischemia-induced neuromigration. For example, newborn neurons can migrate long distances, even from one hemisphere to the other, to reach ischemic brain lesions. In so doing, they appear to rely on more or less stereotypical pathways, which include not only routes observed normally in adult neurogenesis, like the RMS, but also pathways associated with developmental neuromigration. Alternative routes for migration from the adult SVZ include the lateral cortical stream (Nguyen-Ba-Charvet *et al.*, 2004) and ventral migratory mass (De Marchis *et al.*, 2004).

In an effort to address whether neurogenesis induced by cerebral ischemia results in the production of functional neurons, whole-brain ionizing radiation (two 5-Gy doses separated by 7 days) was used to ablate dividing neuronal precursors in the SGZ of guinea pigs, 2 weeks prior to 5 min of global cerebral ischemia (Raber *et al.*, 2004). As had been documented previously (Mizumatsu *et al.*, 2003; Monje *et al.*, 2002; Parent *et al.*, 2002a), this treatment inhibited neurogenesis, although no effect on microvascular morphology or dendritic profiles was observed. One month post-ischemia, the number of BrdU/TuJ1 double-labeled cells in the dentate GCL was reduced by ~80% in irradiated animals. These animals also showed impaired performance on a water–maze task, suggesting that ischemia-induced neurogenesis contributed favorably to outcome in nonirradiated subjects. Others have shown that treatment with cytosine arabinoside (Ara-C) also inhibited SVZ neurogenesis after focal cerebral ischemia produced by MCAO in rats (Arvidsson *et al.*, 2002; Zhang *et al.*, 2004a), although functional outcome was not assessed.

VIII. Neurogenesis in AD

AD is the most common cause of dementia, a syndrome characterized by progressive loss of cognitive function, and affects about 4.5 million people in the United States. Its incidence increases steadily with advancing age. Pathological features of AD comprise cerebral cortical (especially temporoparietal) atrophy from loss of neurons and neuropil, and characteristic histological lesions including extracellular senile (neuritic or amyloid) plaques and intracellular neurofibrillary tangles, which contain β-amyloid (Aβ) and hyperphosphorylated tau proteins, respectively. Most cases of AD are sporadic, but autosomal dominant, familial AD is seen with mutations in β-amyloid precursor protein (β-APP), presenilin-1 (PS1), or presenilin-2 (PS2). How any of these mutations (or the underlying abnormality in sporadic AD) produces the disease is unclear, but toxic effects of cleavage products of β-APP (especially Aβ_{42}) have received most attention. Current treatment for AD is marginally effective in most cases and consists of acetylcholinesterase inhibitors and the use-dependent, NMDA-type glutamate receptor antagonist memantine. Experimental approaches including vaccines directed against Aβ and metal chelators are under investigation. Animal models that express mutant proteins associated with familial AD in humans, such as β-APP and presenilins, exhibit some but not all features of the clinical disease.

One potential approach to treating AD involves using endogenous neuronal precursors to replace lost or damaged cells. Compared to controls, AD brains showed increased expression of immature neuronal marker proteins that signal the birth of new neurons in the hippocampus of AD patients such as DCX, ENCAM, NeuroD, and TUC-4 (Jin *et al.*, 2004c). Expression of DCX and TUC-4 is associated with neurons in the neuroproliferative SGZ of the DG; the physiological destination of these neurons in the GCL; and the CA1 region of Ammon's horn, which is the principal site of hippocampal pathology in AD. These findings suggest that neurogenesis is increased in AD hippocampus, and may generate replacements for neurons lost in the disease.

Studies of neurogenesis in animal models of AD have yielded conflicting results. In one study, 11- to 14-month-old transgenic mice that express APP with the Swedish (APP695[K595N/M596L]) mutation showed reduced numbers of BrdU-, ENCAM-, and BrdU/ENCAM-positive cells in DG or SVZ, consistent with impaired neurogenesis (Haughey *et al.*, 2002). In another study, 24-month-old APP23 (APP751[K670N/M671L]) mice (Sturchler-Pierrat *et al.*, 1997) showed a large increase in BrdU labeling in cerebral neocortex, but BrdU-immunopositive cells were NeuN-immunonegative (Bondolfi *et al.*, 2002), and neither DG nor SVZ was studied. Evidence supporting a role for APP in neurogenesis includes the finding that soluble APP increased EGF-dependent neurogenesis in the SVZ of the adult mouse brain (Caille *et al.*, 2004). PS1 has also been implicated in neurogenesis, in that increased expression of wild-type, but not familial AD mutant, PS1 increases hippocampal neurogenesis (Wen *et al.*, 2002), and environmental enrichment-induced (but not basal) neurogenesis is impaired in DG-SGZ of PS1 knockout mice (Feng *et al.*, 2001).

PDGF-$APP_{Sw,Ind}$ transgenic mice (Hsia *et al.*, 1999) express human APP isoforms APP695, APP751, and APP770 with the Indiana (V717F) and Swedish (K670N/M671L) mutations driven by a platelet-derived growth factor-β (PDGF-β) promoter. $APP_{Sw,Ind}$ mice exhibit extracellular deposits of Aβ that can be detected beginning at 6–9 months of age, as well as synaptic loss and astro- and microgliosis, but neither neurofibrillary tangles nor notable neuronal loss. They differ from APP_{Sw} and APP23 mice in that both of those lines lack the APP Indiana mutation. The effect of AD on neurogenesis can be reproduced in PDGF-$APP_{Sw,Ind}$ mice, which show increased incorporation of BrdU and expression of immature neuronal markers in DG and SVZ (Jin *et al.*, 2004a). These changes, consisting of approximately twofold increases in the number of BrdU-labeled cells, were observed at age 3 months, when neuronal loss and amyloid deposition were not detected. Moreover, introducing an additional Asp→Ala (D664A) mutation in this model, which prevented caspase cleavage at this site, hippocampal synaptic loss (defined by synaptophysin staining), and dentate gyral atrophy, but not Aβ production or amyloid plaque formation, blocked the induction of neurogenesis (Galvan *et al.*, 2006). Because enhanced neurogenesis occurs in both AD and an animal model of AD, it appears to be due to the disease itself, and not confounding clinical factors. Since neurogenesis is increased in PDGF-$APP_{Sw,Ind}$ mice in the absence of neuronal loss, it must be triggered by more subtle disease manifestations such as impaired neurotransmission.

IX. Neurogenesis in HD

HD is an autosomal dominant disorder with complete penetrance and anticipation, which produces choreoathetosis, dementia, and psychosis, usually with onset in middle age. It affects about 30,000 individuals in the United States. The brain in HD shows atrophy of the caudate nucleus, putamen, and cerebral cortex; medium-sized spiny neurons in the caudate and putamen are preferentially affected. In all cases of HD, the disease results from mutations in huntingtin (HTT), a ubiquitously expressed protein found at highest levels in neurons. The responsible mutations are expanded trinucleotide (CAG) repeats that code for N-terminal polyglutamine tracts, which confer a toxic gain of protein function. Mutant forms of HTT are subject to aberrant folding and aggregation, and may be found in intracytoplasmic or intranuclear inclusions. Whether these inclusions promote cell death (by, e.g., complexing the transcription factor binding partner cyclic AMP response element-binding protein (CREB)-binding protein, or CBP), or cell survival (by sequestering potentially toxic mutant HTT), is uncertain. The only available therapy is symptomatic treatment for choreoathetosis or psychosis. The most widely used animal models are transgenic mice that express mutant human HTT or a fragment thereof.

HD might be ameliorated by replacing dead neurons, perhaps achievable by stimulating the proliferation of endogenous neuronal precursors and their migration into damaged regions of the brain. In one human postmortem study, the number of cells that express the cell-proliferation marker PCNA and the immature neuronal marker βIII-tubulin was increased in the subependymal layer near the caudate nucleus in brains of patients who died with HD (Curtis *et al.*, 2003). Considering this, and given that growth factors are neuroprotective in some settings and can also stimulate neurogenesis, HD transgenic R6/2 mice were treated from 8 weeks of age until death with subcutaneous FGF-2 (Jin *et al.*, 2005). In comparison to wild-type littermates, HD transgenic R6/2 mice showed a small increase in the number of BrdU-labeled, Dcx-expressing proliferating cells in the SVZ. FGF-2 further increased the number of proliferating cells in the SVZ by ~30% in wild-type mice and by ~150% in HD transgenic R6/2 mice. FGF-2 also induced the recruitment of new neurons from the SVZ into the neostriatum and cerebral cortex of HD transgenic R6/2 mice. In the striatum, these new neurons were DARPP-32-expressing medium spiny neurons, consistent with the phenotype of neurons lost in HD. FGF-2 was neuroprotective as well, since it blocked cell death induced by mutant expanded HTT in primary striatal cultures. FGF-2 also reduced polyglutamine aggregates, improved motor performance, and extended lifespan by ~20%. Thus, FGF-2 improves neurological deficits and longevity in a transgenic mouse model of HD, and its neuroprotective and neuroproliferative effects may contribute to this improvement.

X. Neurogenesis in PD

PD is a neurodegenerative disorder that is usually sporadic, but may also result from mutations in proteins like α-synuclein, parkin, DJ1, or PTEN-induced kinase 1

(PINK1), or from exposure to toxins like 1-methyl-1,2,3,6-tetrahydropyridine (MPTP). The disease affects ~1.5 million patients in the United States. Clinical features include rigidity, akinesia, tremor, and postural instability, and some patients eventually exhibit dementia as well. The most prominent pathological feature in the PD brain is depigmentation of the substantia nigra pars compacta, which results from selective death of dopaminergic neurons projecting to the caudate and putamen. Other catecholaminergic tracts are also involved. Affected neurons may show inclusions, called Lewy bodies, which stain for α-synuclein and ubiquitin. Nigrostriatal degeneration in PD leads to an increase in the inhibitory output of the striatum. Several approaches to treatment are available, although their efficacy tends to wane as the disease progresses. These include levodopa–carbidopa combinations, dopamine receptor agonists, muscarinic anticholinergics, and monoamine oxidase (MAO) or catechol-*O*-methyl transferase (COMT) inhibitors. Surgical procedures are also available, including pallidotomy and deep brain stimulation. Transplantation of human adult adrenomedullary or human fetal mesencephalic tissue has also been tested, but remains experimental. Animal models of PD include both toxin (6-OHDA, MPTP, paraquat, and rotenone) and transgenic (mutant α-synuclein or parkin) mouse models, although these genetic models are not associated with death of dopaminergic neurons (Fernagut and Chesselet, 2004; Goldberg *et al.*, 2003) and levels of the abnormal proteins they express may not correlate well with clinical involvement in PD (Parkkinen *et al.*, 2005).

Work on neurogenesis in PD has addressed a number of important questions, but has produced conflicting results. As to whether neurogenesis occurs in the adult substantia nigra, the rat substantia nigra was found to contain progenitor cells, identified by labeling with BrdU, that gave rise exclusively to glia *in situ*, but which when cultured in the presence of FGF-2 or FGF-8 *in vitro*, or transplanted into the dentate hilus *in vivo*, produced cells that expressed immature (βIII-tubulin) or mature (NeuN) neuronal markers (Lie *et al.*, 2002). Another group reported that the mouse substantia nigra contained a small number of tyrosine hydroxylase (TH)-expressing cells that could be labeled with BrdU, and which were thought to originate in the SVZ, because they also took up intraventricular DiI (Zhao *et al.*, 2003). However, others could not replicate these results, finding that TH and BrdU were present in adjacent, rather than the same, nigral cells, and that DiI reached this region by retrograde transport along the nigrostriatal tract (Frielingsdorf *et al.*, 2004).

Conflicting results have also been obtained regarding whether animal models of PD, like animal models of other neurodegenerative disorders, stimulate endogenous neurogenesis. Thus, after unilateral injection of 6-OHDA into the medial forebrain bundle of the rat, no BrdU-positive cells that expressed βIII-tubulin, NeuN, or TH could be detected (Lie *et al.*, 2002). Contrarily, administration of 6-OHDA into the substantia nigra and ventral tegmentum, combined with intrastriatal infusion of transforming growth factor-α (TGF-α), led to expansion of the EGF receptor-positive cell population in the rat SVZ, and migration of these cells toward the site of infusion (Fallon *et al.*, 2000). Moreover, some of these migrating cells could be labeled with BrdU, and expressed immature (βIII-tubulin, DCX) or dopaminergic (TH, dopamine (DA) transporter) neuronal markers. Following a single subcutaneous dose of MPTP, the number of TH-positive nigral cells that incorporated BrdU was increased (Zhao *et al.*, 2003). Nevertheless, no such change was observed when 6-OHDA was injected into the medial forebrain bundle (MFB), with or without concomitant administration of BDNF (Frielingsdorf *et al.*, 2004) or TGF-α (Cooper and Isacson, 2004), and the previous result was attributed to failure to distinguish adjacent 6-OHDA-positive and TH-positive cells by three-dimensional confocal analysis (Frielingsdorf *et al.*, 2004).

The discordant results of studies to date suggest that the relationship between neurogenesis and parkinsonism may be complex, and differences in the animal models employed or in the severity or duration of disease may explain some disparities. Dopaminergic input from the substantia nigra to the SVZ has a proneurogenic influence (Baker *et al.*, 2004; Hoglinger *et al.*, 2004), which would be expected to be lost as the disease progresses, and might mask any tendency of PD itself to stimulate neurogenesis. The development of new animal models of PD provides an opportunity to resolve some of these issues. For example, transgenic models of PD are now available (Feany, 2004; Fernagut and Chesselet, 2004), and these could shed light on possible effects of abnormal protein products like α-synuclein on neurogenesis, independent of cell death. Transgenic models may also more closely recapitulate the chronic course of most human cases of PD.

Investigations that involve transplantation of exogenous stem cells or their progeny may also provide important information to help guide studies of endogenous neurogenesis in PD. For example, the survival and neuronal differentiation of mesencephalic stem cells grafted into rat striatum is enhanced in the 6-OHDA-treated, dopamine-depleted striatum (Nishino *et al.*, 2000), suggesting that as in other disease models, the lesion may help to orchestrate its own repair. Unilateral implantation of murine neural stem cells into the midbrain of aged, MPTP-treated mice led to biochemical and functional reconstitution of the injured nigrostriatal pathway, although few of the cells observed were donor derived (Ourednik *et al.*, 2002); this suggests that the beneficial effect of endogenous or exogenous stem cells may not be limited to cell replacement, but might also derive from rescuing injured neurons, perhaps by secreting trophic or protective factors. Finally, studies in mice (Kim *et al.*, 2002) and nonhuman primates (Kim *et al.*, 2002) indicate that cells derived from ES cells, and transplanted into the brains of animals with experimental PD, can adopt phenotypic features of dopaminergic neurons and relieve parkinsonian symptoms.

XI. Neurogenesis in Motor Neuron Disease

ALS is the best-known example of motor neuron disease in which lower motor neurons in the brain stem and anterior horn of the spinal cord and, in some cases, upper motor neurons in motor cortex degenerate, producing weakness, muscle atrophy, and fasciculation. ALS affects ~30,000 persons in the United States and typically leads to death from intercurrent infection within 5 years. The spinal cord shows loss of large neurons in the anterior horns; in cases with upper motor neuron involvement, there is also loss of Betz cells in the motor cortex and degeneration of the corticospinal tracts. Most cases are sporadic, but familial ALS can result from mutations in, for example, Cu/Zn superoxide dismutase (SOD1) or alsin. In some instances, a defect in the trophic effect of VEGF may be involved. Current treatment options are disappointing, consisting of the NMDA-type glutamate receptor antagonist riluzole, which affords little benefit and supportive care. The most popular animal models of ALS are transgenic mice that express an SOD1 mutation, such as G93A, found in some families with the hereditary form of the disease.

Neurogenesis appears to occur in the adult spinal cord, although several studies showing this have been conducted *in vitro* (Kehl *et al.*, 1997; Weiss *et al.*, 1996) or have required transplanting spinal cord precursor cells to neuroproliferative zones of the brain (Shihabuddin *et al.*, 2000). Motor neurons can be produced from human ES cells (Li *et al.*, 2005; Miles *et al.*, 2004; Wichterle *et al.*, 2002), and transplantation of human (Cummings *et al.*, 2005) or murine (Teng *et al.*, 2002) neuronal precursors can attenuate spinal cord injury. Spinal cord neurogenesis has been reported to be increased in rodent models of multiple sclerosis (Danilov *et al.*, 2006) and ALS (Chi *et al.*, 2006). In the latter case, mice expressing a *LacZ* reporter gene driven by the nestin promoter were crossed with mice expressing a mutant superoxide dismutase ($SOD1^{G93A}$) gene found in some pedigrees with familial ALS. The authors found increased proliferation of LacZ-positive cells in the lumbar spinal cord, with migration to the dorsal and then ventral horn. Some of these cells stained with antibodies against neuronal lineage marker proteins such as Hu, Tuj1, and NeuN.

XII. Convergent and Divergent Features of Injury-Induced Neurogenesis

By examining the neurogenesis response to a range of brain injuries, it is possible to discern patterns of both similarity and dissimilarity across pathologies. This information is likely to be useful not only for understanding the basic biology of adult neurogenesis, but possibly also for fashioning therapeutic approaches to

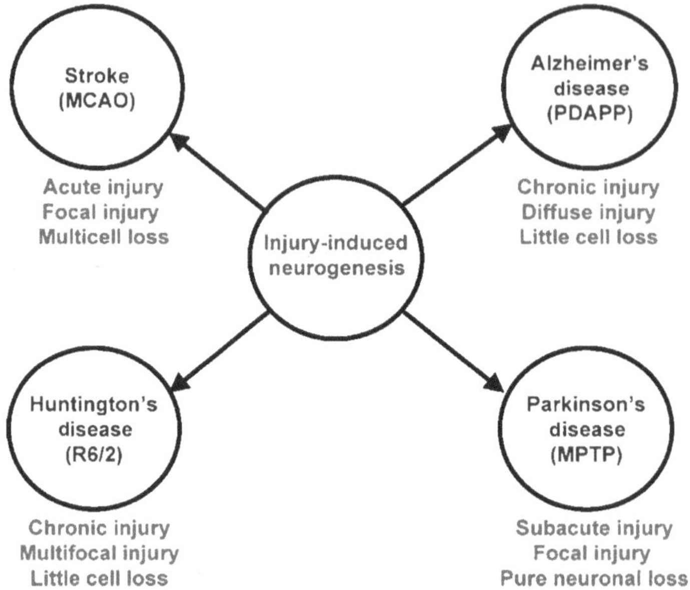

FIG. 3. Interrelationship between studies of neurogenesis in specific neurodegenerative diseases and the principles of injury-induced neurogenesis that they elucidate. (See Color Insert.)

additional disorders. As illustrated in Fig. 3, studies of ischemia have shed light on the bilaterality of neurogenesis induced by unilateral lesions, which may be related to adaptation and repair processes distant from the lesion site. Mouse models of AD have shown that cell death is not a prerequisite for injury-induced neurogenesis, which may instead be triggered by subtler changes in neuronal (perhaps synaptic) function. Transgenic mice bearing a mutation for human HD tell us that growth-factor replacement may be required for neurogenesis in some circumstances, possibly because the disease process produces a local state of growth-factor deficiency. Table II summarizes some of the shared and distinctive features of neurogenesis in these models.

XIII. Unanswered Questions

Work in this field to date points to many basic questions about injury-induced neurogenesis that are unanswered, and which are likely to be important in the eventual design of therapeutic strategies.

TABLE II

Disease-Specific Features of Injury-Induced Neurogenesis

Disease	Mouse model	Spontaneously increased neurogenesis	FGF-2 required for markedly increased neurogenesis	Migration of new neurons to site of injury	Projection of new neurons to appropriate targets	Functional replacement of depleted neurons
Stroke	MCAO	Yes, SVZ > DG	No	Yes, striatum and cortex	?[a]	?
Alzheimer's	$APP_{Sw,Ind}$	Yes, DG > SVZ	No	?	?	?
Huntington's	R6/2	Slight, SVZ	Yes, SVZ	Yes, striatum	Yes, striopallidal	?
Parkinson's	MPTP, 6-OHDA	?	?	?	?	?

[a]?, unknown.

What stimulates injury-induced neurogenesis? Adult neurogenesis is increased in a variety of neurological diseases and animal models of these disorders, but the full spectrum of brain insults capable of eliciting such a response is uncertain. Similarly, the relationship of lesion size, location and character to the magnitude and location of increased neurogenesis is unknown.

How is injury-induced proliferation triggered in target neuronal precursor cells? The nature of injury sensing by neuronal precursor cells has yet to be explored in any detail. In addition, very few studies have addressed the likely heterogeneity of neuronal precursor cell populations within SVZ or SGZ, and whether cell subpopulations respond differently to injury in general or to different types of injury. Similarly, the early molecular steps that translate injury sensing into neuroproliferation have received little attention. Finally, although most pathological stimuli to adult neurogenesis identified thus far seem to increase cell proliferation, it is also possible that other such stimuli may act by inhibiting programmed death of neuronal precursors.

How are the spatial and phenotypic fates of injury-induced new neurons determined? Differentiation and migration of newborn neurons are likely governed by a host of regulatory factors, only a few of which have been identified. The relative contribution of endogenous (cell-autonomous) and exogenous (extracellular) cues in these processes is also unknown.

To what extent does injury-induced neurogenesis yield functional neurons? There are various aspects to this important question, including how fully adult neurogenesis can replicate cells lost in disease, the extent to which these cells can assume normal neuronal functions, and whether newborn neurons can exert beneficial neurotrophic or neuroprotective effects even before they become mature neurons.

Does injury-induced neurogenesis contribute to improved functional outcome? Even if fully functional neurons arise through injury-induced adult neurogenesis, we do not know if they can halt disease progression or restore lost function.

How can endogenous neurogenesis be modified to improve outcome after injury? Ultimately, if functional neurogenesis occurs in response to injury but is insufficient to alter the course of disease, can this process be stimulated to supraphysiological levels by growth factors, drugs, or other treatments so that clinical benefit is achieved?

Acknowledgments

This work is supported by NIH grants NS39912 and NS44921 to D.A.G. and AG21980 to K.J.

References

Altman, J. (1963). Autoradiographic study of cell proliferation in the brains of rats and cats. *Anat. Rec.* **145,** 573–591.

American Heart Association (2004). "Heart Disease and Stroke Statistics—2004 Update." American Heart Association, Dallas, Texas.

Arvidsson, A., Collin, T., Kirik, D., Kokaia, Z., and Lindvall, O. (2002). Neuronal replacement from endogenous precursors in the adult brain after stroke. *Nat. Med.* **8,** 963–970.

Baker, S. A., Baker, K. A., and Hagg, T. (2004). Dopaminergic nigrostriatal projections regulate neural precursor proliferation in the adult mouse subventricular zone. *Eur. J. Neurosci.* **20,** 575–579.

Bayer, S. A., and Altman, J. (1995). Principles of neurogenesis, neuronal migration, and neural circuit formation. *In* "The Rat Nervous System" (G. Paxinos, Ed.), pp. 1079–1098. Academic Press, San Diego, California.

Benraiss, A., Chmielnicki, E., Lerner, K., Roh, D., and Goldman, S. A. (2001). Adenoviral brain-derived neurotrophic factor induces both neostriatal and olfactory neuronal recruitment from endogenous progenitor cells in the adult forebrain. *J. Neurosci.* **21,** 6718–6731.

Bondolfi, L., Calhoun, M., Ermini, F., Kuhn, H. G., Wiederhold, K. H., Walker, L., Staufenbiel, M., and Jucker, M. (2002). Amyloid-associated neuron loss and gliogenesis in the neocortex of amyloid precursor protein transgenic mice. *J. Neurosci.* **22,** 515–522.

Brott, T., and Bogousslavsky, J. (2000). Treatment of acute ischemic stroke. *N. Engl. J. Med.* **343,** 710–722.

Caille, I., Allinquant, B., Dupont, E., Bouillot, C., Langer, A., Muller, U., and Prochiantz, A. (2004). Soluble form of amyloid precursor protein regulates proliferation of progenitors in the adult subventricular zone. *Development* **131,** 2173–2181.

Cameron, H. A., McEwen, B. S., and Gould, E. (1995). Regulation of adult neurogenesis by excitatory input and NMDA receptor activation in the dentate gyrus. *J. Neurosci.* **15,** 4687–4692.

Cameron, H. A., Hazel, T. G., and McKay, R. D. (1998). Regulation of neurogenesis by growth factors and neurotransmitters. *J. Neurobiol.* **36,** 287–306.

Chi, L., Ke, Y., Luo, C., Li, B., Gozal, D., Kalyanaraman, B., and Liu, R. (2006). Motor neuron degeneration promotes neural progenitor cell proliferation, migration and neurogenesis in the spinal cords of ALS Mice. *Stem Cells.* **24,** 34–43.

Cooper, O., and Isacson, O. (2004). Intrastriatal transforming growth factor α delivery to a model of Parkinson's disease induces proliferation and migration of endogenous adult neural progenitor cells without differentiation into dopaminergic neurons. *J. Neurosci.* **24,** 8924–8931.

Cummings, B. J., Uchida, N., Tamaki, S. J., Salazar, D. L., Hooshmand, M., Summers, R., Gage, F. H., and Anderson, A. J. (2005). Human neural stem cells differentiate and promote locomotor recovery in spinal cord-injured mice. *Proc. Natl. Acad. Sci. USA* **102,** 14069–14074.

Cummings, J. L. (2004). Alzheimer's disease. *N. Engl. J. Med.* **351,** 56–67.

Curtis, M. A., Penney, E. B., Pearson, A. G., van Roon-Mom, W. M., Butterworth, N. J., Dragunow, M., Connor, B., and Faull, R. L. (2003). Increased cell proliferation and neurogenesis in the adult human Huntington's disease brain. *Proc. Natl. Acad. Sci. USA* **100,** 9023–9027.

Danilov, A. I., Covacu, R., Moe, M. C., Langmoen, I. A., Johansson, C. B., Olsson, T., and Brundin, L. (2006). Neurogenesis in the adult spinal cord in an experimental model of multiple sclerosis. *Eur. J. Neurosci.* **23,** 394–400.

De Marchis, S., Fasolo, A., and Puche, A. C. (2004). Subventricular zone-derived neuronal progenitors migrate into the subcortical forebrain of postnatal mice. *J. Comp. Neurol.* **476,** 290–300.

Dehay, C., Savatier, P., Cortay, V., and Kennedy, H. (2001). Cell-cycle kinetics of neocortical precursors are influenced by embryonic thalamic axons. *J. Neurosci.* **21,** 201–214.

Ekdahl, C. T., Mohapel, P., Elmer, E., and Lindvall, O. (2001). Caspase inhibitors increase short-term survival of progenitor-cell progeny in the adult rat dentate gyrus following status epilepticus. *Eur. J. Neurosci.* **14,** 937–945.

Fallon, J., Reid, S., Kinyamu, R., Opole, I., Opole, R., Baratta, J., Korc, M., Endo, T. L., Duong, A., Nguyen, G., Karkehabadhi, M., Twardzik, D., *et al.* (2000). *In vivo* induction of massive proliferation, directed migration, and differentiation of neural cells in the adult mammalian brain. *Proc. Natl. Acad. Sci. USA* **97,** 14686–14691.

Feany, M. B. (2004). New genetic insights into Parkinson's disease. *N. Engl. J. Med.* **351,** 1937–1940.

Feng, R., Rampon, C., Tang, Y. P., Shrom, D., Jin, J., Kyin, M., Sopher, B., Miller, M. W., Ware, C. B., Martin, G. M., Kim, S. H., Langdon, R. B., *et al.* (2001). Deficient neurogenesis in forebrain-specific presenilin-1 knockout mice is associated with reduced clearance of hippocampal memory traces. *Neuron* **32,** 911–926.

Fernagut, P. O., and Chesselet, M. F. (2004). Alpha-synuclein and transgenic mouse models. *Neurobiol. Dis.* **17,** 123–130.

Frielingsdorf, H., Schwarz, K., Brundin, P., and Mohapel, P. (2004). No evidence for new dopaminergic neurons in the adult mammalian substantia nigra. *Proc. Natl. Acad. Sci. USA* **101,** 10177–10182.

Galvan, V., Gorostiza, O. H., Banwait, S., Ataie, M., Logvinova, A. V., Carlson, E., Sagi, S. A., Chevallier, N., Jin, K., Greenberg, D. A., and Bredesen, D. E. (2006). Reversal of Alzheimer's-like pathology and behavior in human APP transgenic mice by mutation of Asp664. *Proc. Natl. Acad. Sci. USA* **103,** 7130–7135.

Goldberg, M. S., Fleming, S. M., Palacino, J. J., Cepeda, C., Lam, H. A., Bhatnagar, A., Meloni, E. G., Wu, N., Ackerson, L. C., Klapstein, G. J., Gajendiran, M., Roth, B. L., *et al.* (2003). Parkin-deficient mice exhibit nigrostriatal deficits but not loss of dopaminergic neurons. *J. Biol. Chem.* **278,** 43628–43635.

Gould, E., and Tanapat, P. (1997). Lesion-induced proliferation of neuronal progenitors in the dentate gyrus of the adult rat. *Neuroscience* **80,** 427–436.

Gould, E., Reeves, A. J., Graziano, M. S., and Gross, C. G. (1999). Neurogenesis in the neocortex of adult primates. *Science* **286,** 548–552.

Gu, W., Brannstrom, T., and Wester, P. (2000). Cortical neurogenesis in adult rats after reversible photothrombotic stroke. *J. Cereb. Blood Flow Metab.* **20,** 1166–1173.

Gustafsson, E., Lindvall, O., and Kokaia, Z. (2003). Intraventricular infusion of TrkB-Fc fusion protein promotes ischemia-induced neurogenesis in adult rat dentate gyrus. *Stroke* **34,** 2710–2715.

Haughey, N. J., Nath, A., Chan, S. L., Borchard, A. C., Rao, M. S., and Mattson, M. P. (2002). Disruption of neurogenesis by amyloid β-peptide, and perturbed neural progenitor cell homeostasis, in models of Alzheimer's disease. *J. Neurochem.* **83,** 1509–1524.

Hoehn, M., Kustermann, E., Blunk, J., Wiedermann, D., Trapp, T., Wecker, S., Focking, M., Arnold, H., Hescheler, J., Fleischmann, B. K., Schwindt, W., and Buhrle, C. (2002). Monitoring of implanted stem cell migration *in vivo*: A highly resolved *in vivo* magnetic resonance imaging investigation of experimental stroke in rat. *Proc. Natl. Acad. Sci. USA* **99,** 16267–16272.

Hoglinger, G. U., Rizk, P., Muriel, M. P., Duyckaerts, C., Oertel, W. H., Caille, I., and Hirsch, E. C. (2004). Dopamine depletion impairs precursor cell proliferation in Parkinson disease. *Nat. Neurosci.* **7,** 726–735.

Horner, P. J., and Gage, F. H. (2000). Regenerating the damaged central nervous system. *Nature* **407,** 963–970.

Hsia, A. Y., Masliah, E., McConlogue, L., Yu, G. Q., Tatsuno, G., Hu, K., Kholodenko, D., Malenka, R. C., Nicoll, R. A., and Mucke, L. (1999). Plaque-independent disruption of neural circuits in Alzheimer's disease mouse models. *Proc. Natl. Acad. Sci. USA* **96,** 3228–3233.

Jiang, W., Gu, W., Brännström, T., Rosqvist, R., and Wester, P. (2001). Cortical neurogenesis in adult rats after transient middle cerebral artery occlusion. *Stroke* **32,** 1201–1207.

Jin, K., Minami, M., Lan, J. Q., Mao, X. O., Batteur, S., Simon, R. P., and Greenberg, D. A. (2001). Neurogenesis in dentate subgranular zone and rostral subventricular zone after focal cerebral ischemia in the rat. *Proc. Natl. Acad. Sci. USA* **98,** 4710–4715.

Jin, K., Mao, X. O., Sun, Y., Xie, L., and Greenberg, D. A. (2002a). Stem cell factor stimulates neurogenesis *in vitro* and *in vivo*. *J. Clin. Invest.* **110,** 311–319.

Jin, K., Mao, X. O., Sun, Y., Xie, L., Jin, L., Nishi, E., Klagsbrun, M., and Greenberg, D. A. (2002b). Heparin-binding epidermal growth factor-like growth factor: Hypoxia-inducible expression *in vitro* and stimulation of neurogenesis *in vitro* and *in vivo*. *J. Neurosci.* **22,** 5365–5373.

Jin, K., Zhu, Y., Sun, Y., Mao, X. O., Xie, L., and Greenberg, D. A. (2002c). Vascular endothelial growth factor (VEGF) stimulates neurogenesis *in vitro* and *in vivo*. *Proc. Natl. Acad. Sci. USA* **99,** 11946–11950.

Jin, K., Sun, Y., Xie, L., Peel, A., Mao, X. O., Batteur, S., and Greenberg, D. A. (2003). Directed migration of neuronal precursors into the ischemic cerebral cortex and striatum. *Mol. Cell Neurosci.* **24,** 171–189.

Jin, K., Galvan, V., Xie, L., Mao, X. O., Gorostiza, O. F., Bredesen, D. E., and Greenberg, D. A. (2004a). Enhanced neurogenesis in Alzheimer's disease transgenic (PDGF-$APP_{Sw,Ind}$) mice. *Proc. Natl. Acad. Sci. USA* **101,** 13363–13367.

Jin, K., Mao, X. O., Cottrell, B., Schilling, B., Xie, L., Row, R. H., Sun, Y., Peel, A., Childs, J., Gendeh, G., Gibson, B. W., and Greenberg, D. A. (2004b). Proteomic and immunochemical characterization of a role for stathmin in adult neurogenesis. *FASEB J.* **18,** 287–299.

Jin, K., Peel, A. L., Mao, X. O., Xie, L., Cottrell, B. A., Henshall, D. C., and Greenberg, D. A. (2004c). Increased hippocampal neurogenesis in Alzheimer's disease. *Proc. Natl. Acad. Sci. USA* **101,** 343–347.

Jin, K., Sun, Y., Xie, L., Childs, J., Mao, X. O., and Greenberg, D. A. (2004d). Post-ischemic administration of heparin-binding epidermal growth factor-like growth factor (HB-EGF) reduces infarct size and modifies neurogenesis after focal cerebral ischemia in the rat. *J. Cereb. Blood Flow Metab.* **24,** 399–408.

Jin, K., LaFevre-Bernt, M., Sun, Y., Chen, S., Gafni, J., Crippen, D., Logvinova, A., Ross, C. A., Greenberg, D. A., and Ellerby, L. M. (2005). FGF-2 promotes neurogenesis and neuroprotection and prolongs survival in a transgenic mouse model of Huntington's disease. *Proc. Natl. Acad. Sci. USA* **102,** 18189–18194.

Jin, L. W., Hearn, M. G., Ogburn, C. E., Dang, N., Nochlin, D., Ladiges, W. C., and Martin, G. M. (1998). Transgenic mice over-expressing the C-99 fragment of betaPP with an alpha-secretase site mutation develop a myopathy similar to human inclusion body myositis. *Am. J. Pathol.* **153,** 1679–1686.

Kehl, L. J., Fairbanks, C. A., Laughlin, T. M., and Wilcox, G. L. (1997). Neurogenesis in postnatal rat spinal cord: A study in primary culture. *Science* **276,** 586–589.

Kelly, S., Bliss, T. M., Shah, A. K., Sun, G. H., Ma, M., Foo, W. C., Masel, J., Yenari, M. A., Weissman, I. L., Uchida, N., Palmer, T., and Steinberg, G. K. (2004). Transplanted human fetal neural stem cells survive, migrate, and differentiate in ischemic rat cerebral cortex. *Proc. Natl. Acad. Sci. USA* **101,** 11839–11844.

Kilpatrick, T. J., and Bartlett, P. F. (1993). Cloning and growth of multipotential neural precursors: Requirements for proliferation and differentiation. *Neuron* **10,** 255–265.

Kim, J. H., Auerbach, J. M., Rodriguez-Gomez, J. A., Velasco, I., Gavin, D., Lumelsky, N., Lee, S. H., Nguyen, J., Sanchez-Pernaute, R., Bankiewicz, K., and McKay, R. (2002). Dopamine neurons derived from embryonic stem cells function in an animal model of Parkinson's disease. *Nature* **418,** 50–56.

Kirschenbaum, B., and Goldman, S. A. (1995). Brain-derived neurotrophic factor promotes the survival of neurons arising from the adult rat forebrain subependymal zone. *Proc. Natl. Acad. Sci. USA* **92,** 210–214.

Kornack, D. R., and Rakic, P. (2001). Cell proliferation without neurogenesis in adult primate neocortex. *Science* **294,** 2127–2130.

Kuhn, H. G., Winkler, J., Kempermann, G., Thal, L. J., and Gage, F. H. (1997). Epidermal growth factor and fibroblast growth factor-2 have different effects on neural progenitors in the adult rat brain. *J. Neurosci.* **17,** 5820–5829.

Li, X. J., Du, Z. W., Zarnowska, E. D., Pankratz, M., Hansen, L. O., Pearce, R. A., and Zhang, S. C. (2005). Specification of motoneurons from human embryonic stem cells. *Nat. Biotechnol.* **23,** 215–221.

Lie, D. C., Dziewczapolski, G., Willhoite, A. R., Kaspar, B. K., Shults, C. W., and Gage, F. H. (2002). The adult substantia nigra contains progenitor cells with neurogenic potential. *J. Neurosci.* **22,** 6639–6649.

Liu, J., Solway, K., Messing, R. O., and Sharp, F. R. (1998). Increased neurogenesis in the dentate gyrus after transient global ischemia in gerbils. *J. Neurosci.* **18,** 7768–7778.

Lois, C., and Alvarez-Buylla, A. (1993). Proliferating subventricular zone cells in the adult mammalian forebrain can differentiate into neurons and glia. *Proc. Natl. Acad. Sci. USA* **90,** 2074–2077.

Louissaint, A., Rao, S., Leventhal, C., and Goldman, S. A. (2002). Coordinated interaction of neurogenesis and angiogenesis in the adult songbird brain. *Neuron* **34,** 945–960.

Luskin, M. B. (1993). Restricted proliferation and migration of postnatally generated neurons derived from the forebrain subventricular zone. *Neuron* **11,** 173–189.

Magavi, S. S., Leavitt, B. R., and Macklis, J. D. (2000). Induction of neurogenesis in the neocortex of adult mice. *Nature* **405,** 951–955.

Markakis, E. A., Palmer, T. D., Randolph-Moore, L., Rakic, P., and Gage, F. H. (2004). Novel neuronal phenotypes from neural progenitor cells. *J. Neurosci.* **24,** 2886–2897.

Miles, G. B., Yohn, D. C., Wichterle, H., Jessell, T. M., Rafuse, V. F., and Brownstone, R. M. (2004). Functional properties of motoneurons derived from mouse embryonic stem cells. *J. Neurosci.* **24,** 7848–7858.

Mizumatsu, S., Monje, M. L., Morhardt, D. R., Rola, R., Palmer, T. D., and Fike, J. R. (2003). Extreme sensitivity of adult neurogenesis to low doses of X-irradiation. *Cancer Res.* **63,** 4021–4027.

Monje, M. L., Mizumatsu, S., Fike, J. R., and Palmer, T. D. (2002). Irradiation induces neural precursor-cell dysfunction. *Nat. Med.* **8,** 955–962.

Nakatomi, H., Kuriu, T., Okabe, S., Yamamoto, S., Hatano, O., Kawahara, N., Tamura, A., Kirino, T., and Nakafuku, M. (2002). Regeneration of hippocampal pyramidal neurons after ischemic brain injury by recruitment of endogenous neural progenitors. *Cell* **110,** 429–441.

Nguyen-Ba-Charvet, K. T., Picard-Riera, N., Tessier-Lavigne, M., Baron-Van Evercooren, A., Sotelo, C., and Chedotal, A. (2004). Multiple roles for slits in the control of cell migration in the rostral migratory stream. *J. Neurosci.* **24,** 1497–1506.

Nishino, H., Hida, H., Takei, N., Kumazaki, M., Nakajima, K., and Baba, H. (2000). Mesencephalic neural stem (progenitor) cells develop to dopaminergic neurons more strongly in dopamine-depleted striatum than in intact striatum. *Exp. Neurol.* **164,** 209–214.

O'Kusky, J. R., Ye, P., and D'Ercole, A. J. (2000). Insulin-like growth factor-I promotes neurogenesis and synaptogenesis in the hippocampal dentate gyrus during postnatal development. *J. Neurosci.* **20,** 8435–8442.

Ourednik, J., Ourednik, V., Lynch, W. P., Schachner, M., and Snyder, E. Y. (2002). Neural stem cells display an inherent mechanism for rescuing dysfunctional neurons. *Nat. Biotechnol.* **20,** 1103–1110.

Parent, J. M., Yu, T. W., Leibowitz, R. T., Geschwind, D. H., Sloviter, R. S., and Lowenstein, D. H. (1997). Dentate granule cell neurogenesis is increased by seizures and contributes to aberrant network reorganization in the adult rat hippocampus. *J. Neurosci.* **17,** 3727–3738.

Parent, J. M., Valentin, V. V., and Lowenstein, D. H. (2002a). Prolonged seizures increase proliferating neuroblasts in the adult rat subventricular zone-olfactory bulb pathway. *J. Neurosci.* **22,** 3174–3188.

Parent, J. M., Vexler, Z. S., Gong, C., Derugin, N., and Ferriero, D. M. (2002b). Rat forebrain neurogenesis and striatal neuron replacement after focal stroke. *Ann. Neurol.* **52,** 802–813.

Parkkinen, L., Kauppinen, T., Pirttila, T., Autere, J. M., and Alafuzoff, I. (2005). α-Synuclein pathology does not predict extrapyramidal symptoms or dementia. *Ann. Neurol.* **57,** 82–91.

Pencea, V., Bingaman, K. D., Wiegand, S. J., and Luskin, M. B. (2001). Infusion of brain-derived neurotrophic factor into the lateral ventricle of the adult rat leads to new neurons in the parenchyma of the striatum, septum, thalamus, and hypothalamus. *J. Neurosci.* **21,** 6706–6717.

Pompeiano, M., Blaschke, A. J., Flavell, R. A., Srinivasan, A., and Chun, J. (2000). Decreased apoptosis in proliferative and postmitotic regions of the caspase 3-deficient embryonic central nervous system. *J. Comp. Neurol.* **423,** 1–12.

Raber, J., Fan, Y., Matsumori, Y., Liu, Z., Weinstein, P. R., Fike, J. R., and Liu, J. (2004). Irradiation attenuates neurogenesis and exacerbates ischemia-induced deficits. *Ann. Neurol.* **55,** 381–389.

Ray, J., Peterson, D. A., Schinstine, M., and Gage, F. H. (1993). Proliferation, differentiation, and long-term culture of primary hippocampal neurons. *Proc. Natl. Acad. Sci. USA* **90,** 3602–3606.

Reynolds, B. A., and Weiss, S. (1992). Generation of neurons and astrocytes from isolated cells of the adult mammalian central nervous system. *Science* **255,** 1707–1710.

Romanko, M. J., Rothstein, R. P., and Levison, S. W. (2004). Neural stem cells in the subventricular zone are resilient to hypoxia/ischemia whereas progenitors are vulnerable. *J. Cereb. Blood Flow Metab.* **24,** 814–825.

Rowland, L. P., and Shneider, N. A. (2001). Amyotrophic lateral sclerosis. *N. Engl. J. Med.* **344,** 1688–1700.

Sanai, N., Tramontin, A. D., Quinones-Hinojosa, A., Barbaro, N. M., Gupta, N., Kunwar, S., Lawton, M. T., McDermott, M. W., Parsa, A. T., Manuel-Garcia Verdugo, J., Berger, M. S., and Alvarez-Buylla, A. (2004). Unique astrocyte ribbon in adult human brain contains neural stem cells but lacks chain migration. *Nature* **427,** 740–744.

Shihabuddin, L. S., Horner, P. J., Ray, J., and Gage, F. H. (2000). Adult spinal cord stem cells generate neurons after transplantation in the adult dentate gyrus. *J. Neurosci.* **20,** 8727–8735.

Shingo, T., Sorokan, S. T., Shimazaki, T., and Weiss, S. (2001). Erythropoietin regulates the *in vitro* and *in vivo* production of neuronal progenitors by mammalian forebrain neural stem cells. *J. Neurosci.* **21,** 9733–9743.

Snyder, E. Y., Yoon, C., Flax, J. D., and Macklis, J. D. (1997). Multipotent neural precursors can differentiate toward replacement of neurons undergoing targeted apoptotic degeneration in adult mouse neocortex. *Proc. Natl. Acad. Sci. USA* **94,** 11663–11668.

Song, H. J., Stevens, C. F., and Gage, F. H. (2002). Neural stem cells from adult hippocampus develop essential properties of functional CNS neurons. *Nat. Neurosci.* **5,** 438–445.

Streidter, G. F. (2005). "Principles of Brain Evolution." Sinauer Associates Inc., Sunderland, MA.

Studer, L., Csete, M., Lee, S. H., Kabbani, N., Walikonis, J., Wold, B., and McKay, R. (2000). Enhanced proliferation, survival, and dopaminergic differentiation of CNS precursors in lowered oxygen. *J. Neurosci.* **20,** 7377–7383.

Sturchler-Pierrat, C., Abramowski, D., Duke, M., Wiederhold, K.-H., Mistl, C., Rothacher, S., Ledermann, B., Burki, K., Frey, P., Paganetti, P. A., Waridel, C., Calhoun, M. E., *et al.* (1997). Two amyloid precursor protein transgenic mouse models with Alzheimer disease-like pathology. *Proc. Natl. Acad. Sci. USA* **94,** 13287–13292.

Sun, Y., Jin, K., Xie, L., Childs, J., Mao, X. O., Logvinova, A., and Greenberg, D. A. (2003). VEGF-induced neuroprotection, neurogenesis, and angiogenesis after focal cerebral ischemia. *J. Clin. Invest.* **111,** 1843–1851.

Takagi, Y., Nozaki, K., Takahashi, J., Yodoi, J., Ishikawa, M., and Hashimoto, N. (1999). Proliferation of neuronal precursor cells in the dentate gyrus is accelerated after transient forebrain ischemia in mice. *Brain Res.* **831,** 283–287.

Takasawa, K., Kitagawa, K., Yagita, Y., Sasaki, T., Tanaka, S., Matsushita, K., Ohstuki, T., Miyata, T., Okano, H., Hori, M., and Matsumoto, M. (2002). Increased proliferation of neural progenitor cells but reduced survival of newborn cells in the contralateral hippocampus after focal cerebral ischemia in rats. *J. Cereb. Blood Flow Metab.* **22,** 299–307.

Taupin, P., Ray, J., Fischer, W. H., Suhr, S. T., Hakansson, K., Grubb, A., and Gage, F. H. (2000). FGF-2-responsive neural stem cell proliferation requires CCg, a novel autocrine/paracrine cofactor. *Neuron* **28,** 385–397.

Temple, S. (1989). Division and differentiation of isolated CNS blast cells in microculture. *Nature* **340,** 471–473.

Teng, Y. D., Lavik, E. B., Qu, X., Park, K. I., Ourednik, J., Zurakowski, D., Langer, R., and Snyder, E. Y. (2002). Functional recovery following traumatic spinal cord injury mediated by a unique polymer scaffold seeded with neural stem cells. *Proc. Natl. Acad. Sci. USA* **99,** 3024–3029.

Tonchev, A. B., Yamashima, T., Zhao, L., Okano, H. J., and Okano, H. (2003). Proliferation of neural and neuronal progenitors after global brain ischemia in young adult macaque monkeys. *Mol. Cell Neurosci.* **23,** 292–301.

van Praag, H., Schinder, A. F., Christie, B. R., Toni, N., Palmer, T. D., and Gage, F. H. (2002). Functional neurogenesis in the adult hippocampus. *Nature* **415,** 1030–1034.

Wagner, J. P., Black, I. B., and DiCicco-Bloom, E. (1999). Stimulation of neonatal and adult brain neurogenesis by subcutaneous injection of basic fibroblast growth factor. *J. Neurosci.* **19,** 6006–6016.

Weiss, S., Dunne, C., Hewson, J., Wohl, C., Wheatley, M., Peterson, A. C., and Reynolds, B. A. (1996). Multipotent CNS stem cells are present in the adult mammalian spinal cord and ventricular neuroaxis. *J. Neurosci.* **16,** 7599–7609.

Wen, P. H., Shao, X., Shao, Z., Hof, P. R., Wisniewski, T., Kelley, K., Friedrich, V. L., Jr., Ho, L., Pasinetti, G. M., Shioi, J., Robakis, N. K., and Elder, G. A. (2002). Overexpression of wild type but not an FAD mutant presenilin-1 promotes neurogenesis in the hippocampus of adult mice. *Neurobiol. Dis.* **10,** 8–19.

Wichterle, H., Lieberam, I., Porter, J. A., and Jessell, T. M. (2002). Directed differentiation of embryonic stem cells into motor neurons. *Cell* **110,** 385–397.

Won, S. J., Kim, S. H., Xie, L., Mao, X. O., Jin, K., and Greenberg, D. A. (2006). Reelin-deficient (reeler) mice show impaired adult neurogenesis and increased sensitivity to cerebral ischemia. *Exp. Neurol.* **198,** 250–259.

Yoshimura, S., Takagi, Y., Harada, J., Teramoto, T., Thomas, S. S., Waeber, C., Bakowska, J. C., Breakefield, X. O., and Moskowitz, M. A. (2001). FGF-2 regulation of neurogenesis in adult hippocampus after brain injury. *Proc. Natl. Acad. Sci. USA* **98,** 5874–5879.

Yoshimura, S., Teramoto, T., Whalen, M. J., Irizarry, M. C., Takagi, Y., Qiu, J., Harada, J., Waeber, C., Breakefield, X. O., and Moskowitz, M. A. (2003). FGF-2 regulates neurogenesis and degeneration in the dentate gyrus after traumatic brain injury in mice. *J. Clin. Invest.* **112,** 1202–1210.

Zhang, R., Zhang, Z., Wang, L., Wang, Y., Gousev, A., Zhang, L., Ho, K. L., Morshead, C., and Chopp, M. (2004a). Activated neural stem cells contribute to stroke-induced neurogenesis and neuroblast migration toward the infarct boundary in adult rats. *J. Cereb. Blood Flow Metab.* **24,** 441–448.

Zhang, R., Zhang, Z., Zhang, C., Zhang, L., Robin, A., Wang, Y., Lu, M., and Chopp, M. (2004b). Stroke transiently increases subventricular zone cell division from asymmetric to symmetric and increases neuronal differentiation in the adult rat. *J. Neurosci.* **24,** 5810–5815.

Zhang, R. L., Zhang, Z. G., Zhang, L., and Chopp, M. (2001). Proliferation and differentiation of progenitor cells in the cortex and the subventricular zone in the adult rat after focal cerebral ischemia. *Neuroscience* **105,** 33–41.

Zhang, Z. G., Jiang, Q., Zhang, R., Zhang, L., Wang, L., Arniego, P., Ho, K. L., and Chopp, M. (2003). Magnetic resonance imaging and neurosphere therapy of stroke in rat. *Ann. Neurol.* **53,** 259–263.

Zhao, M., Momma, S., Delfani, K., Carlen, M., Cassidy, R. M., Johansson, C. B., Brismar, H., Shupliakov, O., Frisen, J., and Janson, A. M. (2003). Evidence for neurogenesis in the adult mammalian substantia nigra. *Proc. Natl. Acad. Sci. USA* **100,** 7925–7930.

Zigova, T., Pencea, V., Wiegand, S. J., and Luskin, M. B. (1998). Intraventricular administration of BDNF increases the number of newly generated neurons in the adult olfactory bulb. *Mol. Cell Neurosci.* **11,** 234–245.

SEROTONIN AND BRAIN: EVOLUTION, NEUROPLASTICITY, AND HOMEOSTASIS

Efrain C. Azmitia

Department of Biology and Psychiatry, Center for Neural Science, New York University
New York 10003, USA

I. Introduction
II. Evolution: From Unicellular to Humans
III. Holistic Brain Function Starts at Development
IV. Homeostasis of Brain
V. Clinical Implications of Loss of Homeostasis
References

Neurotransmitters are chemicals released by the presynaptic endings of neurons that have specialized actions through specific receptors on the membrane potential of postsynaptic neurons. However, these chemicals have more complex functions both within the neuron and on its varied cellular targets. These functions include changes in metabolic activities that have consequences on the plasticity of the neurons. Serotonin is an ancient chemical synthesized from an indole-containing precursor, tryptophan. A review of the evolution of this chemical within biological systems helps appreciate its holistic actions on brain homeostasis. In plants serotonin and its allied indole-containing molecules 5-hydroxytryptophan, auxin, and melatonin regulate many of the processes involved in cell differentiation: mitosis, migration, and maturation [Pasternak *et al.* (2005). *J. Exp. Bot.* 56, 1991–2001; Kolar and Machackova (2005). *J. Pineal Res.* 39, 333–341]. In animals, in addition to these trophic properties, 5-hydroxytryptamine (5-HT) participates in most biological functions, especially those associated with limbic and brainstem circuits. The precocious development in the center of the brainstem, its response to a plethora of stimuli and its extensive connection to all areas of the brain provide the framework for 5-HT contribution to holistic functioning of the brain. The fine anatomy of the axons and their exquisite sensitivity to environmental trophic and toxic molecules encourages the dynamics of 5-HT innervation pattern and density. The ability to modify itself by a process called neuroplasticity makes it suited to serve as a regulator in brain homeostasis, and predicts its involvement in many brain disorders, especially those concern with depression and dementia.

DOI: 10.1016/S0074-7742(06)77002-7

0074-7742/07 $35.00

I. Introduction

This chapter presents evidence that the neurotransmitter serotonin (5-hydroxytryptamine, 5-HT) functions as a global factor involved in brain homeostasis. The early appearance of 5-HT axons and their projections throughout the brain and spinal cord occurs prior to the differentiation of most neurons and their participation in functional circuits. The emphasis of the chapter will be on 5-HT's activation of the 5-HT$_{1A}$ receptor to release the glial neurite extension factor, S100B. S100B acts to stabilize the microtubules which form the main framework of the cytoskeleton of neural cells, including neurons and astrocytes.

One of the most dynamic and pervasive neuronal systems is the brainstem raphe serotonergic neurons. The actions of 5-HT on cellular metabolism, movement, and reproduction evolved from its first appearance in aerobic unicellular organisms and plants to its current restricted neuronal localization in the brainstem of humans. In the most primitive organism, 5-HT acts within the cell to regulate cell oxidation largely due to its indole core structure, a unique ring configuration. The indole ring captures light energy and converts it to biological energy by loss of an electron (oxidation). The indole ring is now an oxidizing agent and is reduced by absorbing an electron, usually from a metal ion. The indole ring now functions as antioxidant by acting as a reducing agent that can easily lose this electron. In the cell, serotonin, melatonin, auxin, and many indole-alkaloids act as powerful antioxidants. In addition, 5-HT synthesis by the hydroxylase enzyme directly captures free oxygen and serves the important role of reducing the concentration of this reactive molecule.

In the animal kingdom, while retaining this important antioxidant and differentiating properties, 5-HT begins in crustaceans to influence higher brain functions such as dominance in a social grouping. In the human brain, the 5-HT neurons from the raphe nuclei make connections innervating the entire brain and spinal cord, which builds on its functional impact to include a link in humans to suicide.

The interactions of 5-HT neurons with neuronal and nonneuronal systems are covered with an emphasis on the diversity of the cells receiving released 5-HT: neurons, glial cells, endothelial cells, ependymal cells, and endocrine cells. The diversity of these cellular targets argues for a broad function for 5-HT in brain. The ability of 5-HT to promote brain plasticity and stabilization by acting on the cell cytoskeleton is discussed in the context of homeostasis. The important action of 5-HT on the 5-HT$_{1A}$ receptor in releasing the glial protein S100B is the basis of this regulation of morphological plasticity. The chapter, as a whole, supports the key idea expressed by Cannon in his discussion of homeostasis that "slight instability is the necessary condition for the true stability of the organism." Neuroplasticity is a necessary attribute of a homeostatic system, but early development and global interconnections make this system holistic in scope. The ability to change morphology, stimulate neurogenesis and differentiation, or promote cell survival is influenced by acetylcholine, catecholamines, GABA, EAAs (glutamate and glycine), and neuropeptides. However, only serotonin (5-HT) has the evolutionary and anatomical properties to serve as a global regulator unifying the whole brain into a cohesive biological system.

II. Evolution: From Unicellular to Humans

A delay in our appreciation of the diverse actions of neurotransmitters was the focus on electrophysiology which dominated the study of the brain in the early twentieth century. The chemical substances released by neurons were considered to be mediators of the electrical current across the synapse, thus the word "neurotransmitters." The focus of many early neuroscience studies in the twentieth century was to determine if neurotransmitters were excitatory or inhibitory electrical influences on the membrane potentials of postsynaptic cells. Neurotransmitters by acting on specific receptors were considered to be ionotrophes capable of opening specific ion channels. They traveled in fixed circuits and the notion of neurons and synapses being plastic was not considered until late in the 1950s (Liu and Chambers, 1958).

The reality of the situation is "neurotransmitters" predate the formation of nervous tissue. Serotonin is found in all animals, plants, and most unicellular organisms (Garattini and Valzelli, 1965). It is synthesized from the amino acid tryptophan by the action of two enzymes, tryptophan hydroxylase, and aromatic amino acid decarboxylase (Fig. 1). Tryptophan is synthesized by a variety of enzymes (Zhao and Last, 1996). The creation of the indole structure served an important function in the start of aerobic life on the earth. The conversion of energy (photons) derived from the sun into biological energy requires capturing a light wave and the loss of an electron. Interestingly, the indole ring is the most efficient molecule for doing exactly this and most sensitive to blue light (450 nm) (Borkman and Lerman, 1978). Most proteins are endowed with an intrinsic UV fluorescence because they contain aromatic amino acids, specifically phenylalanine, histamine, tyrosine, and tryptophan. Of these aromatic amino acids, tryptophan has the highest fluorescence quantum yield overshadowing markedly the emissions of the other two. Tryptophan emission maxima in proteins can vary from 332 to 342 nm depending on the protein. Free tryptophan has a characteristic fluorescence emission at 350–360 nm (Borkman and Lerman, 1978). Absorption of blue light waves is able to excite the indole structure so that it loses one of the electrons from its indole ring structure, it becomes oxidized. This single electron begins a directed journey jumping from heavy metals to

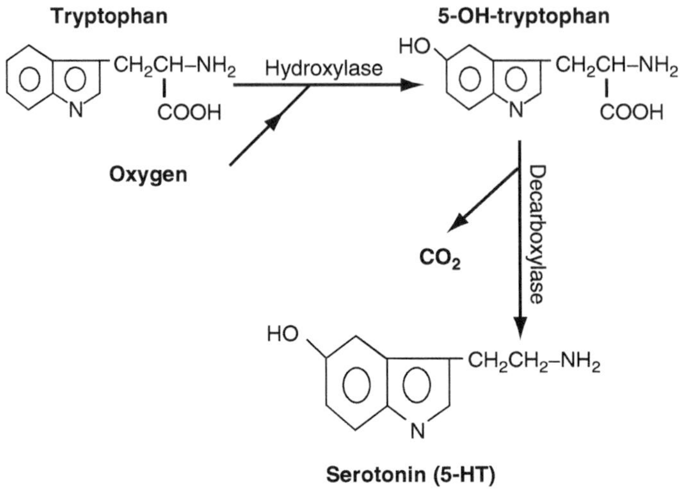

FIG. 1. The synthesis of serotonin from tryptophan involves two enzymes, tryptophan hydroxylase and aromatic amino acid decarboxylase.

finally produce reduced chemical cofactors, such as NADH and NADPH, and generates O_2 from H_2O as a by-product. The production of O_2 is the most relevant for life on the earth, since our atmosphere on the earth contains 20% oxygen, and supports all aerobic organisms. Tryptophan's ability to capture light is used by nearly all proteins (e.g., chlorophyll, rhodopsin, and skin pigment cells) which capture light and convert it to biological energy (Angiolillo and Vanderkooi, 1996). These proteins have tryptophan as the core amino acid for their function. Cells such as blue-green algae, molds, and plants became very adept at producing oxygen. A critical problem developed in these cells: What to do with all the reactive oxidizing agents produced during the generation of oxygen?

One of the solutions for dealing with excess oxidation was to use the biological machinery employed before O_2 was available, and the atmosphere was nearly all CO_2. Anaerobic cells had developed a variety of enzymes for converting CO_2 into biological energy in the form of glucose. The most common process was to produce the sugar, glyceraldehyde-3 phosphate, a three-carbon sugar produced by three molecules of CO_2. The first enzymatic step in this reaction involves the attachment of a molecule of CO_2 to the five-carbon sugar, ribulose bisphosphate (RuBP). The reason this step is emphasized is that the enzyme which catalyzes this initial reaction, and possibly the most abundant enzyme on the earth, is RuBP carboxylase, also known as rubisco. The carboxylase at the very early stages of life on the earth primarily attached to CO_2, however as O_2 levels increased it was shown that this compound could react more favorably with oxygen (Smith, 1976). This was the first enzyme to attach oxygen to a substrate [such as tryptophan to produce 5-hydroxytryptophan (5-HTP)], a process termed hydroxylase because only a single oxygen is used and the other forms water. Furthermore, RuBP carboxylase has the same phosphate-binding site sequence found in tryptophan biosynthetic enzymes (Wilmanns *et al.*, 1991).

The substrates for the primitive hydroxylase enzyme were tryptophan, tyrosine, and phenylalanine, all of which can capture light (Boularand *et al.*, 1998; Grenett *et al.*, 1987; Wiens *et al.*, 1998). The hydroxylase enzyme gave rise to a very large number of complex alkaloids in plants, all of which are potent antioxidants in their own right. As we now know, cellular oxidation is important for cell maturation and division, but excess oxidation results in cell death. The synthesis of pharmaceutically important monoterpenoid indole involves the hydroxylase as well as decarboxylase enzymes (Facchini *et al.*, 2000). 5-HTP, the immediate precursor of serotonin, is formed from tryptophan hydroxylase. This molecule is rapidly converted to serotonin by the ubiquitous working in reverse to function as a decarboxylase. Thus, serotonin is produced from tryptophan by enzymes commonly used in anaerobic organisms before O_2 was formed inside cells. Besides the algae, fungi, and molds, the most efficient generator of O_2 and serotonin is plants. The levels of serotonin inside plants far exceed those seen in the animal brain by 100×; banana skin (40 μg/g) versus rat hippocampus (0.4 μg/g) (Garattini and Valzelli, 1965; Sparks and Slevin, 1985). Interestingly, the immediate precursor of 5-HT, 5-HTP accounts for 20% of the total fresh weight in seeds from *Griffonia simplicifolia*, a tropical shrub of west Africa, which has potent medicinal properties (Lemaire and Adosraku, 2002).

Tryptophan was always a key to life because of its ability to convert solar energy into biological energy. The consequence of this process made tryptophan and its associated molecules involved in all aspects of the organism's life: mitosis, movement, and maturation. As oxygen began to be a major component of the atmosphere of the earth, enzymes that served a central function in conversion of CO_2 into glucose now evolved to hydroxylate many substrates. Hydroxylation leads to 5-HTP and 5-HT as well as to many indole alkaloids used for medicinal purposes today (Fig. 1).

A closer look at plants provides evidence that serotonin and its products, such as melatonin and auxin, serve crucial actions in the life and organization of plants. Plants are complex, multicellular organisms that have specialized cells that function as a unit, a holistic organization. Plants evolved a specialized intracellular organelle, the chloroplast, not only to capture light, but also as the source of tryptophan synthesis. All the enzymes for making tryptophan were localized in these specialized organelles and could only be converted into their mature form when inside the chloroplast (Zhao and Last, 1996). Plants are extremely efficient at capturing light because they were extremely efficient at making tryptophan.

Plants do not have neurons or muscles, but they are nevertheless capable of limited movement by rotating their leaves toward light and sending their roots deep into the soil to capture H_2O and nitrogen. Both the movements of leaves and roots depend on compounds similar to serotonin such as auxin (Ivanchenko *et al.*, 2006). Auxin and other tryptophan derived compounds are transported inside the plant cells and regulate the fast tracing of leaves toward the shifting source of light. The turning of the leaf to its source of energy depends on the rearrangement of the cells cytoskeleton inside the leaf cells. In the root, the emersion into the soil is produced by regulating cell division and maturation. These two forms of producing movement, mitosis, and maturation of plant cells are similar to that seen in unicellular organisms and fungi (Eckert *et al.*, 1999). The actions of serotonin on the cell cytoskeleton and differentiation forecast the actions of serotonin in neuronal development and adult neuroplasticity in mammals (Azmitia, 1999). Receptors for serotonin and other "neurotransmitters" are found in plants. One can assume that receptors are a necessary component for the integration of specialized cells in a multicellular organism, and it appears that their action is concerned with the coordinating metabolic processes.

Animal cells lack chloroplast, an organelle so central to plant photosynthesis and tryptophan synthesis. This lack of a key evolutionary mechanism of life promoted animals to develop a number of traits in order to survive. Animals had to move to capture organism that contained tryptophan and develop specialized cells for extracting O_2 from the atmosphere. 5-HT- and 5-hydroxytryptamine-derived alkaloids are found in sponges, the most primitive form of animal life (Salmoun *et al.*, 2002). This specie does not have a nervous system and feeds by filtration. In hydra, the most primitive animal with a specialized nervous and motor system, 5-HT appears to be localized to sensory cells scattered along the epithelium of the organism (Fig. 2). When 5-HT is applied to hydrazan larvae, the animal undergoes a pronounced metamorphosis in which it develops a variety of specialized cells involved in feeding and movement (McCauley, 1997). This maturation process is triggered by 5-HT release and a receptor action that has protein kinase C (PKC) in its pathway and is blocked by both ketanserine and clozapine. These steps and inhibitors act on human 5-HT receptors (Azmitia, 2001a). When a distinct nervous system is seen, such as in the flatworm (S leucops), serotonin neurons are localized there (Wikgren and Reuter, 1985).

Serotonin in animals is produced in very low quantities because of the limitation of tryptophan and this may explain the very few cells that contain 5-HT. As seen in hydra and flatworms, these specialized 5-HT cells are nevertheless ideally localized and have pronounced actions on the life of the organism. As these organisms evolved a neuronal center for responding to their complex environment, serotonin continues to serve a key part. In Aplysia, there is a primitive brain with only a few rudimentary behaviors mainly concerned with

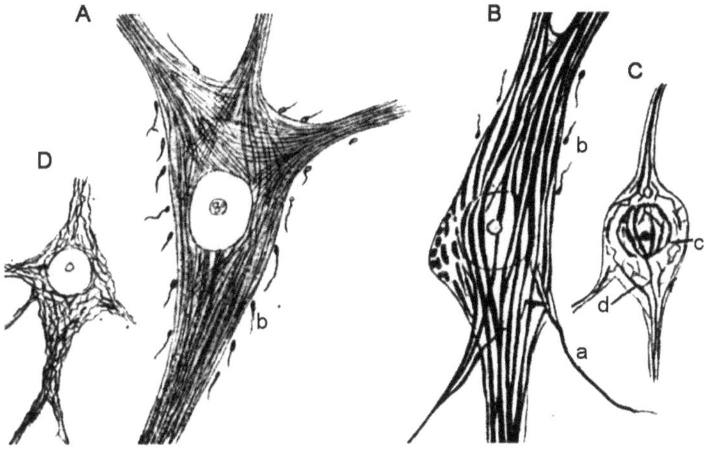

FIG. 2. Cells of lizard spinal cord; illustration demonstrates the changes in neurofibrillar network according to season and temperature. Cajal (1899) used the reduced silver nitrate method. (A and D) Cells of lizard kept in warm or cold for several hours. (B and C) Cells of a lizard in a state of hibernation in lizards. He observed not only was the neuron smaller in size and branching, but there were also fewer connections on its surface.

movement, eating, defense, and reproduction (Marois and Carew, 1997). "The results indicate that the first serotonergic cells emerge at mid-embryogenesis and that a total of five cells make up the entire serotonergic system by hatching. These cells are part of a newly discovered ganglion in Aplysia, called the apical ganglion." The serotonin released from these neurons interacts with specific receptors to increase or decrease the firing rate of its target cells involved in sensory and motor processing. In addition, serotonin changes cAMP and Ca^{2+} levels in its target neurons and influences their transcription rate and modifies cell morphology (Pettigrew *et al.*, 2005). The changes in neuronal morphology are particularly intriguing because they affect neuronal connectivity (Glanzman *et al.*, 1990). 5-HT by increasing cAMP and P-CREB mediates a trophic response that may underlie both maturation and memory formation in this lower animal. Thus, in much the same way as serotonin and its derivatives influence the process and organelles of photosynthesis to move in order to tract the source of light, in animals serotonin influences the morphology of sensory and motor neurons involved in neuronal networking in order to tract the source of relevant stimuli.

As we continue to ascend the animal kingdom, the synthesis of serotonin remains restricted to a few cell types (e.g., mast cells and neurons), involved in promoting differentiation and regulating many key biological functions. Serotonin-producing cells served a defense mechanism (stinging) in coelenterates and in many insects (Horen, 1972; Weiger, 1997). Actions of serotonin on sexual activity and reproduction are seen in Nematodes (Boyle and Yoshino, 2005). In lower animals, serotonin neurons are primarily sensory neurons (activated by external stimuli) and influence food intake, defense withdrawal, and complex locomotor behaviors such as swimming (e.g., in sea urchins, Yaguchi and Katow, 2003). In the ganglia of Annelids, serotonin is first found in interneurons, which permits better regulation of complex behaviors such as swimming (Kristan and Nusbaum, 1982) and possibly learning and memory (Moss *et al.*, 2005). In *Caenorhabditis elegans*, 5-HT is involved in modulating feeding behavior by rapidly altering a chemosensory circuit (Chao *et al.*, 2004). Serotonin in arthropods (lobsters) regulates socially relevant behaviors such as dominance-type posture, offensive tail flicks, and escape responses (Kravitz, 2000). The actions of serotonin thus extend from those of antioxidant through morphogenesis and ascend to being involved in complex behaviors such as position in a social hierarchy. 5-HT-regulated social and mental behaviors increased in number and complexity in vertebrates. In these higher animals, 5-HT continued in its role of a homeostastic regulator in adjusting the dynamic interactions of these many functions.

III. Holistic Brain Function Starts at Development

A few cells in the body are serotonergic, expressing the enzymes tryptophan hydroxylase, aromatic amino acid decarboylase, 5-HT transporter (5-HTT), and the 5-HT_{1A} receptor. These are found in the brainstem, in the enteric nervous system, and in mast cells scattered throughout the body. Serotonin influences cells in all stages of development and in all organs. In the human brain, serotonin neurons are more numerous (>250,000) than in other species and form a tight, small cluster along the midline of the brainstem (Tork, 1990). The projections from these clusters are more restricted than the diffuse projections seen in rodents (Fig. 3). The axons in rats and mice are predominately thin, highly branched, and unmyelinated. In primates, highly myelinated fibers are common compared to the rodents where they are rare (Azmitia and Gannon, 1983) (Fig. 4). Thus, the organization of the serotonin neurons is more evolved in the primate. In Aplysia, a few giant cells contain serotonin, and these cells have dense projections to both sensory and motor centers. In rodents, serotonergic neurons act by mass action, with large number of neurons acting in concert. In the primate brain, the serotonergic cell bodies collect into small clusters and appear to have established more discrete target.

The overall function of serotonin is fairly similar in vertebrates with most biological processes (movement, breathing, reproduction, and temperature regulation) under the strong influence of serotonin. The activity of brainstem serotonergic

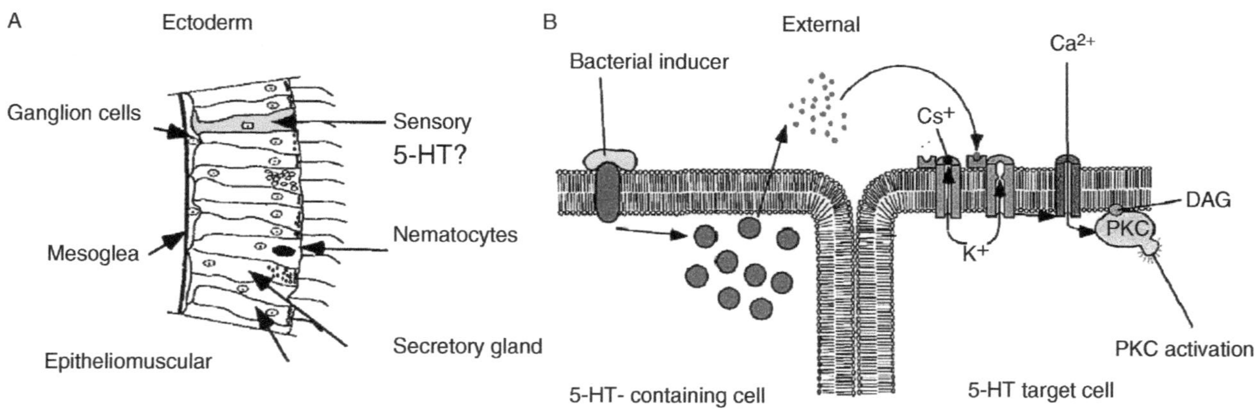

FIG. 3. Proposed 5-HT action in hydra nervous system (modified from McCauley, 1997). (See Color Insert.)

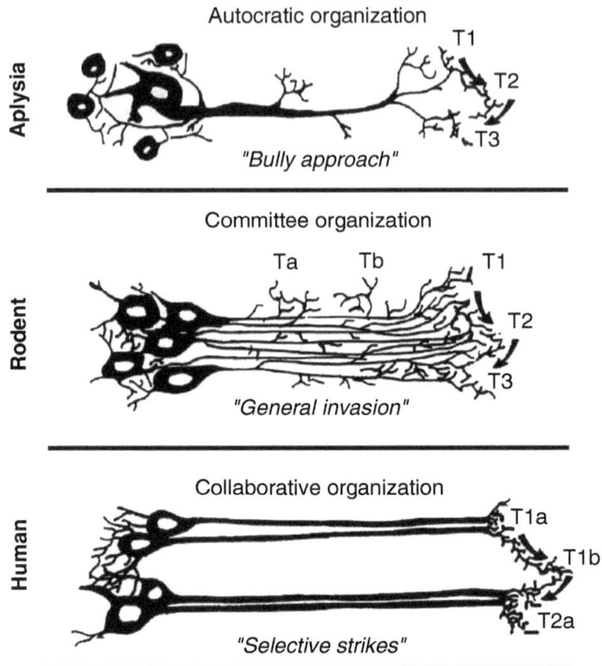

FIG. 4. The progression of 5-HT neurons from Aplysia to humans (Azmitia, 1987).

neurons in rodents and cats has a slow and rhythmic pattern of firing. This pattern produces a constant release of serotonin, ideally suited for the distribution of a trophic molecule rather than a neurotransmitter system involved in point-to-point rapid activity (Jacobs and Azmitia, 1992). In higher animals, serotonin continues to be involved in many behavioral activities, including aggression, sleeping, eating, locomotor activity, attention, learning, memory, sensitization, and sexual activity. It also regulates physiological mechanism such as temperature, feeding, respiration, blood flow and clotting, osmolarity, and hormone secretion. These functions act in concert and there combined function can be described as holistic.

The organization of the 5-HT neurons in the human brain has been extensively reviewed (Azmitia, 1978; Jacobs and Azmitia, 1992; Parent, 1981; Tork, 1990) (Fig. 4). Initially, there are two large groups of serotonergic neurons which appear early in development. The serotonergic phenotypic is induced by the external transcription spacer, Pet-1 (Hendricks *et al.*, 1999). Pet-1 is regulated by the LIM homeodomain and the transcription factors Mashi and Gata-2, after the functional loss of the Nkx2.2 homeodomain. Pet-1 in turn is the transcription factor for 5-HT_{1A} receptor, tryptophan hydroxylase, and the 5-HTT which are expressed around gestational day 12–14 in the mouse brain (first trimester in humans). The two distinct groups of 5-HT neurons appear to have distinct maturational and migrational patterns (Lidov and Molliver, 1982; Wallace and Lauder, 1983). The anterior group (DRM, dosal raphe nucleus; CSD, centralis superior nucleus, pars dorsalis; CSM, centralis superior nucleus pars medianus, supraleminiscal nucleus) projects predominately rostrally to the forebrain, thalamus, and hypothalamus, while the caudal group (RO, nucleus raphe obscurus; NRPa, nucleus raphe pallidus; RM, nucleus raphe magnus; VR, nucleus raphe ventricularis) projects caudally and ventrally to innervate the spinal cord and cerebellum, respectively.

Besides having a number of similar proteins under the control of Pet-1, the various 5-HT neurons share a similar neuronal appearance (large soma, multipolar shape, and largely unmyelinated, highly varicose thin axons) (Cajal, 1899; Scheibel *et al.*, 1975; Steinbusch, 1981). The dendrites from these neurons have spines and are associated with blood vessels and glial cells (Azmitia, 1978). The axons show a propensity to follow myelinated fiber tracts in order to innervate a wide variety of targets by a process termed epiphytic guidance (Azmitia, 1978). Both the dendrites and axons are seen crossing the ependymal layer to enter and exit from the ventricular system. The neurons share a distinct firing pattern which is high during the day when the animals are mobile and very slow at night when the animals are sleeping (Jacobs and Azmitia, 1992). The raphe neurons are sensitive to glucose, pH, blood CO_2 and O_2, and body temperature (Azmitia, 1999; Severson *et al.*, 2003). The serotonergic neurons not only affect the morphology of neurons, but also glial cells (Chang *et al.*, 2005).

Most functions attributed to serotonin have centered on its effects on specific neurons and distinct receptors. For example, electrophysiological studies indicate these neurons are all sensitive to CO_2 (Severson *et al.*, 2003) and may participate in the process of respiration, and correlate with behavioral arousal and may participate in motor activity (Jacobs and Azmitia, 1992). Sensitivity to CO_2 and behavioral arousal can be interpreted to be homeostatic, but not holistic. The next discrete function for serotonin neurons was their involvement in respiration. Using Golgi-stained brainstem material, a close relationship is seen between the raphe reticular neurons and blood vessels. Scheibel *et al.* (1975) wrote nearly three decades ago after they found raphe neurons in contact with blood vessels:

> A chemosensitive role for these raphe elements represents a reasonable extension of putative reticular function. Proactive evidence already available suggests that some brain stem neurons may be sensitive to blood CO_2 levels, and to the osmolarity of the circulating medium. Indeed, it is conceivable that raphe neurons themselves may be sensitive to one or another of these ... given their intimate neurovascular position and their apparent obligatory role in the onset of sleep, they may be capable of detecting circulating substances such as plasma cortisol and ACTH, etc., whose concentrations are time-locked to circadian rhythms and possibly to the shorter rest-activity cycle of Kleitman.

More recently, carbon dioxide chemoreceptors were found in both the medullary and midbrain raphe serotoninergic neurons (Severson *et al.*, 2003). Previously, these chemosensitive cells were believed to be confined to the medulla. This finding is reminiscent of the diurnal rhythm in raphe firing (Jacobs and Azmitia, 1992). The cells contain similar transcription factors to make, release, and detect serotonin (Hendricks *et al.*, 1999) and the fetal innervation of common targets in spinal cord and hippocampus (Azmitia and Whitaker-Azmitia, 1987). Severson *et al.* (2003) suggest their results are relevant to sudden infant death syndrome (SIDS), panic disorder, and migraine headaches. They write that "despite a tendency to study these neurons in relation to only a shared brain function or disease, their highly divergent projections and the homogeneity of their cellular properties (Jacobs and Azmitia, 1992) suggest that there may be a shared function of serotonergic neurons." These studies show a similar function of the midbrain and medullary raphe neurons although these cells develop in different areas of the brainstem, as the rostral and caudal groups seen at gestational day 14 (Scott *et al.*, 2005; Wallace and Lauder, 1983). These shared structural, metabolic, developmental, anatomical, and functional characteristics suggest the raphe neurons, whether located in the midbrain, pons, or medulla, share a holistic action on brain function, irreducible to the sum of its parts (Azmitia, 2004).

IV. Homeostasis of Brain

Historically, many scientists have proposed general (holistic?) functions for serotonin. Brodie and Shore (1957) proposed a holistic metabolic role for serotonin in the neuronal activity of the brain. In their hypothesis, norepinephrine and serotonin modulated opposite systems in the brain based on Hess's (1954) concept of the functional integration of the autonomic system with the central nervous system (CNS). Serotonin was the modulator of the trophotrophic system, which integrates behavioral patterns that are recuperative in nature. This was considered a recessive system, which normally functions during sleep or hibernation. This idea was similar in nature to Cajal's view that sleep represented a time of neuronal rest characterized by a withdrawing of the neuronal connections. This idea comes close to a holistic regulator, the metabolic control may be consistent with a neurotrophic role for 5-HT and NE, and their interactions can be viewed as homeostatic in nature. However, the framework for this system was faulty. Serotonin was considered to function during sleep, although the firing rate of the raphe nuclei is silent during sleep. Furthermore, Brodie and Shore supposed the main interactions were between neurons, and failed to acknowledge the role of nonneuronal cells: glial, ependymal, endothelial, and hormonal. Despite these shortcomings, which mainly reflect the state of neuroscience in 1957, the theories of Brodie and Shore are worth revisiting. A few years after Brodie and Shore (1957), serotonin alone was proposed to be essential for normal mental health (Woolley, 1961). This hypothesis was based on its similar structure to d-lysergic acid diethylamide (LSD), discussed more fully by Whitaker-Azmitia (1999). This idea was not well received and most current mental health professionals do not consider mental health to be a holistic disorder. Mental disorders are thought to be a local dysfunction due to a specific deficit, such as reduced levels of dopamine or serotonin, in a localized region of the brain such as the hippocampus, prefrontal cortex, or cingulate.

We now propose that the globally projecting raphe neurons have the anatomical and functional characteristics to coordinate the physiology of the whole brain. Although specific actions of 5-HT are local, nevertheless the scope is global. The disruption of one group of raphe neurons impacts the system as a whole. The role of 5-HT as an integrating component of neural tissue emphasizes the importance of neuroplasticity. 5-HT neurons show morphological and functional responses to a variety of neuronal and nonneuronal factors. The dynamic view of a plastic 5-HT brainstem system with neurotrophic actions encompasses the concepts of Woolley (1961), Brodie and Shore (1957), and Severson *et al.* (2003). The serotonin neurons evolved from plants as general regulatory system which responds to external stimuli to produce structural changes to meet those signals, be they the source of light or temperature. The system modifies itself to achieve the instability needed for homeostasis. This function of serotonin can be observed in plants and unicellular organisms, long before the advent of neurons. The fluctuations in serotonin levels are broadcast throughout the brain and serve to dynamically integrate and stabilize CNS structure and function. In a previous paper, we introduced the concept that 5-HT raphe neurons might participate in the process of brain homeostasis. The maintenance of a stable nervous system in a dynamic environment is certainly a holistic function since it is difficult to imagine this process being a sum of its component parts. Homeostasis implies not only stability of a given set point or function, but more importantly the dynamic equilibrium seen around that set point.

Proposing 5-HT raphe neurons are involved in homeostasis may be considered a truism, but there are certain implications of this statement that make it interesting to consider. First, a homeostatic regulator needs to sense all the pertinent variables necessary to achieve and maintain an equilibrium. The 5-HT distribution in the brain reaches all areas and includes target cells in the vascular, neuronal, and endocrine systems. The function of 5-HT neurons serves to integrate all cell types in all areas of the brain. The global framework serves to receive and integrate the varied pertinent variables into a holistic unit (Fig. 5). Second, a homeostatic regulator needs to adjust the activity and architecture of the systems involved in equilibrium. 5-HT neurons can produce rapid changes in postsynaptic neuronal firing, glial activity, blood flow, breathing, temperature,

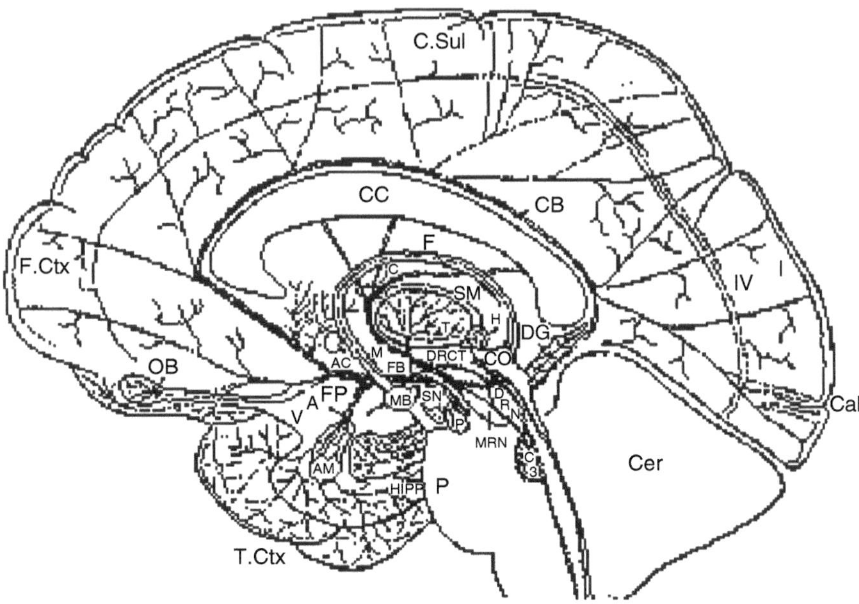

FIG. 5. The 5-HT projections in the human brain (modified from Azmitia and Gannon, 1986).

and hormonal secretion. 5-HT promotes cellular mitosis, migration and maturation of neurons and glial cells, and change how these cellular systems interact. Third, a homeostatic regulator should be able to adjust its own set point to accommodate changes in input to more efficiently reduce fluctuations. 5-HT neurons modify their own cellular architecture in response not only to sensory neuronal inputs, but also to glial cells (Azmitia *et al.*, 1990; Nishi *et al.*, 2000), hormonal levels (Azmitia *et al.*, 1993; Chamas *et al.*, 2004; Cordero *et al.*, 2001), neuropeptides (Davila-Garcia and Azmitia, 1990), and glucose (Martin-Cora *et al.*, 2002). Some of the effects of nutrition can be directly traced to the supply of the essential amino acid tryptophan. "These relationships between precursor availability from the periphery and brain neurotransmitter synthesis may ultimately provide the brain with information about peripheral metabolic state" (Fernstrom, 1977).

The ability to change activity and shape in response to external factors has important implications with respect to homeostasis. It is interesting to quote Professor Walter Cannon, who first elucidated the concept of "fight or flight," and whose research on the PNS and neurotransmission led to the concept of homeostasis. "By an apparent contradiction, it maintains its stability only if it is excitable and capable of modifying itself according to external stimuli, and adjusting its response to the stimulation. In a sense it is stable because it is modifiable—the slight instability is the necessary condition for the true stability of the organism" (Cannon, 1929). We coined the phrase neuronal instability to refer to the tendencies of neuronal cytoskeleton to shrink in the absence of stabilizing molecules such as S100B and glucocorticoids (Azmitia, 2002; Azmitia and Liao, 1994). The morphology of granule neurons in the adult hippocampus decreased when circulating glucocorticoids were removed (Liao *et al.*, 1993). This could be seen by measuring the size of the dentate gyrus and was accompanied by a loss of 5-HT_{1A} receptor mRNA.

The 5-HT_{1A} receptor is an ancient molecule, estimated to be about 800 million years old (Peroutka and Howell, 1994). This is an intronless receptor protein that was mentioned earlier as being induced by PET-1, an external transcription spacer seen in 5-HT neurons prior to their phenotypic differentiation. Thus, the loss of granule neuronal phenotype and 5-HT_{1A} receptor mRNA may be molecularly and evolutionarily linked. The very early developmental expression of the 5-HT_{1A} receptor protein mRNA is seen at fetal day 15 in the rat (Hillion *et al.*, 1993). At this very early period, which is prior to neuronal differentiation in most of the forebrain, the levels of the 5-HT_{1A} receptor mRNA are higher than at any other time in the life of the animal. In the immature cerebellum of the rat, the 5-HT_{1A} receptor protein expression is seen on neurons and astrocytes (Matthiessen *et al.*, 1992). A similar, but lower, glial expression of the 5-HT_{1A} receptor protein was seen in the adult hippocampus (Whitaker-Azmitia *et al.*, 1993). What is the significance of this early expression, and why on both neurons and glial cells?

The action of astrocytes on neuronal survival and differentiation is complex. First, cortical neuronal precursors are believed to form from radial glial cells and astrocytes during early development (Alvarez-Buylla *et al.*, 2001). The astrocytes provide glucose from stored glycogen, a process under the control of the 5-HT_{2A} receptor (Azmitia, 2001a). Glial cells release protein factors involved in neuronal survival (e.g., NGF), attachment (e.g., laminin), and extension (e.g., S100B). These cells contain many different neurotransmitter receptors to control the availability of these factors, and the 5-HT_{1A} receptor is particularly related to S100B release (Ahlemeyer *et al.*, 2000; Eriksen *et al.*, 2002; Whitaker-Azmitia *et al.*, 1990) (Fig. 6). S100B is a neurite extension factor (Azmitia *et al.*, 1990; Kligman and Marshak, 1985). The ability to promote neurite extension is attributed to its ability to prevent the phosphorylation of MAP proteins by PKC (Baudier and Cole, 1988; Sheu *et al.*, 1994). This is an MAP-specific inhibition since S100B does not inhibit PKC phosphorylation of histones (Sheu *et al.*, 1994) and may be related to the ability of S100B to directly interact with MAPs (Donato *et al.*, 1989). The route from the astrocytes where S100B is made to interact with the neuronal cytoskeleton may involve the receptor for advanced glycation endproducts (RAGE) (Hofmann *et al.*, 1999; Rong *et al.*, 2005). RAGE can translocate extracellular S100 into human endothelial cells (Hsieh *et al.*, 2004),

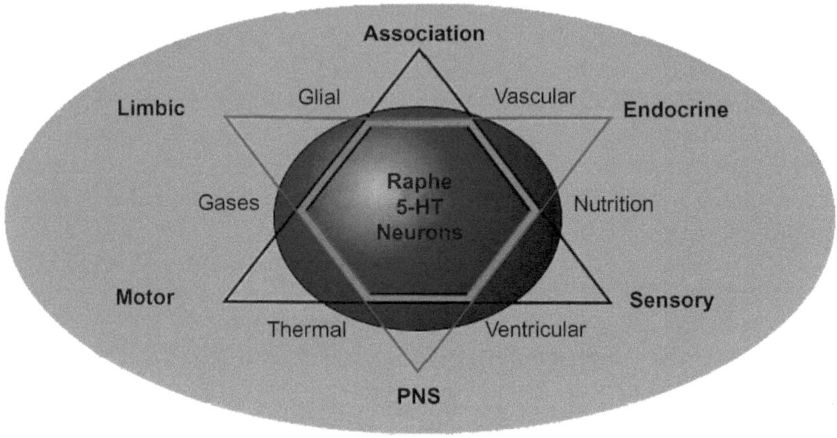

FIG. 6. A schematic representation of 5-HT raphe neuronal interactions with the key systems of the brain (modified from Azmitia, 1999). (See Color Insert.)

and we can assume a similar translocation occurs with neurons that have the RAGE receptor. Interestingly, AGE molecules that are elevated in diabetes block this translocation.

The interactions between the 5-HT_{1A} receptor and S100B thus provide the mechanism for modifying not only the serotonergic neurons themselves but the neurons throughout the brain. The first to show a 5-HT-induced loss of target cell morphology were the late Dr. Okado and his students (Okado *et al.*, 1993). *p*-chlorophenylalanine (PCPA) a serotonin synthesis inhibitor was given to adult chicks for 1 week, and the nonserotonergic axodendritic synapses in the cortex were shown to be dramatically reduced. When the studies were repeated in adult rats, the number of synapses in the hippocampus was significantly reduced and the animals showed memory loss (Matsukawa *et al.*, 1997). 5-HT loss by either PCPA or *para*-chlorophenylalanine produced loss of neuronal dendrites and terminals in adult rat brains (Azmitia *et al.*, 1995; Whitaker-Azmitia *et al.*, 1995), and the injection of a 5-HT_{1A} receptor agonist reversed these losses (Azmitia *et al.*, 1995). The loss of the target morphology not only was reversed by injections of the 5-HT_{1A} agonist, but was shown to involve the glial protein S100B (Eriksen and Druse, 2001; Wilson *et al.*, 1998). These and other changes in adult morphology indicate that 5-HT has a central function in regulating the morphology of adult neurons, either by increasing S100B or by direct action of the 5-HT_{1A} receptor (Azmitia, 2001b). Furthermore, the strong actions of 5-HT during development indicates these trophic actions of serotonin are present at birth and persist throughout life (Mazer *et al.*, 1997; Whitaker-Azmitia, 2005).

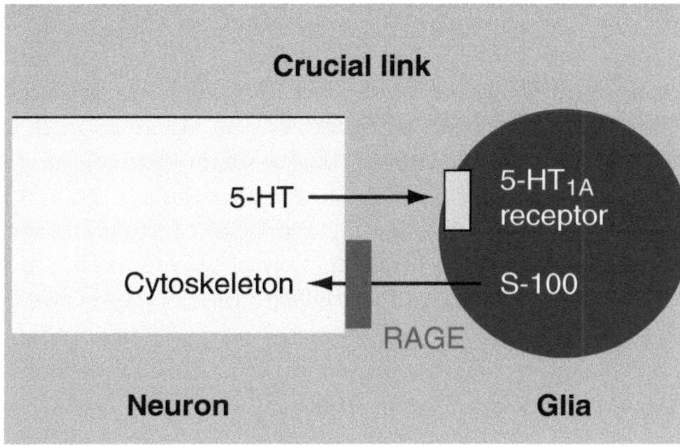

FIG. 7. A schematic showing the relationship between 5-HT neurons and glial cells. (See Color Insert.)

Cannon said: "In a sense it is stable because it is modifiable." Thus, if 5-HT is directly involved in regulating homeostasis of the brain, it produces stability because it functions to release S100B and is also regulated by it (Fig. 7). The direct receptor-mediated stability from 5-HT_{1A} receptor inhibition of C-AMP is developed in a previous review (Azmitia, 1999). 5-HT, as mentioned earlier, is found in all plants and animals. 5-HT from maternal blood begins to bathe the developing fetus from conception, providing a very early start to its functioning as a homeostatic regulator in the dynamic emerging connections of the brain. But what happens when 5-HT is lost? Short-term decreases in 5-HT occur almost every night when we sleep and muscle movement is inhibited (Trulson and Jacobs, 1979). This is consistent with the notion that neuronal connections are unstable and labile during sleep, an idea supported by Cajal that is over 100 years old (Azmitia, 2002). Since serotonin levels fluctuate over the year (Singh, 1964; Wirz-Justice *et al.*, 1977), we can expect, although it has not been shown, that the capacity for learning and memory may show a similar fluctuation. What has been shown to fluctuate is the incidence of suicides (Bjorksten *et al.*, 2005; Dreyer, 1959).

V. Clinical Implications of Loss of Homeostasis

Selye (1956) proposed a unified theory to explain why stress, and a corresponding loss of homeostasis, could impair general health. A few years later, Wooley (1962) suggested a more specific hypothesis that serotonin was the principle factor involved in regulating mental health. His theory was based on his work with serotonin and LSD. He had shown a strong structural similarity between serotonin and LSD, despite the fact he was blind. He also was the first to show that serotonin and LSD both had similar functions in the brain. It is intriguing that Dr. Woolley considered that a single chemical could assume a function so large, so important, as regulating mental health. He did not write of neuroplasticity nor of homeostasis, but his scope was certainly holistic. If 5-HT is a regulator of homeostasis, then a dysfunction of serotonin should have major consequences. One consequence of lowered serotonin is depression. Suicide, the act of an individual voluntarily ending his or her own life, is the ultimate rejection of the evolutionary drive to survive. In suicide brains, 5-HT neurons are smaller in size and number (Underwood *et al.*, 1999).

A decrease in 5-HT levels in the brain is thought to contribute to major depression (Anderson, 2004; Meltzer *et al.*, 2003; Neumeister, 2003; Owens, 2004). The 5-HTT protein is believed to be directly related to the presence of 5-HT fibers, and in the brains of depressed patients there is a reduction in the both levels of 5-HTT (Malison *et al.*, 1998) and 5-HTT-IR axons (Austin *et al.*, 2002). Serotonin-specific reuptake inhibitors (SSRIs) have proved beneficial in a wide variety of affective disorders in addition to depression: panic disorder, obsessive-compulsive disorder, social phobia, and generalized anxiety disorder (Den Boer *et al.*, 2000). All these affective disorders appear responsive to 5-HT dysfunction.

However, if homeostasis is lost, somatic and cognitive diseases, as well as affective disorders are impacted. There are several reports of an underlying disorder of the 5-HT system in neurological disorders: Pick's disease (Sparks and Markesbery, 1991), Parkinson's disease (Halliday *et al.*, 1990; Menza *et al.*, 1999), Alzheimer's disease (AD) (Marksteiner *et al.*, 2003), ischemic heart disease (Stout *et al.*, 2003), spongiform encephalopathies (Fraser *et al.*, 2003), and diffuse Lewy body dementia (DLBD) (Ballard *et al.*, 2002). In DLBD, Lewy bodies occur in the dorsal raphe nucleus (Langlais *et al.*, 1993) and marked reductions of serotonin levels are reported in the striatum (Langlais *et al.*, 1993), neocortex (Ohara *et al.*, 1998), and frontal cortex (Perry *et al.*, 1993). Major depression occurs in at least 30% of individuals suffering from DLBD (Klakta *et al.*, 1996), with an association between the depressive disorder and reduced 5-HTT (Malison *et al.*, 1998).

Neurological disorders all have separate causes and distinct characteristics which can be easily identified by the experienced clinician. Depression, the sign of a 5-HT dysfunction, is considered secondary to the primary neurological disease. This type of thinking is illustrated in the following passage from the seventh edition of "Adam's and Victor's Manual of Neurology." "Moreover, the anatomical substrate of the many diseases causing intellectual decline involve different parts of the cerebral cortex . . .. It is not surprising, therefore, that the dementing diseases may also cause a number of non-cognitive disturbances, such as loss of emotional control, changes in behavior and personality . . ." (Victor and Ropper, 2002, p. 170). This implies the underlying pathology of the dementia and the noncognitive disturbances are the same.

However, if the dementing diseases are causing the affective symptoms, one would assume that cognitive problems would normally precede the noncognitive symptoms. But neurodegenerative disorders often exhibit affective symptoms as an initial presentation of the neurodegenerative disease (Kessing and Andersen, 2004), which can dominate the initial clinical presentation (Ballard *et al.*, 2002). A history of major depression, without specification of episode-related cognitive impairment, appears to be a risk factor for subsequent onset of dementia (Kessing and Andersen, 2004). This is supported by early work with twins, which found that depression and psychiatric illness were risk factors for developing dementia (Wetherell *et al.*, 1999). Depression is also known to be a risk factor for the onset of ADs (Chen *et al.*, 1999; Green *et al.*, 2003; Kokmen *et al.*, 1991; Kral and Emery, 1989). Finally, psychiatric deficits may be improved by treatment, but impaired patients' cognitive functioning often does not completely normalize, especially in the areas of memory, executive function, and information-processing speed (Nebes *et al.*, 2003).

Depression can have a neuropathological consequence. Dr. Varham Haroutunian and colleagues found that the brains of AD patients with a lifetime history of depression showed significantly higher levels of both plaque and tangle formation within the hippocampus than brains of AD patients without a lifetime history of depression (Rapp *et al.*, 2006). This suggests that depression comorbid with AD may act synergistically, while comorbidity of DLBD with AD does not. Could the loss of 5-HT, associated with depression, be a key factor in brain deterioration? Many serotonin researchers over the last 50 years have argued for a larger, more global (holistic) function for serotonin. In unicellular organisms and plants, this may be related to regulation of oxidation. The same appears to be true in mammals (Munoz-Castaneda *et al.*, 2006). These early functions of 5-HT can be envisioned to now include neuroplasticity which disrupts brain homeostasis. Clinically, the eventual result of this holistic dysfunction of the serotonin system is depression. 5-HT loss, associated with depression, leads to increases in the occurrence and severity of neurological disorders. The ultimate breakdown of brain homeostasis results in the destruction of many neurological networks, and in extreme cases, results in suicide, the ultimate rejection of life.

References

Ahlemeyer, B., Beier, H., Semkova, I., Schaper, C., and Krieglstein, J. (2000). S-100beta protects cultured neurons against glutamate- and staurosporine-induced damage and is involved in the antiapoptotic action of the 5 HT(1A)-receptor agonist, Bay x 3702. *Brain Res.* **858,** 121–128.

Alvarez-Buylla, A., García-Verdugo, J. M., and Tramonti, A. D. (2001). A unified hypothesis on the lineage of neural stem cells Nature. *Neuroscience* **2,** 287–293.

Anderson, G. M. (2004). Peripheral and central neurochemical effects of the selective serotonin reuptake inhibitors (SSRIs) in humans and nonhuman primates: Assessing bioeffect and mechanisms of action. *Int. J. Dev. Neurosci.* **22,** 397–404.

Angiolillo, P. J., and Vanderkooi, J. M. (1996). Hydrogen atoms are produced when tryptophan within a protein is irradiated with ultraviolet light. *Photochem. Photobiol.* **64,** 492–495.

Austin, M. C., Whitehead, R. E., Edgar, C. L., Janosky, J. E., and Lewis, D. A. (2002). Localized decrease in serotonin transporter-immunoreactive axons in the prefrontal cortex of depressed subjects committing suicide. *Neuroscience* **114,** 807–815.

Azmitia, E. C. (1978). The serotonin producing neurons of the midbrain median and dorsal raphe nuclei. *In* "The Handbook of Psycho-pharmacology" (L. L. Iversen, S. D. Iversen, and S. Snyder, Eds.), Vol. 9, pp. 233–314. Plenum Press, NY.

Azmitia, E. C. (1987). The primate serotonergic system: Progression towards a collaborative organization. *In* "Psychopharmacology. Third Generation of Progress" (H. Meltzer, Ed.), pp. 61–74. Plenum Press, NY.

Azmitia, E. C. (1999). Serotonin neurons, neuroplasticity, and homeostasis of neural tissue. *Neuropsychopharmacology* **21**(2 Suppl.), 33S–45S.

Azmitia, E. C. (2001a). Modern views on an ancient chemical: Serotonin effects on cell proliferation, maturation, and apoptosis. *Brain Res. Bull.* **56,** 413–424.

Azmitia, E. C. (2001b). Neuronal instability: Implications for Rett's syndrome. *Brain Dev.* **23**(Suppl. 1), S1–S10.

Azmitia, E. C. (2002). Cajal's hypotheses on neurobiones and neurotropic factor match properties of microtubules and S-100β. *Progress in Brain Research* **136,** 87–100.

Azmitia, E. C. (2004). Serotoninergic chemoreceptive neurons: A search for a shared function viewpoint; American Society for Pharmacology and Experimental Therapeutics (ASPET). *Mol. Intervent.* **4,** 18–21.

Azmitia, E. C., and Gannon, P. J. (1983). The ultrastructural localization of serotonin immunoreactivity in myelinated and unmyelinated axons within the medial forebrain bundle of rat and monkey. *J. Neurosci.* **3,** 2083–2090.

Azmitia, E. C., and Gannon, P. J. (1986). The primate serotonergic system: A review of human and animal studies and a report on Macaca fascicularis. *Adv. Neurol.* **43,** 407–468.

Azmitia, E. C., and Liao, B. (1994). Dexamethasone reverses adrenalectomy-induced neuronal de-differentiation in midbrain raphe-hippocampus axis. *Ann. NY Acad. Sci.* **746,** 180–193.

Azmitia, E. C., and Whitaker-Azmitia, P. M. (1987). Target cell stimulation of serotonergic neurons in primary dissociated tissue culture. *Neuroscience* **20,** 47–63.

Azmitia, E. C., Murphy, R. B., and Whitaker-Azmitia, P. M. (1990). MDMA (ecstasy) effects on cultured serotonergic neurons: Evidence for Ca^{++} dependent toxicity linked to release. *Brain Res.* **510,** 97–103.

Azmitia, E. C., Liao, B., and Chen, Y. (1993). Increase of tryptophan hydroxylase enzyme protein by dexamethasone in adrenalectomized rat midbrain. *J. Neurosci.* **13,** 5041–5055.

Azmitia, E. C., Rubinstein, V. J., Strafaci, J. A., Rios, J. C., and Whitaker-Azmitia, P. M. (1995). 5-HT_{1A} agonist and dexamethasone reversal of para-chloroamphetamine induced loss of MAP-2 and synaptophysin immunoreactivity in adult rat brain. *Mol. Brain Res.* **677,** 181–192.

Ballard, C., Johnson, M., Piggott, M., Perry, R., O'Brien, J., Rowan, E., Perry, E., Lantos, P., Cairns, N., and Holmes, C. (2002). A positive association between 5HT re-uptake binding sites and depression in dementia with Lewy bodies. *J. Affect. Disord.* **69,** 219–223.

Baudier, J., and Cole, R. D. (1988). Interactions between the microtubule-associated tau proteins and S100b regulate tau phosphorylation by the Ca^{2+}/calmodulin-dependent protein kinase II. *J. Biol. Chem.* **263,** 5876–5883.

Bjorksten, K. S., Bjerregaard, P., and Kripke, D. F. (2005). Suicides in the midnight sun—a study of seasonality in suicides in West Greenland. *Psychiatry Res.* **133,** 205–213.

Borkman, R. F., and Lerman, S. (1978). Fluorescence spectra of tryptophan residues in human and bovine lens proteins. *Exp. Eye Res.* **26,** 705–713.

Boularand, S., Biguet, N. F., Vidal, B., Veron, M., Mallet, J., Vincent, J. D., Dufour, S., and Vernier, P. (1998). Tyrosine hydroxylase in the european eel (*Anguilla anguilla*): cDNA cloning, brain distribution, and phylogenetic analysis. *J. Neurochem.* **71,** 460–470.

Boyle, J. P., and Yoshino, T. P. (2005). Serotonin-induced muscular activity in Schistosoma mansoni larval stages: Importance of 5-HT transport and role in daughter sporocyst production. *J. Parasitol.* **91,** 542–550.

Brodie, B. B., and Shore, P. A. (1957). A concept for a role of serotonin and norepinephrine as chemical mediators in the brain. *Ann. NY Acad. Sci.* **66,** 631–642.

Cajal, S. R. y. (1899). "Texture of the Nervous System of Man and the Vertebrates." Translation by P. Pasik and T. Pasik (1999). Springer-Verlag, Wien, New York.

Cannon, W. B. (1929). Organization for physiological homeostasis. *Physiol. Rev.* **9,** 399–431.

Chamas, F. M., Underwood, M. D., Arango, V., Serova, L., Kassir, S. A., Mann, J. J., and Sabban, E. L. (2004). Immobilization stress elevates tryptophan hydroxylase mRNA and protein in the rat raphe nuclei. *Biol. Psychiatry* **55,** 278–283.

Chang, M. S., Ariah, L. M., Marks, A., and Azmitia, E. C. (2005). Chronic gliosis induced by loss of S-100B: Knockout mice have enhanced GFAP-immunoreactivity but blunted response to a serotonin challenge. *Brain Res.* **1031,** 1–9.

Chao, M. Y., Komatsu, H., Fukuto, H. S., Dionne, H. M., and Hart, A. C. (2004). Feeding status and serotonin rapidly and reversibly modulate a *Caenorhabditis elegans* chemosensory circuit. *Proc. Natl. Acad. Sci. USA* **101,** 15512–15517.

Chen, P., Ganguli, M., Mulsant, B. H., and DeKosky, S. T. (1999). The temporal relationship between depressive symptoms and dementia: A community-based prospective study. *Arch. Gen. Psychiatry* **56,** 261–266.

Cordero, M. E., Rodríguez, A., Torres, R., and Valenzuela, C. Y. (2001). Human raphe magnus nucleus: A morphometric Golgi-Cox study with emphasis on sex differences. *Brain Res. Dev. Brain Res.* **131,** 85–92.

Davila-Garcia, M. I., and Azmitia, E. C. (1990). Neuropeptides as positive or negative neuronal growth regulatory factors: Effects of ACTH and Leu-enkephalin on cultured serotonergic neurons. *In* "Molecular Aspects of Development and Aging of Nervous System" (J. M. Lauder, A. Privat, E. Giacobini, P. S. Timiras, and A. Vernadakis, Eds.), pp. 75–92. Plenum Press, New York.

Den Boer, J. A., Bosker, F. J., and Slaap, B. R. (2000). Serotonergic drugs in the treatment of depressive and anxiety disorders. *Hum. Psychopharmacol.* **15,** 315–336.

Donato, R., Giambanco, I., and Aisa, M. C. (1989). Molecular interaction of S-100 proteins with microtubule proteins *in vitro*. *J. Neurochem.* **53,** 566–571.

Dreyer, K. (1959). Comparative suicide statistics. II. Death rates from suicide in Denmark since 1921, and seasonal variations since 1835. *Dan. Med. Bull.* **6,** 75–81.

Eckert, S. E., Hoffmann, B., Wanke, C., and Braus, G. H. (1999). Sexual development of Aspergillus nidulans in tryptophan auxotrophic strains. *Arch. Microbiol.* **172,** 157–166.

Eriksen, J. L., and Druse, M. J. (2001). Astrocyte-mediated trophic support of developing serotonin neurons: Effects of ethanol, buspirone, and S100B. *Brain Res. Dev. Brain Res.* **131,** 9–15.

Eriksen, J. L., Gillespie, R., and Druse, M. J. (2002). Effects of ethanol and 5-HT1A agonists on astroglial S100B. *Brain Res. Dev. Brain Res.* **139,** 97–105.

Facchini, P. J., Huber-Allanach, K. L., and Tari, L. W. (2000). Plant aromatic L-amino acid decarboxylases: Evolution, biochemistry, regulation, and metabolic engineering applications. *Phytochemistry* **54,** 121–138.

Fernstrom, J. D. (1977). Effects on the diet on brain neurotransmitters. *Metabolism* **26,** 207–223.

Fraser, E., McDonagh, A. M., Head, M., Bishop, M., Ironside, J. W., and Mann, D. M. (2003). Neuronal and astrocytic responses involving the serotonergic system in human spongiform encephalopathies. *Neuropathol. Appl. Neurobiol.* **29,** 482–495.

Garattini, S., and Valzelli, L. (1965). "Serotonin." Elsevier Publishing Co., Amsterdam, London, New York.

Glanzman, D. L., Kandel, E. R., and Schacher, S. (1990). Target-dependent structural changes accompanying long-term synaptic facilitation in Aplysia neurons. *Science* **249,** 799–802.

Green, R. C., Cupples, L. A., Kurz, A., Auerbach, S., Go, R., Sadovnick, D., Duara, R., Kukull, W. A., Chui, H., Edeki, T., Griffith, P. A., Friedland, R. P., *et al.* (2003). Depression as a risk factor for Alzheimer disease: The MIRAGE Study. *Arch. Neurol.* **60,** 753–759.

Grenett, H. E., Ledley, F. D., Reed, L. L., and Woo, S. L. (1987). Full-length cDNA for rabbit tryptophan hydroxylase: Functional domains and evolution of aromatic amino acid hydroxylases. *Proc. Natl. Acad. Sci. USA* **84,** 5530–5534.

Halliday, G. M., Blumbergs, P. C., Cotton, R. G., Blessing, W. W., and Geffen, L. B. (1990). Loss of brainstem serotonin- and substance P-containing neurons in Parkinson's disease. *Brain Res.* **510,** 104–107.

Hendricks, T., Francis, N., Fyodorov, D., and Deneris, E. S. (1999). The ETS domain factor Pet-1 is an early and precise marker of central serotonin neurons and interacts with a conserved element in serotonergic genes. *J. Neurosci.* **19,** 10348–10356.

Hess, W. R. (1954). "Das Zwischenhirn." Benno Schwabe, Basel.

Hillion, J., Milne-Edwards, J. B., Catelon, J., de Vitry, F., Gros, F., and Hamon, M. (1993). Prenatal developmental expression of rat brain 5-HT1A receptor gene followed by PCR. *Biochem. Biophys. Res. Commun.* **191,** 991–997.

Hofmann, M. A., Drury, S., Fu, C., Qu, W., Taguchi, A., Lu, Y., Avila, C., Kambham, N., Bierhaus, A., Nawroth, P., Neurath, M. F., Slattery, T., *et al.* (1999). RAGE mediates a novel proinflammatory axis: A central cell surface receptor for S100/calgranulin polypeptides. *Cell* **97,** 889–901.

Horen, W. P. (1972). Insect and scorpion sting. *JAMA* **221,** 894–898.

Hsieh, H. L., Schafer, B. W., Weigle, B., and Heizmann, C. W. (2004). S100 protein translocation in response to extracellular S100 is mediated by receptor for advanced glycation endproducts in human endothelial cells. *Biochem. Biophys. Res. Commun.* **316,** 949–959.

Ivanchenko, M. G., Coffeen, W. C., Lomax, T. L., and Dubrovsky, J. G. (2006). Mutations in the Diageotropica (Dgt) gene uncouple patterned cell division during lateral root initiation from proliferative cell division in the pericycle. *Plant J.* **46,** 436–447.

Jacobs, B. L., and Azmitia, E. C. (1992). Structure and function of the brain serotonin system. *Physiol. Rev.* **72,** 163–228.

Kessing, L. V., and Andersen, P. K. (2004). Does the risk of developing dementia increase with the number of episodes in patients with depressive disorder and in patients with bipolar disorder? *J. Neurol. Neurosurg. Psychiatry* **75,** 1662–1666.

Klakta, L. A., Louis, E. D., and Schiffer, R. B. (1996). Psychiatric features in diffuse Lewy body Disease: Findings in 28 pathologically diagnosed cases. *Neurology* **47,** 1148–1152.

Kligman, D., and Marshak, D. R. (1985). Purification and characterization of a neurite extension factor from bovine brain. *Proc. Natl. Acad. Sci. USA* **82,** 7136–7139.

Kokmen, E., Beard, C. M., Chandra, V., Offord, K. P., Schoenberg, B. S., and Ballard, D. J. (1991). Clinical risk factors for Alzheimer's disease: A population-based case-control study. *Neurology* **41,** 1393–1397.

Kolar, J., and Machackova, I. (2005). Melatonin in higher plants: Occurrence and possible functions. *J. Pineal Res.* **39,** 333–341.

Kral, V. A., and Emery, O. B. (1989). Long-term follow-up of depressive pseudodementia of the aged. *Can. J. Psychiatry* **34,** 445–446.

Kravitz, E. A. (2000). Serotonin and aggression: Insights gained from a lobster model system and speculations on the role of amine neurons in a complex behavior. *J. Comp. Physiol. [A]* **186,** 221–238.

Kristan, W. B., Jr., and Nusbaum, M. P. (1982). The dual role of serotonin in leech swimming. *J. Physiol. (Paris)* **78,** 743–747.

Langlais, P. J., Thal, L., Hansen, L., Galasko, D., Alford, M., and Masliah, E. (1993). Neurotransmitters in basal ganglia and cortex of Alzheimer's disease with and without Lewy bodies. *Neurology* **43,** 1927–1934.

Lemaire, P. A., and Adosraku, R. K. (2002). An HPLC method for the direct assay of the serotonin precursor, 5-hydroxytrophan, in seeds of Griffonia simplicifolia. *Phytochem. Anal.* **13,** 333–337.

Liao, B., Miesk, B. H., and Azmitia, E. C. (1993). Loss of 5-HT1A receptor mRNA in the dentate gyrus of the long-term adrenalectomized rat and rapid reversal by dexamethasone. *Mol. Brain Res.* **19,** 328–332.

Lidov, H. G. W., and Molliver, M. E. (1982). An immunohistochemical study of serotonin neuron development in the rat: Ascending pathways and terminal fields. *Brain Res. Bull.* **8,** 389–430.

Liu, C. N., and Chambers, W. W. (1958). Intraspinal sprouting of dorsal root ganglia axons. *Arch. Neurol. Psychiatry* **79,** 46–61.

Malison, R. T., Price, L. H., Berman, R., van Dyck, C. H., Pelton, G. H., Carpenter, L., Sanacora, G., Owens, M. J., Nemeroff, C. B., Rajeevan, N., Baldwin, R. M., Seibyl, J. P., *et al.* (1998). Reduced brain serotonin transporter availability in major depression as measured by [123I]-2 beta-carbomethoxy-3 beta-(4-iodophenyl)tropane and single photon emission computed tomography. *Biol. Psychiatry* **44,** 1090–1098.

Marksteiner, J., Walch, T., Bodner, T., Gurka, P., and Donnemiller, E. (2003). Fluoxetine in Alzheimer's disease with severe obsessive compulsive symptoms and a low density of serotonin transporter sites. *Pharmacopsychiatry* **36,** 207–209.

Marois, R., and Carew, T. J. (1997). Ontogeny of serotonergic neurons in Aplysia californica. *J. Comp. Neurol.* **386,** 477–490.

Martin-Cora, F. J., Fornal, C. A., Metzler, C. W., and Jacobs, B. L. (2002). Insulin-induced hypoglycemia decreases single-unit activity of serotonergic medullary raphe neurons in freely moving cats: Relationship to sympathetic and motor output. *Eur. J. Neurosci.* **16,** 722–734.

Matsukawa, M., Ogawa, M., Nakadate, K., Maeshima, T., Ichitani, Y., Kawai, N., and Okado, N. (1997). Serotonin and acetylcholine are crucial to maintain hippocampal synapses and memory acquisition in rats. *Neurosci. Lett.* **230,** 13–16.

Matthiessen, L., Daval, G., Bailly, Y., Gozlan, H., Hamon, M., and Verge, D. (1992). Quantification of 5-hydroxytryptamine 1A receptors in cerebellum of normal and X-irradiated rats during postnatal development. *Neuroscience* **51,** 475–485.

Mazer, C., Muneyyirci, J., Taheny, K., Raio, N., Borella, A., and Whitaker-Azmitia, P. (1997). Serotonin depletion during synaptogenesis leads to decreased synaptic density and learning deficits in the adult rat: A possible model of neurodevelopmental disorders with cognitive deficits. *Brain Res.* **760,** 68–73.

McCauley, D. W. (1997). Serotonin plays an early role in the metamorphosis of the hydrozoan Phialidium gregarium. *Dev. Biol.* **190,** 229–240.

Meltzer, H. Y., Perry, E., and Jayathilake, K. (2003). Clozapine-induced weight gain predicts improvement in psychopathology. *Schizophr. Res.* **59,** 19–27.

Menza, M. A., Palermo, B., DiPaola, R., Sage, J. I., and Ricketts, M. H. (1999). Depression and anxiety in Parkinson's disease: Possible effect of genetic variation in the serotonin transporter. *J. Geriatr. Psychiatry Neurol.* **12,** 49–52.

Moss, B. L., Fuller, A. D., Sahley, C. L., and Burrell, B. D. (2005). Serotonin modulates axo-axonal coupling between neurons critical for learning in the leech. *J. Neurophysiol.* **94,** 2575–2589.

Munoz-Castaneda, J. R., Montilla, P., Padillo, F. J., Bujalance, I., Munoz, M. C., Muntane, J., and Tunez, I. (2006). Role of serotonin in cerebral oxidative stress in rats. *Acta Neurobiol. Exp.* (*Wars.*) **66,** 1–6.

Nebes, R. D., Pollock, B. G., Houck, P. R., Butters, M. A., Mulsant, B. H., Zmuda, M. D., and Reynolds, C. F., III (2003). Persistence of cognitive impairment in geriatric patients following antidepressant treatment: A randomized, double-blind clinical trial with nortriptyline and paroxetine. *J. Psychiatr. Res.* **37,** 99–108.

Neumeister, A. (2003). Tryptophan depletion, serotonin, and depression: Where do we stand? *Psychopharmacol. Bull.* **37,** 99–115.

Nishi, M., Kawata, M., and Azmitia, E. C. (2000). Trophic interactions between brain-derived neurotrophic factor and s100beta on cultured serotonergic neurons. *Brain Res.* **868,** 113–118.

Ohara, K., Kondo, N., and Ohara, K. (1998). Changes of monoamines in post-mortem brains from patients with diffuse Lewy body disease. *Prog. Neuropsychopharmacol. Biol. Psychiatry* **22,** 311–317.

Okado, N., Cheng, L., Tanatsugu, Y., Hamada, S., and Hamaguchi, K. (1993). Synaptic loss following removal of serotoninergic fibers in newly hatched and adult chickens. *J. Neurobiol.* **24,** 687–698.

Owens, M. J. (2004). Selectivity of antidepressants: From the monoamine hypothesis of depression to the SSRI revolution and beyond. *J. Clin. Psychiatry* **65**(Suppl. 4), 5–10.

Parent, A. (1981). Comparative anatomy of the serotoninergic systems. *J. Physiol.* (*Paris*) **77,** 147–156.

Pasternak, T., Potters, G., Caubergs, R., and Jansen, M. A. (2005). Complementary interactions between oxidative stress and auxins control plant growth responses at plant, organ, and cellular level. *J. Exp. Bot.* **56,** 1991–2001.

Peroutka, S. J., and Howell, T. A. (1994). The molecular evolution of G protein-coupled receptors: Focus on 5-hydroxytryptamine receptors. *Neuropharmacology* **33,** 319–324.

Perry, E. K., Marshall, E., Thompson, P., McKeith, I. G., Collerton, D., Fairbairn, A. F., Ferrier, I. N., Irving, D., and Perry, R. H. (1993). Monoaminergic activities in Lewy body dementia: Relation to hallucinosis and extrapyramidal features. *J. Neural. Transm.* **6,** 167–177.

Pettigrew, D. B., Smolen, P., Baxter, D. A., and Byrne, J. H. (2005). Dynamic properties of regulatory motifs associated with induction of three temporal domains of memory in aplysia. *J. Comput. Neurosci.* **18,** 163–181.

Rapp, M. A., Schnaider-Beeri, M., Grossman, H. T., Sano, M., Perl, D. P., Purohit, D. P., Gorman, J. M., and Haroutunian, V. (2006). Increased hippocampal plaques and tangles in patients with Alzheimer disease with a lifetime history of major depression. *Arch. Gen. Psychiatry* **63,** 161–167.

Rong, L. L., Gooch, C., Szabolcs, M., Herold, K. C., Lalla, E., Hays, A. P., Yan, S. F., Yan, S. S., and Schmidt, A. M. (2005). RAGE: A journey from the complications of diabetes to disorders of the nervous system—striking a fine balance between injury and repair. *Restor. Neurol. Neurosci.* **23,** 355–365.

Salmoun, M., Devijver, C., Daloze, D., Braekman, J. C., and van Soest, R. W. (2002). 5-Hydroxytryptamine-derived alkaloids from two marine sponges of the genus Hyrtios. *J. Nat. Prod.* **65,** 1173–1176.

Scheibel, M. S., Tomiyasu, U., and Scheibek, A. B. (1975). Do raphe nuclei of the reticular formation have a neuroscretory or vascular sensor function? *Exp. Neurol.* **47,** 316–329.

Scott, M. M., Krueger, K. C., and Deneris, E. S. (2005). A differentially autoregulated Pet-1 enhancer region is a critical target of the transcriptional cascade that governs serotonin neuron development. *J. Neurosci.* **25,** 2628–2636.

Selye, H. (1956). "The Stress of Life." McGraw-Hill Book Co., Inc., New York, Toronto, London.

Severson, C. A., Wang, W., Pieribone, V. A., Dohle, C. I., and Richerson, G. B. (2003). Midbrain serotoninergic neurons are central pH chemoreceptors. *Nat. Neurosci.* **6,** 1139–1140.

Sheu, F. S., Azmitia, E. C., Marshak, D. R., Parker, P. J., and Routtenberg, A. (1994). Glial-derived S100b protein selectively inhibits recombinant beta protein kinase C (PKC) phosphorylation of neuron-specific protein F1/GAP43. *Brain Res. Mol. Brain Res.* **21,** 62–66.

Singh, I. (1964). Seasonal variations in the nature of neurotransmitters in a frog vagus-stomach muscle preparation. *Arch. Int. Physiol. Biochim.* **72,** 843–851.

Smith, B. N. (1976). Evolution of C4 photosynthesis in response to changes in carbon and oxygen concentrations in the atmosphere through time. *Biosystems* **8,** 24–32.

Sparks, D. L., and Markesbery, W. R. (1991). Altered serotonergic and cholinergic synaptic markers in Pick's disease. *Arch. Neurol.* **48,** 796–799.

Sparks, D. L., and Slevin, J. T. (1985). Determination of tyrosine, tryptophan and their metabolic derivatives by liquid chromatography-electrochemical detection: Application to post mortem samples from patients with Parkinson's and Alzheimer's disease. *Life Sci.* **36,** 449–457.

Steinbusch, H. W. M. (1981). Distribution of serotonin-immunoreactivity in the central nervous system of the rat-cell bodies and terminals. *Neuroscience* **4,** 557–618.

Stout, S. S., Somerset, W. I., Miller, A., and Musselman, D. L. (2003). Paroxetine use in medically ill patients. *Psychopharmacol. Bull.* **37,** 108–122.

Tork, I. (1990). Anatomy of the Serotonin System. *Ann. NY Acad. Sci.* **600,** 9–36.

Trulson, M. E., and Jacobs, B. L. (1979). Raphe unit activity in freely moving cats: Correlation with level of behavioral arousal. *Brain Res.* **163,** 135–150.

Underwood, M. D., Khaibulina, A. A., Ellis, S. P., Moran, A., Rice, P. M., Mann, J. J., and Arango, V. (1999). Morphometry of the dorsal raphe nucleus serotonergic neurons in suicide victims. *Biol. Psychiatry* **46,** 473–483.

Victor, M., and Ropper, A. H. (2002). "Adam's and Victor's Manual of Neurology," 7th ed. McGraw-Hill, NY.

Wallace, J. A., and Lauder, J. M. (1983). Development of the serotonergic system in the rat embryo: An immunocytochemical study. *Brain Res. Bull.* **10,** 459–479.

Weiger, W. A. (1997). Serotonergic modulation of behaviour: A phylogenetic overview. *Biol. Rev. Camb. Philos. Soc.* **72,** 61–95.

Wetherell, J. L., Gatz, M., Johansson, B., and Pedersen, N. L. (1999). History of depression and other psychiatric illness as risk factors for Alzheimer disease in a twin sample. *Alzheimer Dis. Assoc. Disord.* **13,** 47–52.

Whitaker-Azmitia, P. M. (1999). The discovery of serotonin and its role in neuroscience. *Neuropsychopharmacology* **21**(2S), 2S–8S.

Whitaker-Azmitia, P. M. (2005). Behavioral and cellular consequences of increasing serotonergic activity during brain development: A role in autism? *Int. J. Dev. Neurosci.* **23,** 75–83.

Whitaker-Azmitia, P. M., Murphy, R. B., and Azmitia, E. C. (1990). S-100 protein release from astrocytic glial cells by stimulation of 5-HT1A receptors and regulates the development of serotonergic neurons. *Brain Res.* **528,** 155–158.

Whitaker-Azmitia, P. M., Clarke, C., and Azmitia, E. C. (1993). Localization of 5-HT-1A receptors to astroglial cells in adult rats. *Synapse* **14,** 201–205.

Whitaker-Azmitia, P. M., Borella, A., and Raio, N. (1995). Serotonin depletion in the adult rat causes loss of the dendritic marker MAP-2. A new animal model of schizophrenia? *Neuropsychopharmacology* **12,** 269–272.

Wiens, M., Koziol, C., Batel, R., and Muller, W. E. (1998). Phenylalanine hydroxylase from the sponge Geodia cydonium: Implication for allorecognition and evolution of aromatic amino acid hydroxylases. *Dev. Comp. Immunol.* **22,** 469–478.

Wikgren, M. C., and Reuter, M. (1985). Neuropeptides in a microturbellarian—whole mount immunocytochemistry. *Peptides* **6**(Suppl. 3), 471–475.

Wilmanns, M., Hyde, C. C., Davies, D. R., Kirschner, K., and Jansonius, J. N. (1991). Structural conservation in parallel beta/alpha-barrel enzymes that catalyze three sequential reactions in the pathway of tryptophan biosynthesis. *Biochemistry* **30,** 9161–9169.

Wilson, C. C., Faber, K. M., and Haring, J. H. (1998). Serotonin regulates synaptic connections in the dentate molecular layer of adult rats via 5-HT1a receptors: Evidence for a glial mechanism. *Brain Res.* **782,** 235–239.

Wirz-Justice, A., Lichtsteiner, M., and Feer, H. (1977). Diurnal and seasonal variations in human platelet serotonin in man. *J. Neural Transm.* **41,** 7–15.

Woolley, D. W. (1961). "The Biochemical Basis of Psychosis." John Wiley and Sons, Inc, New York, NY.

Yaguchi, S., and Katow, H. (2003). Expression of tryptophan 5-hydroxylase gene during sea urchin neurogenesis and role of serotonergic nervous system in larval behavior. *J. Comp. Neurol.* **466,** 219–229.

Zhao, J., and Last, R. L. (1996). Coordinate regulation of the tryptophan biosynthetic pathway and indolic phytoalexin accumulation in Arabidopsis. *Plant Cell* **8,** 2235–2244.

THERAPEUTIC APPROACHES TO PROMOTING AXONAL REGENERATION IN THE ADULT MAMMALIAN SPINAL CORD

Sari S. Hannila,[1] Mustafa M. Siddiq,[1] and Marie T. Filbin

Department of Biological Sciences, Hunter College, City University of New York
New York 10021, USA

I. Introduction
II. Growth-Inhibiting Properties of the Glial Scar
III. Inhibition of Axonal Regeneration by CNS Myelin
A. Identification of Myelin-Associated Inhibitors
B. Receptors for Myelin-Associated Inhibitors
C. Lessons from Knockout Mice
D. Mechanisms of Myelin Signal Transduction
E. Role of cAMP in Overcoming Myelin Inhibition
F. The Conditioning Lesion Effect
IV. Cell-Based Strategies to Promote Regeneration
A. Schwann Cell Transplantation
B. Transplantation of OECs
C. Other Methods of Transplantation That Have Been Applied to Spinal Cord Injury
V. Future Directions
References

In the last 25 years, there has been an extraordinary change in our perception of spinal cord injury repair and the resulting scientific advances have been nothing short of revolutionary. It had been believed since the time of Cajal that central nervous system (CNS) axons were incapable of regeneration, but the pioneering work of David and Aguayo overturned this dogma by demonstrating that transected spinal cord axons could regenerate long distances through peripheral nerve grafts. This suggested that the CNS environment played a key role in limiting regeneration, and soon CNS myelin and the glial scar were identified as inhibitors of axonal growth. In this review, we describe several of the pharmacological strategies that have arisen from these findings and how they have been used to encourage axonal regeneration in the injured spinal cord. These include the administration of agents that reverse the inhibitory properties of CNS myelin and the glial scar, the transplantation of cells that create a permissive substrate

[1]These authors contributed equally to this work.

INTERNATIONAL REVIEW OF
NEUROBIOLOGY, VOL. 77
DOI: 10.1016/S0074-7742(06)77003-9

0074-7742/07 $35.00

for axonal growth, and the implantation of artificial scaffolds that have been designed to guide regenerating axons through the injury site.

I. Introduction

> Once development was ended, the founts of growth and regeneration of the axons and dendrites dried up irrevocably. In adult centers the nerve paths are something fixed, ended, immutable. Everything may die, nothing may be regenerated. It is for the science of the future to change, if possible, this harsh decree.
>
> Santiago Ramón y Cajal

At the time these words were written, Cajal had made several fundamental observations that would come to define our perception of axonal regeneration in the central nervous system (CNS). Having devoted most of his career to proving the neuron theory of nervous system structure, Cajal turned his attention to studies of nerve regeneration in the first decade of the twentieth century. With the assistance of his student Francisco Tello, Cajal described the anatomical features of peripheral nerve regeneration in great detail. Together, they demonstrated that regeneration was initiated by axonal sprouting at the proximal end of the severed nerve and that growth cones were present at the tips of regenerating peripheral axons (Cajal, 1928). Transected axons in the brain and spinal cord also displayed growth cones, but after the axons had extended short distances, regeneration was abruptly halted and axonal morphology was dramatically altered. Instead of maintaining a linear trajectory, regenerating CNS axons rapidly became dystrophic and assumed a tortuous, stunted appearance (Cajal, 1928). In addition, the growth cones of these axons were transformed into distorted structures called retraction bulbs (Fig. 1). Cajal described this phenomenon as "abortive regeneration" and it led him to pose the question, "Why, once the reconstructive movement is initiated, do the nerve sprouts lose their energy and suspend their growth?" Tello attempted to address this quandary by performing an experiment in which segments of sciatic nerve were grafted into the injured cerebellum and optic nerve. Using the histological techniques developed by Cajal, he observed axons growing into the grafts, thereby demonstrating for the first time that CNS axons had the capacity to regenerate when provided with a permissive environment (Cajal, 1928). This prompted Cajal to rethink his earlier position and state that their new observations "refute the generally accepted dogma of the essential irregenerability of the central pathways." Seventy years later, a similar experiment would confirm their findings and launch the modern era of CNS regeneration research.

Somewhat ironically, the optimism that Cajal had expressed was overshadowed by his authoritative statements regarding regenerative failure in the CNS, and it was

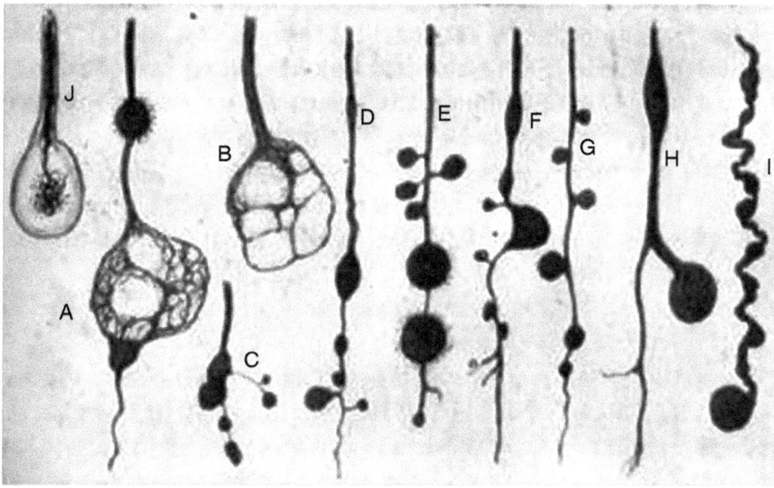

FIG. 1. Diagrams of retraction bulbs observed in the adult mammalian CNS following injury. [Fig. 190, p. 493, from Chapter "Degeneration and Regeneration of the White Matter" from "Cajal's Degeneration & Regeneration of the Nervous System," edited by DeFelipe, Javier and Jones, Edward G. (1991), by permission of Oxford University Press, Inc.]

widely believed in both the scientific and medical communities that CNS regeneration simply was not possible. Consequently, there was little interest in conducting research on CNS repair, but there were a number of prescient studies that produced evidence of innate CNS plasticity. In the dorsal horn of the spinal cord, widespread collateral sprouting of nociceptive dorsal root axons was observed following rhizotomy (Goldberger and Murray, 1974; Liu and Chambers, 1958). Similarly, Anders Björklund demonstrated that lesioned noradrenergic and cholinergic neurons were capable of extending axons over long distances within the brain and spinal cord (Björklund *et al.*, 1971, 1975; Katzman *et al.*, 1971).

The turning point for CNS regeneration research, however, was the publication of two landmark papers from the laboratory of Albert Aguayo. By applying the new technique of horseradish peroxidase (HRP) tracing to Tello's peripheral nerve transplantation experiment, Aguayo's group provided unequivocal proof that regeneration of CNS axons was possible. In the first of these studies, small segments of peripheral nerve were grafted into the injured spinal cord and axons were then retrogradely traced by injection of HRP into the graft (Richardson *et al.*, 1980). Visualization of the tracer revealed substantial axonal growth within the graft, and more importantly, staining of neuronal cell bodies within the spinal cord, which demonstrated that these axons were CNS-derived (Richardson *et al.*, 1980). Samuel David, a postdoctoral fellow in Aguayo's laboratory, then used a highly innovative approach to verify their findings: rather than implant small segments of peripheral nerve, he constructed long sciatic nerve "bridges." The ends of the sciatic nerve grafts were implanted into the medulla oblongata and spinal cord, while the rest of the nerve was positioned extraspinally (David and Aguayo, 1981). The advantage of this approach was that it did not require extensive lesioning of the CNS and therefore did not produce paralysis in the animals, which greatly facilitated their postoperative care. After a postsurgical survival period of 22–30 weeks, HRP was applied to the ends of the transected grafts, which allowed both anterograde and retrograde labeling of axons. The extraspinal application of the tracer and removal of the ipsilateral dorsal root ganglia (DRG) virtually eliminated the possibility of obtaining false-positive results through diffusion of HRP or sprouting of peripheral axons. Large numbers of labeled neurons were observed in the medulla and spinal cord, providing compelling evidence that many of the axons within the grafts had originated in the CNS (David and Aguayo, 1981). It was estimated that some of these axons had extended at least 30 mm, which established that axons in the spinal cord had the capacity to regenerate over long distances (David and Aguayo, 1981). Using similar experimental paradigms, Aguayo's group would later demonstrate that axons in the cerebral cortex (Benfey and Aguayo, 1982) and optic nerve (Vidal-Sanz *et al.*, 1987) could regenerate in a similar manner. The importance of these studies cannot be overstated, as they not only confirmed that CNS axons were capable of regeneration, but also reinforced a concept that has driven virtually all subsequent research in this field: that the CNS environment inhibits axonal regeneration following injury.

II. Growth-Inhibiting Properties of the Glial Scar

The glial scar is one major factor that contributes to the inhibitory environment of the injured CNS, serving as a physical and biochemical barrier to axonal regeneration. Its formation is correlated with expression of inhibitory molecules by reactive astrocytes (Silver and Miller, 2004). CNS injury induces expression of several inhibitory extracellular matrix molecules, including chondroitin sulfate proteoglycan (CSPG), keratin sulfate proteoglycan, NG2, and cytotactin/tenascin (Jones *et al.*, 2002; Martin *et al.*, 2001; McKeon *et al.*, 1991; Silver and Miller, 2004). *In vitro* assays have confirmed that reactive astrocytes up-regulate and secrete proteoglycans that are inhibitory to axonal outgrowth (Canning *et al.*, 1993; Dou and Levine, 1994; McKeon *et al.*, 1991; Smith-Thomas *et al.*, 1994). *In vivo*, CSPGs are rapidly secreted by reactive astrocytes after injury, which leads to increased glial scarring in the brain and spinal cord (Jones *et al.*, 2003a; McKeon *et al.*, 1999; Tang *et al.*, 2003). The mechanism underlying the inhibitory effects of CSPGs appears to be signaling through the Rho/Rho-associated kinase (ROCK) pathway (Borisoff *et al.*, 2003; Dergham *et al.*, 2002; Monnier *et al.*, 2003;

Fig. 2), as inhibition of Rho GTPase enhances neurite outgrowth when primary cortical neurons are plated on CSPG substrates (Dergham *et al.*, 2002). It is believed that the glycosaminoglycan (GAG) side chains are the inhibitory components of proteoglycans, as enzymatic removal of the GAG chains attenuates CSPG inhibitory activity (Fidler *et al.*, 1999; McKeon *et al.*, 1995; Moon *et al.*, 2001; Zuo *et al.*, 1998). *In vivo*, delivery of chondroitinase ABC can promote axonal regeneration and significant functional improvement after spinal cord injury (Bradbury *et al.*, 2002). Administration of a DNA enzyme directed against xylosyltransferase-1, the enzyme responsible for initiating GAG chain synthesis, also enhanced axonal regeneration (Grimpe and Silver, 2004).

III. Inhibition of Axonal Regeneration by CNS Myelin

A. Identification of Myelin-Associated Inhibitors

Cajal was the first to suggest that CNS white matter could block axonal regeneration, and it is now known that myelin contains several inhibitory factors (Fig. 2). In 1988, Martin Schwab's laboratory isolated 35 and 250 kD protein fractions from CNS myelin and showed that these proteins inhibited neurite outgrowth (Caroni and Schwab, 1988a). They subsequently raised a panel of monoclonal antibodies against these fractions and determined that one specific antibody, named IN-1, allowed neurites to grow on a myelin substrate when administered *in vitro* (Caroni and Schwab, 1988b). *In vivo*, administration of IN-1 enhanced regeneration of transected corticospinal axons and recovery of specific reflex and locomotor functions (Bregman *et al.*, 1995; Schnell and Schwab, 1990, 1993).

A protein now known as Nogo was eventually identified as the antigen of the IN-1 antibody (Chen *et al.*, 2000; GrandPre *et al.*, 2000; Prinjha *et al.*, 2000). Nogo exists in three isoforms (A, B, and C) and is a member of the reticulon family of proteins that normally associate with the endoplasmic reticulum. Nogo-A, -B, and -C share a common C-terminus that contains a double lysine endoplasmic reticulum retention signal, a feature that is present in several myelin membrane proteins (Chen *et al.*, 2000). RNA transcripts for all three Nogo isoforms have been detected in several regions of the CNS, including the optic nerve, spinal cord, and cerebral cortex; however, only Nogo-A is enriched in oligodendrocytes (Chen *et al.*, 2000; GrandPre *et al.*, 2000; Prinjha *et al.*, 2000). Recombinant Nogo-A inhibits DRG neurite outgrowth in a dose-dependent manner and this can be neutralized by IN-1. The inhibitory effects of Nogo have been mapped to 66-residue extracellular domain that is shared by all three transcripts, and this domain, deemed Nogo-66, can collapse DRG growth cones (Chen *et al.*, 2000; GrandPre *et al.*, 2000; Prinjha *et al.*, 2000). Nogo-A also has a unique inhibitory domain called amino-Nogo that

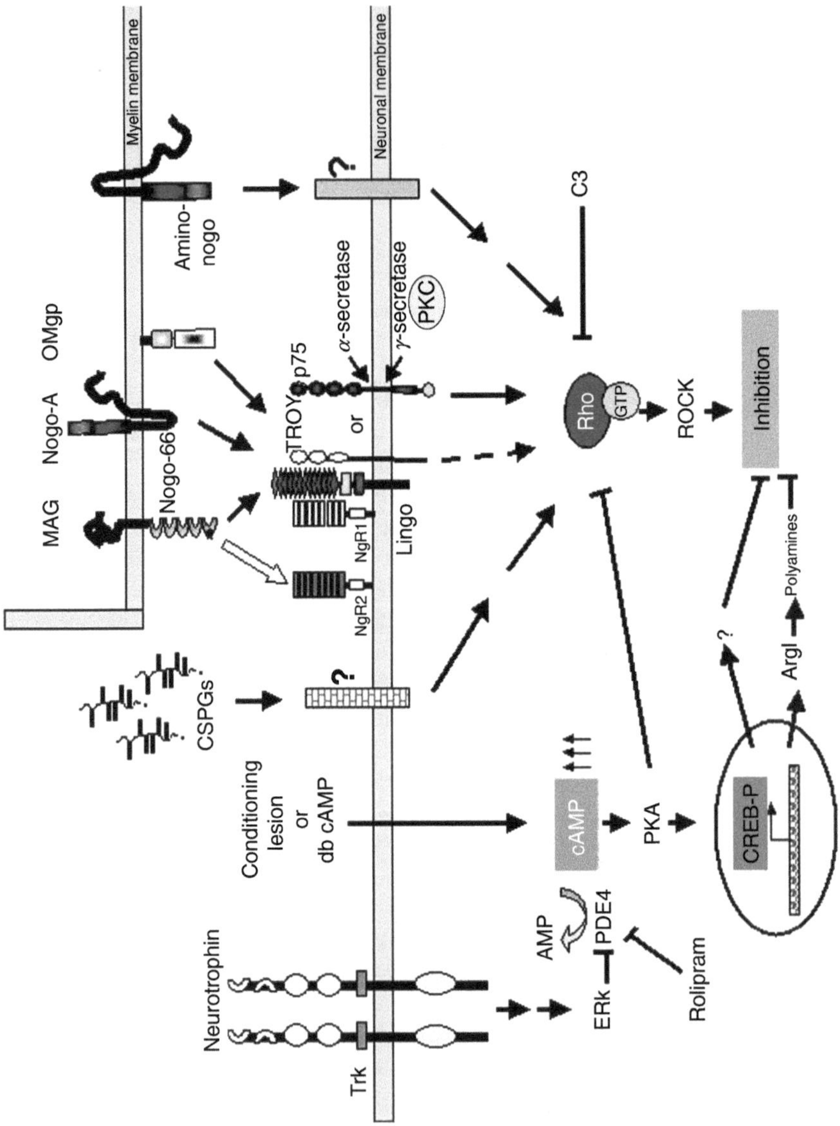

Myelin membrane
Neuronal membrane
MAG
Nogo-A
OMgp
Amino-
nogo
Nogo-66
?
TROY
p75
or
α-secretase
γ-secretase
PKC
NgR1
NgR2
Lingo
CSPGs
?
Conditioning
lesion
or
db cAMP
Neurotrophin
Trk
C3
Rho
GTP
ROCK
Inhibition
cAMP
AMP
PDE4
ERk
Rolipram
PKA
CREB-P
?
Argl
Polyamines

can induce growth cone collapse (Chen *et al.*, 2000; GrandPre *et al.*, 2000; Prinjha *et al.*, 2000). It is believed that Nogo-A has two membrane topologies, and these variants could have different extracellular and intracellular functions (Oertle *et al.*, 2003). Although further work is required to clearly understand the topology of Nogo-A, it is possible that disruption of myelin by injury could expose both Nogo-66 and amino-Nogo (Filbin, 2003; Oertle *et al.*, 2003). Intriguingly, amino-Nogo independently inhibits axonal outgrowth and spreading of fibroblasts (Chen *et al.*, 2000; Fournier *et al.*, 2001), a unique function that is not shared by Nogo-66 (Fournier *et al.*, 2001; Oertle *et al.*, 2003).

Prior to the identification of Nogo-A, myelin-associated glycoprotein (MAG), another well-known myelin protein, was shown to be a potent inhibitor of neurite outgrowth in culture (McKerracher *et al.*, 1994; Mukhopadhyay *et al.*, 1994). MAG is a member of the immunoglobulin (Ig) superfamily and it contains five extracellular Ig-like domains (Salzer *et al.*, 1987, 1990). It is also a sialic acid-binding protein and its first four Ig-like domains are homologous to the Siglec family of sialic acid-binding Ig-like lectins (Kelm *et al.*, 1994). These residues are not critical for its inhibitory function, but could potentiate its effects (Kelm *et al.*, 1994; Tang *et al.*, 1997). Within the first Ig-like domain of MAG, the sialic acid-binding site maps to arginine 118 (R118), a residue that is conserved in other members of the Siglec family (Tang *et al.*, 1997). Mutating R118 eliminates the binding capabilities of the sialic acid residues and the inhibitory function of soluble MAG. Membrane-bound MAG, however, is unaffected by mutation of R118, as neurite outgrowth is strongly inhibited when primary neurons are cocultured with MAG-expressing COS cells. This suggests that there are two distinct sites on MAG that are recognized by neurons, one that binds sialic acid and another that mediates inhibition (Tang *et al.*, 1997).

Intriguingly, MAG seems to be bifunctional, in that it promotes neurite outgrowth from young neurons, but acts as an inhibitor for older neurons (Cai *et al.*, 2001; de Bellard and Filbin, 1999; Johnson *et al.*, 1989; Mukhopadhyay *et al.*, 1994; Turnley and Bartlett, 1998). For instance, DRG neurons are not inhibited by MAG up until P3–4 and these younger neurons also have higher levels of endogenous cyclic AMP (cAMP), whereas in P5 DRG cAMP is barely detected (Cai *et al.*, 2001). In all neuronal subtypes that have been tested to date,

---

FIG. 2. Schematic representation of intracellular signaling pathways involved in mediating and overcoming inhibition of axonal regeneration in the CNS. CSPGs and other components of the glial scar inhibit growth through activation of Rho and its downstream effectors. Myelin inhibitors mediate their effects by binding to the receptor complex composed of NgR1, LINGO-1, and a member of the TNF receptor family ($p75^{NTR}$ or TROY). Activation of this receptor complex culminates in the activation of Rho in a PKC-dependent manner, rearrangement of the actin cytoskeleton, and cessation of axonal growth. Elevation of cAMP leads to activation of PKA, which may inactivate Rho directly, and CREB, which transcribes genes such as ArgI. These events block myelin inhibition and promote axonal regeneration. (See Color Insert.)

neurite outgrowth is initially promoted by MAG, and then the response switches to inhibition during development; however, the age at which the switch occurs varies among different neuronal subtypes (de Bellard *et al.*, 1996; Mukhopadhyay *et al.*, 1994; Turnley and Bartlett, 1998).

In 2002, oligodendrocyte myelin glycoprotein (OMgp), a glycosylphosphatidylinositol (GPI)-linked protein with a leucine-rich repeat (LRR) domain, was identified as a third myelin inhibitor (Wang *et al.*, 2002a). OMgp is a minor component of CNS myelin and it is expressed not only on oligodendrocytes, but also by neurons (Habib *et al.*, 1998; Mikol and Stefansson, 1988; Mikol *et al.*, 1990). Like MAG and Nogo-A, OMgp is a potent inhibitor of neurite outgrowth and induces growth cone collapse (Wang *et al.*, 2002a).

Recently, a fourth myelin-based inhibitor of neurite outgrowth has been described (Benson *et al.*, 2005). Luis Parada's laboratory investigated whether ephrin-B3, a midline repellant for axons of the corticospinal tract (CST) during development, could also inhibit neurite outgrowth in adults. Ephrin-B3 expression was detected using an ephrin-B3 reporter mouse, where reduction in the expression at the midline began at postnatal day 5 (P5). In the white matter of the spinal cord and corpus callosum, expression of ephrin-B3 persisted into adulthood, with high levels of expression around the mature CST (Benson *et al.*, 2005). Using primary cortical and cerebellar granule (CGN) neurons, it was shown that ephrin-B3 did inhibit neurite outgrowth and that this effect was dependent on the p75 neurotrophin receptor ($p75^{NTR}$). Furthermore, myelin inhibition could be reversed by the addition of a soluble recombinant form of either $p75^{NTR}$ or EphA4, the receptor for ephrin-B3.

B. Receptors for Myelin-Associated Inhibitors

In 2001, the first receptor capable of binding myelin inhibitors was cloned from a mouse expression library using a soluble form of Nogo-66 (Fournier *et al.*, 2001). The Nogo receptor (NgR1) is a GPI-linked protein that contains a signal sequence, followed by eight LRR domains and an LRR C-terminal domain (Fournier *et al.*, 2001). Interaction of Nogo-66 with NgR1 was required to induce growth cone collapse (Fournier *et al.*, 2001). Initially, NgR1 was not thought to mediate inhibition by amino-Nogo, but recently, a 24 amino acid sequence of amino-Nogo (Nogo-A-24) was shown to bind to NgR1 with nanomolar affinity (Hu *et al.*, 2005; Oertle *et al.*, 2003). In spite of this ability to bind to NgR1, Nogo-A-24 does not inhibit axonal outgrowth or cell spreading; however, fusion of Nogo-A-24 and Nogo-66 creates a bivalent ligand for NgR1 with substantially enhanced affinity for NgR1 (Oertle *et al.*, 2003). Clearly, further work is required to elucidate the actual topology of Nogo-A and how these two inhibitory regions may interact with NgR1.

One year later, both our laboratory and the laboratory of Steven Strittmatter demonstrated that MAG could also bind to NgR1, and that this binding was independent of the sialic acid residues (Domeniconi *et al.*, 2002; Liu *et al.*, 2002). A soluble form of MAG was able to precipitate NgR1 from primary neurons, and more importantly, MAG binding to NgR1 mediated inhibition of neurite outgrowth. Neutralization of NgR1 function through addition of NgR1 antibody, soluble NgR1, or dominant-negative NgR1 each prevented inhibition of neurite outgrowth by MAG, which confirmed that this receptor was required for MAG inhibition. It was also shown that MAG and Nogo-66 compete for binding to NgR1, suggesting that myelin inhibitors may bind to a common site on the receptor.

When OMgp was identified as an inhibitory component of CNS myelin, the same study demonstrated that it was also capable of binding to NgR1 (Wang *et al.*, 2002a). Expression of exogenous NgR1 conferred responsiveness to OMgp and inhibited neurite outgrowth in neurons that are normally unresponsive to myelin (Wang *et al.*, 2002a). The discovery that NgR1 could bind to three different myelin-associated proteins with no obvious sequence homology came as quite a surprise to the field.

In an attempt to eliminate binding by all three inhibitors to NgR1, the Strittmatter laboratory has devised a competitive antagonist to NgR1 called NEP1-40, which is derived from the N-terminal fragments of Nogo-66. NEP1-40 blocks Nogo-66 or CNS myelin inhibition of neurite outgrowth *in vitro* (GrandPre *et al.*, 2002). More importantly, the delivery of NEP1-40 to rats with spinal cord injury results in significant axonal growth in the CST and improved functional recovery. Since the locations of the binding sites for MAG and OMgp on NgR1 are unknown, it is conceivable that NEP1-40 could prevent the binding of MAG or OMgp as well.

Recently, NgR2, a related molecule of NgR1, was identified as another receptor that can selectively bind MAG (Venkatesh *et al.*, 2005). NgR2 is expressed in the adult rat brain and can mediate inhibition, but unlike what has been previously reported for NgR1, this binding is sialic acid dependent (Venkatesh *et al.*, 2005). This raises the possibility that still other receptors for myelin inhibitors may exist.

Although NgR1 is capable of binding MAG, Nogo-66, and OMgp, it cannot transduce inhibitory signals across the membrane due to its GPI linkage, and this suggested that another transmembrane protein was associating with NgR1. $p75^{NTR}$ was identified as a coreceptor interacting with NgR1 to mediate inhibition (Wang *et al.*, 2002b). $p75^{NTR}$ is a member of the tumor necrosis factor receptor superfamily and it was first identified as a low-affinity neurotrophin receptor (Rabizadeh and Bredesen, 2003). In addition, $p75^{NTR}$ has a death domain and has been shown to mediate programmed cell death (Rabizadeh and Bredesen, 2003). Prior to the discovery that MAG could bind to NgR1, it was already known that p75^{NTR} was the transducer for MAG, since MAG did not inhibit neurite outgrowth from DRG and CGN cultured from mice deficient in p75^{NTR} (p75$^{NTR-/-}$; Yamashita *et al.*, 2002). In the same report, MAG was coimmunoprecipitated with p75^{NTR} from neurons, but they could not show a direct interaction between the two, presumably because of what we know now of NgR1 binding to MAG. p75^{NTR} can be coprecipitated by MAG, Nogo-66, or OMgp and NgR1 was also present in these complexes (Wang *et al.*, 2002b; Wong *et al.*, 2002). The fact that all three known myelin inhibitors bind to NgR1 suggests that just one of these inhibitors would be sufficient for transducing an inhibitory signal and implies that there is functional redundancy in the receptor complex formed by NgR1 and p75^{NTR}.

Studies on the ectodomain crystal structures of NgR1 suggested that the LRR region with its concave exterior surface could contain degenerate ligand-binding sites (He *et al.*, 2003). There is also a deep cleft in the C-terminal base that could provide a site for NgR1 to associate with p75^{NTR} (He *et al.*, 2003). Crystallographic analysis has shown that protein complexes similar to the NgR1/p75^{NTR} complex have some degree of receptor or ligand promiscuity and this is often accomplished by structural adaptation of a highly accessible and hydrophobic surface (DeLano, 2002; He *et al.*, 2003; Kossiakoff and De Vos, 1998). There are two hydrophobic surface "hotspots" within the concave face of NgR1 (He *et al.*, 2003); however, further structural and biochemical analysis is required to better elucidate how NgR1 could bind to MAG, Nogo-66, and OMgp.

In 2004, Mi *et al.* identified LINGO-1 as another component of the NgR1 and p75^{NTR} receptor complex. LINGO-1 is a highly conserved protein that is nervous system specific with a transmembrane domain, 12 LRR motifs, 1 Ig domain and a short cytoplasmic tail that contains a canonical epidermal growth factor receptor (EGFR)-like tyrosine phosphorylation site. The cytoplasmic tail of LINGO-1 suggested it could be involved with signaling, and when a truncated form of LINGO-1 that was incapable of signaling was transfected into P7 CGN, diminished responses to inhibitory substrates were observed. The reverse was also true, in that transfecting CGN with full-length LINGO-1 or using a soluble LINGO-1 made the neurons more responsive to myelin and attenuated neurite outgrowth (Mi *et al.*, 2004).

Soon after p75^{NTR} was identified as the transducing component of the receptor complex, it was noted that expression of p75^{NTR} in the intact mouse CNS was very low (Park *et al.*, 2005). However, it has also been shown that p75^{NTR} expression is significantly increased in both neurons and glia following spinal cord injury (Dubreuil *et al.*, 2003). This indicates that p75^{NTR} likely plays a major role in mediating the inhibitory effects of CNS myelin. Nevertheless, the fact that p75^{NTR} is not widely expressed in the CNS suggested that another transmembrane receptor was associating with NgR1 in these regions. Two labs simultaneously showed that TROY (also known as TAJ) could associate with NgR1 and LINGO-1 to form a functional complex receptor that mediated the inhibitory activity of myelin-based inhibitors (Park *et al.*, 2005; Shao *et al.*, 2005). Like $p75^{NTR}$, TROY is a member of the TNF receptor family and is expressed at high levels in the adult mouse brain (Kojima *et al.*, 2000). CGN and DRG neurons from TROY/TAJ-deficient mice were less inhibited in the presence of OMgp and Nogo-66, and exogenously adding TROY to wild-type neurons *in vitro* reversed neurite outgrowth inhibition. Expression of truncated TROY in rat DRG neurons disinhibited neurite outgrowth in the presence of Nogo-66, while infection of full-length TROY enhanced the inhibition (Park *et al.*, 2005; Shao *et al.*, 2005). Future *in vivo* studies are required with the TROY/TAJ-deficient mice to fully determine the role of this receptor in axonal regeneration.

Recently, activation of the EGFR has been implicated in the inhibitory functions of both myelin and CSPG (Koprivica *et al.*, 2005). Nogo-66, OMgp, and CSPG were shown to phosphorylate EGFR in a calcium-dependent manner, and inhibiting the kinase activity of EGFR promoted neurite outgrowth on myelin, Nogo-66, MAG, and CSPG substrates (Koprivica *et al.*, 2005). *In vivo*, inhibiting EGFR kinase activity does promote regeneration in the optic nerve after crush; however, it was estimated that only a small number of retinal ganglion axons regenerated beyond the lesion site. EGFR did not directly bind myelin inhibitors and it was not coupled to NgR1 or $p75^{NTR}$; however, EGFR activation is NgR1-complex dependent (Koprivica *et al.*, 2005). It was proposed that the EGFR phosphorylation by the myelin inhibitors could be through transactivation of the receptor, by means of signaling pathways downstream of the NgR1.

C. Lessons from Knockout Mice

MAG-deficient ($MAG^{-/-}$) mice were the first to be utilized in regeneration studies. Using $MAG^{-/-}$mice, one study showed that there was only a small amount of spontaneous axonal regeneration in the CST, while another study reported no detectable regeneration after lesioning the optic nerve or spinal cord (Bartsch *et al.*, 1995; Li *et al.*, 1996).

Three Nogo knockouts generated using different strategies have been used for studies of axonal regeneration in the spinal cord (Kim *et al.*, 2003; Simonen *et al.*, 2003; Zheng *et al.*, 2003). Kim *et al.* (2003) used retroviral gene trapping to knockout both *Nogo-A* and *-B* and performed CST lesioning. They observed robust axonal sprouting with long distance growth and improved functional recovery in young adult *Nogo-A/B* knockout mice (Kim *et al.*, 2003). The Schwab group used a more conventional approach to generate *Nogo-A* knockouts, and in response to elimination of Nogo-A, they observed up-regulation of the shorter *Nogo-B* transcript in the CNS (Simonen *et al.*, 2003). However, they did not see extensive regeneration after CST lesioning, and this could potentially be attributed to the up-regulation of Nogo-B. Marc Tessier-Lavigne's laboratory has generated two lines of knockout mice, one lacking *Nogo-A/B* and the other lacking all three isoforms of *Nogo* (Zheng *et al.*, 2003). Unlike Kim *et al.* (2003), they did not observe any spontaneous regeneration after CST injury, suggesting that eliminating Nogo alone is not enough to induce regeneration after spinal cord injury. This supports the redundancy hypothesis, since both MAG and OMgp are intact in all of these knockout mice and can still mediate inhibition through NgR1.

Two different types of *ngr1*-deficient mice ($ngr1^{-/-}$) have also been generated. They first deleted the *ngr1* gene by targeted homologous recombination (Kim *et al.*, 2004). DRG neurons from $ngr1^{-/-}$ mice cannot bind Nogo-66 and the growth cones of these neurons were less sensitive to collapsing in the presence of Nogo-66, MAG, OMgp, and myelin. *In vivo* studies in these mice showed that there was some recovery of motor function after dorsal hemisection or complete transection of the spinal cord with regeneration of some raphespinal and rubrospinal fibers; however, there was no regeneration of CST axons (Kim *et al.*, 2004). Surprisingly, Marc Tessier-Lavigne's laboratory has shown that P7 CGN and P10 DRG from $ngr1^{-/-}$ mice generated by genetic deletion were strongly inhibited by myelin and Nogo-66 and virtually indistinguishable from wild-type neurons (Zheng *et al.*, 2005). In the same report, they utilized $p75NTR^{-/-}$ and observed that DRG neurons from these mice were less inhibited. Neurite outgrowth from $p75^{NTR-/-}$ CGN, however, was inhibited very strongly. In addition, there was a lack of significant regeneration in the CST of either $p75^{NTR-/-}$ or $ngr1^{-/-}$ mice after injury. This study suggests that additional receptors for Nogo-A, MAG, and OMgp may exist. The lack of regeneration observed in these $ngr1^{-/-}$ mice could also be due to the presence of ephrin-B3 in myelin, which binds to EphA2 receptor, or amino-Nogo, whose receptor is unidentified.

It is important to mention the etiological differences that are encountered when rat and mouse models are used in spinal cord injury studies. Within 1 week of a crush injury, the primary lesion sites in mice show far less destruction and cavitation compared to rats (Guth *et al.*, 1999). After 8 weeks, the tissue damage and cavitation in mice have decreased dramatically and healing is observed, but in rats, the primary lesions consist of several huge cavities (Guth *et al.*, 1999). Intriguingly, in rats the administration of steroids is required to attenuate the progressive necrosis and encourage wound healing; however, mice have considerable tissue repair occurring without the addition of steroids. An obvious advantage of utilizing mice is the ability to conduct genetic studies using knockouts, but conversely, the cyst formation and cavitation observed in the rat model is more consistent with what is observed in human spinal cord injury.

D. Mechanisms of Myelin Signal Transduction

Even before the receptor complex for the myelin-based inhibitors was identified, members of the Rho family of small GTPase proteins had been shown to play a role in mediating inhibition (Lehmann *et al.*, 1999). The Rho family of small GTPase proteins is known to regulate the actin cytoskeleton by causing actin polymerization (Jaffe and Hall, 2005). Lisa McKerracher's laboratory showed in PC12 cells that either enzymatic inactivation of Rho or transfection with dominant-negative Rho promoted neurite outgrowth on an inhibitory substrate (Lehmann *et al.*, 1999). Subsequently, several other labs have reported that inactivation of Rho promoted neurite outgrowth *in vitro* on all the myelin-based inhibitors (Nogo-66, MAG, and OMgp) and CSPG (Dergham *et al.*, 2002; Fournier *et al.*, 2003; Niederost *et al.*, 2002; Vinson *et al.*, 2001). This work was further supported *in vivo* when adult rats were treated with C3 enzyme to inactivate Rho following optic nerve crush (Lehmann *et al.*, 1999). This treatment allowed severed axons to transverse the lesion and grow into the distal white matter of the optic nerve (Lehmann *et al.*, 1999). More importantly, administration of C3 enzyme or an inhibitor of Rho's downstream target, ROCK, after spinal cord injury promoted significant axonal regeneration in mice and functional recovery after just 24 hours (Dergham *et al.*, 2002). Although some of this growth and quick functional recovery was attributed to the neuroprotective action of C3, inactivation of Rho seems to be a critical component of promoting regeneration (Dergham *et al.*, 2002; Filbin, 2003).

The last several years have yielded further insight into the signaling events that occur downstream of the NgR1/p75^{NTR}/LINGO-1 receptor complex (Fig. 2). For example, it is now known that interaction of myelin inhibitors with the receptor complex leads to activation of Rho in a protein kinase C (PKC)-dependent manner (Sivasankaran *et al.*, 2004). Activation of Rho is also dependent on p75^{NTR}, as MAG and Nogo-A have been shown to enhance interaction of p75^{NTR} with Rho GDP dissociation inhibitor, an event that allows Rho to assume its active GTP-bound state (Yamashita and Tohyama, 2003). Recently, we showed that MAG binding to primary CGN-induced cleavage of the p75^{NTR} extracellular domain by α-secretase. This was followed by a PKC-dependent γ-secretase cleavage within the transmembrane domain, which releases the intracellular domain into the cytoplasm. These cleavage events are necessary for both activation of Rho and inhibition of neurite outgrowth (Domeniconi *et al.*, 2005). After treatment with pharmacological inhibitors of α- and γ-secretase, CGN are no longer inhibited by MAG and activation of Rho is blocked (Domeniconi *et al.*, 2005). This work suggests that preventing p75^{NTR} cleavage could serve as a means of overcoming inhibition *in vivo*.

E. Role of cAMP in Overcoming Myelin Inhibition

Another approach to overcoming inhibition is to alter the intrinsic state of the neuron so it no longer responds to myelin. Elevation of cAMP overcomes inhibition by MAG and myelin (Cai *et al.*, 1999 Fig. 2), and it is also effective in switching repulsion to attraction in a growth cone-turning assay (Song *et al.*, 1998). There are several different means by which cAMP can be elevated including addition of a nonhydrolyzable, membrane permeable analogue, such as dibutyryl cAMP (dbcAMP), and activating adenylyl cyclase with forskolin (Cai *et al.*, 2001; Kilmer and Carlsen, 1984). Alternatively, elevation of cAMP can be achieved by inhibiting phosphodiesterase 4 (PDE4), the subfamily of enzymes that degrades cAMP in the nervous system (Jin *et al.*, 1999). Rolipram is a specific inhibitor of PDE4, and has been shown to elevate cAMP levels in the spinal cord when administered *in vivo* (Pearse *et al.*, 2004a).

Our lab has further investigated the downstream mechanisms that mediate reversal of inhibition, and it appears that there are two phases to the cAMP effect (Qiu *et al.*, 2002). In the initial phase, PKA is activated after elevating cAMP (Qiu *et al.*, 2002). PKA does interact with the cytoskeleton and could possibly be inactivating Rho (Howe, 2004). This initial up-regulation in cAMP is acute and transcription independent (Filbin, 2003). In the second phase, the effects of cAMP become transcription-dependent (Cai *et al.*, 2002), and one protein that has been shown to be up-regulated by cAMP is Arginase I (ArgI), which is believed to play an anti-apoptotic role in neurons (Cai *et al.*, 2002; Esch *et al.*, 1998). ArgI is a key enzyme in the synthesis of polyamines, which are organic compounds that contain two or more amino groups (Seiler, 2000). It is believed that polyamines play a role in nervous system development (Slotkin and Bartolome, 1986).

High levels of ArgI are expressed in young DRG neurons that are not inhibited by MAG, but when they reach the critical age of P5 there is a decrease in ArgI levels that corresponds with the onset of inhibition by MAG (Cai *et al.*, 2002). Our lab has shown that overexpression of ArgI or exogenous addition of polyamines can overcome inhibition by MAG and myelin. The importance of this pathway is underscored by the fact that inhibiting either ArgI or ornithine decarboxylase, another critical enzyme in the synthesis of polyamines, can abrogate the effects of dbcAMP (Cai *et al.*, 2002). We are currently working to elucidate the molecular mechanism underlying the ability of polyamines to overcome inhibition and their role in spinal cord injury.

Another means of elevating endogenous cAMP levels is by priming with neurotrophins (Cai *et al.*, 1999). Priming is the term used in our laboratory to refer to experiments where neurons are first treated with neurotrophins on a permissive substrate and subsequently exposed to inhibitors such as MAG and myelin. Neurons primed with neurotrophins can overcome inhibition by MAG and myelin (Cai *et al.*, 1999, 2001). By contrast, direct addition of neurotrophins to neurons on inhibitory substrates does not increase neurite outgrowth, but when we add neurotrophins directly in conjunction with pertussis toxin (PTX) inhibition by MAG and myelin are blocked without priming (Cai *et al.*, 1999). PTX inhibits the G_i/G_o protein that can inhibit adenylyl cyclase and decrease intracellular levels of cAMP, and this suggests that MAG and myelin may directly activate G_i/G_o, though the receptor and molecular mechanism behind this remain undefined.

It should be noted that neurotrophins do not commonly act through cAMP signaling (Gao *et al.*, 2003), but instead utilize a dual receptor system consisting of the Trk receptor tyrosine kinases and p75^{NTR} (Teng and Hempstead, 2004). Neurotrophins overcome MAG inhibition by activating the Trk receptors, but not p75^{NTR} (Gao *et al.*, 2003), and can elevate cAMP by extracellular signal-regulated kinase (ERK)-dependent inhibition of PDE4 (Gao *et al.*, 2003).

The PDE4 inhibitor rolipram has proven advantageous for use *in vivo* since it can readily cross the blood–brain barrier (BBB; Krause and Kuhne, 1988). Neurons must be primed with rolipram to overcome inhibition on MAG and myelin *in vitro* (Gao *et al.*, 2003; Nikulina *et al.*, 2004), but delivery of rolipram *in vivo* to P30 rats using osmotic pumps promotes neurite outgrowth when DRG neurons are cultured on MAG (Nikulina *et al.*, 2004). More importantly, subcutaneous delivery of rolipram 2 weeks after a C3/4 hemisection lesion and implantation of embryonic spinal cord tissue increased the number of serotonergic fibers within the transplant (Nikulina *et al.*, 2004). Not only did rolipram promote regeneration, but it also decreased the extent of reactive gliosis as demonstrated by decreased glial fibrillary acidic protein (GFAP) staining in the area of the transplant (Nikulina *et al.*, 2004). When administered in combination with transplantation of Schwann cells or neurotrophin-expressing fibroblasts, rolipram significantly improves axonal regeneration after spinal cord injury (Lu *et al.*, 2004; Pearse *et al.*, 2004a). Clearly rolipram has tremendous therapeutic potential for use in human spinal cord injury because of its ability to readily cross the BBB, promote regeneration, and decrease the extent of glial scarring.

F. The Conditioning Lesion Effect

DRG neurons are the only neurons that have projections in both the CNS and PNS, and this provides a unique opportunity to study the molecular changes that underlie axonal regeneration. Lesioning the peripheral process of a DRG neuron enhances the intrinsic growth state of the neuron, and this is reflected in the increased expression of regeneration-associated genes such as growth-associated protein 43 (GAP-43; Schreyer and Skene, 1993). Conversely, GAP-43 expression does not increase when DRG central processes are lesioned, and these axons fail to regenerate (Schreyer and Skene, 1993). Numerous *in vivo* studies have now demonstrated that regeneration of DRG central axons into a peripheral nerve graft can be increased significantly if the peripheral branches of the neurons are transected at the same time (Chong *et al.*, 1996; Oudega *et al.*, 1994; Richardson and Issa, 1984). This effect is referred to as "conditioning" of the DRG neurons, and it was also shown that optimal regeneration was obtained if the conditioning peripheral lesion was performed 1 week prior to the transection of the axons in the spinal cord (Oudega *et al.*, 1994; Richardson and Issa, 1984). Together, these studies firmly established that a conditioning lesion enhances the regenerative capacity of sensory axons in the CNS.

In 1999, Neumann and Woolf utilized the conditioning lesion paradigm to study regeneration of dorsal column axons, but with one important difference: the experiments were performed in the absence of peripheral nerve grafting. The first group of animals received simultaneous lesions of the sciatic nerve and dorsal column, and axons were anterogradely traced 6–8 weeks later. Animals that did not receive a conditioning lesion showed no signs of axonal regeneration into the lesion site, and this regenerative failure persisted up to 1 year after injury. By contrast, animals that received concomitant central and peripheral lesions displayed abundant growth of axons into and beyond the lesion site. Consistent with earlier findings (Oudega *et al.*, 1994; Richardson and Issa, 1984), performing the conditioning lesion 1 week prior to the dorsal column lesion produced the most extensive regeneration, as axons were observed within the white matter several millimeters beyond the lesion site (Neumann and Woolf, 1999). While the extent of axonal regeneration was impressive, one of the most intriguing aspects of this study was the fact that the conditioning lesion allowed axons to regenerate within the inhibitory environment of the injured CNS. This suggested that a conditioning lesion not only increases the overall growth capacity of DRG neurons but also induces changes at the molecular level that render the neurons unresponsive to myelin inhibitors.

We now know that 24 hours after a conditioning lesion, cAMP levels in the DRG cell bodies have increased twofold and the neurons are no longer inhibited by myelin, while by 1 week post-lesion, cAMP levels have returned to baseline (Qiu *et al.*, 2002). Consistent with our findings with dbcAMP, the enhanced neurite outgrowth on myelin by DRGs 24 hours after a conditioning lesion is PKA-dependent. At 1 week post-lesion, neurite outgrowth on myelin is further enhanced but it is now PKA-independent (Qiu *et al.*, 2002). Also consistent with our earlier observations is the fact that following a conditioning lesion, neurite outgrowth from DRG neurons switches from being transcription-dependent to transcription-independent, and there is substantial elevation of ArgI at 30 hours post-lesion in these neurons (Filbin, 2003). Lastly, injection of dbcAMP into the DRG 1, 2, or 7 days before *ex vivo* transplant on myelin or a dorsal column lesion mimics the conditioning lesioning effect and enhances axonal regeneration (Qiu *et al.*, 2002 Fig. 3;). These important findings demonstrate that directly elevating cAMP in the cell body is sufficient to promote regeneration and reproduce what is observed in the conditioning lesion model (Filbin, 2003).

In 2004, Gao *et al.* demonstrated that cAMP's ability to overcome inhibition by myelin and MAG requires activation of the transcription factor cAMP response element binding protein (CREB). Activation of CREB in response to brain-derived neurotrophic factor (BDNF) and dbcAMP is detected within 5 min of treatment and is maintained for at least 2 hours. Furthermore, activation of CREB is PKA-dependent and leads to enhanced expression of ArgI and polyamines (Gao *et al.*, 2004). Injection of constitutively active CREB into DRGs in adult rats is sufficient to promote regeneration of lesioned dorsal column axons into and beyond the lesion site, which suggests that CREB transcribes a variety of genes that are involved in overcoming myelin inhibition (Gao *et al.*, 2004).

IV. Cell-Based Strategies to Promote Regeneration

The inhibitory properties of CNS myelin have led many researchers to explore cell transplantation as a means of creating a more permissive environment for regeneration following spinal cord injury. In the last 15 years, these approaches have focused on two major cell types: Schwann cells and olfactory ensheathing cells (OECs). Genetically modified fibroblasts, bone marrow stromal cells, embryonic spinal cord tissue, and neural progenitor cells have also emerged as candidates for transplantation in studies of spinal cord injury.

A. Schwann Cell Transplantation

While the injured CNS fails to regenerate, it has been well documented that peripheral nerves have the ability to regenerate fully after injury, and Schwann cells play a key role in this process. Following peripheral nerve transection and Wallerian degeneration, Schwann cells proliferate within the endoneurium and form channels that physically guide regenerating axons to their targets (Fawcett and Keynes, 1990). In addition, Schwann cells respond to injury by up-regulating expression of nerve growth factor (NGF), BDNF, and ciliary neurotrophic factor (CNTF; Heumann *et al.*, 1987; Meyer *et al.*, 1992). These neurotrophic factors stimulate axonal growth and provide trophic support to the injured neurons. Growth-promoting extracellular matrix components and adhesion molecules, such as laminin, L1, neural cell adhesion molecule (N-CAM), and N-cadherin, are also produced by Schwann cells (Bignami *et al.*, 1984; Bixby *et al.*, 1988;

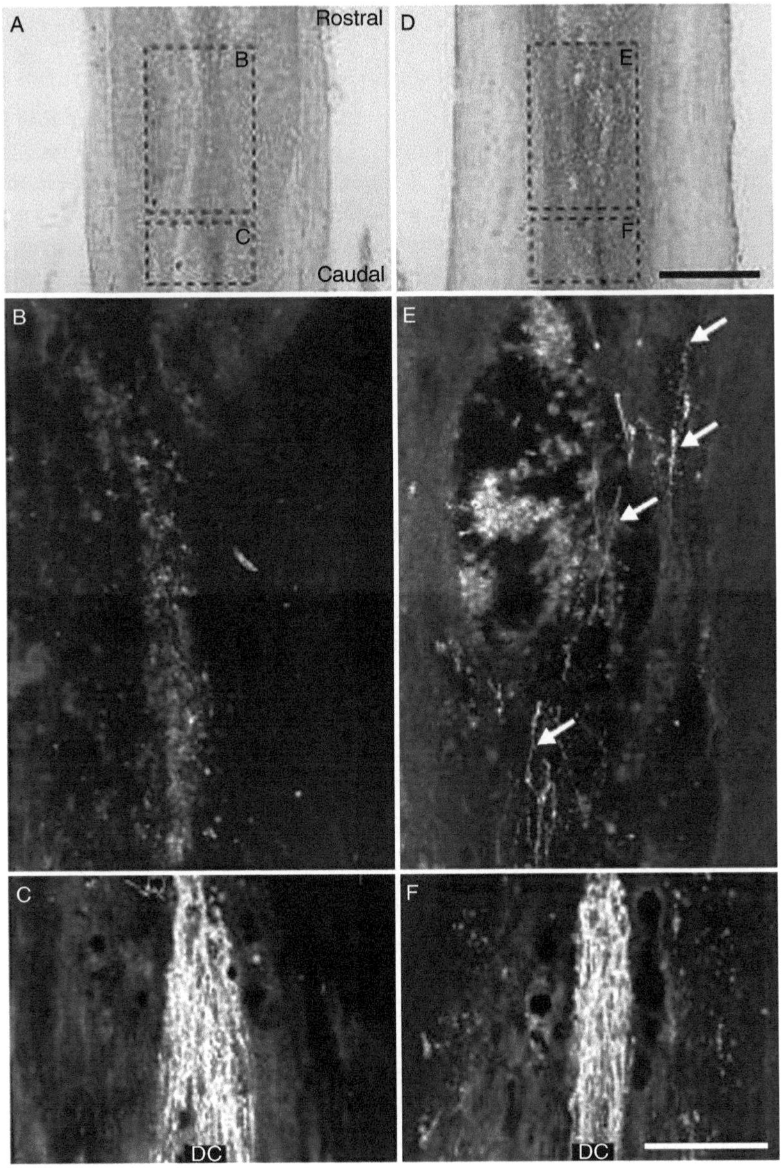

FIG. 3. Regeneration of dorsal column axons induced by injection of dbcAMP. Dorsal column lesions were performed 1 week after injection of saline (A–C) or dbcAMP (D–F) into L5 DRGs. Nissl-stained horizontal sections through the lesion site after injection of saline (A) or dbcAMP (D) depict areas enlarged in darkfield photomicrographs below. Dorsal column axons (DC) were transganglionically labeled with HRP 1–2 weeks after dorsal column lesion and no axons were

Martini and Schachner, 1986). These characteristics underlie the ability of peripheral nerve grafts to support the regeneration of CNS axons (Benfey and Aguayo, 1982; Bixby *et al.*, 1988; David and Aguayo, 1981; Fawcett and Keynes, 1990; Richardson *et al.*, 1980; Vidal-Sanz *et al.*, 1987) and make Schwann cells logical candidates for transplantation in spinal cord injury studies.

One of the first transplantation experiments involved grafting of rat Schwann cells into the demyelinated spinal cords of immune-suppressed mice (Duncan *et al.*, 1981). Remyelination of the dorsal columns was observed over a period of several weeks, demonstrating that Schwann cells purified from peripheral nerve retain their ability to form myelin (Duncan *et al.*, 1981). Several years later, Kromer and Cornbrooks (1985) transplanted Schwann cells grown on a collagen substratum into the brains of rats following a lesion of the septo-hippocampal pathway. The engrafted cells survived within the brain, and regeneration of cholinergic axons across the lesion site was observed within 14 days (Kromer and Cornbrooks, 1985).

Initially, Schwann cells were isolated from neonatal peripheral nerve (Brockes *et al.*, 1979; Montgomery and Robson, 1990, 1993), but in the early 1990s, several laboratories developed techniques for purifying Schwann cells from adult rat and human peripheral nerve, an advance that raised the possibility of autologous Schwann cell transplantation (Morrissey *et al.*, 1991; Rutkowski *et al.*, 1992). In the protocol developed by Richard Bunge, explants of sciatic and phrenic nerve were not mechanically dissociated immediately after dissection; instead, the tissues were repeatedly plated over a period of 5–6 weeks (Morrissey *et al.*, 1991). This allowed contaminating fibroblasts to migrate away from the explants, resulting in a significantly higher yield of Schwann cells (Morrissey *et al.*, 1991). Rutkowski *et al.* (1992) used a similar approach to isolate adult human Schwann cells in which segments of sural nerve were treated with cholera toxin for 7 days prior to dissociation. They also demonstrated that subsequent addition of forskolin and glial growth factor inhibited fibroblast proliferation and produced highly enriched populations of Schwann cells (Rutkowski *et al.*, 1992). Cells purified using these methods, which have become standard protocols for the isolation of Schwann cells, have been widely used for grafting in studies of spinal cord regeneration.

---

present at the midpoint of the lesion site in animals injected with saline (B). In animals that received dbcAMP, axons were observed traversing the lesion site (E). There was abundant labeling of axons caudal to the lesion site in both saline-injected (C) and dbcAMP-injected animals (F), which indicated that axonal labeling was equivalent in both treatment groups. [Reprinted from *Neuron*, 34(6), Qiu, J., Cai, D., Dai, H., McAtee, M., Hoffman, P. N., Bregman, B. S., and Filbin, M. T., Spinal axon regeneration induced by elevation of cyclic AMP, 895–903, copyright 2002, with permission from Elsevier.] (See Color Insert.)

More recent approaches to Schwann cell transplantation have often involved embedding the purified cells within polymer tubes, which then serve as channels for growing axons. Montgomery and Robson (1990, 1993) implanted polycarbonate tubes filled with Schwann cells into the thalamus and after 4–16 weeks, they observed myelinated and unmyelinated axons within the tubes. Thalamic neurons were retrogradely labeled when HRP was applied to the distal ends of the tubes, confirming that these axons had originated in the CNS (Montgomery and Robson, 1990, 1993). This technique was then further refined and applied to spinal cord injury in the laboratory of Mary Bunge. In one of their early studies, 10-mm polymer channels were filled with a suspension of purified Schwann cells and Matrigel, which provided a three-dimensional substrate for axonal growth (Xu *et al.*, 1995a). Adult rats then received a spinal cord injury in which segments T9–T11 were excised, and the rostral end of the spinal cord was inserted into the guidance channel (Xu *et al.*, 1995a). While very few axons were present in channels filled with Matrigel alone, extensive axonal growth was observed within the Schwann cell-filled channels, and ultrastructural analysis showed that many of the axons were myelinated (Xu *et al.*, 1995a). Retrograde tracing revealed that virtually all of the axons originated from DRG neurons and spinal cord interneurons (Xu *et al.*, 1995a). This indicated that regeneration of major descending tracts had not occurred, and this was confirmed by an absence of serotonergic and catecholaminergic markers (Xu *et al.*, 1995a).

This lack of supraspinal regeneration was addressed in a study published that same year in which implantation of Schwann cell guidance channels was combined with administration of BDNF and neurotrophin-3 (NT-3; Xu *et al.*, 1995b). BDNF has been shown to enhance regeneration of rubrospinal axons (Kobayashi *et al.*, 1997), while NT-3 significantly improves regeneration of corticospinal axons after spinal cord injury (Schnell *et al.*, 1994). Infusion of the neurotrophins into the channels produced a significant increase in the regeneration of myelinated axons, which extended for several millimeters within the grafts (Xu *et al.*, 1995b). More importantly, tracing experiments and immunohistochemical staining revealed that many of the ingrowing axons had originated in the raphe and vestibular nuclei of the brainstem, demonstrating that regeneration of supraspinal axons could be enhanced by the combination of Schwann cells and neurotrophins (Xu *et al.*, 1995b). Regeneration of serotonergic and noradrenergic axons was also observed when Schwann cell grafting was combined with administration of methylprednisolone, an anti-inflammatory agent that purportedly reduces secondary damage and improves neurological outcome when administered after spinal cord injury (Chen *et al.*, 1996).

In these studies, only the rostral end of the spinal cord was inserted into the channel; the other end of the channel was capped, preventing contact between the distal spinal cord and the Schwann cell matrix (Chen *et al.*, 1996; Xu *et al.*, 1995b). To test whether Schwann cells could promote regeneration of both ascending and descending axons, the rostral and caudal ends of the transected spinal cord were placed within Schwann cell guidance channels (Xu *et al.*, 1997). The engrafted Schwann cells formed a bridge between the spinal cord stumps and by postsurgical day 30, it was estimated that 17,000 axonal processes had extended into a single graft (Xu *et al.*, 1997). Anterograde and retrograde tracing revealed that the majority of these axons originated from propriospinal neurons within cervical and sacral segments of the spinal cord (Xu *et al.*, 1997). There is no question that this approach produced extensive axonal regeneration, but it is critical to note that it also revealed a problem that is frequently encountered in Schwann cell transplantation studies. Even though the open-ended channels provided continuity between the spinal cord and the Schwann cell grafts, regenerating axons remained within the grafts and very few axons reentered the distal spinal cord (Xu *et al.*, 1997). This lack of long-distance regeneration greatly limits the potential for functional recovery and indicates that Schwann cell transplantation alone is not sufficient to mediate axonal regeneration in the distal spinal cord.

In an attempt to encourage regeneration beyond the Schwann cell grafts, subsequent studies have often used Schwann cell transplantation in combination with neurotrophins and other agents capable of inducing axonal growth. Based on earlier observations showing that infusion of BDNF and NT-3 promotes axonal regeneration into Schwann cell-filled channels (Xu *et al.*, 1995b), several groups have used retroviral vectors to generate genetically modified Schwann cells that express a variety of neurotrophins (Sayers *et al.*, 1998; Tuszynski *et al.*, 1998). When transplanted into the uninjured spinal cord, Schwann cells expressing NGF survived for up to 1 year within the CNS and were able to myelinate CNS axons (Tuszynski *et al.*, 1998). The amount of NGF secreted by these cells was also sufficient to induce collateral sprouting of NGF-responsive axons into the grafts (Tuszynski *et al.*, 1998). In rats with spinal cord injury, transplantation of NGF-expressing Schwann cells led to enhanced regeneration and myelination of axons from the locus coerulus (Weidner *et al.*, 1999). Implantation of BDNF-expressing Schwann cells into the injured spinal cord not only induced growth of reticulospinal and vestibulospinal axons into the grafts, but also allowed these axons to cross from the graft into the distal spinal cord (Menei *et al.*, 1998). A combination of open-ended Schwann cell channels and administration of BDNF and NT-3 also significantly increased axonal regeneration and reentry into the CNS following spinal cord hemisection (Bamber *et al.*, 2001). These findings confirmed that the addition of neurotrophins enhances the capacity of Schwann cells to support axonal growth within the CNS.

Astroglial scarring and myelin inhibitors may also play key roles in preventing axonal reentry into the CNS following Schwann cell transplantation, and several different approaches have been used to overcome these barriers. Immunohistochemical analysis of CSPG expression following Schwann cell transplantation revealed high levels of CSPG immunoreactivity at the boundary between the graft and the caudal spinal cord, which suggests that these molecules may limit the ability of axons to reenter the CNS (Plant *et al.*, 2001). Infusion of chondroitinase ABC into the distal ends of Schwann cell-filled channels greatly reduced CSPG expression and allowed axons to regenerate up to 5 mm within the caudal spinal cord (Chau *et al.*, 2004).

OECs, which will be discussed in detail in the next section, have also been used to facilitate the transition between Schwann cell grafts and the spinal cord. In the first of these studies, implantation of Schwann cell channels was followed by injection of OECs into the rostal and caudal stumps of the spinal cord (Ramón-Cueto *et al.*, 1998). Axons were traced using multiple injections of wheat germ agglutinase (WGA)-HRP at 6 weeks after surgery and it was observed that regenerating axons were able to overcome inhibition by the glial scar, traverse the OEC and Schwann cell grafts, and extend for long distances within myelinated regions of the spinal cord (Ramón-Cueto *et al.*, 1998). This approach has subsequently been used in conjunction with the administration of methylprednisolone and interleukin-10 (Pearse *et al.*, 2004b) or chondroitinase ABC (Fouad *et al.*, 2005) to demonstrate enhanced recovery of locomotor function in spinal cord-injured rats.

As described in the preceding section, increasing intracellular cAMP through administration of dbcAMP or rolipram reverses the inhibitory effects of CNS myelin and enhances axonal regeneration *in vivo* (Nikulina *et al.*, 2004; Qiu *et al.*, 2002). These findings led Pearse *et al.* (2004a) to hypothesize that combining elevation of cAMP levels with Schwann cell transplantation would augment the regenerative response. In this study, adult rats received Schwann cell grafts and intraspinal injections of dbcAMP 1 week after a moderate spinal cord contusion, a model of injury that is considered more clinically relevant than spinal cord transection. Rolipram was delivered subcutaneously either at the time of injury (acute) or at 1 week after the contusion (delayed). Animals that received acute delivery of rolipram combined with Schwann cell grafts and dbcAMP displayed significantly higher levels of cAMP in the CNS at 14 days after injury. This in turn led to enhanced sparing of myelinated axons, greater myelination of spinal cord axons by the engrafted Schwann cells, and an overall increase in the number of axons within the grafts (Fig. 4). The enhanced tissue preservation observed in the spinal cords of animals that received rolipram, dbcAMP, and Schwann cells attests to the neuroprotective effects mediated by this treatment strategy (Fig. 4). While regeneration of corticospinal axons was not observed, there were significant improvements in the regeneration of serotonergic axons across the lesion site and in hindlimb locomotor recovery as measured on the Basso–Beattie–Bresnahan scale (BBB; Basso *et al.*, 1995). These findings demonstrate that elevation of cAMP levels increases the regenerative capacity of CNS axons following spinal cord contusion, and that rolipram and dbcAMP promote significant tissue sparing when administered in conjunction with Schwann cell grafting. The improved axonal regeneration and functional recovery observed in this and other studies confirm that Schwann cell transplantation, particularly

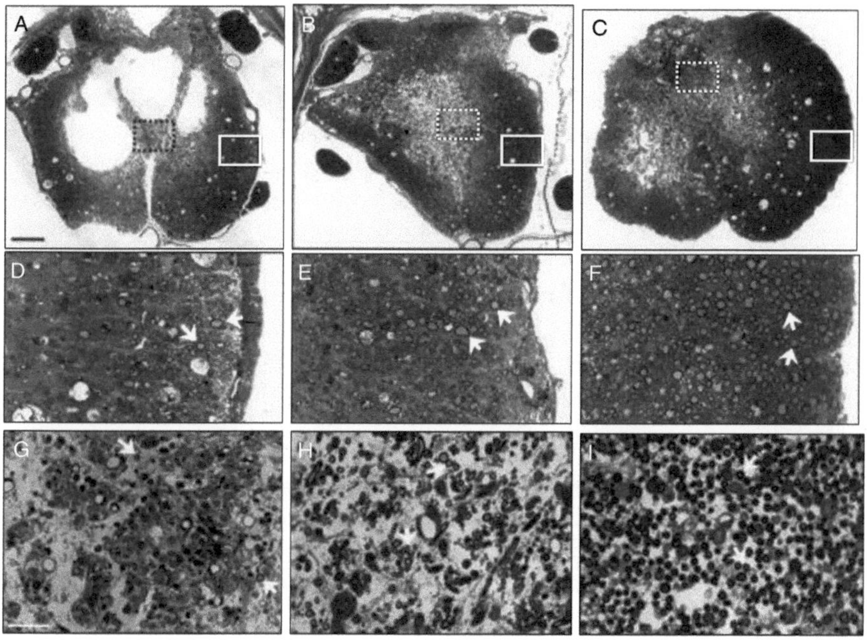

FIG. 4. Schwann cell transplantation in combination with elevation of cAMP promotes tissue sparing and axonal regeneration after spinal cord contusion injury. Transverse sections of control spinal cords at 11 weeks after injury (A) showed extensive cavitation, but in animals that received Schwann cell grafts (B) or Schwann cells plus dbcAMP and rolipram (C) there was greater tissue preservation. Analysis at higher magnification revealed that densities of central and peripheral myelinated axons were significantly increased in animals that received Schwann cells, dbcAMP, and rolipram (F, I) or Schwann cells alone (E, H) compared to control animals (D, G). [Adapted by permission from Macmillan Publishers Ltd.: Pearse, D. D., Pereira, F. C., Marcillo, A. E., Bates, M. L., Berrocal, Y. A., Filbin, M. T., and Bunge, M. B., cAMP and Schwann cells promote axonal growth and functional recovery after spinal cord injury. *Nature Medicine* 10, 610–616, copyright 2004.]

when combined with other agents that reduce inhibition or promote growth, is a viable therapeutic strategy for the treatment of spinal cord injury.

B. Transplantation of OECs

While regenerative failure is commonly observed throughout the brain and spinal cord following injury, there is one area of the CNS that does support axonal regeneration. Neurons in the olfactory epithelium are constantly renewed through a cycle of apoptosis and neurogenesis (Graziadei and Graziadei, 1979a,b; Graziadei and Monti-Graziadei, 1980; Holcomb *et al.*, 1995), and the axons of these neurons form the olfactory nerve (cranial nerve I). The unmyelinated axons of new olfactory neurons extend through the olfactory epithelium to the olfactory bulb, where they form synapses on mitral and tufted cells within the glomeruli (Shipley and Ennis, 1996). Olfactory neurogenesis is also induced by injury, such as transection of the olfactory nerve, and the axons of these newly generated neurons are able to regenerate and reinnervate their normal targets in the olfactory bulb (Doucette *et al.*, 1983; Graziadei and Monti-Graziadei, 1980). Unilateral removal of the olfactory bulb stimulates the differentiation of olfactory neuron precursors as well and the olfactory axons are able to project into the overlying cerebral cortex, where they arborize extensively and form synapses with the dendrites of cortical neurons (Graziadei *et al.*, 1978). These axons are therefore able to enter the CNS and grow relatively unimpeded within this normally hostile environment (Doucette, 1991; Doucette *et al.*, 1983; Graziadei *et al.*, 1978).

The unique ability of olfactory axons to cross the PNS–CNS transitional zone has been attributed to the presence of specialized glial cells known as OECs (Doucette, 1984, 1990, 1991; Raisman, 1985; Ramón-Cueto and Valverde, 1995). OECs surround olfactory axons as they exit the olfactory epithelium and remain associated with the axons as they extend into the CNS and innervate their targets within the olfactory bulb (Doucette, 1984, 1990; Raisman, 1985). It is believed that OECs create a permissive environment for regeneration by physically shielding olfactory axons from myelin and other inhibitors (Ramón-Cueto and Ávila, 1998) and by expressing growth-promoting extracellular matrix molecules (Doucette, 1990). OECs display many features characteristic of Schwann cells, including the expression of adhesion molecules such as L1, laminin, and N-CAM (Doucette, 1990; Franceschini and Barnett, 1996; Miragall *et al.*, 1989), the Schwann cell marker S100 (Franceschini and Barnett, 1996; Pixley, 1992) and $p75^{NTR}$ (Gong *et al.*, 1994; Franceschini and Barnett, 1996; Pixley, 1992). OECs also express and secrete neurotrophic factors such as NGF, BDNF, CNTF, and glial cell line-derived neurotrophic factor (GDNF; Lipson *et al.*, 2003; Woodhall *et al.*, 2001). Interestingly, OECs also have many properties that are normally associated with astrocytes. OECs help to form the glia limitans in the olfactory bulb (Barber and Lindsay, 1982; Doucette, 1984), a function that is performed by astrocytes in other regions of the CNS (Shearer and Fawcett, 2001). Astrocytic proteins such as GFAP and vimentin are also expressed by OECs (Barber and Lindsay, 1982; Franceschini and Barnett, 1996; Pixley, 1992; Schwob *et al.*, 1986). This distinctive combination of phenotypes has made OECs attractive candidates for transplantation in studies of spinal cord injury.

OECs are located in the nerve fiber layer of the olfactory bulb and this structure is a commonly used source of OECs for transplantation. Typically, olfactory bulbs are harvested from either fetal (Devon and Doucette, 1992, 1995) or perinatal rats (Barnett *et al.*, 1993; Chuah and Au, 1993), and the nerve fiber layer is simply peeled away (Chuah and Au, 1993; Devon and Doucette, 1992, 1995). OECs have been obtained from adult animals by isolating the nerve fiber and glomerular layers of the olfactory bulb (Ramón-Cueto and Nieto-Sampedro, 1992). For fetal rats, mechanical dissociation of the tissue and plating of the cell suspension are sufficient to yield a pure culture of OECs as the nerve fiber layer is populated exclusively by OECs at that timepoint (Doucette, 1995). As the animals age, additional purification is required and fluorescence-activated cell sorting for the O4 antigen has been used to produce highly enriched populations of perinatal OECs (Barnett *et al.*, 1993). Adult OECs are prepared in the same manner as fetal OECs and prior to purification, two major populations of OECs can be identified: A-cells and S-cells (Li *et al.*, 1998). While OECs often display features of both astrocytes and Schwann cells (Doucette, 1990), in A-cells a combination of astrocyte and fibroblast characteristics is predominant; these cells express GFAP, vimentin, and fibronectin but not $p75^{NTR}$ (Li *et al.*, 1998; Ramón-Cueto and Nieto-Sampedro, 1992). S-cells closely resemble transplanted Schwann cells and stain positively for $p75^{NTR}$ and S100 but not fibronectin (Li *et al.*, 1998; Ramón-Cueto and Nieto-Sampedro, 1992). Eight days after the initial plating of the cells, adult OECs can be purified by immunopanning with $p75^{NTR}$ antibodies (Ramón-Cueto and Nieto-Sampedro, 1992), and this technique has been used to isolate S-cells for transplantation (Ramón-Cueto and Nieto-Sampedro, 1994; Ramón-Cueto *et al.*, 1998; Richter *et al.*, 2005).

Olfactory axons remain unmyelinated as they extend from the olfactory epithelium to the olfactory bulb, and so, it is generally accepted that OECs do not normally myelinate axons *in vivo* (Boyd *et al.*, 2003; Doucette, 1984, 1990; Raisman, 1985). However, it has been shown *in vitro* that OECs can assume a myelinating phenotype when cocultured with DRG neurons (Devon and Doucette, 1992, 1995). This response is characterized by increased expression of myelin basic protein and P_0, establishment of basal laminae, and formation of myelin sheaths around the neurites (Devon and Doucette, 1992, 1995; Lee *et al.*, 2001). Ultrastructural analysis revealed that the myelin produced by OECs is very similar to PNS myelin (Devon and Doucette, 1992), which suggests that OECs switch to a Schwann cell-like phenotype in the presence of axons. It must also be emphasized that these observations were not the result of Schwann cell contamination as the DRG cultures were pretreated with the antimitotic agent fluorodeoxyuridine to eliminate nonneuronal cells prior to the addition of the OECs (Devon and Doucette, 1992, 1995).

These findings have prompted other groups to use OECs in CNS remyelination studies. Franklin *et al.* (1996) produced areas of focal demyelination in the rat spinal cord by irradiation and injection of ethidium bromide. OECs derived from a clonal cell line were transplanted into the lesion sites and within 3 weeks, large numbers of axons had been remyelinated by OECs expressing P_0 (Franklin *et al.*, 1996). In a similar experiment, extensive remyelination of dorsal column axons was observed following implantation of OECs into the demyelinated spinal cord (Imaizumi *et al.*, 1998). More importantly, electrophysiological recordings showed that the conduction velocities of the remyelinated axons were significantly increased, which indicates that myelination by OECs can restore normal physiological function (Imaizumi *et al.*, 1998). Together, these studies demonstrated that OECs are capable of myelination *in vivo*, which has implications not only for the treatment of spinal cord injury but also demyelinating diseases such as multiple sclerosis.

The first experiments examining the ability of OECs to promote axonal regeneration were performed by Ramón-Cueto and Nieto-Sampedro (1994). In this study, adult rats received a unilateral rhizotomy of the T10 dorsal root, and OECs were injected into the dorsal root entry zone (DREZ) at the level of the injury. Olfactory and dorsal root axons both originate in the PNS and form synapses with CNS neurons, but unlike olfactory axons, dorsal root axons are unable to regenerate across the PNS–CNS transition zone following injury (Cajal, 1928; Carlstedt, 1985). Given their vital role in promoting the growth of olfactory axons, it was hypothesized that OECs may have the ability to improve regeneration of dorsal root axons as well (Ramón-Cueto and Nieto-Sampedro, 1994). By postsurgical day 21, large numbers of regenerating axons were observed within the spinal cords of animals that received OEC transplants (Ramón-Cueto and Nieto-Sampedro, 1994). These axons were immunoreactive for calcitonin gene-related peptide (CGRP) and GAP-43, which confirmed that they were primary afferents originating from DRG neurons (Ramón-Cueto and Nieto-Sampedro, 1994). In a later study involving rhizotomy of multiple dorsal roots, enhanced regeneration of CGRP-positive dorsal root axons was once again observed following implantation of OECs (Navarro *et al.*, 1999). Two months after OEC grafting, these animals showed marked improvements in axonal conduction velocity and recovery of withdrawal reflexes, suggesting that the regenerating axons had formed functional connections with spinal cord neurons (Navarro *et al.*, 1999).

When OECs were transplanted immediately after unilateral electrolytic lesion of the CST, the cells formed a continuous mass within the lesion cavity and migrated into the rostral and caudal ends of the damaged spinal cord (Li *et al.*, 1997, 1998). Axons were anterogradely traced by injecting biotin dextran into the sensorimotor cortex and by postsurgical day 10, numerous CST axons had extended across the lesion site. Bundles of regenerating axons were myelinated by Schwann cell-like OECs, and these processes maintained a linear, unbranched trajectory along the longitudinal axis of the spinal cord. At 2–3 months after surgery, axonal bundles were no longer present and individual axons were myelinated by OECs at a ratio of 1:1, a pattern virtually identical to that of Schwann cell myelination. By this time the axons had grown up to 1 mm into the distal spinal cord and as the axons made the transition from the OEC graft to the CNS, they became myelinated by oligodendrocytes. Recovery of motor function was tested using a forelimb reaching task, and while control rats did not regain use of the affected forelimb, animals that received OEC transplants were capable of performing the task with either forelimb. This indicates that there was significant recovery of fine motor control in these animals (Li *et al.*, 1997, 1998). Histological analysis revealed that OECs extended across the entire lesion site in rats that displayed improved motor function, suggesting that they had enabled functional regeneration of CST axons to occur (Li *et al.*, 1997, 1998).

In collaboration with Mary Bunge, Ramón-Cueto *et al.* (1998)\ demonstrated that implanting OECs at the ends of Schwann cell-filled guidance channels enhanced axonal regeneration in the spinal cord, lending further support to the hypothesis that OECs can facilitate axonal reentry into the CNS. To build on these findings and determine whether OECs alone were capable of supporting regeneration, they injected Hoechst-labeled OECs at multiple sites in the spinal cord following a full spinal cord transection (Ramón-Cueto *et al.*, 2000). The transplanted OECs extended across the lesion site and migrated into the glial scar at the rostral and caudal ends of the spinal cord. Locomotor recovery was assessed from 3 to 7 months after surgery by testing the rats' ability to climb an angled wire grid. Correct paw placement and proprioceptive responses were also measured. Injured rats that received control injections of DMEM showed no recovery of sensorimotor function, but animals with OEC transplants were capable of supporting their body weight and consistently displayed improvements in voluntary hindlimb movement and motor reflexes. At 8 months after surgery, serotonergic and noradrenergic axons were visualized by immunohistochemical staining. These axons crossed the lesion site and grew several millimeters beyond the glial scar in the ventral columns and horns of the distal spinal cord. Regeneration of injured CST axons was also observed in OEC-transplanted animals, but instead of growing within the dorsal funiculus, these axons exited the spinal cord and extended for long distances along the pial surface, accompanied by migrating OECs. Eventually, some CST axons reentered the ventral columns of the spinal cord; however, the majority remained associated with the meninges. There was a strong correlation between functional recovery and the length of regenerated axons (Ramón-Cueto *et al.*, 2000), but in light of these unusual observations, it was unclear as to whether OEC-mediated regeneration of CST axons contributed to the dramatic functional recovery reported in this study.

Recently, a new approach to OEC transplantation has emerged in which OECs isolated from the lamina propria of the olfactory epithelium, or pieces of the olfactory lamina propria (OLP) itself are implanted into the CNS (Lu *et al.*, 2001). Lamina propria OECs (LP-OECs) are the first OECs encountered by olfactory axons as they exit the olfactory epithelium, and they share many phenotypic characteristics with olfactory bulb OECs, including expression of $p75^{NTR}$, S100, and GFAP (Au and Roskams, 2003). LP-OECs also express several unique markers such as CD-44, beta1 integrin, Notch 3, and vascular endothelial growth factor (Au and Roskams, 2003; Carson *et al.*, 2006). The primary motivation for using these cells in spinal cord injury studies has been the fact that the olfactory epithelium is far more accessible in humans compared to the olfactory bulb, raising the possibility of autologous transplantation (Lu *et al.*, 2001, 2002). Unlike olfactory bulb OECs, LP-OECs did not promote regeneration of dorsal root axons following dorsal root rhizotomy (Ramer *et al.*, 2004a), but implantation of LP-OECs or OLP following spinal cord transection did produce regeneration of serotonergic axons and partial recovery of hindlimb function (Lu *et al.*, 2001). It was subsequently shown that delaying transplantation of OLP until 4 weeks after injury had a similar effect on growth of raphespinal axons (Lu *et al.*, 2002). Further characterization of LP-OEC morphology and migration *in vivo* revealed that LP-OECs were capable of migrating longer distances within the spinal cord, and that they interact with astrocytes to reduce glial scar formation and cavitation (Ramer *et al.*, 2004b; Richter *et al.*, 2005). It is believed that these unique abilities contributed to the enhanced axonal sprouting and regeneration that was observed in these studies (Ramer *et al.*, 2004b; Richter *et al.*, 2005).

Although OECs have shown great promise for therapeutic use in spinal cord injury, many questions remain regarding their ability to promote regeneration. One of the most pressing concerns is the lack of reproducibility that has emerged with recent attempts to replicate observations made in earlier studies. For example, contrary to the findings of Devon and Doucette (1992, 1995), Plant *et al.* (2002) found that OECs were unable to myelinate DRG neurites *in vitro* when similar culture conditions were used. This inevitably raises questions as to whether the axonal remyelination that has been observed *in vivo* was truly the result of myelination by OECs. In these studies, it was noted by the authors that the pattern of remyelination strongly resembled that of Schwann cells (Li *et al.*, 1997, 1998) and so, the possibility of myelination by infiltrating Schwann cells must be considered.

Takami *et al.* (2002) compared the abilities of Schwann cells and OECs to promote regeneration following spinal cord contusion. In this clinically relevant model of spinal cord injury, both cell types increased tissue sparing and axonal growth, but regeneration into the distal spinal cord and functional recovery were greater in animals that received Schwann cell transplants (Takami *et al.*, 2002). This suggested that Schwann cells might be a more effective means of promoting functional axonal regeneration following contusion injury (Takami *et al.*, 2002). It should be noted, however, that in a subsequent study that used the same model, functional outcome was greatly improved by delaying OEC transplantation for 1 week, which indicates that immediate OEC transplantation does not elicit an optimal response (Plant *et al.*, 2003).

Lastly, a newly published study by Steward *et al.* (2006) attempted to replicate the OLP transplantation experiments published by Lu *et al.* (2002). In accordance with the published experimental methods, adult rats underwent spinal cord transection and received OLP transplants 30 days later. Recovery of motor function was monitored for 10 weeks, but none of the OLP-transplanted rats showed any significant improvements on the BBB scale. In addition, injections of Fluorogold into the distal spinal cord did not retrogradely label any axons and immunostaining revealed only minimal growth of serotonergic axons beyond the OLP grafts. It was therefore concluded that OLP transplantation do not robustly stimulate axonal regeneration following spinal cord injury (Steward *et al.*, 2006). These results are particularly troubling in light of the fact that human clinical trials using autologous OLP transplantation have already begun (Feron *et al.*, 2005).

Because OECs express proteins that are commonly found in Schwann cells and astrocytes (Doucette, 1990; Franceschini and Barnett, 1996), reliable visualization of the implanted cells is another problem that is regularly encountered in OEC studies. Initially, OECs were labeled with Hoechst or Cell-Tracker dyes (Li *et al.*, 1997, 1998; Ramón-Cueto *et al.*, 1998, 2000), but it has been reported that phagocytosis of the pigments can produce misleading cell survival and migration data (Boyd *et al.*, 2004; Ruitenberg *et al.*, 2002). Adenoviral and retroviral expression of reporter genes such as LacZ (Boyd *et al.*, 2004; Ruitenberg *et al.*, 2002) or using OECs isolated from mice expressing green fluorescent protein (Ramer *et al.*, 2004a,b; Richter *et al.*, 2005) have proven to be more reliable methods of tracking OECs *in vivo*. In a study by Boyd *et al.* (2004), clip compression injury of the spinal cord was followed by delayed transplantation of LacZ-expressing OECs, and staining for β-galactosidase at postsurgical day 21 showed that large numbers of OECs survived within the lesion site. Electron microscopy revealed that the OECs did not form myelin and did not contact axons directly, but rather ensheathed bundles of regenerating axons that were myelinated by Schwann cells. While earlier studies had reported that OECs can remyelinate axons following demyelinating lesions of the spinal cord (Li *et al.*, 1997, 1998), these results provided strong morphological evidence that OECs do not myelinate axons following spinal cord trauma. The lack of interaction between axons and OECs also suggested that these cells might not play a direct role in promoting axonal regeneration (Boyd *et al.*, 2004). It was therefore proposed that OECs help to create a permissive environment for growth by facilitating migration of Schwann cells into the injured spinal cord (Boyd *et al.*, 2004). A new study by Boyd *et al.* (2006) has now identified the actin-binding protein calponin as the first true phenotypic marker for OECs. Proteomic analysis revealed that Schwann cells do not express calponin, but it is found in OECs both *in vivo* and *in vitro* (Boyd *et al.*, 2006). When OECs were transplanted into the spinal cord and immunostained for calponin, it became apparent that the OECs retained the non-myelinating phenotype that they express within the olfactory system (Boyd *et al.*, 2006). This lends further support to the hypothesis that OECs do not promote axonal regeneration directly, but rather enhance growth by providing physical support for infiltrating Schwann cells (Boyd *et al.*, 2006). In light of these findings and the many conflicting reports that have been produced regarding the efficacy of OEC transplantation, the scientific community must thoroughly reevaluate the therapeutic potential of OECs and clearly define their role in mediating functional recovery after spinal cord injury.

C. Other Methods of Transplantation That Have Been Applied to Spinal Cord Injury

Fibroblasts that have been genetically modified to express neurotrophic factors have been generated in an attempt to provide both trophic and physical support to regenerating axons. In one early study, fibroblasts expressing NGF, BDNF, or NT-3 were transplanted into the uninjured spinal cord (Nakahara *et al.*, 1996). Grafts of NGF- and NT-3-expressing cells induced sprouting of nociceptive and catecholaminergic axons, but no effects on motor axon growth were observed, which indicates that these intact axons are less responsive to neurotrophic stimulation (Nakahara *et al.*, 1996). Implantation of NGF-expressing fibroblasts after spinal cord injury elicited a response similar to that seen in the uninjured spinal cord, as extensive sprouting of NGF-responsive axons was observed within the grafts (Grill *et al.*, 1997a; Jones *et al.*, 2003b; Tuszynski *et al.*, 1996). While BDNF-expressing fibroblasts had no effect on axonal plasticity in the uninjured animal (Nakahara *et al.*, 1996), transplantation of these cells after spinal cord hemisection promoted regeneration of rubrospinal axons, which extended within the caudal spinal cord and reached their normal target regions (Liu *et al.*, 1999). Regeneration was confirmed by relesioning of the spinal cord, which eliminated the improved forelimb motor function that had been observed (Liu *et al.*, 1999). Similarly, transplantation of fibroblasts expressing NT-3 enhances survival of axotomized neurons, promotes regeneration of corticospinal axons, and partially improves motor function (Grill *et al.*, 1997b; Himes *et al.*, 2001; Tuszynski *et al.*, 2003). Fibroblast-mediated expression of GDNF within the lesion site induces regeneration of ascending dorsal column axons and migration of Schwann cells into the affected area, which led to remyelination of the ingrowing axons (Blesch and Tuszynski, 2003). Because of their ability to promote growth in several major motor tracts, BDNF and NT-3-expressing fibroblasts have often been used in combination. Studies using a variety of spinal cord injury models have reported that implantation of BDNF and NT-3-expressing fibroblasts enhances axonal regeneration and remyelination, neuronal survival, and recovery of bladder function (McTigue *et al.*, 1998; Mitsui *et al.*, 2005; Tobias *et al.*, 2003). It should be noted that restoration of motor function has been limited in some cases and it has been suggested that cell grafting further damages the spinal cord (Shumsky *et al.*, 2003), but overall, these studies have demonstrated that providing trophic support to injured axons and neurons is a critical component of promoting functional recovery after spinal cord trauma.

Bone marrow is a source of pluripotent hematopoietic stem cells, and it has been shown that marrow stromal cells (MSCs) can differentiate into neurons when transplanted into the CNS (Brazelton *et al.*, 2000; Mezey *et al.*, 2000, 2003). A study by Castro *et al.* (2002) disputed these results, claiming that no differentiation occurred when MSCs were transplanted following injury to the brain. In studies of spinal cord-injured rats, implantation of MSCs 1 week after contusion injury significantly improved functional recovery (Chopp *et al.*, 2000; Hofstetter *et al.*, 2002). The transplanted cells formed tissue bridges that spanned the lesion site, which suggests that they may support axonal regeneration (Hofstetter *et al.*, 2002). MSCs genetically modified to express BDNF support growth of axons into, but not beyond, the site of injury, and as a result, functional recovery was not observed (Lu *et al.*, 2005). An earlier study combined transplantation of MSCs with dbcAMP and NT-3 (Lu *et al.*, 2004). dbcAMP was injected into the L4 DRG prior to injury, while NT-3 was injected into the injury site and caudal spinal cord immediately after dorsal column lesion. MSCs were also transplanted after injury. The goal of this approach was to replicate the conditioning lesion effect *in vivo* by stimulating neuronal cell bodies with dbcAMP, while simultaneously providing trophic support to the transected axons. Animals that received the combination of dbcAMP, NT-3, and MSCs were the only ones that displayed axonal regeneration beyond the lesion site (Fig. 5), but there was no recovery of sensorimotor function as measured by tape-removal, horizontal ladder, and rope-walking tests (Lu *et al.*, 2004). Clearly, the benefits of MSC transplantation have not yet been fully elucidated, but the regeneration observed in this study emphasizes the benefits of combinatorial approaches to spinal cord injury repair.

An alternative to transplanting dissociated cells into the injured spinal cord is implanting segments of embryonic spinal cord tissue, and this has been shown to have many beneficial effects. These include inducing growth of rubrospinal, raphespinal, coerulospinal, and CGRP-expressing nociceptive axons (Bernstein-Goral and Bregman, 1993; Houle and Reier, 1988, 1989; Reier *et al.*, 1986), promoting survival of axotomized neurons (Bregman and Reier, 1986; Bregman *et al.*, 1998; Himes *et al.*, 1994), and reducing glial scar formation (Houle, 1992). Administration of BDNF and NT-3 in conjunction with tissue grafting can further enhance regeneration of axons originating in the brainstem and motor cortex (Bregman *et al.*, 1997; Coumans *et al.*, 2001). One particularly unique aspect of this technique is that it can extend the critical period of plasticity for developing corticospinal axons and thereby enhance their growth following injury (Bregman *et al.*, 1989). Functional recovery is significantly improved when embryonic tissue is implanted into neonatal and adult rats after spinal cord overhemisection or complete transection (Bregman *et al.*, 1993; Diener and

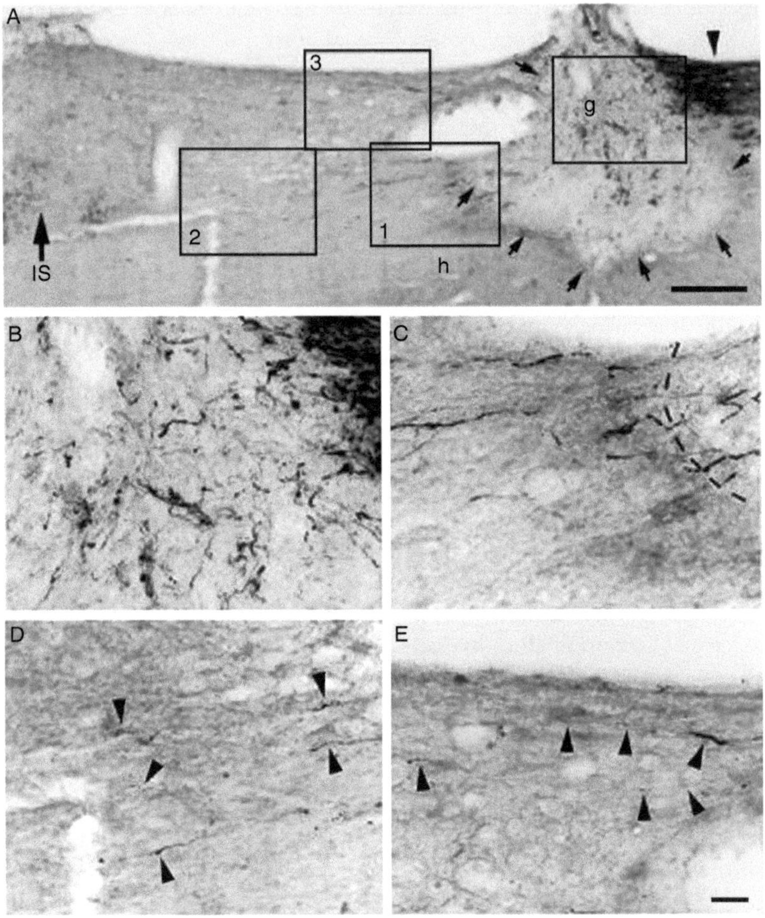

FIG. 5. Axonal regeneration induced by a combination of elevated cAMP, NT-3, and MSC transplantation. At 1–3 months after dorsal column lesion, animals that received intraganglionic injection of dbcAMP, MSC grafts (g), and intraspinal injection of NT-3 (IS) displayed regeneration of axons within and beyond the lesion site (A). Axons labeled with cholera toxin B are present within the MSC graft (B), and reenter the host tissue (h) rostral to the graft (C). These axons extend up to 0.7 mm beyond the lesion site (D) in both the ventral and dorsal (E) aspects of the spinal cord. [Copyright 2004 by the Society for Neuroscience.]

Bregman, 1998a,b; Miya *et al.*, 1997), and it was later shown that delaying transplantation of embryonic tissue until 2–4 weeks after injury further improves motor function (Coumans *et al.*, 2001). This approach has proven to be a very valuable experimental model, but it would be difficult to implement a similar strategy in human spinal cord injury due to the ethical questions that would inevitably be raised by the use of human fetal tissue.

The newest candidates for transplantation in spinal cord injury are multipotent neural progenitor cells, which can be derived from embryonic stem (ES) cells or isolated from embryonic CNS tissue (Cao *et al.*, 2002a; Gage, 2000; McDonald *et al.*, 2004). The cells can then be expanded by treatment with fibroblast growth factor-2 or EGF, and differentiated by withdrawal of these mitogenic factors (Gage, 2000; Ray *et al.*, 1993; Reynolds and Weiss, 1992; Reynolds *et al.*, 1992; Richards *et al.*, 1992). The addition of platelet-derived growth factor or retinoic acid induces ES-cells to differentiate into oligodendrocytes and these cells are capable of remyelinating axons when implanted after chemical demyelination or spinal cord contusion (Brüstle *et al.*, 1999; Keirstead *et al.*, 2005; Liu *et al.*, 2000). It has been shown that while ES cells were able to differentiate into both neurons and glia *in vitro*, only glial cells were produced when undifferentiated ES cells were transplanted following spinal cord injury (Cao *et al.*, 2001). In addition, the majority of these cells displayed an astrocytic phenotype, which suggested that differentiation prior to transplantation was required to obtain oligodendrocytes (Cao *et al.*, 2001). Infection of embryonic glial-restricted precursor cells with retroviruses expressing the multineurotrophin D15A substantially increased the percentage of oligodendrocytes, and by 6 weeks after transplantation into the contused spinal cord, these cells expressed myelin basic protein (Cao *et al.*, 2005). This led to axonal remyelination, restoration of motor-evoked potentials, and recovery of hindlimb motor function (Cao *et al.*, 2005). Neuronal-restricted precursor cells differentiate into neurons expressing βIII-tubulin, GABA, and glutamate within the uninjured spinal cord, but curiously, they fail to do so when implanted following injury (Cao *et al.*, 2002b). This finding illustrates the difficulty of replacing damaged neurons, but neuronal differentiation has been observed when mouse ES cells were transplanted into the spinal cord 9 days after a contusion injury (McDonald *et al.*, 1999). Animals that received these transplants showed improvements in weight bearing and motor coordination (McDonald *et al.*, 1999). More recently, Nakamura *et al.* (2005) demonstrated that embryonic neurospheres differentiated into neurons and oligodendrocytes within the spinal cord, and that regeneration of corticospinal, raphespinal, and coerulospinal axons was consequently enhanced. Motor reflexes and locomotion were also improved, indicating that functional recovery had occurred (Nakamura *et al.*, 2005).

Neural precursor cells are also present in the adult CNS, particularly in regions with high rates of neurogenesis such as the subventricular zone and hippocampus (Gage, 2000; Gage *et al.*, 1995; Reynolds and Weiss, 1992; Weiss *et al.*, 1996). In an attempt to avoid the ethical concerns raised by the use of fetal tissue, these cells are now being isolated for use in transplantation. Oka *et al.* (2004) have obtained neural precursor cells from the subventricular zone in mature nonhuman primates and transplanted these cells into the demyelinated spinal cord. They observed substantial remyelination of CNS axons within 3 weeks, which suggests that autologous transplantation of adult neural precursor cells could be an effective treatment for demyelinating disorders (Oka *et al.*, 2004). A newly published study from the laboratory of Michael Fehlings has now tested the ability of adult neural progenitor cells to promote functional recovery after clip compression injury of the spinal cord (Karimi-Abdolrezaee *et al.*, 2006). Axonal regeneration was not assessed, but delayed transplantation of neural progenitor cells did produce axonal remyelination by precursor-derived oligodendrocytes and significant improvements in motor function as measured by grid walking tests and the BBB scale (Karimi-Abdolrezaee *et al.*, 2006). This field is still in its infancy, but it is clear from these results that neural progenitor cells hold great promise for the treatment of spinal cord injury.

V. Future Directions

As interest in spinal cord injury research has grown, therapeutic approaches have become increasingly innovative. Advances in biomedical engineering have led to the creation of artificial scaffolds designed to support axonal regeneration when implanted into the spinal cord. Unlike the nondegradable polycarbonate and polyvinylchloride tubes used in earlier studies of Schwann cell transplantation (Montgomery and Robson, 1990, 1993; Xu *et al.*, 1995a,b, 1997), these scaffolds are composed of resorbable, nonimmunogenic polymers such as polylactic acid, polyhydroxybutyrate, or agarose (Moore *et al.*, 2006; Novikov *et al.*, 2002; Oudega *et al.*, 2001; Stokols and Tuszynski, 2006; Teng *et al.*, 2002). The biodegradable nature of these materials means that they can be engineered to deliver neurotrophins, and it has been shown that agarose scaffolds containing BDNF can effectively promote axonal regeneration following an aspiration lesion of the spinal cord (Stokols and Tuszynski, 2006). It has been hypothesized that the microscopic channels present within the scaffolds may allow the organization of spinal cord tracts to be preserved, and this is partially supported by the observation that the regenerating axons maintained a linear trajectory (Stokols and Tuszynski, 2006). The channels can also be seeded with Schwann cells or neural stem cells to further enhance axonal growth and guidance (Hurtado *et al.*, 2006; Moore *et al.*, 2006; Oudega *et al.*, 2001; Teng *et al.*, 2002). Animals that underwent spinal cord hemisection displayed regeneration of corticospinal axons into the caudal spinal cord and recovery of hindlimb function following implantation of scaffolds containing neural stem cells (Teng *et al.*, 2002). In studies using scaffolds seeded with Schwann cells, only modest axonal regeneration and functional recovery have been reported and this has been attributed to collapse of the scaffold and poor survival of Schwann cells within the channels (Hurtado *et al.*, 2006; Oudega *et al.*, 2001). Clearly this technique is not yet optimized, but as the procedures for manufacturing and implanting these scaffolds become more refined, the therapeutic value of these devices will undoubtedly increase.

The last 15 years have seen unprecedented advances in the field of spinal cord injury repair, particularly in the identification of factors that limit axonal regeneration. As a result, the scientific community no longer asks whether regeneration is possible, but rather how regeneration can be achieved, and new strategies aimed at accomplishing this goal continue to emerge. It was recently shown that expression of retinoic acid receptor $\beta 2$ increases levels of cAMP within DRG and allows axons to regenerate through the dorsal root entry zone *in vivo* (Wong *et al.*, 2006). Administration of interleukin-6 after dorsal column lesion also reverses myelin inhibition and enhances axonal regeneration in the spinal cord (Cafferty *et al.*, 2004; Cao *et al.*, 2006). These two agents join the many others that are capable of promoting regeneration, but it has become clear that no single drug will provide a cure for spinal cord injury. This realization has led to increased collaboration between researchers and the development of new combinatorial therapies that target multiple barriers to regeneration. As described herein, the results of recent combinatorial studies have been extremely encouraging and there is every reason to believe that this approach will eventually yield effective treatments for spinal cord injury. Developing these treatments is the task that now falls to the science of the future.

References

Au, E., and Roskams, A. J. (2003). Olfactory ensheathing cells of the lamina propria *in vivo* and *in vitro*. *Glia* **41,** 224–236.

Bamber, N. I., Li, H., Lu, X., Oudega, M., Aebischer, P., and Xu, X. M. (2001). Neurotrophins BDNF and NT-3 promote axonal re-entry into the distal host spinal cord through Schwann cell-seeded mini-channels. *Eur. J. Neurosci.* **13,** 257–268.

Barber, P. C., and Lindsay, R. M. (1982). Schwann cells of the olfactory nerves contain glial fibrillary acidic protein and resemble astrocytes. *Neuroscience* **7,** 3077–3090.

Barnett, S. C., Hutchins, A. M., and Noble, M. (1993). Purification of olfactory nerve ensheathing cells from the olfactory bulb. *Dev. Biol.* **155,** 337–350.

Bartsch, U., Bandtlow, C. E., Schnell, L., Bartsch, S., Spillmann, A. A., Rubin, B. P., Hillenbrand, R., Montag, D., Schwab, M. E., and Schachner, M. (1995). Lack of evidence that myelin-associated glycoprotein is a major inhibitor of axonal regeneration in the CNS. *Neuron* **15,** 1375–1381.

Basso, D. M., Beattie, M. S., and Bresnahan, J. C. (1995). A sensitive and reliable locomotor rating scale for open field testing in rats. *J. Neurotrauma* **12,** 1–21.

Benfey, M., and Aguayo, A. J. (1982). Extensive elongation of axons from rat brain into peripheral nerve grafts. *Nature* **296,** 150–152.

Benson, M. D., Romero, M. I., Lush, M. E., Lu, Q. R., Henkemeyer, M., and Parada, L. F. (2005). Ephrin-B3 is a myelin-based inhibitor of neurite outgrowth. *Proc. Natl. Acad. Sci. USA* **102,** 10694–10699.

Bernstein-Goral, H., and Bregman, B. S. (1993). Spinal cord transplants support the regeneration of axotomized neurons after spinal cord lesions at birth: A quantitative double-labeling study. *Exp. Neurol.* **123,** 118–132.

Bignami, A., Chi, N. H., and Dahl, D. (1984). Laminin in rat sciatic nerve undergoing Wallerian degeneration. Immunofluorescence study with laminin and neurofilament antisera. *J. Neuropathol. Exp. Neurol.* **43,** 94–103.

Bixby, J. L., Lilien, J., and Reichardt, L. F. (1988). Identification of the major proteins that promote neuronal process outgrowth on Schwann cells *in vitro*. *J. Cell Biol.* **107,** 353–361.

Björklund, A., Katzman, R., Stenevi, U., and West, K. A. (1971). Development and growth of axonal sprouts from noradrenaline and 5-hydroxytryptamine neurones in the rat spinal cord. *Brain Res.* **31,** 21–33.

Björklund, A., Johansson, B., Stenevi, U., and Svendgaard, N. A. (1975). Re-establishment of functional connections by regenerating central adrenergic and cholinergic axons. *Nature* **253,** 446–448.

Blesch, A., and Tuszynski, M. H. (2003). Cellular GDNF delivery promotes growth of motor and dorsal column sensory axons after partial and complete spinal cord transections and induces remyelination. *J. Comp. Neurol.* **467,** 403–417.

Borisoff, J. F., Chan, C. C., Hiebert, G. W., Oschipok, L., Robertson, G. S., Zamboni, R., Steeves, J. D., and Tetzlaff, W. (2003). Suppression of Rho-kinase activity promotes axonal growth on inhibitory CNS substrates. *Mol. Cell. Neurosci.* **22,** 405–416.

Boyd, J. G., Skihar, V., Kawaja, M., and Doucette, R. (2003). Olfactory ensheathing cells: Historical perspective and therapeutic potential. *Anat. Rec.* **271,** 49–60.

Boyd, J. G., Lee, J., Skihar, V., Doucette, R., and Kawaja, M. D. (2004). LacZ-expressing olfactory ensheathing cells do not associate with myelinated axons after implantation into the compressed spinal cord. *Proc. Natl. Acad. Sci. USA* **101,** 2162–2166.

Boyd, J. G., Jahed, A., McDonald, T. G., Krol, K. M., Van Eyk, J. E., Doucette, R., and Kawaja, M. D. (2006). Proteomic evaluation reveals that olfactory ensheathing cells but not Schwann cells express calponin. *Glia* **53,** 434–440.

Bradbury, E. J., Moon, L. D., Popat, R. J., King, V. R., Bennett, G. S., Patel, P. N., Fawcett, J. W., and McMahon, S. B. (2002). Chondroitinase ABC promotes functional recovery after spinal cord injury. *Nature* **416,** 636–640.

Brazelton, T. R., Rossi, F. M., Keshet, G. I., and Blau, H. M. (2000). From marrow to brain: Expression of neuronal phenotypes in adult mice. *Science* **290,** 1775–1779.

Bregman, B. S., and Reier, P. J. (1986). Neural tissue transplants rescue axotomized rubrospinal cells from retrograde death. *J. Comp. Neurol.* **244,** 86–95.

Bregman, B. S., Kunkel-Bagden, E., McAtee, M., and O'Neill, A. (1989). Extension of the critical period for developmental plasticity of the corticospinal pathway. *J. Comp. Neurol.* **282,** 355–370.

Bregman, B. S., Kunkel-Bagden, E., Reier, P. J., Dai, H. N., McAtee, M., and Gao, D. (1993). Recovery of function after spinal cord injury: Mechanisms underlying transplant-mediated recovery of function differ after spinal cord injury in newborn and adult rats. *Exp. Neurol.* **123,** 3–16.

Bregman, B. S., Kunkel-Bagden, E., Schnell, L., Dai, H. N., Gao, D., and Schwab, M. E. (1995). Recovery from spinal cord injury mediated by antibodies to neurite growth inhibitors. *Nature* **378,** 498–501.

Bregman, B. S., McAtee, M., Dai, H. N., and Kuhn, P. L. (1997). Neurotrophic factors increase axonal growth after spinal cord injury and transplantation in the adult rat. *Exp. Neurol.* **148,** 475–494.

Bregman, B. S., Broude, E., McAtee, M., and Kelley, M. S. (1998). Transplants and neurotrophic factors prevent atrophy of mature CNS neurons after spinal cord injury. *Exp. Neurol.* **149,** 13–27.
Brockes, J. P., Fields, K. L., and Raff, M. C. (1979). Studies on cultured rat Schwann cells. I. Establishment of purified populations from cultures of peripheral nerve. *Brain Res.* **165,** 105–118.
Brüstle, O., Jones, K. N., Learish, R. D., Karram, K., Choudhary, K., Wiestler, O. D., Duncan, I. D., and McKay, R. D. (1999). Embryonic stem cell-derived glial precursors: A source of myelinating transplants. *Science* **285,** 754–756.
Cafferty, W. B., Gardiner, N. J., Das, P., Qiu, J., McMahon, S. B., and Thompson, S. W. (2004). Conditioning injury-induced spinal axon regeneration fails in interleukin-6 knock-out mice. *J. Neurosci.* **24,** 4432–4443.
Cai, D., Shen, Y., De Bellard, M., Tang, S., and Filbin, M. T. (1999). Prior exposure to neurotrophins blocks inhibition of axonal regeneration by MAG and myelin via a cAMP-dependent mechanism. *Neuron* **22,** 89–101.
Cai, D., Qiu, J., Cao, Z., McAtee, M., Bregman, B. S., and Filbin, M. T. (2001). Neuronal cyclic AMP controls the developmental loss in ability of axons to regenerate. *J. Neurosci.* **21,** 4731–4739.
Cai, D., Deng, K., Mellado, W., Lee, J., Ratan, R. R., and Filbin, M. T. (2002). Arginase I and polyamines act downstream from cyclic AMP in overcoming inhibition of axonal growth MAG and myelin *in vitro*. *Neuron* **35,** 711–719.
Cajal, S. R. y. (1928). "Degeneration and Regeneration of the Nervous System" (J. DeFelipe and E. G. Jones, Eds.). Oxford University Press, Oxford.
Canning, D. R., McKeon, R. J., DeWitt, D. A., Perry, G., Wujek, J. R., Frederickson, R. C., and Silver, J. (1993). beta-Amyloid of Alzheimer's disease induces reactive gliosis that inhibits axonal outgrowth. *Exp. Neurol.* **124,** 289–298.
Cao, Q., Benton, R. L., and Whittemore, S. R. (2002a). Stem cell repair of central nervous system injury. *J. Neurosci. Res.* **68,** 501–510.
Cao, Q., Xu, X. M., Devries, W. H., Enzmann, G. U., Ping, P., Tsoulfas, P., Wood, P. M., Bunge, M. B., and Whittemore, S. R. (2005). Functional recovery in traumatic spinal cord injury after transplantation of multineurotrophin-expressing glial-restricted precursor cells. *J. Neurosci.* **25,** 6947–6957.
Cao, Q. L., Zhang, Y. P., Howard, R. M., Walters, W. M., Tsoulfas, P., and Whittemore, S. R. (2001). Pluripotent stem cells engrafted into the normal or lesioned adult rat spinal cord are restricted to a glial lineage. *Exp. Neurol.* **167,** 48–58.
Cao, Q. L., Howard, R. M., Dennison, J. B., and Whittemore, S. R. (2002b). Differentiation of engrafted neuronal-restricted precursor cells is inhibited in the traumatically injured spinal cord. *Exp. Neurol.* **177,** 349–359.
Cao, Z., Gao, Y., Bryson, J. B., Hou, J., Chaudry, N., Siddiq, M., Martinez, J., Spencer, T., Carmel, J., Hart, R. B., and Filbin, M. T. (2006). The cytokine, IL-6, is sufficient but not necessary to mimic the peripheral conditioning lesion effect on axonal growth. *J. Neurosci.* **261,** 5565–5573.
Carlstedt, T. (1985). Regenerating axons form nerve terminals at astrocytes. *Brain Res.* **347,** 188–191.
Caroni, P., and Schwab, M. E. (1988a). Two membrane protein fractions from rat central myelin with inhibitory properties for neurite growth and fibroblast spreading. *J. Cell Biol.* **106,** 1281–1288.
Caroni, P., and Schwab, M. E. (1988b). Antibody against myelin-associated inhibitor of neurite growth neutralizes nonpermissive substrate properties of CNS white matter. *Neuron* **1,** 85–96.
Carson, C., Murdoch, B., and Roskams, A. J. (2006). Notch 2 and Notch 1/3 segregate to neuronal and glial lineages of the developing olfactory epithelium. *Dev. Dyn.* **235**(6), 1678–1688.
Castro, R. F., Jackson, K. A., Goodell, M. A., Robertson, C. S., Liu, H., and Shine, H. D. (2002). Failure of bone marrow cells to transdifferentiate into neural cells *in vivo*. *Science* **297,** 1299.

Chau, C. H., Shum, D. K., Li, H., Pei, J., Lui, Y. Y., Wirthlin, L., Chan, Y. S., and Xu, X. M. (2004). Chondroitinase ABC enhances axonal regrowth through Schwann cell-seeded guidance channels after spinal cord injury. *FASEB J.* **18,** 194–196.

Chen, A., Xu, X. M., Kleitman, N., and Bunge, M. B. (1996). Methylprednisolone administration improves axonal regeneration into Schwann cell grafts in transected adult rat thoracic spinal cord. *Exp. Neurol.* **138,** 261–276.

Chen, M. S., Huber, A. B., van der Haar, M. E., Frank, M., Schnell, L., Spillmann, A. A., Christ, F., and Schwab, M. E. (2000). Nogo-A is a myelin-associated neurite outgrowth inhibitor and an antigen for monoclonal antibody IN-1. *Nature* **403,** 434–439.

Chong, M. S., Woolf, C. J., Turmaine, M., Emson, P. C., and Anderson, P. N. (1996). Intrinsic versus extrinsic factors in determining the regeneration of the central processes of rat dorsal root ganglion neurons: The influence of a peripheral nerve graft. *J. Comp. Neurol.* **370,** 97–104.

Chopp, M., Zhang, X. H., Li, Y., Wang, L., Chen, J., Lu, D., Lu, M., and Rosenblum, M. (2000). Spinal cord injury in rat: Treatment with bone marrow stromal cell transplantation. *Neuroreport* **11,** 3001–3005.

Chuah, M. I., and Au, C. (1993). Cultures of ensheathing cells from neonatal rat olfactory bulbs. *Brain Res.* **601,** 213–220.

Coumans, J. V., Lin, T. T., Dai, H. N., MacArthur, L., McAtee, M., Nash, C., and Bregman, B. S. (2001). Axonal regeneration and functional recovery after complete spinal cord transection in rats by delayed treatment with transplants and neurotrophins. *J. Neurosci.* **21,** 9334–9344.

David, S., and Aguayo, A. J. (1981). Axonal elongation into peripheral nervous system "bridges" after central nervous system injury in adult rats. *Science* **214,** 931–933.

de Bellard, M. E., and Filbin, M. T. (Bellard 1999). Myelin-associated glycoprotein, MAG, selectively binds several neuronal proteins. *J. Neurosci. Res.* **56,** 213–218.

de Bellard, M. E., Tang, S., Mukhopadhyay, G., Shen, Y. J., and Filbin, M. T. (Bellard 1996). Myelin-associated glycoprotein inhibits axonal regeneration from a variety of neurons via interaction with a sialoglycoprotein. *Mol. Cell. Neurosci.* **7,** 89–101.

DeLano, W. L. (2002). Unraveling hot spots in binding interfaces: Progress and challenges. *Curr. Opin. Struct. Biol.* **12,** 14–20.

Dergham, P., Ellezam, B., Essagian, C., Avedissian, H., Lubell, W. D., and McKerracher, L. (2002). Rho signaling pathway targeted to promote spinal cord repair. *J. Neurosci.* **22,** 6570–6577.

Devon, R., and Doucette, R. (1992). Olfactory ensheathing cells myelinate dorsal root ganglion neurites. *Brain Res.* **589,** 175–179.

Devon, R., and Doucette, R. (1995). Olfactory ensheathing cells do not require L-ascorbic acid *in vitro* to assemble a basal lamina or to myelinate dorsal root ganglion neurites. *Brain Res.* **688,** 223–229.

Diener, P. S., and Bregman, B. S. (1998a). Fetal spinal cord transplants support the development of target reaching and coordinated postural adjustments after neonatal cervical spinal cord injury. *J. Neurosci.* **18,** 763–778.

Diener, P. S., and Bregman, B. S. (1998b). Fetal spinal cord transplants support growth of supraspinal and segmental projections after cervical spinal cord hemisection in the neonatal rat. *J. Neurosci.* **18,** 779–793.

Domeniconi, M., Cao, Z., Spencer, T., Sivasankaran, R., Wang, K., Nikulina, E., Kimura, N., Cai, H., Deng, K., Gao, Y., He, Z., and Filbin, M. (2002). Myelin-associated glycoprotein interacts with the Nogo66 receptor to inhibit neurite outgrowth. *Neuron* **35,** 283–290.

Domeniconi, M., Zampieri, N., Spencer, T., Hilaire, M., Mellado, W., Chao, M. V., and Filbin, M. T. (2005). MAG induces regulated intramembrane proteolysis of the p75 neurotrophin receptor to inhibit neurite outgrowth. *Neuron* **46,** 849–855.

Dou, C. L., and Levine, J. M. (1994). Inhibition of neurite growth by the NG2 chondroitin sulfate proteoglycan. *J. Neurosci.* **14,** 7616–7628.

Doucette, J. R. (1984). The glial cells in the nerve fiber layer of the rat olfactory bulb. *Anat. Rec.* **210,** 385–391.

Doucette, J. R., Kiernan, J. A., and Flumerfelt, B. A. (1983). The re-innervation of olfactory glomeruli following transection of primary olfactory axons in the central or peripheral nervous system. *J. Anat.* **137,** 1–19.

Doucette, R. (1990). Glial influences on axonal growth in the primary olfactory system. *Glia* **3,** 433–449.

Doucette, R. (1991). PNS-CNS transitional zone of the first cranial nerve. *J. Comp. Neurol.* **312,** 451–466.

Doucette, R. (1995). Olfactory ensheathing cells: Potential for glial cell transplantation into areas of CNS injury. *Histol. Histopathol.* **10,** 503–507.

Dubreuil, C. I., Winton, M. J., and McKerracher, L. (2003). Rho activation patterns after spinal cord injury and the role of activated Rho in apoptosis in the central nervous system. *J. Cell Biol.* **162,** 233–243.

Duncan, I. D., Aguayo, A. J., Bunge, R. P., and Wood, P. M. (1981). Transplantation of rat Schwann cells grown in tissue culture into the mouse spinal cord. *J. Neurol. Sci.* **49,** 241–252.

Esch, F., Lin, K. I., Hills, A., Zaman, K., Baraban, J. M., Chatterjee, S., Rubin, L., Ash, D. E., and Ratan, R. R. (1998). Purification of a multipotent antideath activity from bovine liver and its identification as arginase: Nitric oxide-independent inhibition of neuronal apoptosis. *J. Neurosci.* **18,** 4083–4095.

Fawcett, J. W., and Keynes, R. J. (1990). Peripheral nerve regeneration. *Annu. Rev. Neurosci.* **13,** 43–60.

Feron, F., Perry, C., Cochrane, J., Licina, P., Nowitzke, A., Urquhart, S., Geraghty, T., and Mackay-Sim, A. (2005). Autologous olfactory ensheathing cell transplantation in human spinal cord injury. *Brain* **128,** 2951–2960.

Fidler, P. S., Schuette, K., Asher, R. A., Dobbertin, A., Thornton, S. R., Calle-Patino, Y., Muir, E., Levine, J. M., Geller, H. M., Rogers, J. H., Faissner, A., and Fawcett, J. W. (1999). Comparing astrocytic cell lines that are inhibitory or permissive for axon growth: The major axon-inhibitory proteoglycan is NG2. *J. Neurosci.* **19,** 8778–8788.

Filbin, M. T. (2003). Myelin-associated inhibitors of axonal regeneration in the adult mammalian CNS. *Nat. Rev. Neurosci.* **4,** 703–713.

Fouad, K., Schnell, L., Bunge, M. B., Schwab, M. E., Liebscher, T., and Pearse, D. D. (2005). Combining Schwann cell bridges and olfactory-ensheathing glia grafts with chondroitinase promotes locomotor recovery after complete transection of the spinal cord. *J. Neurosci.* **25,** 1169–1178.

Fournier, A. E., GrandPre, T., and Strittmatter, S. M. (2001). Identification of a receptor mediating Nogo-66 inhibition of axonal regeneration. *Nature* **409,** 341–346.

Fournier, A. E., Takizawa, B. T., and Strittmatter, S. M. (2003). Rho kinase inhibition enhances axonal regeneration in the injured CNS. *J. Neurosci.* **23,** 1416–1423.

Franceschini, I. A., and Barnett, S. C. (1996). Low-affinity NGF-receptor and E-N-CAM expression define two types of olfactory nerve ensheathing cells that share a common lineage. *Dev. Biol.* **173,** 327–343.

Franklin, R. J. M., Gilson, J. M., Franceschini, I. A., and Barnett, S. C. (1996). Schwann cell-like myelination following transplantation of an olfactory bulb-ensheathing cell line into areas of demyelination in the adult CNS. *Glia* **17,** 217–224.

Gage, F. H. (2000). Mammalian neural stem cells. *Science* **287,** 1433–1438.

Gage, F. H., Coates, P. W., Palmer, T. D., Kuhn, H. G., Fisher, L. J., Suhonen, J. O., Peterson, D. A., Suhr, S. T., and Ray, J. (1995). Survival and differentiation of adult neuronal progenitor cells transplanted to the adult brain. *Proc. Natl. Acad. Sci. USA* **92,** 11879–11883.

Gao, Y., Nikulina, E., Mellado, W., and Filbin, M. T. (2003). Neurotrophins elevate cAMP to reach a threshold required to overcome inhibition by MAG through extracellular signal-regulated kinase-dependent inhibition of phosphodiesterase. *J. Neurosci.* **23,** 11770–11777.

Gao, Y., Deng, K., Hou, J., Bryson, J. B., Barco, A., Nikulina, E., Spencer, T., Mellado, W., Kandel, E. R., and Filbin, M. T. (2004). Activated CREB is sufficient to overcome inhibitors in myelin and promote spinal axon regeneration *in vivo*. *Neuron* **44,** 609–621.

Goldberger, M. E., and Murray, M. (1974). Restitution of function and collateral sprouting in the cat spinal cord: The deafferented animal. *J. Comp. Neurol.* **158,** 37–53.

Gong, Q., Bailey, M. S., Pixley, S. K., Ennis, M., Liu, W., and Shipley, M. T. (1994). Localization and regulation of low affinity nerve growth factor receptor expression in the rat olfactory system during development and regeneration. *J. Comp. Neurol.* **344,** 336–348.

GrandPre, T., Nakamura, F., Vartanian, T., and Strittmatter, S. M. (2000). Identification of the Nogo inhibitor of axon regeneration as a Reticulon protein. *Nature* **403,** 439–444.

GrandPre, T., Li, S., and Strittmatter, S. M. (2002). Nogo-66 receptor antagonist peptide promotes axonal regeneration. *Nature* **417,** 547–551.

Graziadei, P. P., and Graziadei, G. A. (1979a). Neurogenesis and neuron regeneration in the olfactory system of mammals. I. Morphological aspects of differentiation and structural organization of the olfactory sensory neurons. *J. Neurocytol.* **8,** 1–18.

Graziadei, P. P., and Graziadei, G. A. (1979b). Neurogenesis and neuron regeneration in the olfactory system of mammals. II. Degeneration and reconstitution of the olfactory sensory neurons after axotomy. *J. Neurocytol.* **8,** 197–213.

Graziadei, P. P., and Monti-Graziadei, G. A. (1980). Neurogenesis and neuron regeneration in the olfactory system of mammals. III. Deafferentation and reinnervation of the olfactory bulb following section of the fila olfactoria in rat. *J. Neurocytol.* **9,** 145–162.

Graziadei, P. P., Levine, R. R., and Graziadei, G. A. (1978). Regeneration of olfactory axons and synapse formation in the forebrain after bulbectomy in neonatal mice. *Proc. Natl. Acad. Sci. USA* **75,** 5230–5234.

Grill, R. J., Blesch, A., and Tuszynski, M. H. (1997a). Robust growth of chronically injured spinal cord axons induced by grafts of genetically modified NGF-secreting cells. *Exp. Neurol.* **148,** 444–452.

Grill, R., Murai, K., Blesch, A., Gage, F. H., and Tuszynski, M. H. (1997b). Cellular delivery of neurotrophin-3 promotes corticospinal axonal growth and partial functional recovery after spinal cord injury. *J. Neurosci.* **17,** 5560–5572.

Grimpe, B., and Silver, J. (2004). A novel DNA enzyme reduces glycosaminoglycan chains in the glial scar and allows microtransplanted dorsal root ganglia axons to regenerate beyond lesions in the spinal cord. *J. Neurosci.* **24,** 1393–1397.

Guth, L., Zhang, Z., and Steward, O. (1999). The unique histopathological responses of the injured spinal cord. Implications for neuroprotective therapy. *Ann. N. Y. Acad. Sci.* **890,** 366–384.

Habib, A. A., Marton, L. S., Allwardt, B., Gulcher, J. R., Mikol, D. D., Hognason, T., Chattopadhyay, N., and Stefansson, K. (1998). Expression of the oligodendrocyte-myelin glycoprotein by neurons in the mouse central nervous system. *J. Neurochem.* **70,** 1704–1711.

He, X. L., Bazan, J. F., McDermott, G., Park, J. B., Wang, K., Tessier-Lavigne, M., He, Z., and Garcia, K. C. (2003). Structure of the Nogo receptor ectodomain: A recognition module implicated in myelin inhibition. *Neuron* **38,** 177–185.

Heumann, R., Korsching, S., Bandtlow, C., and Thoenen, H. (1987). Changes of nerve growth factor synthesis in nonneuronal cells in response to sciatic nerve transection. *J. Cell Biol.* **104,** 1623–1631.

Himes, B. T., Goldberger, M. E., and Tessler, A. (1994). Grafts of fetal central nervous system tissue rescue axotomized Clarke's nucleus neurons in adult and neonatal operates. *J. Comp. Neurol.* **339,** 117–131.

Himes, B. T., Liu, Y., Solowska, J. M., Snyder, E. Y., Fischer, I., and Tessler, A. (2001). Transplants of cells genetically modified to express neurotrophin-3 rescue axotomized Clarke's nucleus neurons after spinal cord hemisection in adult rats. *J. Neurosci. Res.* **65,** 549–564.

Hofstetter, C. P., Schwarz, E. J., Hess, D., Widenfalk, J., El Manira, A., Prockop, D. J., and Olson, L. (2002). Marrow stromal cells form guiding strands in the injured spinal cord and promote recovery. *Proc. Natl. Acad. Sci. USA* **99,** 2199–2204.

Holcomb, J. D., Mumm, J. S., and Calof, A. L. (1995). Apoptosis in the neuronal lineage of the mouse olfactory epithelium: Regulation *in vivo* and *in vitro*. *Dev. Biol.* **172,** 307–323.

Houle, J. (1992). The structural integrity of glial scar tissue associated with a chronic spinal cord lesion can be altered by transplanted fetal spinal cord tissue. *J. Neurosci. Res.* **31,** 120–130.

Houle, J. D., and Reier, P. J. (1988). Transplantation of fetal spinal cord tissue into the chronically injured adult rat spinal cord. *J. Comp. Neurol.* **269,** 535–547.

Houle, J. D., and Reier, P. J. (1989). Regrowth of calcitonin gene-related peptide (CGRP) immunoreactive axons from the chronically injured rat spinal cord into fetal spinal cord tissue transplants. *Neurosci. Lett.* **103,** 253–258.

Howe, A. K. (2004). Regulation of actin-based cell migration by cAMP/PKA. *Biochim. Biophys. Acta* **1692,** 159–174.

Hu, F., Liu, B. P., Budel, S., Liao, J., Chin, J., Fournier, A., and Strittmatter, S. M. (2005). Nogo-A interacts with the Nogo-66 receptor through multiple sites to create an isoform-selective subnanomolar agonist. *J. Neurosci.* **25,** 5298–5304.

Hurtado, A., Moon, L. D., Maquet, V., Blits, B., Jerome, R., and Oudega, M. (2006). Poly (D,L-lactic acid) macroporous guidance scaffolds seeded with Schwann cells genetically modified to secrete a bi-functional neurotrophin implanted in the completely transected adult rat thoracic spinal cord. *Biomaterials* **27,** 430–442.

Imaizumi, T., Lankford, K. L., Waxman, S. G., Greer, C. A., and Kocsis, J. D. (1998). Transplanted olfactory ensheathing cells remyelinate and enhance axonal conduction in the demyelinated dorsal columns of the rat spinal cord. *J. Neurosci.* **18,** 6176–6185.

Jaffe, A. B., and Hall, A. (2005). Rho GTPases: Biochemistry and biology. *Annu. Rev. Cell. Dev. Biol.* **21,** 247–269.

Jin, S. L., Richard, F. J., Kuo, W. P., D'Ercole, A. J., and Conti, M. (1999). Impaired growth and fertility of cAMP-specific phosphodiesterase PDE4D-deficient mice. *Proc. Natl. Acad. Sci. USA* **96,** 11998–12003.

Johnson, P. W., Abramow-Newerly, W., Seilheimer, B., Sadoul, R., Tropak, M. B., Arquint, M., Dunn, R. J., Schachner, M., and Roder, J. C. (1989). Recombinant myelin-associated glycoprotein confers neural adhesion and neurite outgrowth function. *Neuron* **3,** 377–385.

Jones, L. L., Yamaguchi, Y., Stallcup, W. B., and Tuszynski, M. H. (2002). NG2 is a major chondroitin sulfate proteoglycan produced after spinal cord injury and is expressed by macrophages and oligodendrocyte progenitors. *J. Neurosci.* **22,** 2792–2803.

Jones, L. L., Margolis, R. U., and Tuszynski, M. H. (2003a). The chondroitin sulfate proteoglycans neurocan, brevican, phosphacan, and versican are differentially regulated following spinal cord injury. *Exp. Neurol.* **182,** 399–411.

Jones, L. L., Sajed, D., and Tuszynski, M. H. (2003b). Axonal regeneration through regions of chondroitin sulfate proteoglycan deposition after spinal cord injury: A balance of permissiveness and inhibition. *J. Neurosci.* **23,** 9276–9288.

Karimi-Abdolrezaee, S., Eftekharpour, E., Wang, J., Morshead, C. M., and Fehlings, M. G. (2006). Delayed transplantation of adult neural precursor cells promotes remyelination and functional neurological recovery after spinal cord injury. *J. Neurosci.* **26,** 3377–3389.

Katzman, R., Björklund, A., Owman, C., Stenevi, U., and West, K. A. (1971). Evidence for regenerative axon sprouting of central catecholamine neurons in the rat mesencephalon following electrolytic lesions. *Brain Res.* **25,** 579–596.

Keirstead, H. S., Nistor, G., Bernal, G., Totoiu, M., Cloutier, F., Sharp, K., and Steward, O. (2005). Human embryonic stem cell-derived oligodendrocyte progenitor cell transplants remyelinate and restore locomotion after spinal cord injury. *J. Neurosci.* **25,** 4694–4705.

Kelm, S., Pelz, A., Schauer, R., Filbin, M. T., Tang, S., de Bellard, M. E., Schnaar, R. L., Mahoney, J. A., Hartnell, A., Bradfield, P., and Crocker, P. R. (1994). Sialoadhesin, myelin-associated glycoprotein and CD22 define a new family of sialic acid-dependent adhesion molecules of the immunoglobulin superfamily. *Curr. Biol.* **4,** 965–972.

Kilmer, S. L., and Carlsen, R. C. (1984). Forskolin activation of adenylate cyclase *in vivo* stimulates nerve regeneration. *Nature* **307,** 455–457.

Kim, J. E., Li, S., GrandPre, T., Qiu, D., and Strittmatter, S. M. (2003). Axon regeneration in young adult mice lacking Nogo-A/B. *Neuron* **38,** 187–199.

Kim, J. E., Liu, B. P., Park, J. H., and Strittmatter, S. M. (2004). Nogo-66 receptor prevents raphespinal and rubrospinal axon regeneration and limits functional recovery from spinal cord injury. *Neuron* **44,** 439–451.

Kobayashi, N. R., Fan, D. P., Giehl, K. M., Bedard, A. M., Wiegand, S. J., and Tetzlaff, W. (1997). BDNF and NT-4/5 prevent atrophy of rat rubrospinal neurons after cervical axotomy, stimulate GAP-43 and Talpha1-tubulin mRNA expression, and promote axonal regeneration. *J. Neurosci.* **17,** 9583–9595.

Kojima, T., Morikawa, Y., Copeland, N. G., Gilbert, D. J., Jenkins, N. A., Senba, E., and Kitamura, T. (2000). TROY, a newly identified member of the tumor necrosis factor receptor superfamily, exhibits a homology with Edar and is expressed in embryonic skin and hair follicles. *J. Biol. Chem.* **275,** 20742–20747.

Koprivica, V., Cho, K. S., Park, J. B., Yiu, G., Atwal, J., Gore, B., Kim, J. A., Lin, E., Tessier-Lavigne, M., Chen, D. F., and He, Z. (2005). EGFR activation mediates inhibition of axon regeneration by myelin and chondroitin sulfate proteoglycans. *Science* **310,** 106–110.

Kossiakoff, A. A., and De Vos, A. M. (1998). Structural basis for cytokine hormone-receptor recognition and receptor activation. *Adv. Protein Chem.* **52,** 67–108.

Krause, W., and Kuhne, G. (1988). Pharmacokinetics of rolipram in the rhesus and cynomolgus monkeys, the rat and the rabbit. Studies on species differences. *Xenobiotica* **18,** 561–571.

Kromer, L. F., and Cornbrooks, C. J. (1985). Transplants of Schwann cell cultures promote axonal regeneration in the adult mammalian brain. *Proc. Natl. Acad. Sci. USA* **82,** 6330–6334.

Lee, M. J., Calle, E., Brennan, A., Ahmed, S., Sviderskaya, E., Jessen, K. R., and Mirsky, R. (2001). In early development of the rat mRNA for the major myelin protein P(0) is expressed in nonsensory areas of the embryonic inner ear, notochord, enteric nervous system, and olfactory ensheathing cells. *Dev. Dyn.* **222,** 40–51.

Lehmann, M., Fournier, A., Selles-Navarro, I., Dergham, P., Sebok, A., Leclerc, N., Tigyi, G., and McKerracher, L. (1999). Inactivation of Rho signaling pathway promotes CNS axon regeneration. *J. Neurosci.* **19,** 7537–7547.

Li, M., Shibata, A., Li, C., Braun, P. E., McKerracher, L., Roder, J., Kater, S. B., and David, S. (1996). Myelin-associated glycoprotein inhibits neurite/axon growth and causes growth cone collapse. *J. Neurosci. Res.* **46,** 404–414.

Li, Y., Field, P. M., and Raisman, G. (1997). Repair of adult rat corticospinal tract by transplants of olfactory ensheathing cells. *Science* **277,** 2000–2002.

Li, Y., Field, P. M., and Raisman, G. (1998). Regeneration of adult rat corticospinal axons induced by transplanted olfactory ensheathing cells. *J. Neurosci.* **18,** 10514–10524.

Lipson, A. C., Widenfalk, J., Lindqvist, E., Ebendal, T., and Olson, L. (2003). Neurotrophic properties of olfactory ensheathing glia. *Exp. Neurol.* **180,** 167–171.

Liu, C. N., and Chambers, W. W. (1958). Intraspinal sprouting of dorsal root axons; development of new collaterals and preterminals following partial denervation of the spinal cord in the cat. *Arch. Neurol. Psychiatry* **79,** 46–61.

Liu, Y., Kim, D., Himes, B. T., Chow, S. Y., Schallert, T., Murray, M., Tessler, A., and Fischer, I. (1999). Transplants of fibroblasts genetically modified to express BDNF promote regeneration of adult rat rubrospinal axons and recovery of forelimb function. *J. Neurosci.* **19,** 4370–4387.

Liu, S., Qu, Y., Stewart, T. J., Howard, M. J., Chakrabortty, S., Holekamp, T. F., and McDonald, J. W. (2000). Embryonic stem cells differentiate into oligodendrocytes and myelinate in culture and after spinal cord transplantation. *Proc. Natl. Acad. Sci. USA* **97,** 6126–6131.

Liu, B. P., Fournier, A., GrandPre, T., and Strittmatter, S. M. (2002). Myelin-associated glycoprotein as a functional ligand for the Nogo-66 receptor. *Science* **297,** 1190–1193.

Lu, J., Feron, F., Ho, S. M., Mackay-Sim, A., and Waite, P. M. (2001). Transplantation of nasal olfactory tissue promotes partial recovery in paraplegic adult rats. *Brain Res.* **889,** 344–357.

Lu, J., Feron, F., Mackay-Sim, A., and Waite, P. M. (2002). Olfactory ensheathing cells promote locomotor recovery after delayed transplantation into transected spinal cord. *Brain* **125,** 14–21.

Lu, P., Yang, H., Jones, L. L., Filbin, M. T., and Tuszynski, M. H. (2004). Combinatorial therapy with neurotrophins and cAMP promotes axonal regeneration beyond sites of spinal cord injury. *J. Neurosci.* **24,** 6402–6409.

Lu, P., Jones, L. L., and Tuszynski, M. H. (2005). BDNF-expressing marrow stromal cells support extensive axonal growth at sites of spinal cord injury. *Exp. Neurol.* **191,** 344–360.

Martin, S., Levine, A. K., Chen, Z. J., Ughrin, Y., and Levine, J. M. (2001). Deposition of the NG2 proteoglycan at nodes of Ranvier in the peripheral nervous system. *J. Neurosci.* **21,** 8119–8128.

Martini, R., and Schachner, M. (1986). Immunoelectron microscopic localization of neural cell adhesion molecules (L1, N-CAM, and MAG) and their shared carbohydrate epitope and myelin basic protein in developing sciatic nerve. *J. Cell Biol.* **103,** 2439–2448.

McDonald, J. W., Liu, X. Z., Qu, Y., Liu, S., Mickey, S. K., Turetsky, D., Gottlieb, D. I., and Choi, D. W. (1999). Transplanted embryonic stem cells survive, differentiate and promote recovery in injured rat spinal cord. *Nat. Med.* **5,** 1410–1412.

McDonald, J. W., Becker, D., Holekamp, T. F., Howard, M., Liu, S., Lu, A., Lu, J., Platik, M. M., Qu, Y., Stewart, T., and Vadivelu, S. (2004). Repair of the injured spinal cord and the potential of embryonic stem cell transplantation. *J. Neurotrauma* **21,** 383–393.

McKeon, R. J., Schreiber, R. C., Rudge, J. S., and Silver, J. (1991). Reduction of neurite outgrowth in a model of glial scarring following CNS injury is correlated with the expression of inhibitory molecules on reactive astrocytes. *J. Neurosci.* **11,** 3398–3411.

McKeon, R. J., Hoke, A., and Silver, J. (1995). Injury-induced proteoglycans inhibit the potential for laminin-mediated axon growth on astrocytic scars. *Exp. Neurol.* **136,** 32–43.

McKeon, R. J., Jurynec, M. J., and Buck, C. R. (1999). The chondroitin sulfate proteoglycans neurocan and phosphacan are expressed by reactive astrocytes in the chronic CNS glial scar. *J. Neurosci.* **19,** 10778–10788.

McKerracher, L., David, S., Jackson, D. L., Kottis, V., Dunn, R. J., and Braun, P. E. (1994). Identification of myelin-associated glycoprotein as a major myelin-derived inhibitor of neurite growth. *Neuron* **13,** 805–811.

McTigue, D. M., Horner, P. J., Stokes, B. T., and Gage, F. H. (1998). Neurotrophin-3 and brain-derived neurotrophic factor induce oligodendrocyte proliferation and myelination of regenerating axons in the contused adult rat spinal cord. *J. Neurosci.* **18,** 5354–5365.

Menei, P., Montero-Menei, C., Whittemore, S. R., Bunge, R. P., and Bunge, M. B. (1998). Schwann cells genetically modified to secrete human BDNF promote enhanced axonal regrowth across transected adult rat spinal cord. *Eur. J. Neurosci.* **10,** 607–621.

Meyer, M., Matsuoka, I., Wetmore, C., Olson, L., and Thoenen, H. (1992). Enhanced synthesis of brain-derived neurotrophic factor in the lesioned peripheral nerve: Different mechanisms are responsible for the regulation of BDNF and NGF mRNA. *J. Cell Biol.* **119,** 45–54.

Mezey, E., Chandross, K. J., Harta, G., Maki, R. A., and McKercher, S. R. (2000). Turning blood into brain: Cells bearing neuronal antigens generated *in vivo* from bone marrow. *Science* **290,** 1779–1782.

Mezey, E., Key, S., Vogelsang, G., Szalayova, I., Lange, G. D., and Crain, B. (2003). Transplanted bone marrow generates new neurons in human brains. *Proc. Natl. Acad. Sci. USA* **100,** 1364–1369.

Mi, S., Lee, X., Shao, Z., Thill, G., Ji, B., Relton, J., Levesque, M., Allaire, N., Perrin, S., Sands, B., Crowell, T., Cate, R. L., *et al.* (2004). LINGO-1 is a component of the Nogo-66 receptor/p75 signaling complex. *Nat. Neurosci.* **7,** 221–228.

Mikol, D. D., and Stefansson, K. (1988). A phosphatidylinositol-linked peanut agglutinin-binding glycoprotein in central nervous system myelin and on oligodendrocytes. *J. Cell Biol.* **106,** 1273–1279.

Mikol, D. D., Gulcher, J. R., and Stefansson, K. (1990). The oligodendrocyte-myelin glycoprotein belongs to a distinct family of proteins and contains the HNK-1 carbohydrate. *J. Cell Biol.* **110,** 471–479.

Miragall, F., Kadmon, G., and Schachner, M. (1989). Expression of L1 and N-CAM cell adhesion molecules during development of the mouse olfactory system. *Dev. Biol.* **135,** 272–286.

Mitsui, T., Fischer, I., Shumsky, J. S., and Murray, M. (2005). Transplants of fibroblasts expressing BDNF and NT-3 promote recovery of bladder and hindlimb function following spinal contusion injury in rats. *Exp. Neurol.* **194,** 410–431.

Miya, D., Giszter, S., Mori, F., Adipudi, V., Tessler, A., and Murray, M. (1997). Fetal transplants alter the development of function after spinal cord transection in newborn rats. *J. Neurosci.* **17,** 4856–4872.

Monnier, P. P., Sierra, A., Schwab, J. M., Henke-Fahle, S., and Mueller, B. K. (2003). The Rho/ROCK pathway mediates neurite growth-inhibitory activity associated with the chondroitin sulfate proteoglycans of the CNS glial scar. *Mol. Cell. Neurosci.* **22,** 319–330.

Montgomery, C. T., and Robson, J. A. (1990). New method of transplanting purified glial cells into the brain. *J. Neurosci. Methods* **32,** 135–141.

Montgomery, C. T., and Robson, J. A. (1993). Implants of cultured Schwann cells support axonal growth in the central nervous system of adult rats. *Exp. Neurol.* **122,** 107–124.

Moon, L. D., Asher, R. A., Rhodes, K. E., and Fawcett, J. W. (2001). Regeneration of CNS axons back to their target following treatment of adult rat brain with chondroitinase ABC. *Nat. Neurosci.* **4,** 465–466.

Moore, M. J., Friedman, J. A., Lewellyn, E. B., Mantila, S. M., Krych, A. J., Ameenuddin, S., Knight, A. M., Lu, L., Currier, B. L., Spinner, R. J., Marsh, R. W., Windebank, A. J., *et al.* (2006). Multiple-channel scaffolds to promote spinal cord axon regeneration. *Biomaterials* **27,** 419–429.

Morrissey, T. K., Kleitman, N., and Bunge, R. P. (1991). Isolation and functional characterization of Schwann cells derived from adult peripheral nerve. *J. Neurosci.* **11,** 2433–2442.

Mukhopadhyay, G., Doherty, P., Walsh, F. S., Crocker, P. R., and Filbin, M. T. (1994). A novel role for myelin-associated glycoprotein as an inhibitor of axonal regeneration. *Neuron* **13,** 757–767.

Nakahara, Y., Gage, F. H., and Tuszynski, M. H. (1996). Grafts of fibroblasts genetically modified to secrete NGF, BDNF, NT-3, or basic FGF elicit differential responses in the adult spinal cord. *Cell Transplant.* **5,** 191–204.

Nakamura, M., Okano, H., Toyama, Y., Dai, H. N., Finn, T. P., and Bregman, B. S. (2005). Transplantation of embryonic spinal cord-derived neurospheres support growth of supraspinal projections and functional recovery after spinal cord injury in the neonatal rat. *J. Neurosci. Res.* **81,** 457–468.

Navarro, X., Valero, A., Gudino, G., Fores, J., Rodriguez, F. J., Verdu, E., Pascual, R., Cuadras, J., and Nieto-Sampedro, M. (1999). Ensheathing glia transplants promote dorsal root regeneration and spinal reflex restitution after multiple lumbar rhizotomy. *Ann. Neurol.* **45,** 207–215.

Neumann, S., and Woolf, C. J. (1999). Regeneration of dorsal column fibers into and beyond the lesion site following adult spinal cord injury. *Neuron* **23,** 83–91.

Niederost, B., Oertle, T., Fritsche, J., McKinney, R. A., and Bandtlow, C. E. (2002). Nogo-A and myelin-associated glycoprotein mediate neurite growth inhibition by antagonistic regulation of RhoA and Rac1. *J. Neurosci.* **22,** 10368–10376.

Nikulina, E., Tidwell, J. L., Dai, H. N., Bregman, B. S., and Filbin, M. T. (2004). The phosphodiesterase inhibitor rolipram delivered after a spinal cord lesion promotes axonal regeneration and functional recovery. *Proc. Natl. Acad. Sci. USA* **101,** 8786–8790.

Novikov, L. N., Novikova, L. N., Mosahebi, A., Wiberg, M., Terenghi, G., and Kellerth, J. O. (2002). A novel biodegradable implant for neuronal rescue and regeneration after spinal cord injury. *Biomaterials* **23,** 3369–3376.

Oertle, T., van der Haar, M. E., Bandtlow, C. E., Robeva, A., Burfeind, P., Buss, A., Huber, A. B., Simonen, M., Schnell, L., Brosamle, C., Kaupmann, K., Vallon, R., *et al.* (2003). Nogo-A inhibits neurite outgrowth and cell spreading with three discrete regions. *J. Neurosci.* **23,** 5393–5406.

Oka, S., Honmou, O., Akiyama, Y., Sasaki, M., Houkin, K., Hashi, K., and Kocsis, J. D. (2004). Autologous transplantation of expanded neural precursor cells into the demyelinated monkey spinal cord. *Brain Res.* **1030,** 94–102.

Oudega, M., Varon, S., and Hagg, T. (1994). Regeneration of adult rat sensory axons into intraspinal nerve grafts: Promoting effects of conditioning lesion and graft predegeneration. *Exp. Neurol.* **129,** 194–206.

Oudega, M., Gautier, S. E., Chapon, P., Fragoso, M., Bates, M. L., Parel, J. M., and Bunge, M. B. (2001). Axonal regeneration into Schwann cell grafts within resorbable poly(alpha-hydroxyacid) guidance channels in the adult rat spinal cord. *Biomaterials* **22,** 1125–1136.

Park, J. B., Yiu, G., Kaneko, S., Wang, J., Chang, J., He, X. L., Garcia, K. C., and He, Z. (2005). A TNF receptor family member, TROY, is a coreceptor with Nogo receptor in mediating the inhibitory activity of myelin inhibitors. *Neuron* **45,** 345–351.

Pearse, D. D., Pereira, F. C., Marcillo, A. E., Bates, M. L., Berrocal, Y. A., Filbin, M. T., and Bunge, M. B. (2004a). cAMP and Schwann cells promote axonal growth and functional recovery after spinal cord injury. *Nat. Med.* **10,** 610–616.

Pearse, D. D., Marcillo, A. E., Oudega, M., Lynch, M. P., Wood, P. M., and Bunge, M. B. (2004b). Transplantation of Schwann cells and olfactory ensheathing glia after spinal cord injury: Does pretreatment with methylprednisolone and interleukin-10 enhance recovery? *J. Neurotrauma* **21,** 1223–1239.

Pixley, S. K. (1992). The olfactory nerve contains two populations of glia, identified both *in vivo* and *in vitro*. *Glia* **5,** 269–284.

Plant, G. W., Bates, M. L., and Bunge, M. B. (2001). Inhibitory proteoglycan immunoreactivity is higher at the caudal than the rostral Schwann cell graft-transected spinal cord interface. *Mol. Cell. Neurosci.* **17,** 471–487.

Plant, G. W., Currier, P. F., Cuervo, E. P., Bates, M. L., Pressman, Y., Bunge, M. B., and Wood, P. M. (2002). Purified adult ensheathing glia fail to myelinate axons under culture conditions that enable Schwann cells to form myelin. *J. Neurosci.* **22,** 6083–6091.

Plant, G. W., Christensen, C. L., Oudega, M., and Bunge, M. B. (2003). Delayed transplantation of olfactory ensheathing glia promotes sparing/regeneration of supraspinal axons in the contused adult rat spinal cord. *J. Neurotrauma* **20,** 1–16.

Prinjha, R., Moore, S. E., Vinson, M., Blake, S., Morrow, R., Christie, G., Michalovich, D., Simmons, D. L., and Walsh, F. S. (2000). Inhibitor of neurite outgrowth in humans. *Nature* **403,** 383–384.

Qiu, J., Cai, D., Dai, H., McAtee, M., Hoffman, P. N., Bregman, B. S., and Filbin, M. T. (2002). Spinal axon regeneration induced by elevation of cyclic AMP. *Neuron* **34,** 895–903.

Rabizadeh, S., and Bredesen, D. E. (2003). Ten years on: Mediation of cell death by the common neurotrophin receptor p75(NTR). *Cytokine Growth Factor Rev.* **14,** 225–239.

Raisman, G. (1985). Specialized neuroglial arrangement may explain the capacity of vomeronasal axons to reinnervate central neurons. *Neuroscience* **14,** 237–254.

Ramer, L. M., Richter, M. W., Roskams, A. J., Tetzlaff, W., and Ramer, M. S. (2004a). Peripherally-derived olfactory ensheathing cells do not promote primary afferent regeneration following dorsal root injury. *Glia* **47,** 189–206.

Ramer, L. M., Au, E., Richter, M. W., Liu, J., Tetzlaff, W., and Roskams, A. J. (2004b). Peripheral olfactory ensheathing cells reduce scar and cavity formation and promote regeneration after spinal cord injury. *J. Comp. Neurol.* **473,** 1–15.

Ramón-Cueto, A., and Ávila, J. (1998). Olfactory ensheathing glia: Properties and function. *Brain Res. Bull.* **46,** 175–187.

Ramón-Cueto, A., and Nieto-Sampedro, M. (1992). Glial cells from adult rat olfactory bulb: Immunocytochemical properties of pure cultures of ensheathing cells. *Neuroscience* **47,** 213–220.

Ramón-Cueto, A., and Nieto-Sampedro, M. (1994). Regeneration into the spinal cord of transected dorsal root axons is promoted by ensheathing glia transplants. *Exp. Neurol.* **127,** 232–244.

Ramón-Cueto, A., and Valverde, F. (1995). Olfactory bulb ensheathing glia: A unique cell type with axonal growth-promoting properties. *Glia* **14,** 163–173.

Ramón-Cueto, A., Plant, G. W., Ávila, J., and Bunge, M. B. (1998). Long-distance axonal regeneration in the transected adult rat spinal cord is promoted by olfactory ensheathing glia transplants. *J. Neurosci.* **18,** 3803–3815.

Ramón-Cueto, A., Cordero, M. I., Santos-Benito, F. F., and Ávila, J. (2000). Functional recovery of paraplegic rats and motor axon regeneration in their spinal cords by olfactory ensheathing glia. *Neuron* **25,** 425–435.

Ray, J., Peterson, D. A., Schinstine, M., and Gage, F. H. (1993). Proliferation, differentiation, and long-term culture of primary hippocampal neurons. *Proc. Natl. Acad. Sci. USA* **90,** 3602–3606.

Reier, P. J., Bregman, B. S., and Wujek, J. R. (1986). Intraspinal transplantation of embryonic spinal cord tissue in neonatal and adult rats. *J. Comp. Neurol.* **247,** 275–296.

Reynolds, B. A., and Weiss, S. (1992). Generation of neurons and astrocytes from isolated cells of the adult mammalian central nervous system. *Science* **255,** 1707–1710.

Reynolds, B. A., Tetzlaff, W., and Weiss, S. (1992). A multipotent EGF-responsive striatal embryonic progenitor cell produces neurons and astrocytes. *J. Neurosci.* **12,** 4565–4574.

Richards, L. J., Kilpatrick, T. J., and Bartlett, P. F. (1992). *De novo* generation of neuronal cells from the adult mouse brain. *Proc. Natl. Acad. Sci. USA* **89,** 8591–8595.

Richardson, P. M., and Issa, V. M. (1984). Peripheral injury enhances central regeneration of primary sensory neurones. *Nature* **309,** 791–793.

Richardson, P. M., McGuinness, U. M., and Aguayo, A. J. (1980). Axons from CNS neurons regenerate into PNS grafts. *Nature* **284,** 264–265.

Richter, M. W., Fletcher, P. A., Liu, J., Tetzlaff, W., and Roskams, A. J. (2005). Lamina propria and olfactory bulb ensheathing cells exhibit differential integration and migration and promote differential axon sprouting in the lesioned spinal cord. *J. Neurosci.* **25,** 10700–10711.

Ruitenberg, M. J., Plant, G. W., Christensen, C. L., Blits, B., Niclou, S. P., Harvey, A. R., Boer, G. J., and Verhaagen, J. (2002). Viral vector-mediated gene expression in olfactory ensheathing glia implants in the lesioned rat spinal cord. *Gene Ther.* **9,** 135–146.

Rutkowski, J. L., Tennekoon, G. I., and McGillicuddy, J. E. (1992). Selective culture of mitotically active human Schwann cells from adult sural nerves. *Ann. Neurol.* **31,** 580–586.

Salzer, J. L., Holmes, W. P., and Colman, D. R. (1987). The amino acid sequences of the myelin-associated glycoproteins: Homology to the immunoglobulin gene superfamily. *J. Cell Biol.* **104,** 957–965.

Salzer, J. L., Pedraza, L., Brown, M., Struyk, A., Afar, D., and Bell, J. (1990). Structure and function of the myelin-associated glycoproteins. *Ann. N. Y. Acad. Sci.* **605,** 302–312.

Sayers, S. T., Khan, N., Ahmed, Y., Shahid, R., and Khan, T. (1998). Preparation of brain-derived neurotrophic factor- and neurotrophin-3-secreting Schwann cells by infection with a retroviral vector. *J. Mol. Neurosci.* **10,** 143–160.

Schnell, L., and Schwab, M. E. (1990). Axonal regeneration in the rat spinal cord produced by an antibody against myelin-associated neurite growth inhibitors. *Nature* **343,** 269–272.

Schnell, L., and Schwab, M. E. (1993). Sprouting and regeneration of lesioned corticospinal tract fibres in the adult rat spinal cord. *Eur. J. Neurosci.* **5,** 1156–1171.

Schnell, L., Schneider, R., Kolbeck, R., Barde, Y. A., and Schwab, M. E. (1994). Neurotrophin-3 enhances sprouting of corticospinal tract during development and after adult spinal cord lesion. *Nature* **367,** 170–173.

Schreyer, D. J., and Skene, J. H. (1993). Injury-associated induction of GAP-43 expression displays axon branch specificity in rat dorsal root ganglion neurons. *J. Neurobiol.* **24,** 959–970.

Schwob, J. E., Farber, N. B., and Gottlieb, D. I. (1986). Neurons of the olfactory epithelium in adult rats contain vimentin. *J. Neurosci.* **6,** 208–217.

Seiler, N. (2000). Oxidation of polyamines and brain injury. *Neurochem. Res.* **25,** 471–490.

Shao, Z., Browning, J. L., Lee, X., Scott, M. L., Shulga-Morskaya, S., Allaire, N., Thill, G., Levesque, M., Sah, D., McCoy, J. M., Murray, B., Jung, V., *et al.* (2005). TAJ/TROY, an orphan TNF receptor family member, binds Nogo-66 receptor 1 and regulates axonal regeneration. *Neuron* **45,** 353–359.

Shearer, M. C., and Fawcett, J. W. (2001). The astrocyte/meningeal cell interface—a barrier to successful nerve regeneration? *Cell Tissue Res.* **305,** 267–273.

Shipley, M. T., and Ennis, M. (1996). Functional organization of olfactory system. *J. Neurobiol.* **30,** 123–176.

Shumsky, J. S., Tobias, C. A., Tumolo, M., Long, W. D., Giszter, S. F., and Murray, M. (2003). Delayed transplantation of fibroblasts genetically modified to secrete BDNF and NT-3 into a spinal cord injury site is associated with limited recovery of function. *Exp. Neurol.* **184,** 114–130.

Silver, J., and Miller, J. H. (2004). Regeneration beyond the glial scar. *Nat. Rev. Neurosci.* **5,** 146–156.

Simonen, M., Pedersen, V., Weinmann, O., Schnell, L., Buss, A., Ledermann, B., Christ, F., Sansig, G., van der Putten, H., and Schwab, M. E. (2003). Systemic deletion of the myelin-associated outgrowth inhibitor Nogo-A improves regenerative and plastic responses after spinal cord injury. *Neuron* **38,** 201–211.

Sivasankaran, R., Pei, J., Wang, K. C., Zhang, Y. P., Shields, C. B., Xu, X. M., and He, Z. (2004). PKC mediates inhibitory effects of myelin and chondroitin sulfate proteoglycans on axonal regeneration. *Nat. Neurosci.* **7,** 261–268.

Slotkin, T. A., and Bartolome, J. (1986). Role of ornithine decarboxylase and the polyamines in nervous system development: A review. *Brain Res. Bull.* **17,** 307–320.

Smith-Thomas, L. C., Fok-Seang, J., Stevens, J., Du, J. S., Muir, E., Faissner, A., Geller, H. M., Rogers, J. H., and Fawcett, J. W. (1994). An inhibitor of neurite outgrowth produced by astrocytes. *J. Cell Sci.* **107,** 1687–1695.

Song, H., Ming, G., He, Z., Lehmann, M., McKerracher, L., Tessier-Lavigne, M., and Poo, M. (1998). Conversion of neuronal growth cone responses from repulsion to attraction by cyclic nucleotides. *Science* **281,** 1515–1518.

Steward, O., Sharp, K., Selvan, G., Hadden, A., Hofstadter, M., Au, E., and Roskams, J. (2006). A re-assessment of the consequences of delayed transplantation of olfactory lamina propria following complete spinal cord transection in rats. *Exp. Neurol.* **198**(2), 483–499.

Stokols, S., and Tuszynski, M. H. (2006). Freeze-dried agarose scaffolds with uniaxial channels stimulate and guide linear axonal growth following spinal cord injury. *Biomaterials* **27,** 443–451.

Takami, T., Oudega, M., Bates, M. L., Wood, P. M., Kleitman, N., and Bunge, M. B. (2002). Schwann cell but not olfactory ensheathing glia transplants improve hindlimb locomotor performance in the moderately contused adult rat thoracic spinal cord. *J. Neurosci.* **22,** 6670–6681.

Tang, S., Shen, Y. J., DeBellard, M. E., Mukhopadhyay, G., Salzer, J. L., Crocker, P. R., and Filbin, M. T. (1997). Myelin-associated glycoprotein interacts with neurons via a sialic acid binding site at ARG118 and a distinct neurite inhibition site. *J. Cell Biol.* **138,** 1355–1366.

Tang, X., Davies, J. E., and Davies, S. J. (2003). Changes in distribution, cell associations, and protein expression levels of NG**2,** neurocan, phosphacan, brevican, versican V**2,** and tenascin-C during acute to chronic maturation of spinal cord scar tissue. *J. Neurosci. Res.* **71,** 427–444.

Teng, K. K., and Hempstead, B. L. (2004). Neurotrophins and their receptors: Signaling trios in complex biological systems. *Cell. Mol. Life Sci.* **61,** 35–48.

Teng, Y. D., Lavik, E. B., Qu, X., Park, K. I., Ourednik, J., Zurakowski, D., Langer, R., and Snyder, E. Y. (2002). Functional recovery following traumatic spinal cord injury mediated by a unique polymer scaffold seeded with neural stem cells. *Proc. Natl. Acad. Sci. USA* **99,** 3024–3029.

Tobias, C. A., Shumsky, J. S., Shibata, M., Tuszynski, M. H., Fischer, I., Tessler, A., and Murray, M. (2003). Delayed grafting of BDNF and NT-3 producing fibroblasts into the injured spinal cord stimulates sprouting, partially rescues axotomized red nucleus neurons from loss and atrophy, and provides limited regeneration. *Exp. Neurol.* **184,** 97–113.

Turnley, A. M., and Bartlett, P. F. (1998). MAG and MOG enhance neurite outgrowth of embryonic mouse spinal cord neurons. *Neuroreport* **9,** 1987–1990.

Tuszynski, M. H., Gabriel, K., Gage, F. H., Suhr, S., Meyer, S., and Rosetti, A. (1996). Nerve growth factor delivery by gene transfer induces differential outgrowth of sensory, motor, and noradrenergic neurites after adult spinal cord injury. *Exp. Neurol.* **137,** 157–173.

Tuszynski, M. H., Weidner, N., McCormack, M., Miller, I., Powell, H., and Conner, J. (1998). Grafts of genetically modified Schwann cells to the spinal cord: Survival, axon growth, and myelination. *Cell Transplant.* **7,** 187–196.

Tuszynski, M. H., Grill, R., Jones, L. L., Brant, A., Blesch, A., Low, K., Lacroix, S., and Lu, P. (2003). NT-3 gene delivery elicits growth of chronically injured corticospinal axons and modestly improves functional deficits after chronic scar resection. *Exp. Neurol.* **181,** 47–56.

Venkatesh, K., Chivatakarn, O., Lee, H., Joshi, P. S., Kantor, D. B., Newman, B. A., Mage, R., Rader, C., and Giger, R. J. (2005). The Nogo-66 receptor homolog NgR2 is a sialic acid-dependent receptor selective for myelin-associated glycoprotein. *J. Neurosci.* **25,** 808–822.

Vidal-Sanz, M., Bray, G. M., Villegas-Perez, M. P., Thanos, S., and Aguayo, A. J. (1987). Axonal regeneration and synapse formation in the superior colliculus by retinal ganglion cells in the adult rat. *J. Neurosci.* **7,** 2894–2909.

Vinson, M., Strijbos, P. J., Rowles, A., Facci, L., Moore, S. E., Simmons, D. L., and Walsh, F. S. (2001). Myelin-associated glycoprotein interacts with ganglioside GT1b. A mechanism for neurite outgrowth inhibition. *J. Biol. Chem.* **276,** 20280–20285.

Wang, K. C., Koprivica, V., Kim, J. A., Sivasankaran, R., Guo, Y., Neve, R. L., and He, Z. (2002a). Oligodendrocyte-myelin glycoprotein is a Nogo receptor ligand that inhibits neurite outgrowth. *Nature* **417,** 941–944.

Wang, K. C., Kim, J. A., Sivasankaran, R., Segal, R., and He, Z. (2002b). P75 interacts with the Nogo receptor as a co-receptor for Nogo, MAG and OMgp. *Nature* **420,** 74–78.

Weidner, N., Blesch, A., Grill, R. J., and Tuszynski, M. H. (1999). Nerve growth factor-hypersecreting Schwann cell grafts augment and guide spinal cord axonal growth and remyelinate central nervous system axons in a phenotypically appropriate manner that correlates with expression of L1. *J. Comp. Neurol.* **413,** 495–506.

Weiss, S., Dunne, C., Hewson, J., Wohl, C., Wheatley, M., Peterson, A. C., and Reynolds, B. A. (1996). Multipotent CNS stem cells are present in the adult mammalian spinal cord and ventricular neuroaxis. *J. Neurosci.* **16,** 7599–7609.

Wong, L. F., Yip, P. K., Battaglia, A., Grist, J., Corcoran, J., Maden, M., Azzouz, M., Kingsman, S. M., Kingsman, A. J., Mazarakis, N. D., and McMahon, S. B. (2006). Retinoic acid receptor beta2 promotes functional regeneration of sensory axons in the spinal cord. *Nat. Neurosci.* **9,** 243–250.

Wong, S. T., Henley, J. R., Kanning, K. C., Huang, K. H., Bothwell, M., and Poo, M. M. (2002). A p75(NTR) and Nogo receptor complex mediates repulsive signaling by myelin-associated glycoprotein. *Nat. Neurosci.* **5,** 1302–1308.

Woodhall, E., West, A. K., and Chuah, M. I. (2001). Cultured olfactory ensheathing cells express nerve growth factor, brain-derived neurotrophic factor, glia cell line-derived neurotrophic factor and their receptors. *Brain Res. Mol. Brain Res.* **88,** 203–213.

Xu, X. M., Guenard, V., Kleitman, N., and Bunge, M. B. (1995a). Axonal regeneration into Schwann cell-seeded guidance channels grafted into transected adult rat spinal cord. *J. Comp. Neurol.* **351,** 145–160.

Xu, X. M., Guenard, V., Kleitman, N., Aebischer, P., and Bunge, M. B. (1995b). A combination of BDNF and NT-3 promotes supraspinal axonal regeneration into Schwann cell grafts in adult rat thoracic spinal cord. *Exp. Neurol.* **134,** 261–272.

Xu, X. M., Chen, A., Guenard, V., Kleitman, N., and Bunge, M. B. (1997). Bridging Schwann cell transplants promote axonal regeneration from both the rostral and caudal stumps of transected adult rat spinal cord. *J. Neurocytol.* **26,** 1–16.

Yamashita, T., and Tohyama, M. (2003). The p75 receptor acts as a displacement factor that releases Rho from Rho-GDI. *Nat. Neurosci.* **6,** 461–467.

Yamashita, T., Higuchi, H., and Tohyama, M. (2002). The p75 receptor transduces the signal from myelin-associated glycoprotein to Rho. *J. Cell Biol.* **157,** 565–570.

Zheng, B., Ho, C., Li, S., Keirstead, H., Steward, O., and Tessier-Lavigne, M. (2003). Lack of enhanced spinal regeneration in Nogo-deficient mice. *Neuron* **38,** 213–224.

Zheng, B., Atwal, J., Ho, C., Case, L., He, X. L., Garcia, K. C., Steward, O., and Tessier-Lavigne, M. (2005). Genetic deletion of the Nogo receptor does not reduce neurite inhibition *in vitro* or promote corticospinal tract regeneration *in vivo*. *Proc. Natl. Acad. Sci. USA* **102,** 1205–1210.

Zuo, J., Neubauer, D., Dyess, K., Ferguson, T. A., and Muir, D. (1998). Degradation of chondroitin sulfate proteoglycan enhances the neurite-promoting potential of spinal cord tissue. *Exp. Neurol.* **154,** 654–662.

EVIDENCE FOR NEUROPROTECTIVE EFFECTS OF ANTIPSYCHOTIC DRUGS: IMPLICATIONS FOR THE PATHOPHYSIOLOGY AND TREATMENT OF SCHIZOPHRENIA

Xin-Min Li and Haiyun Xu

Neuropsychiatry Research Unit, Department of Psychiatry, University of Saskatchewan
Saskatoon, SK, S7N 5E4, Canada

I. Introduction
II. Altered Levels of NTs and Their Receptors by Antipsychotic Drugs
 A. NTs and Their Receptors
 B. Altered Levels of NGF in the Brain by Antipsychotics
 C. Altered Levels of BDNF and Trk Receptors in Brains of Patients with Schizophrenia
 D. Effects of Antipsychotic Drugs on the Expression of BDNF and Trk Receptors in Animal Brains
 E. Decreased Expression of p?75^{NTR} by Antipsychotic Drugs
III. The Regulation of Neurogenesis by Antipsychotic Drugs
 A. Changed Cell Proliferation by Haloperidol
 B. Increased Brain Neurogenesis by Atypical Antipsychotic Drugs
 C. Factors Involved in the Regulation of Neurogenesis by Antipsychotics
IV. Neuroprotective Effects of Antipsychotic Drugs
 A. Cytoprotective/Cytotrophic Effects of Antipsychotic Drugs in Cell Cultures
 B. Protective Effects of Antipsychotic Drugs in Animals
 C. Effects of Atypical Antipsychotics on the Activity of HPA Axis and the Stress-Induced Changes in the Hippocampus
 D. Combinations of Atypical Antipsychotics and Antidepressants
References

Schizophrenia is a severe psychiatric disorder, and its etiology, pathophysiology, and mechanism of antipsychotic treatment remain to be elucidated. Although the first neuroleptic drugs (chlorpromazine, reserpine, and haloperidol) have revolutionized both the theory and the therapy of schizophrenia, the dopamine hypothesis is being challenged by recent advances in schizophrenia research. This chapter examines recent advances in the exploration of neural mechanisms of antipsychotic drugs, putting emphases on *in vitro* and *in vivo* evidence of their neuroprotective effects. Specifically, the chapter deals with the effects of antipsychotic drugs on neurotrophins and their receptors in the brain, the regulation of brain neurogenesis by antipsychotics, as well as the cytoprotective/neurotrophic effects of atypical antipsychotics in cell culture and animal studies. In addition, data from clinical research reports are included, which support the neuroprotective hypothesis

INTERNATIONAL REVIEW OF
NEUROBIOLOGY, VOL. 77
DOI: 10.1016/S0074-7742(06)77004-0

0074-7742/07 $35.00

of antipsychotic drugs. All the data considered in the chapter, taken together, should provide new insights into the principal mechanisms of action of antipsychotic drugs and the pathophysiology of schizophrenia.

I. Introduction

Schizophrenia is one of the most severe psychiatric disorders, affecting about 1% of the population. It is characterized by: positive symptoms (hallucinations, delusions, and disorganization), negative symptoms (social withdrawal and lack of emotion), cognitive impairment (poor learning and reasoning ability), and depressive symptoms (Andreasen, 1995; Andreasen *et al.*, 1994; Lublin *et al.*, 2005). In addition, some patients may show aggressive or suicidal tendencies (Keck *et al.*, 2000; Perenyi and Forlano, 2005; Sharif, 1998).

A large number of drugs belonging to different chemical classes have shown therapeutic efficacy in patients with schizophrenia, and have been defined as antipsychotic drugs. These drugs are categorized into typical (conventional) and atypical (novel) antipsychotics. Typical antipsychotics, represented by chlorpromazine and haloperidol, are effective in reducing positive symptoms. Atypical antipsychotics, including clozapine, olanzapine, quetiapine, and risperidone, are not only effective in treating the positive symptoms of schizophrenia, but also reduce negative symptoms and cognitive deficits (Breier, 1999; Lindenmayer, 1995; Meltzer *et al.*, 1994; Weiss *et al.*, 2002).

Antipsychotic drugs block multiple amine receptors in the brain. Typical antipsychotics are highly D_2 receptor selective, whereas atypical antipsychotics are broad-acting agents with affinity for many neuroreceptors. Of particular interest is their stronger affinity for 5-HT_{2A} receptors than for D_2 receptors. The blocking of D_2 receptors in mesolimbic areas is believed to be responsible for the therapeutic effects of antipsychotic drugs on the positive symptoms of patients with schizophrenia (Ellenbroek, 1993; Lyne *et al.*, 2004; Meltzer *et al.*, 1999; Weiss *et al.*, 2002). However, the mechanisms underlying the improvement of negative and cognitive symptoms by atypical antipsychotic drugs remain to be elucidated.

The purpose of this chapter is to review recent advances in exploring the neural mechanisms of antipsychotic drugs, with emphases on *in vitro* and *in vivo* evidence for their neuroprotective effects. In Section II, we describe the effects of antipsychotic drugs on expression of the neurotrophins (NTs) and their receptors in the brain, with a special emphasis on the modulation of brain-derived neurotrophic factor (BDNF) by the drugs, because BDNF is mainly expressed and distributed in brain neurons. In Section III, we delineate the regulation of brain neurogenesis by antipsychotics, emphasizing hippocampal neurogenesis, which is highly relevant to learning and memory functions of the brain.

Section IV highlights the cytoprotective and neurotrophic effects of atypical antipsychotics in cell culture and animal studies, and also lists combinations of atypical antipsychotics and antidepressants used in animal studies and clinical practice. The data provide new insights into the principal mechanisms of action of antipsychotic drugs, as well as the pathophysiology of schizophrenia.

II. Altered Levels of NTs and Their Receptors by Antipsychotic Drugs

A. NTs and Their Receptors

NTs are growth factors that act directly on neurons to support their growth, differentiation, and survival (Yuen and Mobley, 1996). The NT family includes nerve growth factor (NGF), BDNF, NT-3, and NT-4. NTs bind to the tyrosine kinase (Trk) family of receptors (including TrkA, TrkB, and TrkC) and to the p75 NT receptor ($p75^{NTR}$), a member of the tumor necrosis factor receptor family (Huang and Reichardt, 2001). NGF signals primarily through TrkA, BDNF and NT-4 through TrkB, and NT-3 through TrkC. All of the NTs bind to $p75^{NTR}$ with similar low affinities. The Trk receptors are responsible for most of the survival and growth properties of NTs. The $p75^{NTR}$ is the first identified dependence receptor. In the presence of ligand, it enhances NT-mediated Trk receptor activity leading to survival, differentiation, or migration. In the absence of a ligand, it initiates or amplifies a signal for programmed cell death (Mehlen and Bredesen, 2004; Roux and Barker, 2002).

B. Altered Levels of NGF in the Brain by Antipsychotics

Pilot studies on NGF levels in patients with schizophrenia and on effects of antipsychotic drugs on NGF levels in the brain showed that acute haloperidol treatment (10 mg/kg) in adult mice decreased the levels of NGF protein in the hypothalamus (Alleva *et al.*, 1996). These studies reported that NGF plasma levels in patients with schizophrenia are lower compared to healthy controls (Bersani *et al.*, 2000, 2004), pointing to a potential role of NTs, particularly NGF, in the pathogenesis and treatment of schizophrenia.

Other studies found that chronic antipsychotic treatment with haloperidol or risperidone increased NGF levels in the hypothalamus of rats and decreased NGF levels in the striatum and hippocampus (Angelucci *et al.*, 2000a). Similarly, haloperidol (in drinking water for 45 days, 1 or 2 mg/kg/day) reduced NGF immunoreactivity in the rat hippocampal dentate gyrus (DG), CA1, and CA3 areas (Parikh *et al.*, 2004b). In mice, however, haloperidol induced NGF mRNA expression in a dose-dependent fashion in the hippocampus, piriform cortex, striatum, and nucleus accumbens. Another D_2 antagonist, (−)-sulpiride, showed the same effect. The haloperidol- and (−)-sulpiride-induced NGF mRNA expression attained a maximum level 120 min after injection and returned to control levels 24 hours later (Ozaki *et al.*, 1999). The same laboratory also reported that chronic administration of haloperidol or (−)-sulpiride increased NGF protein levels in the same brain regions (Ozaki, 2000).

In recent studies, effects of the atypical antipsychotic olanzapine on NGF levels in the brain were observed. It was shown that olanzapine (5 or 10 mg/kg, given in drinking water for 45 days) increased levels of NGF in the hippocampus (Parikh *et al.*, 2004a). In another study, olanzapine (3 or 15 mg/kg) was administered for 29 days and shown to increase NGF levels in the hippocampus, occipital cortex, and hypothalamus (Angelucci *et al.*, 2005). Given that schizophrenia is considered to be a developmental disorder of the cerebral cortex (Raedler *et al.*, 1998), NGF plays a crucial role in the survival and function of forebrain cholinergic neurons (Mobley *et al.*, 1986) and mesencephalic dopaminergic neurons (Chaturvedi *et al.*, 2006), and that NGF is involved in mechanisms regulating neuronal death (McCollum and Estus, 2004), determining the role of NGF in the pathogenesis and treatment of schizophrenia is of significance theoretically as well as for clinical practice.

C. Altered Levels of BDNF and Trk Receptors in Brains of Patients with Schizophrenia

BDNF has a wide distribution in the brain. It is synthesized predominantly in neurons and is expressed at highest levels in the hippocampus and the cerebral cortex (Ernfors *et al.*, 1990; Hofer *et al.*, 1990; Katoh-Semba *et al.*, 1997; Wetmore *et al.*, 1990). In addition to providing neurotrophic support for cholinergic neurons, BDNF has been proposed to have a potential role in promoting the functioning and survival of dopaminergic, GABAergic, noradrenergic, and serotonergic neurons (for review see Connor and Dragunow, 1998; Nagtegaal *et al.*, 1998). The expression of both the BDNF gene and its corresponding protein has been shown to be regulated by a number of factors, including neuronal activity, seizures, restraint stress, neurotransmitter actions, and second messenger cascades such as cAMP (Condorelli *et al.*, 1994; Humpel *et al.*, 1993; Lindefors *et al.*, 1995; Nibuya *et al.*, 1995, 1996; Smith *et al.*, 1995; Vaidya *et al.*, 1997; Wetmore *et al.*, 1994; Xu *et al.*, 2004).

Takahashi *et al.* (2000) determined levels of NTs and their receptors in the postmortem brains of schizophrenic patients and normal control subjects. Among the NTs examined, levels of BDNF were elevated specifically in the anterior cingulate cortex and hippocampus of schizophrenic patients. This observation led to further studies to determine the levels of BDNF and its receptors in patients with schizophrenia. In one study, the BDNF content in cortical areas of brains of schizophrenic patients was significantly higher than in controls. In the hippocampus, the opposite effect was observed: the BDNF concentrations were significantly lower in schizophrenia cases than in normal controls (Durany *et al.*, 2001). In contrast, other studies detected a significant reduction in BDNF mRNA and protein in the dorsolateral prefrontal cortex (Weickert *et al.*, 2003), and in the prefrontal cortex (Hashimoto *et al.*, 2005) of patients with schizophrenia compared to normal individuals, and showed higher BDNF immunoreactivity in the hippocampal formation in schizophrenic patients (Iritani *et al.*, 2003). Serum BDNF levels in patients with schizophrenia were also inconsistent, while some studies showed that BDNF levels were significantly reduced in the serum of schizophrenic patients (Pirildar *et al.*, 2004; Toyooka *et al.*, 2002), another study reported that there were no significant differences in serum BDNF levels among antipsychotic-naïve and medicated patients and normal controls (Shimizu *et al.*, 2003).

TrkB and TrkC receptor levels in brains of schizophrenic patients were also altered. In most studies (Hashimoto *et al.*, 2005; Iritani *et al.*, 2003; Takahashi *et al.*, 2000), the expression of TrkB receptor was reduced in the hippocampus and the prefrontal cortex, but one study showed that TrkB immunoreactivity in the hippocampal formation of schizophrenic patients was more intense (Iritani *et al.*, 2003). TrkC mRNA levels in the frontal and prefrontal cortices of patients with schizophrenia were lower as compared to controls (Schramm *et al.*, 1998; Weickert *et al.*, 2005).

The data described above suggest that the expression of BDNF, TrkB, and TrkC in the brain is implicated in the pathophysiology of schizophrenia. However, it is not known whether antipsychotic treatment is related to these changes. In addressing this question, animal studies have provided meaningful insight described below.

D. Effects of Antipsychotic Drugs on the Expression of BDNF and Trk Receptors in Animal Brains

Chronic administration (29 days) of haloperidol and risperidone triggered a decrease in BDNF levels in the prefrontal cortex, occipital cortex, and hippocampus of rats. Haloperidol also significantly decreased TrkB immunoreactivity in neurons of the hippocampus, substantia nigra, and ventral tegmental area, whereas the effects of risperidone were not statistically significant (Angelucci *et al.*, 2000b). These effects of haloperidol on BDNF and TrkB were further confirmed in a recent animal study in which chronic exposure to haloperidol (2 mg/kg/day for 45 days) significantly reduced the levels of BDNF and TrkB receptors in the hippocampus (Parikh *et al.*, 2004a). Acute administration of haloperidol also produced evident effects on basal levels of BDNF in the brain. Three days of haloperidol administration (0.5 mg/kg) dramatically decreased BDNF immunostaining in the neurons and fibers of the prefrontal cortex, hippocampus, extended amygdala, and ventral tegmental area (Dawson *et al.*, 2001). Similarly, following 3 days of haloperidol treatment (1 mg/kg), the density of BDNF-immunoreactive fibers and cells were reduced in most forebrain and midbrain areas, including the cortex and amygdala. In the ventral pallidum, BDNF immunoreactivity was reduced to a level almost below detection. The treatments markedly reduced tyrosine hydroxylase (TH)-immunoreactive labeling throughout the ventral pallidum and induced fiber and terminal degeneration in this brain area (Meredith *et al.*, 2004).

The effects of atypical antipsychotic drugs on the expression of BDNF in the brain differed in various studies. In one study, olanzapine (3 or 15 mg/kg, for 29 days) decreased BDNF in the hippocampus, frontal cortex, and occipital cortex (Angelucci *et al.*, 2005), while in another study, olanzapine (10 mg/kg, for 45 days) did not alter BDNF levels in the hippocampus (Parikh *et al.*, 2004a). The latter study also showed that posttreatment with olanzapine markedly restored the haloperidol-induced reductions in both BDNF and TrkB receptors in the hippocampus.

In some studies, effects of antipsychotic drugs on the expression of BDNF mRNA in the brain were measured. A single dose or 28 days of clozapine (10 mg/kg) and haloperidol (0.5 and 1 mg/kg) reduced BDNF mRNA levels in the hippocampus of rats, but did not lower its expression in the prefrontal cortex (Lipska *et al.*, 2001). Our team, however, found that chronic administration (28 days) of clozapine (10 mg/kg) and olanzapine (2.7 mg/kg) up-regulated BDNF mRNA expression in the hippocampus, whereas haloperidol (1 mg/kg) decreased BDNF mRNA levels in the same brain region (Bai *et al.*, 2003; Fig. 1). In another animal study, we confirmed that chronic administration (19 days) of haloperidol decreased hippocampal BDNF mRNA in comparison to controls (Chlan-Fourney *et al.*, 2002). Furthermore, in the same study we observed a dose-dependent regulation of BDNF mRNA levels by atypical antipsychotics. Specifically, a higher dose of risperidone (4.11 mg/kg) significantly decreased BDNF mRNA levels in the hippocampus, whereas a lower (0.066 mg/kg) and an intermediate (0.67 mg/kg) dose of the drug were without effect. A lower dose of clozapine (1.3 mg/kg) increased BDNF mRNA levels, whereas a higher dose (27 mg/kg) produced opposite effects (Chlan-Fourney *et al.*, 2002; Table I). This dose-dependent effect provided a reasonable explanation to those inconsistent results described above. With the same, it can be explained why neither acute nor chronic clozapine treatment affected the expression of BDNF mRNA in any brain area investigated by Linden *et al.* (2000).

All of the data described above strongly suggest that antipsychotic drugs do affect the expression of BDNF and TrkB and TrkC receptors in the brain. Directions of the

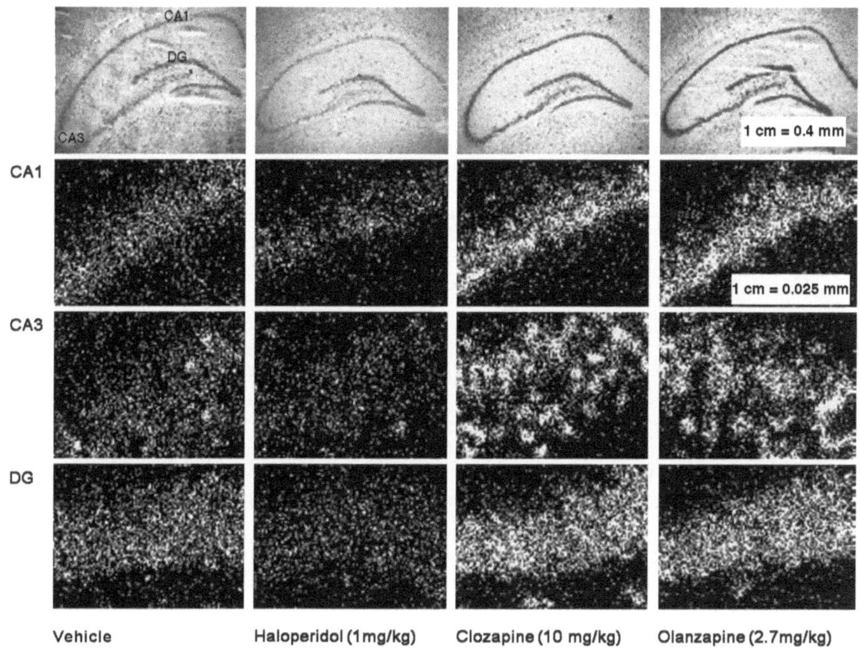

FIG. 1. Animals were given daily injections of vehicle, haloperidol, clozapine, or olanzapine for 28 days and were sacrificed 18 hours after the last injection. Brain sections were hybridized with 35S-labeled BDNF RNA probes and dipped into photographic emulsion. Low- magnification (2.5×) photomicrographs show BDNF mRNA expression in the hippocampus (top row). High-magnification (40×) photomicrographs show detailed changes in silver grain density in the CA1, CA3, and DG regions compared with the control.

regulation may be related to the dose of the drugs used, the experimental paradigms, brain regions, animal species, and a number of other variables.

E. Decreased Expression of p75^{NTR} by Antipsychotic Drugs

In an earlier *in vivo* study evaluating effects of antipsychotic drugs on levels of NGF and its receptor, Alberch *et al.* (1991) administered haloperidol (2 mg/kg, s.c.) to pregnant Sprague-Dawley rats daily for 11 days beginning on E10, and found a decrease in NGF receptor mRNA in the neonate forebrain. Furthermore, they showed a decrease in the K_d value (74%) and the B_{max} value (63%) of the low-affinity binding sites, but did not find any changes in the high-affinity binding sites. In 1999, we examined effects of atypical antipsychotics on p75^{NTR} mRNA levels in PC12 cells (Li *et al.*, 1999). Using the PC12 cell *in vitro* model, we showed

TABLE I

THE EFFECTS OF CHRONIC ANTIPSYCHOTIC ADMINISTRATION ON BDNF MRNA BY *IN SITU* HYBRIDIZATION ANALYSES[a]

Treatment[b]	Brain region		
	Dentate	CA1	CA3
Control	100.00 ± 6.1	100 ± 3.0	100 ± 1.3
Haloperidol	**65.00 ± 7.3	*85 ± 3.7	*77 ± 4.6
Risperidone low	99.00 ± 12.2	93 ± 2.9	99 ± 7.6
Risperidone intermed[c]	107.00 ± 9.1	96 ± 1.1	99 ± 6.2
Risperidone high	*81.00 ± 2.4	96 ± 8.0	92 ± 5.2
Clozapine low	#115.00 ± 5.5	113 ± 8.0	102 ± 5.1
Clozapine intermed	101.00 ± 11.1	104 ± 5.0	96 ± 3.1
Clozapine high	#86.00 ± 5.8	103 ± 3.9	95 ± 4.2

[a]Measured as dpm/mm^2, expressed in the Table as percentage control ± SEM.

[b]Doses: rats received either vehicle (0.8% glacial acetic acid in 5% glucose), haloperidol (1 mg/kg i.p.), risperidone (0.066, 0.67, and 4.11 mg/kg i.p.), or clozapine (1.3 , 9.0, and 27.0 mg/kg i.p.) daily for 19 days ($n = 4$).

[c]Intermediate dose.

*Significantly different vs control, $p < 0.05$; **vs CON, $p < 0.01$; #clozapine high vs clozapine low, $p < 0.05$.

that olanzapine decreased p75^{NTR} mRNA levels in PC12 cells. This effect held true for other atypical antipsychotics, including clozapine, quetiapine, and risperidone (Bai *et al.*, 2002). In a more recent study, haloperidol showed the same action of down-regulating the expression of p75^{NTR} in PC12 cells (Bai *et al.*, 2006). These results suggest that both haloperidol and atypical antipsychotic drugs have the same action of decreasing p75^{NTR} mRNA expression. As mentioned before, p75^{NTR} is a common NT receptor and has been shown to be capable of mediating programmed cell death (Rabizadeh and Bredesen, 2003). Therefore, down-regulation of p75^{NTR} mRNA may have cytoprotective potential.

III. The Regulation of Neurogenesis by Antipsychotic Drugs

Neurogenesis is a process of generating functionally integrated neurons from progenitor cells (Ming and Song, 2005). In most mammals, active neurogenesis occurs throughout life in the subventricular zone (SVZ) of the lateral ventricles and in the subgranular zone (SGZ) of DG. Newborn neuronal cells in the SGZ migrate to the granular cell layer of DG, integrate into the existing hippocampal circuitry (Cameron and McKay, 2001; van Praag *et al.*, 2002), and play important roles in learning and memory formation (Gould *et al.*, 1999; Shors, 2004; Shors *et al.*, 2001).

The newborn neurons can be detected and identified by using exogenous and inherent markers. The most commonly used exogenous markers are [H^3]-thymidine and bromodeoxyuridine (BrdU), both of which can be incorporated into newly synthesized DNA and then passed on to cell progeny. The incorporated [H^3]-thymidine marker can be detected autoradiographically (Rogers and Lietman, 1977), while the BrdU marker can be detected with immunohistochemistry and allows for both phenotypic analysis and stereological quantification of newborn cells. In addition, some specific inherent markers, such as PSA-NCAM (polysialylated neural cell adhesion molecule), Tuji (β-tubulin isoform III), CRMP (collaspin response-mediated protein 4), and DCX (doublecortin), have been used for identifying immature neurons (Ming and Song, 2005).

A. Changed Cell Proliferation by Haloperidol

A number of studies have been done to explore the teratogenic potential of neuroleptic drugs based on evidence that substances which act as neurotransmitters in the developed brain may also function as neurohumoral agents involved in the regulation of both brain growth and development (Hohmann and Coyle, 1988; Lauder, 1983; Patel and Lewis, 1988; Whitaker-Azmitia and Azmitia, 1989; Williams *et al.*, 1992). It was hypothesized that haloperidol might have effects on cell proliferation as it had been believed to act primarily on receptors of dopamine which might exert a trophic effect on early brain development. Indeed, *in vivo* and *in vitro* data showed that haloperidol produced a significant depression on the rate of [^3H]-thymidine incorporation into the DNA of brain cells (Backhouse *et al.*, 1982; Barochovsky and Patel, 1982; Holson *et al.*, 1994; Jurand and Martin, 1990; Williams *et al.*, 1992). In these studies, the doses of haloperidol were in the range of 10–20 mg/kg for acute or subchronic maternal exposure (Backhouse *et al.*, 1982; Holson *et al.*, 1994), 2.5–5 mg/kg for chronic prenatal exposure (Williams *et al.*, 1992), or 1–1000 μM for slice culture (Barochovsky and Patel, 1982; Holson *et al.*, 1994). At these relatively high doses, it is difficult to estimate human reproductive risks from these animal data for human therapeutic doses. Other caveats are: (1) the growth stunting produced by prenatal neuroleptic exposure was not restricted to dopamine-rich areas of the brain (Holson *et al.*, 1994) and (2) although agonists and antagonists both showed similar inhibitory effects, the action of agonists was reversed by the appropriate antagonists (Barochovsky and Patel, 1982).

In male gerbils, Dawirs *et al.* (1998) reported that four doses of haloperidol increased the mean individual mitotic activity in SGZ by about 75%. Following this study, investigators were not able to replicate the same effect in animals that received subchronic or chronic haloperidol treatment (Schmitt *et al.*, 2004;

Wakade *et al.*, 2002). *In vitro* studies, however, showed that D_2 antagonists, including haloperidol, blocked the inhibition of cell proliferation by dopamine (An *et al.*, 2003; Arita *et al.*, 1998). Therefore, the failure to establish a clear link between haloperidol and cell proliferation can be attributed to the use of different protocols to detect cell genesis that is sensitive to mixed or different populations of proliferating and maturing cell types. With this consideration, a more recent study (Kippin *et al.*, 2005) examined the effect of haloperidol on neural stem cells (NSCs), the ultimate precursors for adult cell genesis. The authors showed that haloperidol increased NSC numbers, resulting in more progenitors and more new neurons and glias in the adult rat brain (Fig. 2). They suggested that this effect was done by antagonizing dopamine at D_2 receptors on NSCs, as D_2 receptor stimulation *in vitro* inhibited NSC proliferation, which was reversed by haloperidol. Furthermore, $D_2^{+/+}$ mice that received haloperidol had increased numbers of neurospheres, but $D_2^{-/-}$ mice that received haloperidol did not show an increase in the numbers of NSCs.

B. Increased Brain Neurogenesis by Atypical Antipsychotic Drugs

Wakade *et al.* (2002) reported that both olanzapine and risperidone induced a significant increase in $BrdU^{(+)}$ cells in SVZ. The atypical antipsychotics stimulated a two- to threefold increase of $BrdU^{(+)}$ cells when compared to controls. The expression of a neuronal marker, NeuN, in $BrdU^{(+)}$ cells in rats treated with olanzapine or risperidone also suggested that these compounds might modulate *in vivo* differentiation of neuronal progenitor cells even within a day of BrdU injection. In another study, chronic (21 days) olanzapine administration increased the number of newborn cells in DG of adult rats. The treatment also increased the number of proliferating cells in the prelimbic cortex (Kodama *et al.*, 2004). Similarly, olanzapine increased both the total number and density of BrdU-labeled cells in the prefrontal cortex and dorsal striatum (Wang *et al.*, 2004), and increased cell proliferation of PC12 cells (Lu *et al.*, 2004). Clozapine, another atypical antipsychotic, in a low dose (0.5 mg/kg) and chronic treatment (28 days) paradigm, increased the number of $BrdU^{(+)}$ cells in DG by twofold (Halim *et al.*, 2004). These results suggest that atypical antipsychotic drugs may increase brain neurogenesis.

C. Factors Involved in the Regulation of Neurogenesis by Antipsychotics

Although the mechanism by which antipsychotic drugs stimulate neurogenesis is not clear, current data suggest the involvement of the following factors in the regulation of brain neurogenesis by antipsychotic drugs.

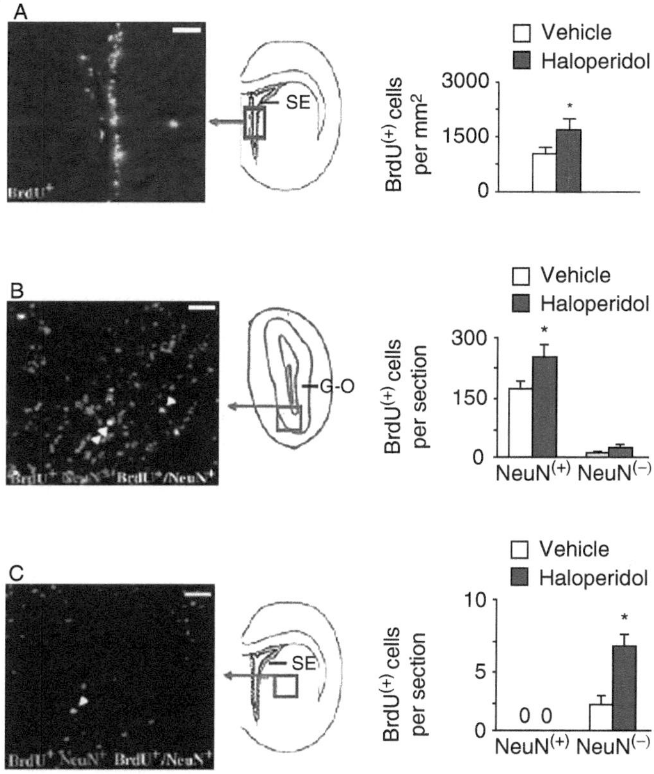

FIG. 2. Haloperidol increases the number of CP cells in the subependyma, neurogenesis in the olfactory bulbs, and gliogenesis in the striatum. (A) Left, photomicrograph of progenitor cells after haloperidol treatment. Right, haloperidol increased $BrdU^{(+)}$ cell numbers in lateral ventricle subependyma. (B) Left, photomicrograph of neurogenesis in the olfactory bulb. Right, haloperidol increased $BrdU^{(+)}/NeuN^{(+)}$ cell numbers in the olfactory bulbs. (C) Left, photomicrograph of nonneuronal cell genesis in the striatum. Right, haloperidol increased $BrdU^{(+)}/NeuN^{(-)}$ cell numbers in the striatum. Single and double arrowheads indicate $BrdU^{(+)}/NeuN^{(-)}$ and $BrdU^{(+)}/NeuN^{(+)}$ cells, respectively. All data are presented as mean ± SEM. Scale bars, 50 μm. $*p < 0.05$.

1. D_2 receptors. By binding to D_2 receptors, antipsychotics may block the inhibition of cell proliferation by dopamine (An *et al.*, 2003; Arita *et al.*, 1998; Kippin *et al.*, 2005).

2. $5\text{-}HT_{1A}$ receptors. In recent studies, atypical antipsychotic drugs have shown their capacity of occupying and activating the $5\text{-}HT_{1A}$ receptors (Chou *et al.*, 2003; Chung *et al.*, 2004; Cussac *et al.*, 2002; Diaz-Mataix *et al.*, 2006; Newman-Tancredi *et al.*, 1998, 2001). Injections of the $5\text{-}HT_{1A}$ agonists increased the numbers of the BrdU-labeled cells in the rat DG (Santarelli *et al.*, 2003). In adult rats, three different antagonists of the $5\text{-}HT_{1A}$ receptors (NAN-190, *p*-MPPI, and WAY-100635)

reduced the numbers of newborn cells in DG by about 30% (Radley and Jacobs, 2002), whereas the activation of 5-HT_{1A} heteroreceptors produced similar increases in the number of BrdU-labeled cells (Banasr *et al.*, 2004).

3. BDNF is another important contributor. In support of this notion, incubation of cultured progenitor cells with BDNF was reported to increase the differentiation of cells into neurons (Palmer *et al.*, 1997). There are also *in vivo* studies showing that increasing BDNF levels via different approaches lead to increased numbers of new neurons in the brain (Pencea *et al.*, 2001; Scharfman *et al.*, 2005). By affecting BDNF levels in the brain, antipsychotic drugs may regulate brain neurogenesis.

4. The cAMP response element-binding protein (CREB). Previous studies have shown that increased cell proliferation is accompanied by phosphorylation of CREB in granule cells of DG (Nakagawa *et al.*, 2002). Furthermore, activation of CREB stimulates neurogenesis in the adult DG (Zhu *et al.*, 2004) and increases the differentiation of newborn cells into neurons (Palmer *et al.*, 1997). In line with these previous results, we found that quetiapine administration (for 7 or 21 days) resulted in parallel increases in BrdU-labeled cells and phosphorylated-$CREB^{(+)}$ cells in DG of rats, suggesting that this drug might activate CREB, which in turn stimulates hippocampal neurogenesis (Luo *et al.*, 2005). Indeed, in hippocampal neuron cultures, the levels of CREB phosphorylation significantly increased after treatment with haloperidol (50 nM/liter) and risperidone (50 nM/liter) (Yang *et al.*, 2004).

In addition to the above factors, some atypical antipsychotic drugs have unique actions, which may contribute to enhanced neurogenesis. For example, acute or chronic administration of clozapine produced a selective increase of fibroblast growth factor-2 (FGF-2) mRNA and protein in the striatum (Riva *et al.*, 1999). The FGF-2 is believed to increase neurogenesis, exemplified by the findings that subcutaneous administration of FGF-2 to newborn rat pups promotes neuroblast proliferation in regions still undergoing neurogenesis, including the hippocampus and the granule layer of the cerebellum (Tao *et al.*, 1997), and that chronic infusion of FGF-2 into the lateral ventricle of adult rats results in an increase in the population of proliferating precursors in SVZ and a subsequent increase in the number of neurons migrating from the SVZ to the olfactory bulb (Kuhn *et al.*, 1997).

IV. Neuroprotective Effects of Antipsychotic Drugs

A. Cytoprotective/Cytotrophic Effects of Antipsychotic Drugs in Cell Cultures

Ueda *et al.* (1997) reported that the phenothiazine derivatives chlorpromazine, promethazine, and trifluoperazine inhibited toxicities of β-amyloid peptide [$A\beta_{25-35}$] in primary cultures of rat cortical neurons and PC12 cells. Chlorpromazine dose-dependently inhibited the $A\beta_{25-35}$-induced decrease in cell viability and Ca^{2+} uptake in both types of cell cultures. Since promethazine, which has the phenothiazine structure without neuroleptic effects (Nyback and Sedvall, 1976), also reduced $A\beta_{25-35}$ cytotoxicity, the protective effects of phenothiazine derivatives could be attributed to structure specificity, rather than to neuroleptic effects (Ueda *et al.*, 1997). This extrapolation is applicable to haloperidol, which did not show cytoprotective effects (Qing *et al.*, 2003), even induced cytotoxicities in recent *in vitro* studies (Galili-Mosberg *et al.*, 2000; Sagara, 1998; Ukai *et al.*, 2004).

Given the demonstrated action of $A\beta_{25-35}$ to produce reactive oxygen species (ROS), which in turn increase Ca^{2+}, thereby inducing neurotoxicity (Harris *et al.*, 1995; Hensley *et al.*, 1994), the cytoprotective effects of chlorpromazine and trifluoperazine might be the reflection of their antioxidant potential. Indeed, in lipid peroxidation induced in microsomal membranes, chlorpromazine and trifluoperazine acted as very good antioxidants, and in aqueous solution they appeared particularly effective in the bleaching of crocin (Dalla Libera *et al.*, 1998). In the same study, clozapine also showed antioxidant properties, which were more evident in a hydrophobic environment.

Our team has carried out a series of *in vitro* experiments in recent years and the results further demonstrated that atypical antipsychotics have antioxidant capacity. We found that the atypical antipsychotics clozapine, olanzapine, quetiapine, and risperidone increased the gene expression of superoxide dismutase (SOD_1), while they decreased $p75^{NTR}$ mRNA levels in PC12 cells and prevented the cells from death after serum withdrawal (Bai *et al.*, 2002; Li *et al.*, 1999). These drugs also reduced the *N*-methyl-4-phenylpyridinium ion (MPP^+)-induced apoptosis and DNA fragmentation in PC12 cultures (Qing *et al.*, 2003) and decreased cell death induced by hydrogen peroxide H_2O_2 or $A\beta_{25-35}$ in PC12 cultures (Wei *et al.*, 2003a,b). Consistent results were also reported by other investigators, who revealed that olanzapine enhanced the survival of PC12 cells, SH-SY5Y cells, and 3T3 preadipocytes, and protected neuronal cells from death induced by serum and glutamine deprivation, β-amyloid peptide, and fluphenazine (Lu *et al.*, 2004). The cytotoxic insults mentioned above have in common the capability of causing oxidative stress, as indicated by overproduction of H_2O_2 and changes in antioxidant enzymes following the treatments (Lee *et al.*, 2002; Qing *et al.*, 2003; Wei *et al.*, 2003a; Xiao *et al.*, 2002). Therefore, it may be concluded that these atypical antipsychotics have a common antioxidant action responsible for their cytoprotective effects in the circumstances described above. In support of this notion, one of our recent studies showed that olanzapine and quetiapine prevented PC12 cells from $A\beta_{25-35}$-induced apoptosis (Fig. 3) and the overproduction of intracellular ROS (Fig. 4); attenuated $A\beta_{25-35}$-induced changes in activities of the antioxidant enzymes SOD_1, catalase, and glutathione peroxidase (Fig. 5); and blocked $A\beta_{25-35}$-induced decrease in mitochondrial membrane potential in PC12 cells (Wang *et al.*, 2005; Fig. 6).

Olanzapine, quetiapine, and clozapine, also enhanced neurite outgrowth induced by NGF in PC12 cells, increased the number of cells bearing neurites, the length of primary neurites, and the size of the cell body of NGF-differentiated PC12 cells, and induced sprouting of neurite-like processes in PC12 cells in the absence of NGF, suggesting that these drugs have neurotrophic potential (Lu and Dwyer, 2005). These effects appear to involve the phosphatidylinositol 3-kinase (PI3K)/Akt and extracellular signal-regulated kinase (ERK) signal transduction pathways that play

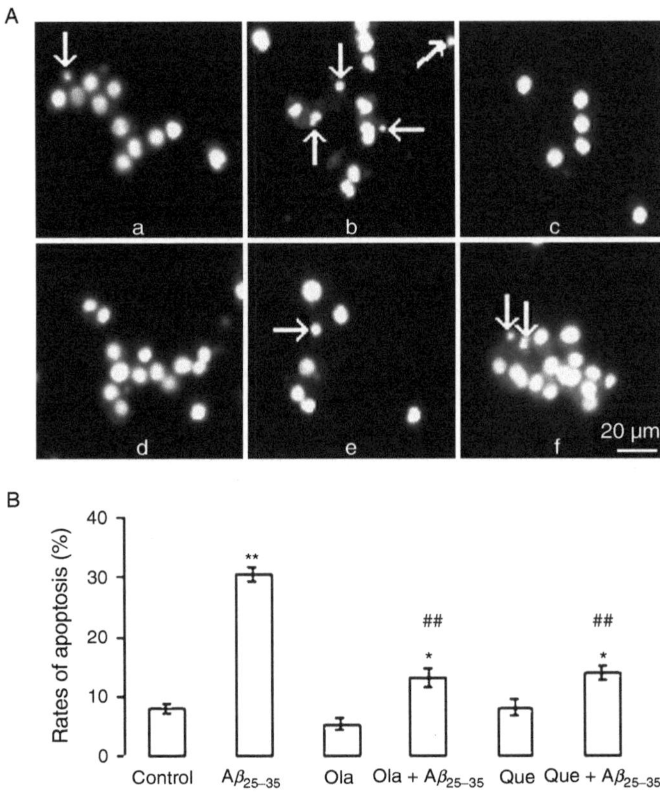

FIG. 3. Olanzapine and quetiapine diminished the $A\beta_{25-35}$-induced apoptosis. Cells were pretreated with vehicle Dulbecco's modified eagle's medium (DMEM), 50 μM of olanzapine or quetiapine for 24 hours, followed by exposing to 0 or 20 μM of $A\beta_{25-35}$ in the absence or presence of 50 μM of olanzapine or quetiapine for another 24 hours. (A) Photographs a–f are from control, $A\beta_{25-35}$, olanzapine, olanzapine + $A\beta_{25-35}$, quetiapine, and quetiapine + $A\beta_{25-35}$, respectively. Arrows indicate apoptotic cells with condensed nuclei. (B) The results of quantitative analyses. Data are presented as means ± SEM. $*p < 0.05$, $**p < 0.01$ compared with control. $^{\#\#}p < 0.01$ compared with cultures exposed to $A\beta_{25-35}$. Ola, olanzapine; Que, quetiapine; $A\beta$, $A\beta_{25-35}$.

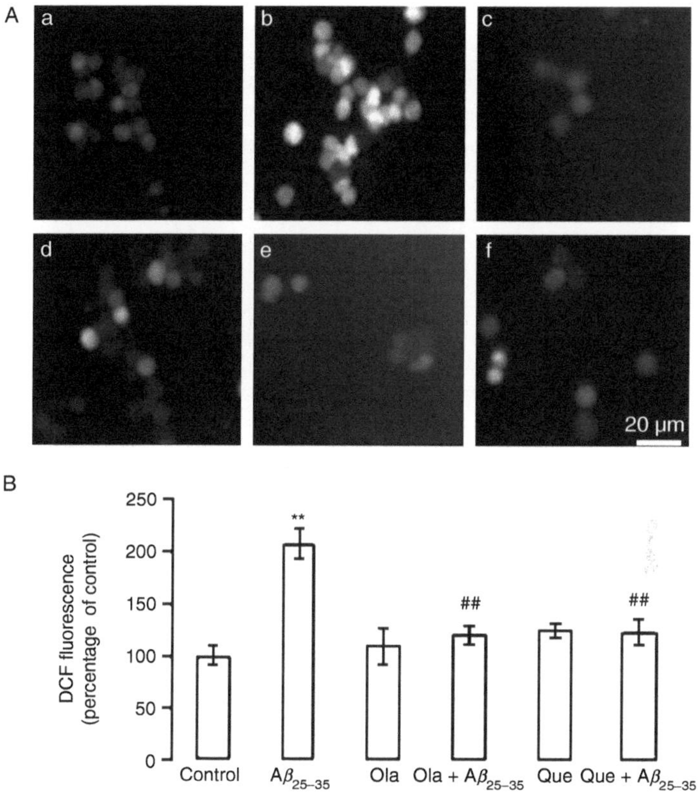

FIG. 4. Olanzapine and quetiapine prevented the $A\beta_{25-35}$-induced overproduction of intracellular ROS. Cells were pretreated with vehicle (DMEM) or 50 μM of olanzapine or quetiapine for 24 hours, followed by exposing to 0 or 20 μM of $A\beta_{25-35}$ in the absence or presence of 50 μM of olanzapine or quetiapine for another 6 hours. Then, intracellular ROS accumulation was measured. (A) Photographs a–f were from control, $A\beta_{25-35}$, olanzapine, olanzapine + $A\beta_{25-35}$, quetiapine, and quetiapine + $A\beta_{25-35}$, respectively. (B) The results of quantitative analyses. Data are presented as means ± SEM. $^{**}p < 0.01$ compared with control. $^{\#\#}p < 0.01$ compared with cultures exposed to $A\beta_{25-35}$ alone.

important roles in neuronal cell growth, survival, and differentiation (Ciani *et al.*, 2002; Dudek *et al.*, 1997; Lawlor and Alessi, 2001; Vaudry *et al.*, 2002). In line with this explanation, atypical antipsychotics stimulated phosphorylation of Akt and ERK (Cussac *et al.*, 2002; Lu *et al.*, 2004); olanzapine, quetiapine, and clozapine significantly enhanced ERK phosphorylation induced by NGF in PC12 cells; and PI3K inhibitor LY294002 decreased the positive effects of antipsychotics on NGF-induced neurite outgrowth (Lu and Dwyer, 2005). In addition, the selective 5-HT_{1A} receptor antagonist WAY-100635 abolished the phosphorylation of ERK by

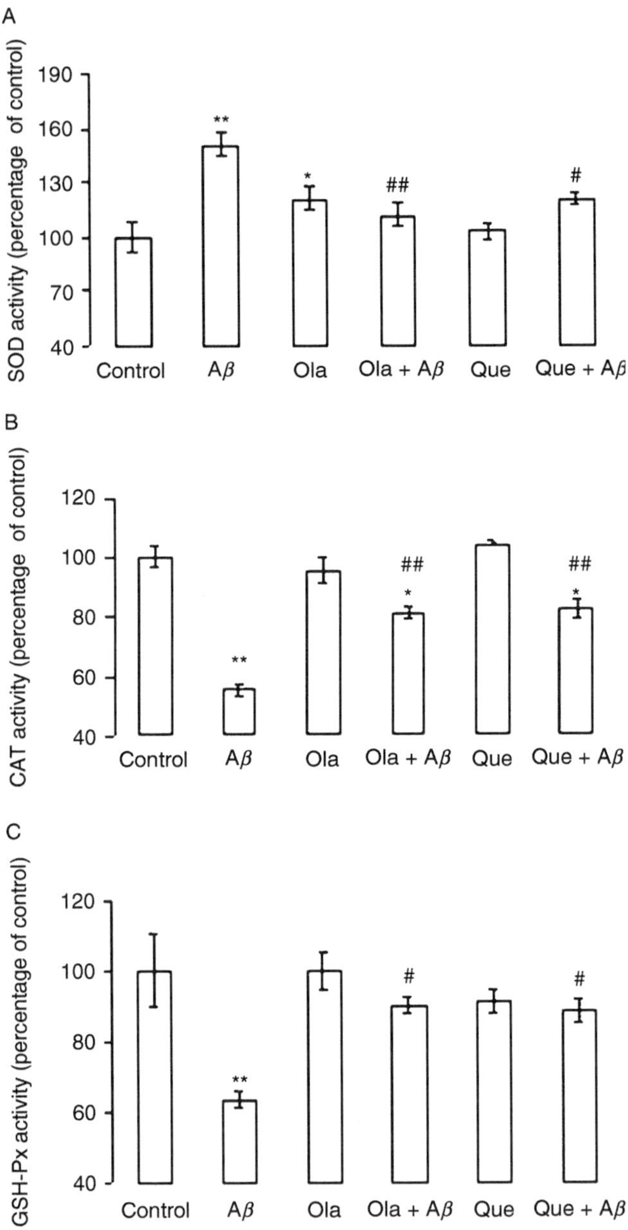

FIG. 5. Olanzapine and quetiapine prevented the $A\beta_{25-35}$-induced changes in antioxidant enzymes SOD (A), CAT (B), and GSH-Px (C). Cells were pretreated with vehicle (DMEM) or 50 μM of olanzapine or quetiapine for 24 hours, followed by exposure to 0 or 20 μM of $A\beta_{25-35}$ in the absence

atypical antipsychotics, suggesting the involvement of the receptors in these actions (Cussac *et al.*, 2002). Indeed, all of these atypical antipsychotic drugs have relative high affinities for the 5-HT_{1A} receptors, as discussed before.

B. Protective Effects of Antipsychotic Drugs in Animals

Chlorpromazine is the first neuroleptic drug introduced into clinical practice that has revolutionized both the theory and the therapy of schizophrenia, and has been known for its cytoprotective effects on cell injury. Results from earlier studies show that this drug effectively prevented cell necrosis in experimentally induced ischemic liver and heart disease (Chien *et al.*, 1979; Farber *et al.*, 1978; Saville *et al.*, 1988), markedly lessened the degree of structural and functional impairment seen in mercuric chloride or *N*-(3,5-dichlorophenyl) succinimide (NDPS) (an agricultural fungicide)-induced acute renal failure in rats, and increased the rate of

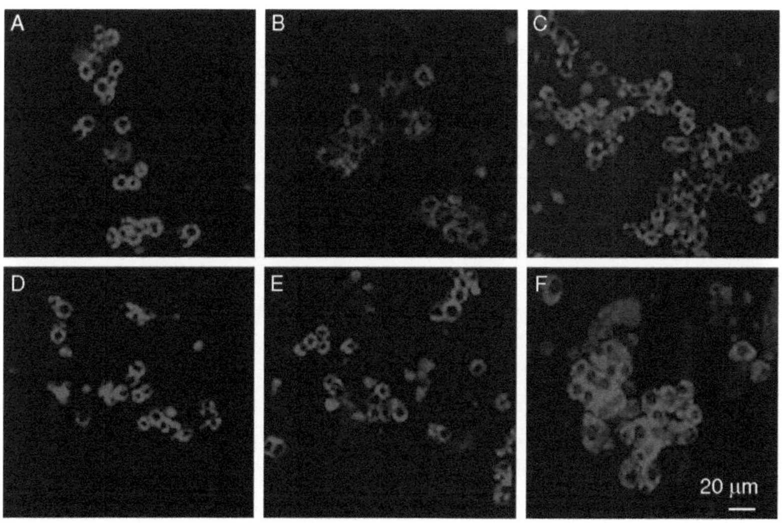

FIG. 6. Olanzapine and quetiapine prevented the $A\beta_{25-35}$-induced decrease in mitochondrial membrane potential (DCm). Cells were pretreated with vehicle (DMEM), 50 μM of olanzapine or quetiapine for 24 hours, followed by exposure to 0 or 20 μM of $A\beta_{25-35}$ in the absence or presence of 50 μM of olanzapine or quetiapine for another 12 hours. Photographs a–f are from control, $A\beta_{25-35}$, olanzapine, olanzapine + $A\beta_{25-35}$, quetiapine, and quetiapine + $A\beta_{25-35}$, respectively. (See Color Insert.)

---

or presence of 50 μM of olanzapine or quetiapine for another 16 hours. Data are presented as means ± SEM. *$p < 0.05$, **$p < 0.01$ compared with control. $^{\#}p < 0.05$, $^{\#\#}p < 0.01$ compared with cultures exposed to $A\beta_{25-35}$ alone.

recovery (Dobyan and Bulger, 1984; Rankin *et al.*, 1991). The protective effects of chlorpromazine have been attributed to its ability to antagonize Ca^{2+} influx into cytoplasm and/or intracellular Ca^{2+}-dependent processes (Chien *et al.*, 1979; Rankin *et al.*, 1991; Saville *et al.*, 1988). This explanation is consistent with the understanding of the cytoprotective effects of chlorpromazine and trifluoperazine in cultured cells that suffered from oxidative stress, as discussed above.

The cytoprotective effects of chlorpromazine in peripheral tissues seemed to be nonsignificant, as patients with schizophrenia have no neuronal necrosis in their brains. However, recent advances in schizophrenia research suggest that the pathophysiology of this disorder involves abnormal neuronal formation, apoptosis, oxidative damage, and neurodegeneration, leading to dysfunctional neural synaptic activity, which results in cognitive impairment (Dakhale *et al.*, 2004; Jann, 2004; Yao *et al.*, 2001). In fact, cognitive impairment, including deficits in attention and memory function, has been considered to be one of the core features of schizophrenia (Weinberger and Gallhofer, 1997). Numerous clinical and multicenter trials using open-label conditions and real-world situations have demonstrated consistent improvement in various cognitive symptoms with atypical antipsychotics (reviewed by Jann, 2004). In light of the above, one may speculate that atypical antipsychotics restore or help recover pathological/pathophysiological changes in the brain, and thus exert beneficial effects on cognitive functions of patients with schizophrenia. In the following paragraphs, we will describe some of the recent studies relevant to this proposal.

It has been found that phencyclidine [PCP, a noncompetitive antagonist of the *N*-methyl-*D*-aspartate (NMDA) subtype of glutamate receptor] and various other NMDA receptor antagonists including ketamine and MK-801 (dizocilpine), which trigger psychotic reactions in humans, induce cytopathological changes in cerebrocortical neurons in the adult rat brain (Corso *et al.*, 1997; Ellison, 1994; Ellison and Switzer, 1993; Fix *et al.*, 1993; Olney *et al.*, 1989). Recent studies have shown that atypical antipsychotic drugs produced protective effects on the NMDA receptor antagonist-induced changes in animals. For example, the dizocilpine-induced delayed alteration impairment in rats was prevented by pretreatment with clozapine (Hauber, 1993). Also clozapine protected against NMDA-antagonist neurotoxicity reaction in rats (Farber *et al.*, 1993; Hashimoto *et al.*, 2000). Although traditional antipsychotics (e.g., loxapine, haloperidol, and thioridazine) showed similar effects, their neuroprotective action was not attributed to the drugs' binding affinity to dopamine receptors, as sulpiride, an agent that binds to dopamine receptors only, did not significantly reduce MK-801 neurotoxicity at doses up to 30 mg/kg. However, DTG [di(2-tolyl)guanidine] and rimcazole, which bind to σ receptors without appreciable binding to dopamine receptors (Largent *et al.*, 1988; Weber *et al.*, 1986), both prevented MK-801-induced cerebrocortical neurotoxicity (Farber *et al.*, 1993).

Similar to clozapine, other atypical antipsychotic drugs, such as olanzapine, quetiapine, risperidone, and ziprasidone, also prevented PCP or MK-801 neurotoxicity in rats (Dickerson and Sharp, 2006; Farber *et al.*, 1996; Fumagalli *et al.*, 2004; He *et al.*, 2006). These drugs not only have affinity to D_2 receptors, but also have effects on other receptors, including other dopamine (D_1, D_3, D_4), serotonin (5-HT_{2A}, 5-HT_{1A}, and 5-$HT_{3,6}$ receptors), and alpha-1 adrenergic receptors (Miyamoto *et al.*, 2005; Richelson, 1999; Tatsumi *et al.*, 1999). Since dopamine receptors are unlikely to be involved in neuroprotection, other receptors are worthy of consideration. As indicated by the following findings: (1) 5-$HT_{2A/2C}$ agonists (±)-1-(2,5-dimethoxy-4-bromophenyl)-2-aminopropane (DOB), (±)-1-(2,5-dimethoxy-4-methylphenyl)-2-aminopropane (DOM), (±)-1-(2,5-dimethoxy-4-iodophenyl)-2-aminopropane (DOI), and d-lysergic acid diethylamide (LSD) all dose-dependently prevented MK-801 neurotoxicity; (2) lisuride, a selective 5-HT_{2A} agonist, potently protected against MK-801's neurotoxic action; and (3) SDZ SER-802, a selective 5-HT_{2C} antagonist was ineffective in reversing DOI's blocking action (Farber *et al.*, 1998), 5-HT_{2A} receptors are likely involved in the neuroprotective actions of atypical antipsychotics.

In addition, atypical antipsychotics also showed neuroprotective actions against other neurotoxic effects induced by amphetamine, kainic acid, okadaic acid, NMDA, or surgical lesions. Risperidone, for example, was shown to reverse memory deficits in kainic acid-lesioned rats (Csernansky *et al.*, 2001). Similarly, quetiapine was effective in ameliorating deficits in contextual fear conditioning and cued fear conditioning in kainic acid-treated animals (Martin *et al.*, 2005). In a recent study, we found that chronic pretreatment of rats with olanzapine effectively attenuated methamphetamine-induced neurotoxicity that includes the methamphetamine-induced mortality, hyperthermia, and decreases in TH and Bcl-2 immunostaining in the caudate putamen (He *et al.*, 2004). In another study, chronic administration of quetiapine significantly attenuated the *dl*-amphetamine-induced hyperthermia and the anxiety-like behavioral changes in the light/dark box and in the open field tests (He *et al.*, 2005a). Acute injections of clozapine to pretrained rats improved memory in animals with fimbria-fornix lesions (Addy *et al.*, 2005). In the absence of pretraining, chronic administration of clozapine improved performance in the delayed spatial alternation task in rats with direct damage (by infusion of NMDA) to the hippocampus (Bardgett *et al.*, 2006). The above data suggest that these atypical antipsychotic drugs have primary cognition-enhancing effects in animals with neuropathy in the brain.

The cognition-enhancing effects of atypical antipsychotics in animals with neuropathy in the brain suggest that these drugs may exert direct protection on brain lesions. Evidence supporting this notion has been emerging. For instance, at a pharmacological dose, clozapine and olanzapine blocked the neuropathological changes (neuronal vacuolization) and the expression of Fos-like protein and *c-fos* mRNA induced in the rat retrosplenial cortex by administration of dizocilpine (Fujimura *et al.*, 2000). In a follow-up study, this protective effect on the dizocilpine-induced neuronal vacuolization was confirmed and extended to zotepine, another atypical antipsychotic (Okamura *et al.*, 2003). In a recent study, we showed that olanzapine significantly attenuated the spatial memory impairment (Fig. 7) and hippocampal cell death induced by okadaic acid (He *et al.*, 2005b; Fig. 8). Moreover, clozapine, ziprasidone, and aripiprazole decreased kainic acid-induced striatal lesion volumes in mice (Cosi *et al.*, 2005).

The aforementioned protective and therapeutic effects of atypical antipsychotic drugs in animals with neuropathy in the brain and behavioral changes are of significance to patients with schizophrenia, since neuropathy has been observed in brains of schizophrenic patients (Antonova *et al.*, 2004; Arnold, 2000; Csernansky *et al.*, 2002; Knable *et al.*, 2004). Explicitly, these protective effects may contribute to therapeutic effects of these drugs on negative and cognitive symptoms in patients with schizophrenia.

C. Effects of Atypical Antipsychotics on the Activity of HPA Axis and the Stress-Induced Changes in the Hippocampus

Many psychiatric illnesses are known to be associated with cortisol dysregulation (Corcoran *et al.*, 2001). There has been evidence linking the dysfunction of the hypothalamic–pituitary–adrenocortical (HPA) axis to schizophrenia (Kaneko *et al.*, 1992), indicated by higher basal cortisol levels and/or the cortisol/dehydroepiandrosterone ratio in patients with schizophrenia than in normal controls (Lammers *et al.*, 1995; Ritsner *et al.*, 2004, 2005; Ryan *et al.*, 2004; Taherianfard and Shariaty, 2004). Moreover, atypical antipsychotics have been shown to be effective in treating corticosteroid-induced psychosis (DeSilva *et al.*, 2002; Siddiqui *et al.*, 2005), reduce cortisol excretion in healthy subjects (Cohrs *et al.*, 2004; Meier *et al.*, 2005), and reverse increased cortisol levels in patients with schizophrenia while improving patients' symptoms (Breier *et al.*, 1993; Mann *et al.*, 2006; Wik, 1995; Zhang *et al.*, 2005).

High levels of cortisol and stress have been demonstrated to be harmful to the brain, especially to the hippocampus, in animal and human studies. Patients with schizophrenia showed smaller hippocampi than control normal subjects (Wright *et al.*, 2000). Postmortem studies also demonstrated abnormal cytoarchitecture, lower cell counts, disorientation of pyramidal cells, and smaller neurons in the hippocampi of patients with schizophrenia (Jonsson *et al.*, 1999; Weinberger, 1999). Likewise, functional imaging showed reduced hippocampal activation in schizophrenic patients during verbal episodic memory retrieval and odor discrimination (Heckers *et al.*, 1998; Malaspina *et al.*, 1998).

A plausible inference from current knowledge, as described above, is that atypical antipsychotics may protect the hippocampus against stress insults. In support of this notion, we found that chronic administration of quetiapine prevented the decrease in

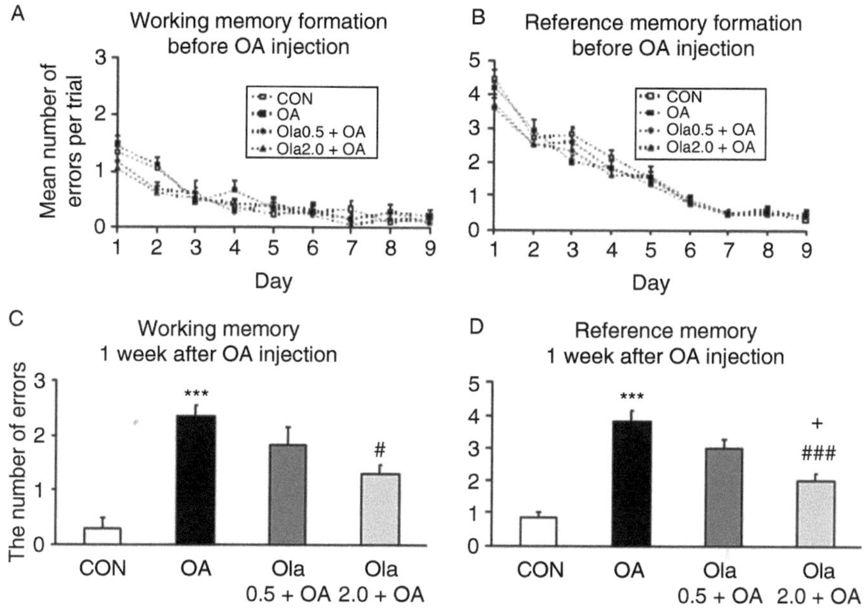

FIG. 7. Olanzapine did not affect the spatial working (A) and reference (B) memory formation before okadaic acid (OA) or saline microinjection in the radial arm maze task in rats. However, olanzapine significantly attenuated the OA-induced impairment in animals' working (C) and reference (D) memory, which was measured by the radial arm maze task 1 week after the microinjection of OA or saline into the right hippocampus. Results are expressed as means ± SEM ($n = 6$–7 in each group). $^{***}p < 0.001$ vs control (CON), $^{\#}p < 0.05$ and $^{\#\#\#}p < 0.001$ vs OA, and $^{+}p < 0.05$ vs Ola0.5 + OA.

hippocampal levels of BDNF (Xu *et al.*, 2002), and reversed the suppression of hippocampal neurogenesis caused by repeated restraint stress (Luo *et al.*, 2004). In addition, chronic administration of quetiapine prevented the stress-induced decrease in hippocampal levels of heme oxygenase-2 (HO-2; Chen *et al.*, 2005), which is an isoform of HO, and could be down-regulated by prolonged exposure to the adrenal glucocorticoids in rats (Weber *et al.*, 1994). Olanzapine, other atypical antipsychotic, showed similar effects of accelerating the recovery processes of changed BDNF and Bcl-2 levels in hippocampi of stressed rats (Lu *et al.*, 2004). Further studies are ongoing in our laboratory to extend the current results and explore possible mechanisms responsible for these neuroprotective effects.

D. Combinations of Atypical Antipsychotics and Antidepressants

In clinical practice, the atypical antipsychotics clozapine, olanzapine, quetiapine, and risperidone were added to antidepressant treatment in cases of depression that had not responded to antidepressant treatment alone. For example, in

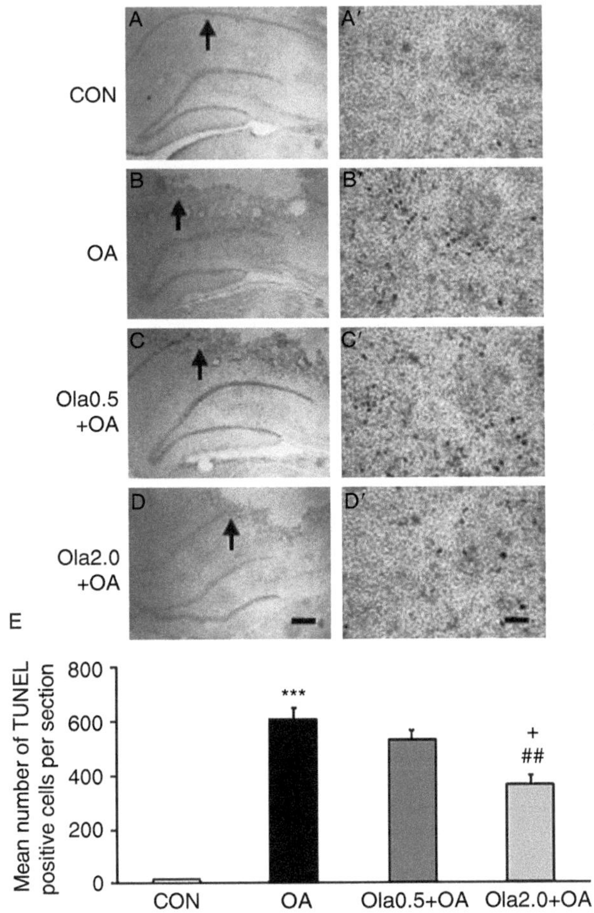

FIG. 8. (A–D, A′–D′) Representative photomicrographs of TUNEL staining in the injected hippocampus of the rats in the CON (A, A′), OA (B, B′), Ola0.5 + OA (C, C′), and Ola2.0 + OA (D, D′) groups. The high magnification of right photomicrographs (A–D′) were picked up from A–D, respectively. The arrows on the low-magnification panels indicate the location of the high-magnification images. In the hippocampi of OA-injected groups (B, B′; C, C′; D, D′), the TUNEL-positive cells are visible in different frequency, whereas there are almost no TUNEL-positive cells in the hippocampus of the control rat (A, A′). The scale bar represents 300 μm in A–D, and represents 30 mm in A′–D′. (E) Quantitative analysis of the effect of olanzapine on the OA-induced increase of TUNEL-positive cells in the injected hippocampus. The number of TUNEL-positive cells in the hippocampus was counted at 400× magnification. Results are expressed as means ± SEM ($n = 6$–7 in each group). $^{***}p < 0.001$ vs CON, $^{\#\#}p < 0.01$ vs OA, and $^{+}p < 0.05$ vs Ola0.5 + OA.

the case of a 48-year-old woman with recurrent depressive disorder, resistant to antidepressants and to electroconvulsive therapy, a marked improvement was achieved when a combination treatment of clozapine and maprotiline (an antidepressant) was commenced (Hrdlicka, 2002). There are many more clinical reports on the addition of olanzapine for treatment-resistant depression (Pitchot and Ansseau, 2001), psychotic major depressive disorder (Barbee *et al.*, 2004; Matthews *et al.*, 2002), and posttraumatic stress disorder (Stein *et al.*, 2002). Similar reports have also been published on quetiapine (Barbee *et al.*, 2004; Calabrese *et al.*, 2005) and risperidone (Barbee *et al.*, 2004; Ghaemi *et al.*, 2004; Hirose and Ashby, 2002; Tani *et al.*, 2004). In fact, accumulating evidence suggests that these atypical antipsychotics produce antidepressant effects (Ghaemi *et al.*, 2000; Ishak *et al.*, 2004; Klein *et al.*, 2004; McIntyre and Katzman, 2003; McIntyre *et al.*, 2004; Sajatovic *et al.*, 2002). Atypical antipsychotic drugs have also been used in other mental disorders such as Alzheimer's disease (AD; De Deyn *et al.*, 2004; Fujikawa *et al.*, 2004; Negron and Reichman, 2000) and treatment-refractory obsessive-compulsive disorder (OCD; Adityanjee and Schulz, 2002; Bogan *et al.*, 2005; Carey *et al.*, 2005; Denys *et al.*, 2002, 2004).

Although determining the precise mechanism of the synergism of the combination of atypical antipsychotics with antidepressants is a daunting challenge, the clinical efficacy of atypical antipsychotic drugs in schizophrenia, depression, and mood disorders, all of which have their own unique pathogenesis, suggests that these drugs may share some common targets at the molecular levels. There is increasing evidence suggesting that mood stabilizers and antidepressants have neuroprotective properties (Bachmann *et al.*, 2005; Cavus and Duman, 2003; Chuang, 2004; Duman, 1998; Duman *et al.*, 1997; Haynes *et al.*, 2004). Recent studies, as described in this chapter, also suggest that antipsychotic drugs may have neuroprotective capacity. Therefore, it may be postulated that a combination of atypical antipsychotics and antidepressants enhance their neuroprotective effects. In support of this reasoning, Maragnoli *et al.* (2004) reported that a combination of fluoxetine and olanzapine produced higher levels of FGF-2, a neurotrophic molecule that is highly expressed in the mature brain and is protective for a wide range of neurons (Dono, 2003), in prefrontal cortex than either drug alone. In a more recent study, we found that the combination of quetiapine and venlafaxine produced a synergistic effect in preventing the stress-induced decrease in hippocampal levels of HO-2 (Chen *et al.*, 2005). The same combination synergistically prevented the stress-induced decrease in hippocampal levels of BDNF and neurogenesis (Xu *et al.*, 2006). Hopefully, these results will promote further research, since neuropathies in the prefrontal cortex and hippocampus are common features of depression, schizophrenia, and AD, and in turn, these disorders show similar negative and cognitive behavioral changes (Keshavan *et al.*, 2005; Krystal *et al.*, 2001; McIlroy and Craig, 2004; Miyamoto *et al.*, 2003; Shinosaki *et al.*, 2000; Vaidya and Duman, 2001; Zubenko, 1997).

Acknowledgments

We are grateful to Canadian Institutes of Health Research, Canadian Psychiatry Research Foundation, Saskatchewan Health Research Foundation, Schizophrenia Society of Saskatchewan, and Royal University Hospital Foundation for their grants support. The authors would like to give thanks to Drs. Augusto V. Juorio, Sergey Fedoroff, and Steve Richardson for their helpful comments during the preparation of this chapter, and to Gabriel Stegeman and Yvonne Wilkinson for their help in editing the chapter.

References

Addy, N. A., Pocivavsek, A., and Levin, E. D. (2005). Reversal of clozapine effects on working memory in rats with fimbria-fornix lesions. *Neuropsychopharmacology* **30,** 1121–1127.

Adityanjee, and Schulz, S. C. (2002). Clinical use of quetiapine in disease states other than schizophrenia. *J. Clin. Psychiatry* **63**(Suppl. 13), 32–38.

Alberch, J., Brito, B., Notario, V., and Castro, R. (1991). Prenatal haloperidol treatment decreases nerve growth factor receptor and mRNA in neonate rat forebrain. *Neurosci. Lett.* **131,** 228–232.

Alleva, E., Della Seta, D., Cirulli, F., and Aloe, L. (1996). Haloperidol treatment decreases nerve growth factor levels in the hypothalamus of adult mice. *Prog. Neuropsychopharmacol. Biol. Psychiatry* **20,** 483–489.

An, J. J., Cho, S. R., Jeong, D. W., Park, K. W., Ahn, Y. S., and Baik, J. H. (2003). Anti-proliferative effects and cell death mediated by two isoforms of dopamine D2 receptors in pituitary tumor cells. *Mol. Cell. Endocrinol.* **206,** 49–62.

Andreasen, N. C. (1995). Symptoms, signs, and diagnosis of schizophrenia. *Lancet* **346,** 477–481.

Andreasen, N. C., Nopoulos, P., Schultz, S., Miller, D., Gupta, S., Swayze, V., and Flaum, M. (1994). Positive and negative symptoms of schizophrenia: Past, present, and future. *Acta Psychiatr. Scand.* **384**(Suppl.), 51–59.

Angelucci, F., Aloe, L., Gruber, S. H., Fiore, M., and Mathe, A. A. (2000a). Chronic antipsychotic treatment selectively alters nerve growth factor and neuropeptide Y immunoreactivity and the distribution of choline acetyl transferase in rat brain regions. *Int. J. Neuropsychopharmacol.* **3,** 13–25.

Angelucci, F., Mathe, A. A., and Aloe, L. (2000b). Brain-derived neurotrophic factor and tyrosine kinase receptor TrkB in rat brain are significantly altered after haloperidol and risperidone administration. *J. Neurosci. Res.* **60,** 783–794.

Angelucci, F., Aloe, L., Iannitelli, A., Gruber, S. H., and Mathe, A. A. (2005). Effect of chronic olanzapine treatment on nerve growth factor and brain-derived neurotrophic factor in the rat brain. *Eur. Neuropsychopharmacol.* **15,** 311–317.

Antonova, E., Sharma, T., Morris, R., and Kumari, V. (2004). The relationship between brain structure and neurocognition in schizophrenia: A selective review. *Schizophr. Res.* **70,** 117–145.

Arita, J., Hashi, A., Hoshi, K., Mazawa, S., and Suzuki, S. (1998). D2 dopamine-receptor-mediated inhibition of proliferation of rat lactotropes in culture is accompanied by changes in cell shape. *Neuroendocrinology* **68,** 163–171.

Arnold, S. E. (2000). Cellular and molecular neuropathology of the parahippocampal region in schizophrenia. *Ann. NY Acad. Sci.* **911,** 275–292.

Bachmann, R. F., Schloesser, R. J., Gould, T. D., and Manji, H. K. (2005). Mood stabilizers target cellular plasticity and resilience cascades: Implications for the development of novel therapeutics. *Mol. Neurobiol.* **32,** 173–202.

Backhouse, B., Barochovsky, O., Malik, C., Patel, A. J., and Lewis, P. D. (1982). Effects of haloperidol on cell proliferation in the early postnatal rat brain. *Neuropathol. Appl. Neurobiol.* **8,** 109–116.

Bai, O., Xu, H., and Li, X. M. (2006). Haloperidol and atypical antipsychotics share a same action of decreasing p75NTR mRNA levels in PC12 cells. *Life Sci.* **79,** 570–574.

Bai, O., Wei, Z., Lu, W., Bowen, R., Keegan, D., and Li, X. M. (2002). Protective effects of atypical antipsychotic drugs on PC12 cells after serum withdrawal. *J. Neurosci. Res.* **69,** 278–283.

Bai, O., Chlan-Fourney, J., Bowen, R., Keegan, D., and Li, X. M. (2003). Expression of brain-derived neurotrophic factor mRNA in rat hippocampus after treatment with antipsychotic drugs. *J. Neurosci. Res.* **71,** 127–131.

Banasr, M., Hery, M., Printemps, R., and Daszuta, A. (2004). Serotonin-induced increases in adult cell proliferation and neurogenesis are mediated through different and common 5-HT receptor subtypes in the dentate gyrus and the subventricular zone. *Neuropsychopharmacology* **29,** 450–460.

Barbee, J. G., Conrad, E. J., and Jamhour, N. J. (2004). The effectiveness of olanzapine, risperidone, quetiapine, and ziprasidone as augmentation agents in treatment-resistant major depressive disorder. *J. Clin. Psychiatry* **65,** 975–981.

Bardgett, M. E., Griffith, M. S., Foltz, R. F., Hopkins, J. A., Massie, C. M., and O'Connell, S. M. (2006). The effects of clozapine on delayed spatial alternation deficits in rats with hippocampal damage. *Neurobiol. Learn. Mem.* **85,** 86–94.

Barochovsky, O., and Patel, A. J. (1982). Effect of central nervous system acting drugs on brain cell replication *in vitro*. *Neurochem. Res.* **7,** 1059–1074.

Bersani, G., Iannitelli, A., Fiore, M., Angelucci, F., and Aloe, L. (2000). Data and hypotheses on the role of nerve growth factor and other neurotrophins in psychiatric disorders. *Med. Hypotheses* **55,** 199–207.

Bersani, G., Iannitelli, A., Massoni, E., Garavini, A., Grilli, A., Di Giannantonio, M., Conti, C. M., and Pancheri, P. (2004). Ultradian variation of nerve growth factor plasma levels in healthy and schizophrenic subjects. *Int. J. Immunopathol. Pharmacol.* **17,** 367–372.

Bogan, A. M., Koran, L. M., Chuong, H. W., Vapnik, T., and Bystritsky, A. (2005). Quetiapine augmentation in obsessive-compulsive disorder resistant to serotonin reuptake inhibitors: An open-label study. *J. Clin. Psychiatry* **66,** 73–79.

Breier, A. (1999). Cognitive deficit in schizophrenia and its neurochemical basis. *Br. J. Psychiatry* **174**(Suppl. 37), 16–18.

Breier, A., Kirkpatrick, B., and Buchanan, R. W. (1993). Clozapine attenuates meta-chlorophenyl-piperazine (mCPP)-induced plasma cortisol increases in schizophrenia. *Biol. Psychiatry* **34,** 492–494.

Calabrese, J. R., Keck, P. E., Jr., Macfadden, W., Minkwitz, M., Ketter, T. A., Weisler, R. H., Cutler, A. J., McCoy, R., Wilson, E., and Mullen, J. (2005). A randomized, double-blind, placebo-controlled trial of quetiapine in the treatment of bipolar I or II depression. *Am. J. Psychiatry* **162,** 1351–1360.

Cameron, H. A., and McKay, R. D. (2001). Adult neurogenesis produces a large pool of new granule cells in the dentate gyrus. *J. Comp. Neurol.* **435,** 406–417.

Carey, P. D., Vythilingum, B., Seedat, S., Muller, J. E., van Ameringen, M., and Stein, D. J. (2005). Quetiapine augmentation of SRIs in treatment refractory obsessive-compulsive disorder: A double-blind, randomised, placebo-controlled study [ISRCTN83050762]. *BMC Psychiatry* **5,** 5.

Cavus, I., and Duman, R. S. (2003). Influence of estradiol, stress, and 5-HT2A agonist treatment on brain-derived neurotrophic factor expression in female rats. *Biol. Psychiatry* **54,** 59–69.

Chaturvedi, R. K., Shukla, S., Seth, K., and Agrawal, A. K. (2006). Nerve growth factor increases survival of dopaminergic graft, rescue nigral dopaminergic neurons and restores functional deficits in rat model of Parkinson's disease. *Neurosci. Lett.* **398,** 44–49.

Chen, Z., Xu, H., Haimanot, S., Li, X., and Li, X. M. (2005). Quetiapine and venlafaxine synergically regulate heme oxygenase-2 protein expression in hippocampal formations of stressed rats. *Neurosci. Lett.* **389,** 173–177.

Chien, K. R., Peau, R. G., and Farber, J. L. (1979). Ischemic myocardial cell injury. Prevention by chlorpromazine of an accelerated phospholipid degradation and associated membrane dysfunction. *Am. J. Pathol.* **97,** 505–529.

Chlan-Fourney, J., Ashe, P., Nylen, K., Juorio, A. V., and Li, X. M. (2002). Differential regulation of hippocampal BDNF mRNA by typical and atypical antipsychotic administration. *Brain Res.* **954,** 11–20.

Chou, Y. H., Halldin, C., and Farde, L. (2003). Occupancy of 5-HT1A receptors by clozapine in the primate brain: A PET study. *Psychopharmacology (Berl.)* **166,** 234–240.

Chuang, D. M. (2004). Neuroprotective and neurotrophic actions of the mood stabilizer lithium: Can it be used to treat neurodegenerative diseases? *Crit. Rev. Neurobiol.* **16,** 83–90.

Chung, Y. C., Li, Z., Dai, J., Meltzer, H. Y., and Ichikawa, J. (2004). Clozapine increases both acetylcholine and dopamine release in rat ventral hippocampus: Role of 5-HT1A receptor agonism. *Brain Res.* **1023,** 54–63.

Ciani, E., Virgili, M., and Contestabile, A. (2002). Akt pathway mediates a cGMP-dependent survival role of nitric oxide in cerebellar granule neurones. *J. Neurochem.* **81,** 218–228.

Cohrs, S., Pohlmann, K., Guan, Z., Jordan, W., Meier, A., Huether, G., Ruther, E., and Rodenbeck, A. (2004). Quetiapine reduces nocturnal urinary cortisol excretion in healthy subjects. *Psychopharmacology (Berl.)* **174,** 414–420.

Condorelli, D. F., Dell'Albani, P., Mudo, G., Timmusk, T., and Belluardo, N. (1994). Expression of neurotrophins and their receptors in primary astroglial cultures: Induction by cyclic AMP-elevating agents. *J. Neurochem.* **63,** 509–516.

Connor, B., and Dragunow, M. (1998). The role of neuronal growth factors in neurodegenerative disorders of the human brain. *Brain Res. Brain Res. Rev.* **27,** 1–39.

Corcoran, C., Gallitano, A., Leitman, D., and Malaspina, D. (2001). The neurobiology of the stress cascade and its potential relevance for schizophrenia. *J. Psychiatr. Pract.* **7,** 3–14.

Corso, T. D., Sesma, M. A., Tenkova, T. I., Der, T. C., Wozniak, D. F., Farber, N. B., and Olney, J. W. (1997). Multifocal brain damage induced by phencyclidine is augmented by pilocarpine. *Brain Res.* **752,** 1–14.

Cosi, C., Waget, A., Rollet, K., Tesori, V., and Newman-Tancredi, A. (2005). Clozapine, ziprasidone and aripiprazole but not haloperidol protect against kainic acid-induced lesion of the striatum in mice, *in vivo*: Role of 5-HT1A receptor activation. *Brain Res.* **1043,** 32–41.

Csernansky, J. G., Humphrey, W. M., Jefferson, A., and Bardgett, M. E. (2001). Risperidone restores memory deficits in rats with excitotoxin-induced hippocampal neuropathy. Hawaii Poster presented at the 40th Meeting of the ACNP, Waikoloa, Hawaii. December 9–13, 2001.

Csernansky, J. G., Mahmoud, R., Brenner, R., and Risperidone-USA-79 Study Group (2002). A comparison of risperidone and haloperidol for the prevention of relapse in patients with schizophrenia. *N. Engl. J. Med.* **346,** 16–22.

Cussac, D., Duqueyroix, D., Newman-Tancredi, A., and Millan, M. J. (2002). Stimulation by antipsychotic agents of mitogen-activated protein kinase (MAPK) coupled to cloned, human (h) serotonin (5-HT)(1A) receptors. *Psychopharmacology (Berl.)* **162,** 168–177.

Dakhale, G., Khanzode, S., Khanzode, S., Saoji, A., Khobragade, L., and Turankar, A. (2004). Oxidative damage and schizophrenia: The potential benefit by atypical antipsychotics. *Neuropsychobiology* **49,** 205–209.

Dalla Libera, A., Scutari, G., Boscolo, R., Rigobello, M. P., and Bindoli, A. (1998). Antioxidant properties of clozapine and related neuroleptics. *Free Radic. Res.* **29,** 151–157.

Dawirs, R. R., Hildebrandt, K., and Teuchert-Noodt, G. (1998). Adult treatment with haloperidol increases dentate granule cell proliferation in the gerbil hippocampus. *J. Neural. Transm.* **105,** 317–327.

Dawson, N. M., Hamid, E. H., Egan, M. F., and Meredith, G. E. (2001). Changes in the pattern of brain-derived neurotrophic factor immunoreactivity in the rat brain after acute and subchronic haloperidol treatment. *Synapse* **39,** 70–81.

De Deyn, P. P., Carrasco, M. M., Deberdt, W., Jeandel, C., Hay, D. P., Feldman, P. D., Young, C. A., Lehman, D. L., and Breier, A. (2004). Olanzapine versus placebo in the treatment of psychosis with or without associated behavioral disturbances in patients with Alzheimer's disease. *Int. J. Geriatr. Psychiatry* **19,** 115–126.

Denys, D., van Megen, H., and Westenberg, H. (2002). Quetiapine addition to serotonin reuptake inhibitor treatment in patients with treatment-refractory obsessive-compulsive disorder: An open-label study. *J. Clin. Psychiatry* **63,** 700–703.

Denys, D., de Geus, F., van Megen, H. J., and Westenberg, H. G. (2004). A double-blind, randomized, placebo-controlled trial of quetiapine addition in patients with obsessive-compulsive disorder refractory to serotonin reuptake inhibitors. *J. Clin. Psychiatry* **65,** 1040–1048.

DeSilva, C. C., Nurse, M. C., and Vokey, K. (2002). Steroid-induced psychosis treated with risperidone. *Can. J. Psychiatry* **47,** 388–389.

Diaz-Mataix, L., Artigas, F., and Celada, P. (2006). Activation of pyramidal cells in rat medial prefrontal cortex projecting to ventral tegmental area by a 5-HT(1A) receptor agonist. *Eur. Neuropsychopharmacol.* **16,** 288–296.

Dickerson, J., and Sharp, F. R. (2006). Atypical antipsychotics and a Src kinase inhibitor (PP1) prevent cortical injury produced by the psychomimetic, noncompetitive NMDA receptor antagonist MK-801. *Neuropsychopharmacol.* **31,** 1420–1430.

Dobyan, D. C., and Bulger, R. E. (1984). Partial protection by chlorpromazine in mercuric chloride-induced acute renal failure in rats. *Lab. Invest.* **50,** 578–586.

Dono, R. (2003). Fibroblast growth factors as regulators of central nervous system development and function. *Am. J. Physiol. Regul. Integr. Comp. Physiol.* **284,** R867–R881.

Dudek, H., Datta, S. R., Franke, T. F., Birnbaum, M. J., Yao, R., Cooper, G. M., Segal, R. A., Kaplan, D. R., and Greenberg, M. E. (1997). Regulation of neuronal survival by the serine-threonine protein kinase Akt. *Science.* **275,** 661–665.

Duman, R. S. (1998). Novel therapeutic approaches beyond the serotonin receptor. *Biol. Psychiatry* **44,** 324–335.

Duman, R. S., Heninger, G. R., and Nestler, E. J. (1997). A molecular and cellular theory of depression. *Arch. Gen. Psychiatry* **54,** 597–606.

Durany, N., Michel, T., Zochling, R., Boissl, K. W., Cruz-Sanchez, F. F., Riederer, P., and Thome, J. (2001). Brain-derived neurotrophic factor and neurotrophin 3 in schizophrenic psychoses. *Schizophr. Res.* **52,** 79–86.

Ellenbroek, B. A. (1993). Treatment of schizophrenia: A clinical and preclinical evaluation of neuroleptic drugs. *Pharmacol. Ther.* **57,** 1–78.

Ellison, G. (1994). Stimulant-induced psychosis, the dopamine theory of schizophrenia, and the habenula. *Brain Res. Brain Res. Rev.* **19,** 223–239.

Ellison, G., and Switzer, R. C., III (1993). Dissimilar patterns of degeneration in brain following four different addictive stimulants. *Neuroreport* **5,** 17–20.

Ernfors, P., Ibanez, C. F., Ebendal, T., Olson, L., and Persson, H. (1990). Molecular cloning and neurotrophic activities of a protein with structural similarities to nerve growth factor: Developmental and topographical expression in the brain. *Proc. Natl. Acad. Sci. USA* **87,** 5454–5458.

Farber, J. L., Martin, J. T., and Chien, K. R. (1978). Irreversible ischemic cell injury. Prevention by chlorpromazine of the aggregation of the intramembranous particles of rat liver plasma membranes. *Am. J. Pathol.* **92,** 713–732.

Farber, N. B., Price, M. T., Labruyere, J., Nemnich, J., St. Peter, H., Wozniak, D. F., and Olney, J. W. (1993). Antipsychotic drugs block phencyclidine receptor-mediated neurotoxicity. *Biol. Psychiatry* **34,** 119–121.

Farber, N. B., Foster, J., Duhan, N. L., and Olney, J. W. (1996). Olanzapine and fluperlapine mimic clozapine in preventing MK-801 neurotoxicity. *Schizophr. Res.* **21,** 33–37.

Farber, N. B., Hanslick, J., Kirby, C., McWilliams, L., and Olney, J. W. (1998). Serotonergic agents that activate 5HT2A receptors prevent NMDA antagonist neurotoxicity. *Neuropsychopharmacology* **18,** 57–62.

Fix, A. S., Horn, J. W., Wightman, K. A., Johnson, C. A., Long, G. G., Storts, R. W., Farber, N., Wozniak, D. F., and Olney, J. W. (1993). Neuronal vacuolization and necrosis induced by the noncompetitive N-methyl-D-aspartate (NMDA) antagonist MK(+)801 (dizocilpine maleate): A light and electron microscopic evaluation of the rat retrosplenial cortex. *Exp. Neurol.* **123,** 204–215.

Fujikawa, T., Takahashi, T., Kinoshita, A., Kajiyama, H., Kurata, A., Yamashita, H., and Yamawaki, S. (2004). Quetiapine treatment for behavioral and psychological symptoms in patients with senile dementia of Alzheimer type. *Neuropsychobiology* **49,** 201–204.

Fujimura, M., Hashimoto, K., and Yamagami, K. (2000). Effects of antipsychotic drugs on neurotoxicity, expression of fos-like protein and c-fos mRNA in the retrosplenial cortex after administration of dizocilpine. *Eur. J. Pharmacol.* **398,** 1–10.

Fumagalli, F., Molteni, R., Bedogni, F., Gennarelli, M., Perez, J., Racagni, G., and Riva, M. A. (2004). Quetiapine regulates FGF-2 and BDNF expression in the hippocampus of animals treated with MK-801. *Neuroreport* **15,** 2109–2112.

Galili-Mosberg, R., Gil-Ad, I., Weizman, A., Melamed, E., and Offen, D. (2000). Haloperidol-induced neurotoxicity—possible implications for tardive dyskinesia. *J. Neural. Transm.* **107,** 479–490.

Ghaemi, S. N., Cherry, E. L., Katzow, J. A., and Goodwin, F. K. (2000). Does olanzapine have antidepressant properties? A retrospective preliminary study. *Bipolar Disord.* **2,** 196–199.

Ghaemi, S. N., Hsu, D. J., Rosenquist, K. J., Katzow, J. J., and Goodwin, F. K. (2004). Long-term observational comparison of risperidone and olanzapine in bipolar disorder. *Ann. Clin. Psychiatry* **16,** 69–73.

Gould, E., Tanapat, P., Hastings, N. B., and Shors, T. J. (1999). Neurogenesis in adulthood: A possible role in learning. *Trends Cogn. Sci.* **3,** 186–192.

Halim, N. D., Weickert, C. S., McClintock, B. W., Weinberger, D. R., and Lipska, B. K. (2004). Effects of chronic haloperidol and clozapine treatment on neurogenesis in the adult rat hippocampus. *Neuropsychopharmacology* **29,** 1063–1069.

Harris, M. E., Carney, J. M., Cole, P. S., Hensley, K., Howard, B. J., Martin, L., Bummer, P., Wang, Y., Pedigo, N. W., Jr., and Butterfield, D. A. (1995). Beta-Amyloid peptide-derived, oxygen-dependent free radicals inhibit glutamate uptake in cultured astrocytes: Implications for Alzheimer's disease. *Neuroreport* **6,** 1875–1879.

Hashimoto, K., Fujimura, M., and Yamagami, K. (2000). Dizocilpine-induced neuropathological changes in rat retrosplenial cortex are reversed by subsequent clozapine treatment. *Life Sci.* **66,** 1071–1078.

Hashimoto, T., Bergen, S. E., Nguyen, Q. L., Xu, B., Monteggia, L. M., Pierri, J. N., Sun, Z., Sampson, A. R., and Lewis, D. A. (2005). Relationship of brain-derived neurotrophic factor and its receptor TrkB to altered inhibitory prefrontal circuitry in schizophrenia. *J. Neurosci.* **25,** 372–383.

Hauber, W. (1993). Clozapine improves dizocilpine-induced delayed alteration impairment in rats. *J. Neural. Transm. Gen. Sect.* **94,** 223–233.

Haynes, L. E., Barber, D., and Mitchell, I. J. (2004). Chronic antidepressant medication attenuates dexamethasone-induced neuronal death and sublethal neuronal damage in the hippocampus and striatum. *Brain Res.* **1026,** 157–167.

He, J., Xu, H., Yang, Y., Zhang, X., and Li, X. M. (2004). Neuroprotective effects of olanzapine on methamphetamine-induced neurotoxicity are associated with an inhibition of hyperthermia and prevention of Bcl-2 decrease in rats. *Brain Res.* **1018,** 186–192.

He, J., Xu, H., Yang, Y., Zhang, X., and Li, X. M. (2005a). Chronic administration of quetiapine alleviates the anxiety-like behavioural changes induced by neurotoxic regimen of *dl*-amphetamine in rats. *Behav. Brain Res.* **160,** 178–187.

He, J., Yang, Y., Xu, H., Zhang, X., and Li, X. M. (2005b). Olanzapine attenuates the okadaic acid-induced spatial memory impairment and hippocampal cell death in rats. *Neuropsychopharmacology* **30,** 1511–1520.

He, J., Xu, H., Yang, Y., Rajakumar, D., Li, X., and Li, X. M. (2006). The effects of chronic administration of quetiapine on the phencyclidine-induced reference memory impairment and decrease of Bcl-X(L)/Bax ratio in the posterior cingulate cortex in rats. *Behav. Brain Res.* **168,** 236–242.

Heckers, S., Rauch, S. L., Goff, D., Savage, C. R., Schacter, D. L., Fischman, A. J., and Alpert, N. M. (1998). Impaired recruitment of the hippocampus during conscious recollection in schizophrenia. *Nat. Neurosci.* **1,** 318–323.

Hensley, K., Carney, J. M., Mattson, M. P., Aksenova, M., Harris, M., Wu, J. F., Floyd, R. A., and Butterfield, D. A. (1994). A model for beta-amyloid aggregation and neurotoxicity based on free radical generation by the peptide: Relevance to Alzheimer disease. *Proc. Natl. Acad. Sci. USA* **91,** 3270–3274.

Hirose, S., and Ashby, C. R., Jr. (2002). An open pilot study combining risperidone and a selective serotonin reuptake inhibitor as initial antidepressant therapy. *J. Clin. Psychiatry* **63,** 733–736.

Hofer, M., Pagliusi, S. R., Hohn, A., Leibrock, J., and Barde, Y. A. (1990). Regional distribution of brain-derived neurotrophic factor mRNA in the adult mouse brain. *EMBO J.* **9,** 2459–2464.

Hohmann, C. F., and Coyle, J. T. (1988). Long-term effects of basal forebrain lesions on cholinergic, noradrenergic and serotonergic markers in mouse neocortex. *Brain Res. Bull.* **21,** 13–20.

Holson, R. R., Webb, P. J., Grafton, T. F., and Hansen, D. K. (1994). Prenatal neuroleptic exposure and growth stunting in the rat: An *in vivo* and *in vitro* examination of sensitive periods and possible mechanisms. *Teratology* **50,** 125–136.

Hrdlicka, M. (2002). Combination of clozapine and maprotiline in refractory psychotic depression. *Eur. Psychiatry* **17,** 484.

Huang, E. J., and Reichardt, L. F. (2001). Neurotrophins: Roles in neuronal development and function. *Annu. Rev. Neurosci.* **24,** 677–736.

Humpel, C., Wetmore, C., and Olson, L. (1993). Regulation of brain-derived neurotrophic factor messenger RNA and protein at the cellular level in pentylenetetrazol-induced epileptic seizures. *Neuroscience* **53,** 909–918.

Iritani, S., Niizato, K., Nawa, H., Ikeda, K., and Emson, P. C. (2003). Immunohistochemical study of brain-derived neurotrophic factor and its receptor, TrkB, in the hippocampal formation of schizophrenic brains. *Prog. Neuropsychopharmacol. Biol. Psychiatry* **27,** 801–807.

Ishak, W. W., Rapaport, M. H., and Gotto, J. G. (2004). The effectiveness of atypical antipsychotic medications in depressive disorders. *Curr. Psychiatry Rep.* **6,** 422–424.

Jann, M. W. (2004). Implications for atypical antipsychotics in the treatment of schizophrenia: Neurocognition effects and a neuroprotective hypothesis. *Pharmacotherapy* **24,** 1759–1783.

Jonsson, S. A., Luts, A., Guldberg-Kjaer, N., and Ohman, R. (1999). Pyramidal neuron size in the hippocampus of schizophrenics correlates with total cell count and degree of cell disarray. *Eur. Arch. Psychiatry Clin. Neurosci.* **249,** 169–173.

Jurand, A., and Martin, L. V. (1990). Teratogenic potential of two neurotropic drugs, haloperidol and dextromoramide, tested on mouse embryos. *Teratology* **42,** 45–54.

Kaneko, M., Yokoyama, F., Hoshino, Y., Takahagi, K., Murata, S., Watanabe, M., and Kumashiro, H. (1992). Hypothalamic-pituitary-adrenal axis function in chronic schizophrenia: Association with clinical features. *Neuropsychobiology* **25,** 1–7.

Katoh-Semba, R., Takeuchi, I. K., Semba, R., and Kato, K. (1997). Distribution of brain-derived neurotrophic factor in rats and its changes with development in the brain. *J. Neurochem.* **69,** 34–42.

Keck, P. E., Jr., Strakowski, S. M., and McElroy, S. L. (2000). The efficacy of atypical antipsychotics in the treatment of depressive symptoms, hostility, and suicidality in patients with schizophrenia. *J. Clin. Psychiatry* **61**(Suppl. 3), 4–9.

Keshavan, M. S., Berger, G., Zipursky, R. B., Wood, S. J., and Pantelis, C. (2005). Neurobiology of early psychosis. *Br. J. Psychiatry* **187**(Suppl. 48), s8–s18.

Kippin, T. E., Kapur, S., and van der Kooy, D. (2005). Dopamine specifically inhibits forebrain neural stem cell proliferation, suggesting a novel effect of antipsychotic drugs. *J. Neurosci.* **25,** 5815–5823.

Klein, N., Sacher, J., Wallner, H., Tauscher, J., and Kasper, S. (2004). Therapy of treatment resistant depression: Focus on the management of TRD with atypical antipsychotics. *CNS Spectr.* **9,** 823–832.

Knable, M. B., Barci, B. M., Webster, M. J., Meador-Woodruff, J., and Torrey, E. F., Stanley Neuropathology Consortium (2004). Molecular abnormalities of the hippocampus in severe psychiatric illness: Postmortem findings from the Stanley Neuropathology Consortium. *Mol. Psychiatry* **9,** 609–620.

Kodama, M., Fujioka, T., and Duman, R. S. (2004). Chronic olanzapine or fluoxetine administration increases cell proliferation in hippocampus and prefrontal cortex of adult rat. *Biol. Psychiatry* **56,** 570–580.

Krystal, J. H., D'Souza, D. C., Sanacora, G., Goddard, A. W., and Charney, D. S. (2001). Current perspectives on the pathophysiology of schizophrenia, depression, and anxiety disorders. *Med. Clin. North Am.* **85,** 559–577.

Kuhn, H. G., Winkler, J., Kempermann, G., Thal, L. J., and Gage, F. H. (1997). Epidermal growth factor and fibroblast growth factor-2 have different effects on neural progenitors in the adult rat brain. *J. Neurosci.* **17,** 5820–5829.

Lammers, C. H., Garcia-Borreguero, D., Schmider, J., Gotthardt, U., Dettling, M., Holsboer, F., and Heuser, I. J. (1995). Combined dexamethasone/corticotropin-releasing hormone test in patients with schizophrenia and in normal controls: II. *Biol. Psychiatry* **38,** 803–807.

Largent, B. L., Wikstrom, H., Snowman, A. M., and Snyder, S. H. (1988). Novel antipsychotic drugs share high affinity for sigma receptors. *Eur. J. Pharmacol.* **155,** 345–347.

Lauder, J. M. (1983). Hormonal and humoral influences on brain development. *Psychoneuroendocrinology* **8,** 121–155.

Lawlor, M. A., and Alessi, D. R. (2001). PKB/Akt: A key mediator of cell proliferation, survival and insulin responses? *J. Cell Sci.* **114,** 2903–2910.

Lee, J., Kim, M. S., Park, C., Lim, Y. S., Lee, I., Moon, B. S., Lee, H. S., and Park, R. (2002). Protective effects of Debo on serum-deprived apoptosis of PC12 cells via inhibition of H_2O_2 generation and caspase 3-like protease activity. *Immunopharmacol. Immunotoxicol.* **24,** 227–243.

Li, X. M., Chlan-Fourney, J., Juorio, A. V., Bennett, V. L., Shrikhande, S., Keegan, D. L., Qi, J., and Boulton, A. A. (1999). Differential effects of olanzapine on the gene expression of superoxide dismutase and the low affinity nerve growth factor receptor. *J. Neurosci. Res.* **56,** 72–75.

Lindefors, N., Brodin, E., and Metsis, M. (1995). Spatiotemporal selective effects on brain-derived neurotrophic factor and trkB messenger RNA in rat hippocampus by electroconvulsive shock. *Neuroscience* **65,** 661–670.

Linden, A. M., Vaisanen, J., Lakso, M., Nawa, H., Wong, G., and Castren, E. (2000). Expression of neurotrophins BDNF and NT-3, and their receptors in rat brain after administration of antipsychotic and psychotrophic agents. *J. Mol. Neurosci.* **14,** 27–37.

Lindenmayer, J. P. (1995). New pharmacotherapeutic modalities for negative symptoms in psychosis. *Acta Psychiatr. Scand. Suppl.* **388,** 15–19.

Lipska, B. K., Khaing, Z. Z., Weickert, C. S., and Weinberger, D. R. (2001). BDNF mRNA expression in rat hippocampus and prefrontal cortex: Effects of neonatal ventral hippocampal damage and antipsychotic drugs. *Eur. J. Neurosci.* **14,** 135–144.

Lu, X. H., and Dwyer, D. S. (2005). Second-generation antipsychotic drugs, olanzapine, quetiapine, and clozapine enhance neurite outgrowth in PC12 cells via PI3K/AKT, ERK, and pertussis toxin-sensitive pathways. *J. Mol. Neurosci.* **27,** 43–64.

Lu, X. H., Bradley, R. J., and Dwyer, D. S. (2004). Olanzapine produces trophic effects *in vitro* and stimulates phosphorylation of Akt/PKB, ERK1/2, and the mitogen-activated protein kinase p38. *Brain Res.* **1011,** 58–68.

Luo, C., Xu, H., and Li, X. M. (2004). Post-stress changes in BDNF and Bcl-2 immuno-reactivities in hippocampal neurons: Effect of chronic administration of olanzapine. *Brain Res.* **1025,** 194–202.

Luo, C., Xu, H., and Li, X. M. (2005). Quetiapine reverses the suppression of hippocampal neurogenesis caused by repeated restraint stress. *Brain Res.* **1063,** 32–39.

Lyne, J., Kelly, B. D., and O'Connor, W. T. (2004). Schizophrenia: A review of neuropharmacology. *Ir. J. Med. Sci.* **173,** 155–159.

Malaspina, D., Perera, G. M., Lignelli, A., Marshall, R. S., Esser, P. D., Storer, S., Furman, V., Wray, A. D., Coleman, E., Gorman, J. M., and Van Heertum, R. L. (1998). SPECT imaging of odor identification in schizophrenia. *Psychiatry Res.* **82,** 53–61.

Mann, K., Rossbach, W., Muller, M. J., Muller-Siecheneder, F., Pott, T., Linde, I., Dittmann, R. W., and Hiemke, C. (2006). Nocturnal hormone profiles in patients with schizophrenia treated with olanzapine. *Psychoneuroendocrinology* **31,** 256–264.

Maragnoli, M. E., Fumagalli, F., Gennarelli, M., Racagni, G., and Riva, M. A. (2004). Fluoxetine and olanzapine have synergistic effects in the modulation of fibroblast growth factor 2 expression within the rat brain. *Biol. Psychiatry* **55,** 1095–1102.

Martin, M. V., Dong, H., Bertchume, A., and Csernansky, J. G. (2005). Low dose quetiapine reverses deficits in contextual and cued fear conditioning in rats with excitotoxin-induced hippocampal neuropathy. *Pharmacol. Biochem. Behav.* **82,** 263–269.

Matthews, J. D., Bottonari, K. A., Polania, L. M., Mischoulon, D., Dording, C. M., Irvin, R., and Fava, M. (2002). An open study of olanzapine and fluoxetine for psychotic major depressive disorder: Interim analyses. *J. Clin. Psychiatry* **63,** 1164–1170.

McCollum, A. T., and Estus, S. (2004). NGF acts via p75 low-affinity neurotrophin receptor and calpain inhibition to reduce UV neurotoxicity. *J. Neurosci. Res.* **77,** 552–564.

McIlroy, S., and Craig, D. (2004). Neurobiology and genetics of behavioural syndromes of Alzheimer's disease. *Curr. Alzheimer Res.* **1,** 135–142.

McIntyre, R., and Katzman, M. (2003). The role of atypical antipsychotics in bipolar depression and anxiety disorders. *Bipolar Disord.* **5**(Suppl. 2), 20–35.

McIntyre, R. S., Mancini, D. A., Lin, P., and Jordan, J. (2004). Treating bipolar disorder. Evidence-based guidelines for family medicine. *Can. Fam. Physician* **50,** 388–394.

Mehlen, P., and Bredesen, D. E. (2004). The dependence receptor hypothesis. *Apoptosis* **9,** 37–49.

Meier, A., Neumann, A. C., Jordan, W., Huether, G., Rodenbeck, A., Ruther, E., and Cohrs, S. (2005). Ziprasidone decreases cortisol excretion in healthy subjects. *Br. J. Clin. Pharmacol.* **60,** 330–336.

Meltzer, H. Y., Lee, M. A., and Ranjan, R. (1994). Recent advances in the pharmacotherapy of schizophrenia. *Acta Psychiatr. Scand. Suppl.* **384,** 95–101.

Meltzer, H. Y., Park, S., and Kessler, R. (1999). Cognition, schizophrenia, and the atypical antipsychotic drugs. *Proc. Natl. Acad. Sci. USA* **96,** 13591–13593.

Meredith, G. E., Switzer, R. C., III, and Napier, T. C. (2004). Short-term, D2 receptor blockade induces synaptic degeneration, reduces levels of tyrosine hydroxylase and brain-derived neurotrophic factor, and enhances D2-mediated firing in the ventral pallidum. *Brain Res.* **995,** 14–22.

Ming, G. L., and Song, H. (2005). Adult neurogenesis in the mammalian central nervous system. *Annu. Rev. Neurosci.* **28,** 223–250.

Miyamoto, S., LaMantia, A. S., Duncan, G. E., Sullivan, P., Gilmore, J. H., and Lieberman, J. A. (2003). Recent advances in the neurobiology of schizophrenia. *Mol. Interv.* **3,** 27–39.

Miyamoto, S., Duncan, G. E., Marx, C. E., and Lieberman, J. A. (2005). Treatments for schizophrenia: A critical review of pharmacology and mechanisms of action of antipsychotic drugs. *Mol. Psychiatry* **10,** 79–104.

Mobley, W. C., Rutkowski, J. L., Tennekoon, G. I., Gemski, J., Buchanan, K., and Johnston, M. V. (1986). Nerve growth factor increases choline acetyltransferase activity in developing basal forebrain neurons. *Brain Res.* **387,** 53–62.

Nagtegaal, I. D., Lakke, E. A., and Marani, E. (1998). Trophic and tropic factors in the development of the central nervous system. *Arch. Physiol. Biochem.* **106,** 161–202.

Nakagawa, S., Kim, J. E., Lee, R., Malberg, J. E., Chen, J., Steffen, C., Zhang, Y. J., Nestler, E. J., and Duman, R. S. (2002). Regulation of neurogenesis in adult mouse hippocampus by cAMP and the cAMP response element-binding protein. *J. Neurosci.* **22,** 3673–3682.

Negron, A. E., and Reichman, W. E. (2000). Risperidone in the treatment of patients with Alzheimer's disease with negative symptoms. *Int. Psychogeriatr.* **12,** 527–536.

Newman-Tancredi, A., Gavaudan, S., Conte, C., Chaput, C., Touzard, M., Verriele, L., Audinot, V., and Millan, M. J. (1998). Agonist and antagonist actions of antipsychotic agents at 5-HT1A receptors: A [35S]GTPgammaS binding study. *Eur. J. Pharmacol.* **355,** 245–256.

Newman-Tancredi, A., Verriele, L., Touzard, M., and Millan, M. J. (2001). Efficacy of antipsychotic agents at human 5-HT(1A) receptors determined by [3H]WAY100,635 binding affinity ratios: Relationship to efficacy for G-protein activation. *Eur. J. Pharmacol.* **428,** 177–184.

Nibuya, M., Morinobu, S., and Duman, R. S. (1995). Regulation of BDNF and trkB mRNA in rat brain by chronic electroconvulsive seizure and antidepressant drug treatments. *J. Neurosci.* **15,** 7539–7547.

Nibuya, M., Nestler, E. J., and Duman, R. S. (1996). Chronic antidepressant administration increases the expression of cAMP response element binding protein (CREB) in rat hippocampus. *J. Neurosci.* **16,** 2365–2372.

Nyback, H., and Sedvall, G. (1976). Effects of neuroleptics on striatal dopamine synthesized from 14C-tyrosine. *Pharmacol. Ther. [B]* **2,** 49–63.

Okamura, N., Hashimoto, K., Kanahara, N., Shimizu, E., Kumakiri, C., Komatsu, N., and Iyo, M. (2003). Protective effect of the antipsychotic drug zotepine on dizocilpine-induced neuropathological changes in rat retrosplenial cortex. *Eur. J. Pharmacol.* **461,** 93–98.

Olney, J. W., Labruyere, J., and Price, M. T. (1989). Pathological changes induced in cerebrocortical neurons by phencyclidine and related drugs. *Science* **244,** 1360–1362.

Ozaki, T. (2000). Comparative effects of dopamine D(1) and D(2) receptor antagonists on nerve growth factor protein induction. *Eur. J. Pharmacol.* **402,** 39–44.

Ozaki, T., Mui, K., and Yamagami, S. (1999). Comparison of the effects of dopamine D1 and D2 receptor antagonists on nerve growth factor mRNA expression. *Eur. J. Pharmacol.* **369,** 133–143.

Palmer, T. D., Takahashi, J., and Gage, F. H. (1997). The adult rat hippocampus contains primordial neural stem cells. *Mol. Cell Neurosci.* **8,** 389–404.

Parikh, V., Khan, M. M., Terry, A., and Mahadik, S. P. (2004a). Differential effects of typical and atypical antipsychotics on nerve growth factor and choline acetyltransferase expression in the cortex and nucleus basalis of rats. *J. Psychiatr. Res.* **38,** 521–529.

Parikh, V., Terry, A. V., Khan, M. M., and Mahadik, S. P. (2004b). Modulation of nerve growth factor and choline acetyltransferase expression in rat hippocampus after chronic exposure to haloperidol, risperidone, and olanzapine. *Psychopharmacology (Berl.)* **172,** 365–374.

Patel, A. J., and Lewis, P. D. (1988). Brain cell acquisition and neurotropic drugs with special reference to functional teratogenesis. *Prog. Brain Res.* **73,** 389–403.

Pencea, V., Bingaman, K. D., Wiegand, S. J., and Luskin, M. B. (2001). Infusion of brain-derived neurotrophic factor into the lateral ventricle of the adult rat leads to new neurons in the parenchyma of the striatum, septum, thalamus, and hypothalamus. *J. Neurosci.* **21,** 6706–6717.

Perenyi, A., and Forlano, R. (2005). Suicide in schizophrenia. *Neuropsychopharmacol. Hung.* **7,** 107–117.

Pirildar, S., Gonul, A. S., Taneli, F., and Akdeniz, F. (2004). Low serum levels of brain-derived neurotrophic factor in patients with schizophrenia do not elevate after antipsychotic treatment. *Prog. Neuropsychopharmacol. Biol. Psychiatry* **28,** 709–713.

Pitchot, W., and Ansseau, M. (2001). Addition of olanzapine for treatment-resistant depression. *Am. J. Psychiatry* **158,** 1737–1738.

Qing, H., Xu, H., Wei, Z., Gibson, D., and Li, X. M. (2003). The ability of atypical antipsychotic drugs vs. haloperidol to protect PC12 cells against MPP^+-induced apoptosis. *Eur. J. Neurosci.* **17,** 1563–1570.

Rabizadeh, S., and Bredesen, D. E. (2003). Ten years on: Mediation of cell death by the common neurotrophin receptor p75(NTR). *Cytokine Growth Factor Rev.* **14,** 225–239.

Radley, J. J., and Jacobs, B. L. (2002). 5-HT1A receptor antagonist administration decreases cell proliferation in the dentate gyrus. *Brain Res.* **955,** 264–267.

Raedler, T. J., Knable, M. B., and Weinberger, D. R. (1998). Schizophrenia as a developmental disorder of the cerebral cortex. *Curr. Opin. Neurobiol.* **8,** 157–161.

Rankin, G. O., Valentovic, M. A., Teets, V. J., Nicoll, D. W., Sutherland, C. H., and Brown, P. I. (1991). Effect of calcium antagonism by nifedipine and chlorpromazine on acute N-(3,5-dichlorophenyl)succinimide-induced nephrotoxicity in Fischer 344 rats. *Toxicology* **66,** 19–34.

Richelson, E. (1999). Receptor pharmacology of neuroleptics: Relation to clinical effects. *J. Clin. Psychiatry* **60**(Suppl. 10), 5–14.

Ritsner, M., Maayan, R., Gibel, A., Strous, R. D., Modai, I., and Weizman, A. (2004). Elevation of the cortisol/dehydroepiandrosterone ratio in schizophrenia patients. *Eur. Neuropsychopharmacol.* **14,** 267–273.

Ritsner, M., Gibel, A., Maayan, R., Ratner, Y., Ram, E., Biadsy, H., Modai, I., and Weizman, A. (2005). Cortisol/dehydroepiandrosterone ratio and responses to antipsychotic treatment in schizophrenia. *Neuropsychopharmacology* **30,** 1913–1922.

Riva, M. A., Molteni, R., Tascedda, F., Massironi, A., and Racagni, G. (1999). Selective modulation of fibroblast growth factor-2 expression in the rat brain by the atypical antipsychotic clozapine. *Neuropharmacology* **38,** 1075–1082.

Rogers, H. J., and Lietman, P. S. (1977). Effect of folate and folinate on 3H-thymidine incorporation by transforming human lymphocytes *in vitro*. *Experientia* **33,** 671–672.

Roux, P. P., and Barker, P. A. (2002). Neurotrophin signaling through the p75 neurotrophin receptor. *Prog. Neurobiol.* **67,** 203–233.

Ryan, M. C., Sharifi, N., Condren, R., and Thakore, J. H. (2004). Evidence of basal pituitary-adrenal overactivity in first episode, drug naive patients with schizophrenia. *Psychoneuroendocrinology* **29,** 1065–1070.

Sagara, Y. (1998). Induction of reactive oxygen species in neurons by haloperidol. *J. Neurochem.* **71,** 1002–1012.

Sajatovic, M., Mullen, J. A., and Sweitzer, D. E. (2002). Efficacy of quetiapine and risperidone against depressive symptoms in outpatients with psychosis. *J. Clin. Psychiatry* **63,** 1156–1163.

Santarelli, L., Saxe, M., Gross, C., Surget, A., Battaglia, F., Dulawa, S., Weisstaub, N., Lee, J., Duman, R., Arancio, O., Belzung, C., and Hen, R. (2003). Requirement of hippocampal neurogenesis for the behavioral effects of antidepressants. *Science* **301,** 805–809.

Saville, J. G., Davidson, C. P., D'Adrea, G. H., Born, C. K., and Hamrick, M. E. (1988). Inhibition of acetaminophen hepatotoxicity by chlorpromazine in fed and fasted mice. *Biochem. Pharmacol.* **37,** 2467–2471.

Scharfman, H., Goodman, J., Macleod, A., Phani, S., Antonelli, C., and Croll, S. (2005). Increased neurogenesis and the ectopic granule cells after intrahippocampal BDNF infusion in adult rats. *Exp. Neurol.* **192,** 348–356.

Schmitt, A., Weber, S., Jatzko, A., Braus, D. F., and Henn, F. A. (2004). Hippocampal volume and cell proliferation after acute and chronic clozapine or haloperidol treatment. *J. Neural. Transm.* **111,** 91–100.

Schramm, M., Falkai, P., Feldmann, N., Knable, M. B., and Bayer, T. A. (1998). Reduced tyrosine kinase receptor C mRNA levels in the frontal cortex of patients with schizophrenia. *Neurosci. Lett.* **257,** 65–68.

Sharif, Z. A. (1998). Treatment refractory schizophrenia: How should we proceed? *Psychiatr. Q.* **69,** 263–281.

Shimizu, E., Hashimoto, K., Watanabe, H., Komatsu, N., Okamura, N., Koike, K., Shinoda, N., Nakazato, M., Kumakiri, C., Okada, S., and Iyo, M. (2003). Serum brain-derived neurotrophic factor (BDNF) levels in schizophrenia are indistinguishable from controls. *Neurosci. Lett.* **351,** 111–114.

Shinosaki, K., Nishikawa, T., and Takeda, M. (2000). Neurobiological basis of behavioral and psychological symptoms in dementia of the Alzheimer type. *Psychiatry Clin. Neurosci.* **54,** 611–620.

Shors, T. J. (2004). Memory traces of trace memories: Neurogenesis, synaptogenesis and awareness. *Trends Neurosci.* **27,** 250–256.

Shors, T. J., Miesegaes, G., Beylin, A., Zhao, M., Rydel, T., and Gould, E. (2001). Neurogenesis in the adult is involved in the formation of trace memories. *Nature* **410,** 372–376.

Siddiqui, Z., Ramaswamy, S., and Petty, F. (2005). Quetiapine therapy for corticosteroid-induced mania. *Can. J. Psychiatry* **50,** 77–78.

Smith, M. A., Makino, S., Kvetnansky, R., and Post, R. M. (1995). Stress and glucocorticoids affect the expression of brain-derived neurotrophic factor and neurotrophin-3 mRNAs in the hippocampus. *J. Neurosci.* **15,** 1768–1777.

Stein, M. B., Kline, N. A., and Matloff, J. L. (2002). Adjunctive olanzapine for SSRI-resistant combat-related PTSD: A double-blind, placebo-controlled study. *Am. J. Psychiatry* **159,** 1777–1779.

Taherianfard, M., and Shariaty, M. (2004). Evaluation of serum steroid hormones in schizophrenic patients. *Indian J. Med. Sci.* **58,** 3–9.

Takahashi, M., Shirakawa, O., Toyooka, K., Kitamura, N., Hashimoto, T., Maeda, K., Koizumi, S., Wakabayashi, K., Takahashi, H., Someya, T., and Nawa, H. (2000). Abnormal expression of brain-derived neurotrophic factor and its receptor in the corticolimbic system of schizophrenic patients. *Mol. Psychiatry* **5,** 293–300.

Tani, K., Takei, N., Kawai, M., Suzuki, K., Sekine, Y., Toyoda, T., Minabe, Y., and Mori, N. (2004). Augmentation of milnacipran by risperidone in treatment for major depression. *Int. J. Neuropsychopharmacol.* **7,** 55–58.

Tao, Y., Black, I. B., and DiCicco-Bloom, E. (1997). *In vivo* neurogenesis is inhibited by neutralizing antibodies to basic fibroblast growth factor. *J. Neurobiol.* **33,** 289–296.

Tatsumi, M., Jansen, K., Blakely, R. D., and Richelson, E. (1999). Pharmacological profile of neuroleptics at human monoamine transporters. *Eur. J. Pharmacol.* **368,** 277–283.

Toyooka, K., Asama, K., Watanabe, Y., Muratake, T., Takahashi, M., Someya, T., and Nawa, H. (2002). Decreased levels of brain-derived neurotrophic factor in serum of chronic schizophrenic patients. *Psychiatry Res.* **110,** 249–257.

Ueda, K., Yagami, T., Asakura, K., and Kawasaki, K. (1997). Chlorpromazine reduces toxicity and Ca^{2+} uptake induced by amyloid beta protein (25–35) *in vitro*. *Brain Res.* **748,** 184–188.

Ukai, W., Ozawa, H., Tateno, M., Hashimoto, E., and Saito, T. (2004). Neurotoxic potential of haloperidol in comparison with risperidone: Implication of Akt-mediated signal changes by haloperidol. *J. Neural. Transm.* **111,** 667–681.

Vaidya, V. A., and Duman, R. S. (2001). Depresssion—emerging insights from neurobiology. *Br. Med. Bull.* **57,** 61–79.

Vaidya, V. A., Marek, G. J., Aghajanian, G. K., and Duman, R. S. (1997). 5-HT2A receptor-mediated regulation of brain-derived neurotrophic factor mRNA in the hippocampus and the neocortex. *J. Neurosci.* **17,** 2785–2795.

van Praag, H., Schinder, A. F., Christie, B. R., Toni, N., Palmer, T. D., and Gage, F. H. (2002). Functional neurogenesis in the adult hippocampus. *Nature* **415,** 1030–1034.

Vaudry, D., Stork, P. J., Lazarovici, P., and Eiden, L. E. (2002). Signaling pathways for PC12 cell differentiation: Making the right connections. *Science* **296,** 1648–1649.

Wakade, C. G., Mahadik, S. P., Waller, J. L., and Chiu, F. C. (2002). Atypical neuroleptics stimulate neurogenesis in adult rat brain. *J. Neurosci. Res.* **69,** 72–79.

Wang, H., Xu, H., Dyck, L., and Li, X. M. (2005). Atypical antipsychotics protect PC12 cells against amyloid-induced apoptosis by inhibiting oxidative stress. *J. Neurosci. Res.* **81,** 572–580.

Wang, H. D., Dunnavant, F. D., Jarman, T., and Deutch, A. Y. (2004). Effects of antipsychotic drugs on neurogenesis in the forebrain of the adult rat. *Neuropsychopharmacology* **29,** 1230–1238.

Weber, C. M., Eke, B. C., and Maines, M. D. (1994). Corticosterone regulates heme oxygenase-2 and NO synthase transcription and protein expression in rat brain. *J. Neurochem.* **63,** 953–962.

Weber, E., Sonders, M., Quarum, M., McLean, S., Pou, S., and Keana, J. F. (1986). 1,3-Di(2-[5–3H]tolyl) guanidine: A selective ligand that labels sigma-type receptors for psychotomimetic opiates and antipsychotic drugs. *Proc. Natl. Acad. Sci. USA* **83,** 8784–8788.

Wei, Z., Bai, O., Richardson, J. S., Mousseau, D. D., and Li, X. M. (2003a). Olanzapine protects PC12 cells from oxidative stress induced by hydrogen Peroxide. *J. Neurosci. Res.* **73,** 364–368.

Wei, Z., Mousseau, D. D., Richardson, J. S., Dyck, L. E., and Li, X. M. (2003b). Atypical antipsychotics attenuate neurotoxicity of β-amyloid (25–35) by modulating Bax and Bcl- $X_{L/S}$ expression and localization. *J. Neurosci. Res.* **74,** 942–947.

Weickert, C. S., Hyde, T. M., Lipska, B. K., Herman, M. M., Weinberger, D. R., and Kleinman, J. E. (2003). Reduced brain-derived neurotrophic factor in prefrontal cortex of patients with schizophrenia. *Mol. Psychiatry* **8,** 592–610.

Weickert, C. S., Ligons, D. L., Romanczyk, T., Ungaro, G., Hyde, T. M., Herman, M. M., Weinberger, D. R., and Kleinman, J. E. (2005). Reductions in neurotrophin receptor mRNAs in the prefrontal cortex of patients with schizophrenia. *Mol. Psychiatry* **10,** 637–650.

Weinberger, D. R. (1999). Cell biology of the hippocampal formation in schizophrenia. *Biol. Psychiatry* **45,** 395–402.

Weinberger, D. R., and Gallhofer, B. (1997). Cognitive function in schizophrenia. *Int. Clin. Psychopharmacol.* **12**(Suppl. 4), S29–S36.

Weiss, E. M., Bilder, R. M., and Fleischhacker, W. W. (2002). The effects of second-generation antipsychotics on cognitive functioning and psychosocial outcome in schizophrenia. *Psychopharmacology (Berl.)* **162,** 11–17.

Wetmore, C., Ernfors, P., Persson, H., and Olson, L. (1990). Localization of brain-derived neurotrophic factor mRNA to neurons in the brain by in situ hybridization. *Exp. Neurol.* **109,** 141–152.

Wetmore, C., Olson, L., and Bean, A. J. (1994). Regulation of brain-derived neurotrophic factor (BDNF) expression and release from hippocampal neurons is mediated by non-NMDA type glutamate receptors. *J. Neurosci.* **14,** 1688–1700.

Whitaker-Azmitia, P. M., and Azmitia, E. C. (1989). Stimulation of astroglial serotonin receptors produces culture media which regulates growth of serotonergic neurons. *Brain Res.* **497,** 80–85.

Wik, G. (1995). Effects of neuroleptic treatment on cortisol and 3-methoxy-4-hydroxyphenylethyl glycol levels in blood. *J. Endocrinol.* **144,** 425–429.

Williams, R., Ali, S. F., Scalzo, F. M., Soliman, K., and Holson, R. R. (1992). Prenatal haloperidol exposure: Effects on brain weights and caudate neurotransmitter levels in rats. *Brain Res. Bull.* **29,** 449–458.

Wright, I. C., Rabe-Hesketh, S., Woodruff, P. W., David, A. S., Murray, R. M., and Bullmore, E. T. (2000). Meta-analysis of regional brain volumes in schizophrenia. *Am. J. Psychiatry* **157,** 16–25.

Xiao, X. Q., Zhang, H. Y., and Tang, X. C. (2002). Huperzine A attenuates amyloid beta-peptide fragment 25–35-induced apoptosis in rat cortical neurons via inhibiting reactive oxygen species formation and caspase-3 activation. *J. Neurosci. Res.* **67,** 30–36.

Xu, H., Qing, H., Lu, W., Keegan, D., Richardson, J. S., and Li, X. M. (2002). Quetiapine attenuates the immobilization stress-induced decrease of brain-derived neurotrophic factor expression in rat hippocampus. *Neurosci. Lett.* **321,** 65–68.

Xu, H., Luo, C., Richardson, J. S., and Li, X. M. (2004). Recovery of hippocampal cell proliferation and BDNF levels, both of which are reduced by repeated restraint stress, is accelerated by chronic venlafaxine. *Pharmacogenomics J.* **4,** 322–331.

Xu, H., Chen, Z., He, J., Haimanot, S., Li, X., Dyck, L., and Li, X. M. (2006). Synergetic effects of quetiapine and venlafaxine in preventing the chronic restraint stress-induced decrease in cell proliferation and BDNF expression in rat hippocampus. *Hippocampus* **16**(6), 551–559.

Yang, B. H., Son, H., Kim, S. H., Nam, J. H., Choi, J. H., and Lee, J. S. (2004). Phosphorylation of ERK and CREB in cultured hippocampal neurons after haloperidol and risperidone administration. *Psychiatry Clin. Neurosci.* **58,** 262–267.

Yao, J. K., Reddy, R. D., and van Kammen, D. P. (2001). Oxidative damage and schizophrenia: An overview of the evidence and its therapeutic implications. *CNS Drugs* **15,** 287–310.

Yuen, E. C., and Mobley, W. C. (1996). Therapeutic potential of neurotrophic factors for neurological disorders. *Ann. Neurol.* **40,** 346–354.

Zhang, X. Y., Zhou, D. F., Cao, L. Y., Wu, G. Y., and Shen, Y. C. (2005). Cortisol and cytokines in chronic and treatment-resistant patients with schizophrenia: Association with psychopathology and response to antipsychotics. *Neuropsychopharmacology* **30,** 1532–1538.

Zhu, D. Y., Lau, L., Liu, S. H., Wei, J. S., and Lu, Y. M. (2004). Activation of cAMP-response-element-binding protein (CREB) after focal cerebral ischemia stimulates neurogenesis in the adult dentate gyrus. *Proc. Natl. Acad. Sci. USA* **101,** 9453–9457.

Zubenko, G. S. (1997). Molecular neurobiology of Alzheimer's disease (syndrome?). *Harv. Rev. Psychiatry* **5,** 177–213.

NEUROGENESIS AND NEUROENHANCEMENT IN THE PATHOPHYSIOLOGY AND TREATMENT OF BIPOLAR DISORDER

Robert J. Schloesser, Guang Chen, and Husseini K. Manji

Laboratory of Molecular Pathophysiology, Mood and Anxiety Disorders Research Program
National Institute of Mental Health, Bethesda, Maryland 20892, USA

I. Introduction
II. Clinical Evidence for a Role of Neuroplasticity in BPD
 A. Course and Precipitating Factors of the Illness
 B. HPA Axis and BPD
 C. Imaging and Activational Studies
 D. Magnetic Resonance Spectroscopy
 E. Postmortem Brain Analyses
III. Preclinical *In Vivo* and *In Vitro* Studies
 A. Cell Proliferation and Differentiation
 B. Stress Decreases Adult Hippocampal Neurogenesis
 C. Antidepressants Action on Adult Hippocampal Neurogenesis
 D. Mood Stabilizers
 E. Atypical Antipsychotics
 F. Functional/Behavioral Considerations
 G. Cellular Plasticity in Other Brain Areas
 H. Neurite Growth, Synaptic Mechanisms, and Dendritic Remodeling
 I. GSK-3 Inhibition and Wnt Pathway
 J. ERK Pathway
 K. Cell Death
IV. Conclusions
 References

In recent years, clinical research findings indicate that mood disorders, such as bipolar disorder (BPD), are associated with abnormal structure and function of the central nervous system (CNS). Dysfunctions in neuroplastic processes are implicated as the underlying cause for these impairments. "Neuroplasticity" subsumes diverse processes such as dendritic remodeling, axonal sprouting, neurite extension, alterations in synaptic strength, and even neurogenesis. By these active mechanisms the brain reorganizes itself in response to environmental stimuli. Psychotropic drugs commonly used to treat mood disorders modulate several different forms of neuroplasticity, including neurogenesis. This chapter will discuss the role of neuroplasticity in the pathophysiology and treatment of BPD by reviewing literature from both clinical and preclinical arenas. Clinical

DOI: 10.1016/S0074-7742(06)77005-2 0074-7742/07 $35.00

data clearly support the role of abnormal CNS function and structure in BPD. Preclinical studies parallel these findings by providing us with possible mechanisms that underlie the abnormalities found in clinical research. Understanding both of them together as a whole will help generate hypotheses to further our studies in BPD and mood disorders in general.

I. Introduction

The ability to adapt to complex environmental stimuli, not anticipated by genetic programming, is vital for higher organisms and their social groups. Learning and memory, homeostasis of complex physiology, as well as advantageous adjustments of behavior rely on the ability to change in response to external and internal influences. The term neuroplasticity addresses biological correlates for these types of adaptive mechanisms found in the CNS. Neuroplasticity enables the brain to reorganize itself by forming new neural connections and altering the attributes of existing ones through modifications in synaptic strength, variations in neurotransmitter release, and modeling of axon and dendrite architecture. In addition, even in adulthood the brain still produces new neurons and glial cells derived from pluripotent precursors. Qualitative, quantitative, and temporal characteristics of stimuli determine the nature of these adaptations. For instance, chronic or repeated stimuli lead to different effects than acute stimuli or a single stimulus (Hyman and Nestler, 1996) and changes in the brain due to certain stimuli are dependent on the developmental phase in which they occur.

In recent years, clinical research as well as cellular and molecular biology associated mood disorders with structural and functional impairments in several regions of the CNS. Furthermore, psychotropic drugs commonly used to treat these conditions can modulate neuroplasticity. The goal of this chapter is to discuss the role of neuroplasticity in the pathophysiology and treatment of bipolar disorder (BPD). We will first examine the nature of BPD's course. Then we will look at biological correlates found in clinical imaging and postmortem pathological research. Finally, we will explore preclinical research on cellular and synaptic plasticity and their related underlying molecular mechanisms.

BPD is a severe, chronic illness that affects about 1–6% of the general population (Judd and Akiskal, 2003). Patients afflicted with this condition typically alternate between episodes of depression and mania. Depressed mood, anhedonia, and despair, taken together with cognitive impairments, changes in psychomotor behavior, and neurovegetative symptoms, including distinctive disturbances of sleep, characterize bipolar depression. Patients in manic stages present a hyperaroused state with either euphoric or dysphoric mood, motor activity elevations, racing thoughts, impaired judgment, and decreased sleep as well as a decreased need for sleep. During the chronic course of the disease many patients experience episode relapse, residual symptoms, and impairments in cognitive and psychosocial functions. Suicide is the cause of death in 10–15% of patients. BPD frequently manifests with many other deleterious nonpsychiatric conditions such as cardiovascular disease, diabetes mellitus, and obesity (Kupfer, 2005). The World Health Organization has listed BPD as a leading cause of disability worldwide and has projected a greater impact for the future (Murray and Lopez, 1997). The cost associated with disability and premature death due to BPD comes close to tens of billions of dollars annually in the United States alone.

Patients with BPD are generally treated with a class of medications referred to as "mood stabilizers." The prototypic agent of this class is lithium. More recently, a variety of anticonvulsant agents, most notably valproic acid, have been effectively used as mood stabilizers (Bowden *et al.*, 1994; Pope *et al.*, 1991). Mood stabilizers are efficacious for the treatment of mania (Macritchie *et al.*, 2003; Poolsup *et al.*, 2000) and may also possess antidepressant properties. In addition, they exert prophylactic effects in preventing the relapse of manic or depressive episodes. However, a great number of BPD patients do not adequately respond to mood stabilizer treatment plus unfortunately medication adherence is often quite poor due to side effects (Nemeroff, 2003). Even in individuals who do respond and comply with the treatment regimen, relapse is common (Keck and Manji, 2002). Therefore, further research on the underlying causes and treatment of this disorder needs to be pursued. The role of neuroplasticity in BPD is an avenue of investigation with great potential and is likely to lead to additional, improved treatments of the disease.

II. Clinical Evidence for a Role of Neuroplasticity in BPD

A. Course and Precipitating Factors of the Illness

The pathophysiology of BPD is likely based on complex interactions of multiple susceptibility and protective genes and environmental influences. Family and adoption studies show the importance of genetics (Craddock and Jones, 1999; Smoller and Finn, 2003), but psychosocial environment clearly plays a significant role (Brown and Harris, 1989). Current data suggest that high levels of stress can increase the risk of a new episode (Ellicott *et al.*, 1990; Malkoff-Schwartz *et al.*, 1998, 2000) or the severity of symptoms (Johnson, 2005; Johnson *et al.*, 2000). Early psychosocial stress (i.e., abuse during childhood) can cause an adverse course of the illness (Leverich *et al.*, 2002). Furthermore, individuals with BPD seem to sensitize to stress after repeated episodes of depression or mania (Hlastala *et al.*, 2000) and relapse becomes more frequent with the number of prior hospitalizations (Post, 1992).

A growing number of studies indicate significant impairments of cognitive and psychosocial functioning even during remission. The course of the illness impacts the severity of neuropsychological dysfunction, whereas the degree of depressive symptoms determines social impairments (Fagiolini *et al.*, 2005; Robinson and Ferrier, 2006). These findings along with the descriptions of stressor and episode sensitization suggest biological residues left by the disease that are likely evidence for impairments in neuroplasticity.

B. HPA Axis and BPD

As implied by the precipitating factor of stress, there is considerable clinical evidence for dysregulation of the stress response in BPD. Many studies focus on abnormalities of the hypothalamic pituitary adrenal (HPA) axis. The HPA system is the central endocrine system for the regulation of stress. Stress is defined as the body's nonspecific response to any demand, physical or psychological. During stress several cortical and limbic brain areas cause the hypothalamus to secret corticotrophin-releasing hormone (CRH). CRH acts on the pituitary to stimulate adrenocorticotropin hormone (ACTH) release into the bloodstream. ACTH triggers the production and secretion of cortisol, a glucocorticoid, by the adrenal glands. Two nuclear receptors subsequently mediate glucocorticoid effects, the high-affinity type I receptor or mineralocorticoid receptor (MR), and the low-affinity type II or glucocorticoid receptor (GR). Negative feedback ensures HPA axis homeostasis. This is achieved by suppression of ACTH from the pituitary and the regulation of the hypothalamus and other brain areas by cortisol.

Many clinical studies report HPA hyperactivity in patients with mood disorders such as unipolar depression or BPD. In bipolar patients, hyperactivity of the HPA axis has been demonstrated and is found in particular in patients with depressive and mixed episodes (Gold *et al.*, 1984; Goodwin and Jamison, 1990; Vieta *et al.*, 1997). HPA axis function is frequently assessed by measuring changes in cortisol, ACTH, and CRH in response to dexamethasone (DEX) alone or in combination with CRH. DEX, a synthetic glucocorticoid, has high affinity to GR receptors and in normal individuals suppresses the release of ACTH and cortisol. Reports of nonsuppression to DEX in both unipolar and BPD as well as enhanced response to CRH (Vieta *et al.*, 1999) suggest primary abnormalities in HPA axis inhibition (Zhou *et al.*, 1987). Additionally, magnetic resonance imaging (MRI) studies show decreases in pituitary volume in patients with BPD (Sassi *et al.*, 2001). The potential hyperactivation of the HPA axis in mood disorders might be causative for the regional brain atrophy observed in many patients with BPD (see imaging in Section II.C). To what extent these findings represent the sequelae of the biochemical changes (e.g., in glucocorticoid levels) accompanying repeated affective episodes remains to be elucidated. The notion that changes in glucocorticoid levels correlate with brain atrophy is also supported by observations that chronic stress or glucocorticoid administration produce atrophy and cell death in hippocampal neurons of rodents and primates. Another consideration is that chronic excessive secretion of glucocorticoids, as seen in patients with ACTH- or cortisol-secreting tumors (Cushings syndrome), has been associated with mania, depression, and cognitive dysfunction (Sonino and Fava, 2001). The incidence of manic or hypomanic episodes during the course of Cushings syndrome is about 30% (Haskett, 1985; Hudson *et al.*, 1987) and current MRI studies reveal reduced hippocampal volumes in patients with Cushings syndrome (Starkman *et al.*, 1999).

C. Imaging and Activational Studies

MRI, fMRI, PET, and SPECT technologies enabled researchers over the last two to three decades to collect data on volumes, flow, and metabolism rates as well as activational differences in human BPD patients compared to healthy volunteers. These measurable changes in the brain are somewhat indicative of changes in neuroplasticity, lending further support to a role for neuroplasticity in the etiology of BPD.

Regarding the brain in its entirety, gray matter volume is not different in BPD (Brambilla *et al.*, 2001; Schlaepfer *et al.*, 1994; Zipursky *et al.*, 1997); however, many studies found reductions in specific regions. The majority of the differences found were in the frontal lobes of the cerebral cortex. Some studies found volume reductions in the left subgenual region 24 (SG24), a structure in the anterior gyrus ventral to the genu of the corpus callosum (Drevets *et al.*, 1997), but others have not (Brambilla *et al.*, 2002). Other studies indicate reductions of gray matter in three other important areas of the frontal cortex, the left dorsolateral prefrontal cortex (DLPFC) (Brambilla *et al.*, 2002), the ventral PFC, and the orbital PFC (Frangou *et al.*, 2002).

Temporal lobe structures such as the hippocampus and amygdala have not been as thoroughly researched as the frontal lobes. No volume differences related to BPD have been observed in the hippocampus (Altshuler *et al.*, 2000; Swayze *et al.*, 1992); however, in one study on monozygotic twins discordant for BPD a reduction of volume in the right hippocampus was indicated in the affected twin (Noga *et al.*, 2001). In contrast, the amygdala data are convoluted. One study indicated reductions in volume on the left (Pearlson *et al.*, 1997), another found no difference (Swayze *et al.*, 1992), and many others found an enlargement of both left and right or just the left (Altshuler *et al.*, 1998; Brambilla *et al.*, 2002, 2003; Frangou *et al.*, 2002; Strakowski *et al.*, 1999).

With regards to white matter measurements, BPD patients appear to have a reduced amount of white matter (Brambilla *et al.*, 2001; Strakowski *et al.*, 1993; Zipursky *et al.*, 1997) and increases in white matter hyperintensities (WMH). WMH are defined as bright spots in T2-weighted images that are considered to be an indication of vascular degeneration. WMH are typically found in normal aging as well as in cardiovascular diseases; however, they are seen in BPD patients of all ages, even in adolescents (Botteron *et al.*, 1995; Lyoo *et al.*, 2002; Pillai *et al.*, 2002). In mood disorders this seems specific for BPD, in unipolar depression it is only seen in elderly patients (Soares and Mann, 1997). The relevance of these WMH is unknown, but they are correlated with the severity or chronicity of BPD (Moore *et al.*, 2001).

Resting state studies using 18F-FDG-PET have been utilized to measure glucose metabolism in depressed BPD patients. Two studies highlight two different areas: one study found a decrease in the left dorsal anterolateral PFC, whereas the other study reported decreases in the right insula, claustrum, anteroventral caudate/ putamen, and temporal cortex, but higher metabolism in the anterior cingulate (Baxter *et al.*, 1989; Dunn *et al.*, 2002). Another study using similar methods reported an increase in glucose metabolism in the amygdala in BPD-depressed patients (Drevets *et al.*, 2002). This study also demonstrated that mood stabilizers normalize glucose metabolism. Other resting state studies using 99m-hexamethylpropyleneamine oxime (HMPAO) SPECT report a decrease in CBF in the PFC, limbic, and paralimbic areas (Ito *et al.*, 1996). Many studies have parallel findings in manic patients (Migliorelli *et al.*, 1993; Rubin *et al.*, 1995) with the exception of one study that found an increase in the temporal lobe (O'Connell *et al.*, 1995). Likewise, using a different SPECT method resulted in the observation of an increased perfusion of the superior anterior cingulate and the left orbitofrontal cortex (OFC) after lithium withdrawal and relapse in BPD patients (Goodwin *et al.*, 1997).

Brain activation studies report decreased right rostral and orbital PFC activation in manic patients undergoing a word generation task (Blumberg *et al.*, 1999) and reduction of regional cerebral blood flow (rCBF), when compared to controls, in the prefrontal but not the left temporal regions of the cortex while being given a verbal fluency paradigm. Studies using fMRI in BPD patients performing the color-word Stroup task show a smaller response in the right ventral PFC during manic episodes in comparison to a smaller response in the left ventral PFC during a depressed episode. Fearful facial affect recognition paradigms report reduced activation in the DLPFC and increased amygdala activation in BPD patients (Yurgelun-Todd *et al.*, 2000). Similarly, other studies using happy stimuli find a decreased striatal response in depressed patients and an increased striatal response in mania (Blumberg *et al.*, 2000; Keedwell *et al.*, 2005).

D. Magnetic Resonance Spectroscopy

Magnetic resonance spectroscopy (MRS) is based on the same principles as MRI but can extract *in vivo* biochemical information from body tissue. *N*-acetyl aspartate (NAA), a marker for neuronal viability, shows a decrease in the PFC

(Cecil *et al.*, 2002; Winsberg *et al.*, 2000) and the hippocampus (Bertolino *et al.*, 2003) of BPD patients. Choline-containing compounds (Cho) signal has been thought to reflect neuronal membrane metabolism. Studies indicate BPD patients have decreased signals in the OFC (Cecil *et al.*, 2002), increased signals in the basal ganglia (Kato *et al.*, 1996), and a reduction in the right anterior cingulate cortex (ACC) (Moore *et al.*, 2000a,b).

Treatments of BPD influence MRS findings. Lithium increases NAA levels in various cortical regions, but valproic acid does not (Moore *et al.*, 2000a,b; Silverstone *et al.*, 2002). Lithium and valproic acid reduce myoinositol concentrations, important in the phosphoinositol (PI) second messenger system (Section III.H) (Moore *et al.*, 1999; Silverstone *et al.*, 2002). Additionally, other important modulators of second messenger systems such as phosphor-monoesters (PME) and phosphodiesters (PDE) appear to be altered in BPD as well. PME are decreased in temporal and frontal regions of the brain (Deicken *et al.*, 1995a,b; Kato *et al.*, 1994a,b), while PDE is increased in the frontal lobes (Deicken *et al.*, 1995a,b).

E. Postmortem Brain Analyses

Postmortem studies have concentrated greatly on the DLPFC and SG24 areas in the brain due to the high number of findings in imaging studies over the years. In the DLPFC, there is a reduction of neuronal density in glutamatergic (Rajkowska *et al.*, 2001) and GABAergic neurons (Beasley *et al.*, 2002). Additionally, glial cell density is greatly reduced which correlates with a reduction in volume (Rajkowska *et al.*, 2001). A significant reduction of GFAP labeling in the PFC has been reported (Johnston-Wilson *et al.*, 2000). In the SG24, similar reductions in glial density and number (but not size) have been found (Ongur *et al.*, 1998) as well as reductions of neuronal densities (Bouras *et al.*, 2001), particularly a reduction of interneurons (Benes *et al.*, 2001a,b).

Studies in postmortem tissue implicate other areas of the brain such as the anterior cingulate (BA 24B), suggesting that it may be a site of abnormalities in synaptic plasticity resulting in a decrease of the synaptic proteins: synaptophysin, complexin 2, and GAP43, in BPD patients (Eastwood and Harrison, 2001). Corresponding with those decreases another researcher finds decreases in spine density in the subiculum of the hippocampus, which is known for being a target for anterior cingulate axons (Rosoklija *et al.*, 2000). Hippocampal reductions of GABAergic interneurons have also been found (Benes and Berretta, 2001; Heckers *et al.*, 2002). Other brain areas such as the amygdala and the raphe also present reduced numbers of glia or neurons, respectively (Baumann *et al.*, 2002; Bowley *et al.*, 2002). However, the locus coeruleus, best known for its role in norepinephrine release, shows an increase in neuronal numbers in BPD patients (Baumann *et al.*, 2002).

Based on neuropsychological findings in humans and animals, two systems for emotional perception and regulation of the affective state have been proposed (Phillips *et al.*, 2003). A ventral system that is important to identify emotional significance and a dorsal system that is used for executive functioning (i.e., selective attention, planning, and effortful regulation of affective states). The ventral system includes the amygdala, insula, ventral striatum, ventral anterior cingulate gyrus, and prefrontal cortex. The dorsal system encompasses the hippocampus, dorsal ACC, and prefrontal cortex. The pattern of structural and functional abnormalities reviewed above could help explain the prominent mood swings, emotional lability, and distractability observed in BPD. However, a more thorough understanding of the neurobiology underlying these abnormalities requires research in a preclinical setting in order to allow analysis and manipulation on a cellular and molecular level. Throughout the rest of the chapter we review preclinical research focusing on the role of cellular and synaptic plasticity in the pathophysiology and treatment of BPD.

III. Preclinical *In Vivo* and *In Vitro* Studies

The imaging and postmortem neuropathological findings reviewed above strongly suggest functional and cellular impairments related to BPD pathophysiology. The data also suggest that mood stabilizers bring about their salutary actions by attenuating or reversing these defects. Repair of existing damage requires mechanisms that increase trophic support or cellular proliferation and improve the resilience of neuronal cells against harmful events. Preclinical research provides further insight into these neuroplastic mechanisms. Animal and cell culture models allow a close look at the proposed pathophysiology that can be useful in discovering potential molecular targets of mood stabilizers.

A. Cell Proliferation and Differentiation

Although the adult CNS contains predominantly fully differentiated neurons and glial cells, it also continues to give birth to new immature progenitors throughout life and into old age. These proliferating cells can be detected in subcortical white matter (Nunes *et al.*, 2003) and cortical regions of the brain (Goldman and Sim, 2005). There are two germinal zones that contain self-renewing, multipotent neural stem cells, the subventricular zone (SVZ) of the lateral ventricles and the subgranular zone (SGZ) of the hippocampal dentate gyrus (DG) (Fig. 1). Multipotency describes the ability to differentiate into all three neural lineages: neurons, astrocytes, and oligodendrocytes. New born cells from the SVZ migrate into the olfactory bulb where they differentiate into new neurons.

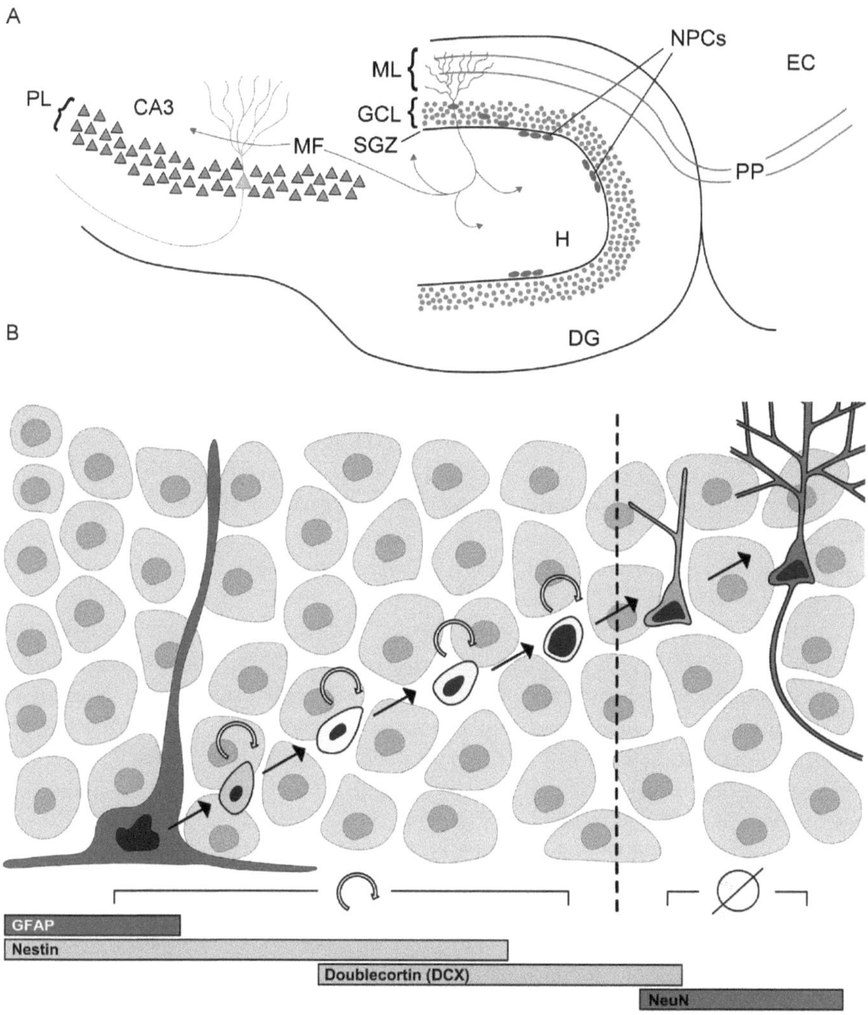

FIG. 1. Neurogenesis in the adult hippocampus.

Cells from the SGZ of the hippocampal DG migrate into the hippocampal granule cell layer where they preferentially differentiate into granule cell neurons that subsequently develop dendrites and extend their axons toward the CA3 region of the hippocampus. During their differentiation these cells express several markers that make it possible to distinguish different stages of maturation (Fig. 1). The putative stem cell expresses both the glial cell marker GFAP and the stem cell marker nestin. Young migrating neuroblasts express doublecortin. Postmitotic young neurons express the markers calbindin and calretinin and finally become positive for the neuron-specific marker NeuN (Kempermann *et al.*, 2004). Newly born neurons are very plastic as measured by their high sensitivity to the induction of long-term potentiation (LTP) (Schmidt-Hieber *et al.*, 2004), but once matured their electrophysiology is indistinguishable from older granule cells (van Praag *et al.*, 2002). In the field of mood disorder research, adult hippocampal neurogenesis has drawn a lot of attention due to the fact that it is modulated by environment, stress, and psychotropic drugs. Additionally, the hippocampus is the region with the highest expression of GRs (Lopez *et al.*, 1999) and has been the most studied area in adaptive plasticity. This is due to the well-defined and easily studied neuronal populations of the hippocampus.

B. Stress Decreases Adult Hippocampal Neurogenesis

Different types of acute and chronic stressors provoke depressive-like behavior in animal models. Most of these stressors can decrease hippocampal progenitor cell proliferation (Table I) (Czeh *et al.*, 2001, 2002; Gould *et al.*, 1998; Tanapat *et al.*, 1998; van der Hart *et al.*, 2002). Quality, quantity, and chronological characteristics of stressors are important factors controlling the permanence of these effects. Rats subjected to chronic and intense uncontrollable stress in adulthood show prolonged inhibition of cell proliferation leading to a prolonged decline in new neuron production (Czeh *et al.*, 2002; Heine *et al.*, 2004; Malberg and Duman, 2003; Pham *et al.*, 2003; Westenbroek *et al.*, 2004). In contrast, decreased proliferation due to acute stress recovers within a 24-hour period (Heine *et al.*, 2004) and active avoidance learning, which is stressful but nonetheless controllable, does not affect DG cell proliferation (Van der Borght *et al.*, 2005). This is very similar to the role of negative life events in human affective disorders where the controllability of the employed stressor is important to outcome measurements of mood changes. Stress during development can have long-lasting effects on adult hippocampal neurogenesis. In pregnant animals, stress induces a decrease of proliferation in the DG of the offspring which extends into adulthood (Lemaire *et al.*, 2000, 2006). Interestingly, postnatal stimulation of pups from prenataly stressed mothers can oppose this effect (Lemaire *et al.*, 2006), while maternal separation stress during the early postnatal period inhibits cell proliferation later in life (Mirescu *et al.*, 2004).

HPA axis activation and glucocorticoid levels are important mediators of stress-related decreases in adult neurogenesis. Inhibition of the HPA axis by surgical (Cameron and Gould, 1994; Cameron and McKay, 1999; Gould *et al.*, 1992; Tanapat *et al.*, 2001) or pharmacological (Alonso *et al.*, 2004) means increases cell proliferation and neurogenesis in the DG. In contrast, administration of corticosteroids suppresses

TABLE I
MOOD STABILIZER EFFECTS ON CELL PROLIFERATION AND DIFFERENTIATION[a]

Drug	Model	Findings	Proposed mechanism	References
In vivo				
Lithium	Mouse (C57/Bl6)	Increase in new born cells, no change in neuronal differentiation	Bcl-2 increase	Chen *et al.* (2000)
Lithium	Rat (SD)	Increase in new born cells, increase in neuronal differentiation	ERK pathway	Kim *et al.* (2004), Son *et al.* (2003)
Lithum	Rat (SD) old age	Increase in new born cells, no mature neuronal phenotype		Yu *et al.* (2003)
VPA	Mouse (C57/Bl6)	No difference in proliferation, increase in new born cells after 6-week survival	ERK pathway	Hao *et al.* (2004)
VPA	Rat (F344)	Decrease in proliferation, increase in neuronal differentiation	HDAC inhibition	Hsieh *et al.* (2004)
In vitro				
Lithium	CGC, CC	Increase in proliferation		Hashimoto *et al.* (2003)
VPA	aHPC	Decrease in proliferation, increase in neuronal differentiation	HDAC inhibition	Hsieh *et al.* (2004)
VPA	reNSC, heNSC	Increase in neuronal differentiation, GABAergic phenotype		Laeng *et al.* (2004)

[a]SD, Sprague-Dawley; F344, Fischer 344; CGC, cerebellar granule cell culture; CC, cerebral cortical culture; aHPC, adult hippocampal progenitor cell culture; reNSC, rat embryonic neural stem cells; heNSC, human embryonic neural stem cells.

adult neurogenesis (Cameron and Gould, 1994; Gould *et al.*, 1991; Wong and Herbert, 2006). It is not fully understood through which mechanisms glucocorticoids decrease neurogenesis. Direct effects through GRs (Garcia *et al.*, 2004) as well as indirect effects mediated by the surrounding environment are both possible. Increased glucocorticoids do stimulate glutamate release and alter NMDA receptor expression which in turn decreases granule cell precursors proliferation (Cameron *et al.*, 1998). Furthermore, glucocorticoids negatively modulate expression of neurotrophic factors such as brain-derived neurotrophic factor (BDNF) (Schaaf *et al.*, 2000). BDNF plays an important role in adult neurogenesis as it increases proliferation and modulates survival of new neurons.

Several drugs of abuse, such as alcohol and nicotine, but also morphine and psychostimulants, decrease adult hippocampal neurogenesis (Eisch and Harburg, 2006).

C. Antidepressants Action on Adult Hippocampal Neurogenesis

In contrast to stress and glucocorticoids, many classes of antidepressant drugs increase neurogenesis in the adult hippocampus. In humans, antidepressants require several weeks of treatment in order to reduce symptoms, and interestingly, a similar timeframe is required to increase neurogenesis (Malberg *et al.*, 2000; Manev *et al.*, 2001; Nakagawa *et al.*, 2002; Santarelli *et al.*, 2003). Moreover, chronic antidepressant treatment blocks the effects of stress on hippocampal neurogenesis (Alonso *et al.*, 2004; Czeh *et al.*, 2001; Lee *et al.*, 2001; Malberg and Duman, 2003; van der Hart *et al.*, 2002). However, there is one study that indicates antidepressant treatment not only increases proliferation but also increases apoptosis of granule cells in the DG. Thus, antidepressant treatment might accelerate the turnover of cells rather than just increasing the total number of granule cells (Sairanen *et al.*, 2005). Other nonpharmacological antidepressant treatments similarly affect neurogenesis. Electroconvulsive seizures (ECS), an efficacious treatment against major depression as well as bipolar depression (Grunhaus *et al.*, 2002), increase the number of new neurons in rodent models; however, the changes with ECS occur more rapidly than with antidepressant drugs (Madsen *et al.*, 2000; Malberg *et al.*, 2000). The data on sleep deprivation, another treatment known to reduce depressive symptoms, is less consistent. Some studies show a decrease in cell proliferation after forced deprivation of sleep (Guzman-Marin *et al.*, 2005; Hairston *et al.*, 2005); however, one study does find an increase in proliferation (Grassi Zucconi *et al.*, 2006). These contradictory results might easily be explained by the confound of stress that is inevitable in sleep deprivation studies.

The mechanisms by which antidepressants bring about these effects are not well understood. Modulation of serotonin (5-HT) transmission by several antidepressant drugs seems to be one important factor. Inhibition of 5-HT synthesis, *in vivo*, decreases the number of newborn cells in the SVZ and the DG (Brezun and Daszuta, 1999). Conversely, 5-HT-producing neurons grafted into the hippocampus increase cell proliferation (Brezun and Daszuta, 2000). 5-HT_{1A} receptors in particular appear to have a critical role in mediating these effects where several specific antagonists have been tested and shown to decrease neurogenesis (Radley and Jacobs, 2002), while agonists appear to increase cell proliferation (Banasr *et al.*, 2004). Furthermore, 5-HT1A knockout mice show a decreased rate of adult neurogenesis (Santarelli *et al.*, 2003).

Other proposed mechanisms through which antidepressants increase neurogenesis involve the up-regulation of BDNF and other growth factors.

Stress decreases the levels of BDNF expression in the hippocampus while several, but not all, antidepressant treatments including ECS increase BDNF (for review see Duman and Monteggia, 2006). Chronic BDNF infusion into the hippocampus increases neurogenesis; however, this treatment also induces seizures, which can also increase neurogenesis (Madsen *et al.*, 2000; Radley and Jacobs, 2003). Interestingly, a recent report found that the physiological diurnal rhythm of glucocorticoids is necessary for the selective serotonin reuptake inhibitor fluoxetine to stimulate neurogenesis (Huang and Herbert, 2006), thus reiterating the involvement of the HPA axis in adult hippocampal neurogenesis.

D. Mood Stabilizers

Most studies report that chronic lithium treatment increases adult hippocampal neurogenesis in rodents (Chen *et al.*, 2000; Son *et al.*, 2003); however, one study in aged animals found an increase in new born cells, but these cells did not obtain a mature neuronal phenotype after 28-day survival (Yu *et al.*, 2003). Pharmacological doses of VPA increase the number of newborn neurons in the hippocampus of adult mice (Hao *et al.*, 2004); however, in another study using a higher dose of VPA, a decreased proliferation of progenitor cells was seen (Hsieh *et al.*, 2004). Lithium stimulates the proliferation of progenitor cells in rat primary neuronal cultures and additionally prevents the loss of proliferation induced by glutamate, glucocorticoids, and haloperidol (Hashimoto *et al.*, 2003). However, it is still unclear whether mood stabilizers increase neurogenesis *in vivo* through changes in proliferation or prolongations of cell survival. Additionally, both lithium and VPA selectively induce neuronal differentiation. Neural stem cells, derived from embryonic hippocampi, show an increase in neuronal as opposed to glial differentiation when treated with lithium. Likewise, lithium- and VPA-treated animals have both been shown to induce neuronal differentiation of progenitor cells *in vivo* (Hsieh *et al.*, 2004; Kim *et al.*, 2004). This acceleration in differentiation and maturation could lead to more efficient incorporation of newborn cells into the neural circuitry. The specific types of neurons these treatments are influencing are still unidentified. One study found that, *in vitro*, VPA treatment of human and rat embryonic stem cell cultures produce a tenfold increase of immunostaining for GABA and the GABA-synthesizing enzyme GAD65/67 (Laeng *et al.*, 2004); however, whether mood stabilizers induce a GABAergic phenotype of new born neurons *in vivo* is not known. Although most newly generated neurons in the DG are excitatory glutamatergic granule cells, GABAergic interneurons derived from adult progenitor cells have been found (Liu *et al.*, 2003). Moreover, researchers finding neurogenesis in the rat cortex have proposed these cells to be distinctively GABAergic interneurons (Dayer *et al.*, 2005) and adult excitatory granule cells can take on a GABAergic phenotype when treated with BDNF or after induction of seizures (Gomez-Lira *et al.*, 2002, 2005). These data along with the human postmortem findings indicating decreased hippocampal GABAergic interneurons in subjects with BPD (Heckers *et al.*, 2002) give us good reason to further evaluate the role of GABAergic neurogenesis in BPD.

In contrast to classical antidepressant drugs, mood stabilizers do not appear to have major direct interactions with cell surface receptors and are likely to exert their effects directly or indirectly through modulation of intracellular targets. Chronic lithium and valproate treatment raise levels of the anti-apoptotic protein Bcl-2 in many regions of the adult brain, including the frontal cortex (Chen *et al.*, 1999), the striatum, and the DG of the hippocampus (Manji *et al.*, 2000). Bcl-2 attenuates apoptotic cell death through several molecular mechanisms, such as the sequestering death-driving caspases, and by preventing the release of mitochondrial apoptogenic factors such as calcium and "cytochrome C." Bcl-2 acts on mitochondria to stabilize membrane integrity and to prevent opening of the permeability transition pore, and protects neurons from various insults both *in vitro* and *in vivo* (Adams and Cory, 1998; Bruckheimer *et al.*, 1998). Besides these well-known anti-apoptotic properties, Bcl-2 accelerates maturation of neuroblasts (Middleton *et al.*, 1998; Suzuki and Tsutomi, 1998) and regulates neuronal differentiation (Zhang *et al.*, 1996). Neural stem and progenitor cells express Bcl-2 and the major neurogenic zones of the adult brain have higher levels of Bcl-2 than nonneurogenic zones (Bernier *et al.*, 2002). Mice overexpressing Bcl-2 in immature neurons show reduced apoptotic cell death, higher production of new granule neurons, and a larger pool of differentiating progenitor cells, but interestingly, no changes in progenitor proliferation (Kuhn *et al.*, 2005). Another molecule related to Bcl-2 is Bcl-2-associated athanogene (BAG-1). BAG-1 is up-regulated by both lithium and valproic acid. BAG-1 can potentiate the anti-apoptotic actions of Bcl-2, but is also similar to Bcl-2 in that it is essential for the differentiation of neuronal cells (Gotz *et al.*, 2005). BAG-1 is also involved in the translocation of GRs to the nucleus where they regulate transcription. Mood stabilizers can attenuate this translocation through modulating BAG-1. Given the important role of glucocorticoids in modulating adult neurogenesis, this mechanism could increase the resilience of new born cells against deleterious effects induced by high levels of glucocorticoids.

Similar to antidepressants, chronic mood stabilizer treatment can increase BDNF levels in rodent brains (Einat *et al.*, 2003; Fukumoto *et al.*, 2001). Furthermore, both lithium and valproate activate the extracellular signal-regulated kinase (ERK) signaling pathway (Section III.J), one of the major pathways activated by neurotrophic factors (Einat *et al.*, 2003; Yuan *et al.*, 2001). ERK increases the activation of several transcription factors. One of these transcription factors is CREB. CREB activation increases Bcl-2 and BDNF expression and may be involved in the mechanism of antidepressant action (Duman, 2002).

E. Atypical Antipsychotics

Clinical reports show that atypical antipsychotics are effective treatment choices for BPD (Ghaemi and Goodwin, 1999). Atypical but not typical antipsychotics have antidepressant effects in behavioral models for depressive-like behavior (Weiner *et al.*, 2003). Similar to antidepressant and mood stabilizers, chronic treatment with antipsychotics increases proliferation of adult neural stem cells *in vitro* and *in vivo*. Some argue that this effect is specific to atypical antipsychotics (Halim *et al.*, 2004; Malberg *et al.*, 2000; Wakade *et al.*, 2002; Wang *et al.*, 2004), while others find a similar result with the typical antipsychotic haloperidol and explain the increase through a blockade of D2 receptors which are expressed on neural stem cells (Kippin *et al.*, 2005).

F. Functional/Behavioral Considerations

Along with its role in memory consolidation, some include the hippocampus as part of a regulatory system for affective states and emotional-related responses in physiology and behavior (Phillips *et al.*, 2003). The hippocampus partly inhibits stress and the HPA axis via its connections to the hypothalamus (Lopez *et al.*, 1999). Observations of impaired extinction in animals with hippocampal lesions generated theories highlighting behavioral inhibition of defensive behavior, and hippocampal lesioned animals display reduced anxiety levels in several behavioral tests (Davidson and Jarrard, 2004).

The functional role of adult neurogenesis in the hippocampus is not yet understood. Existing studies suggest that adult born neurons can contribute to some forms of hippocampal-dependent learning and memory (for review see Leuner *et al.*, 2006), whereas inhibition of adult neurogenesis leads to deficits in spatial memory task performance weeks after training (Snyder *et al.*, 2005). In the context of emotional behavior many studies found several forms of stress leading to both behavioral despair and a decrease in neurogenesis, but these findings are merely correlative rather than causative. Not all studies find correlations between adult hippocampal neurogenesis and depressive-like behavior. For instance, in one study the rate of hippocampal neurogenesis was not reduced in rats bred for depressive-like behavior (Henn and Vollmayr, 2004). Therefore, it is not clear whether the changes seen in adult hippocampal neurogenesis are causative for responses to stress or just epiphenomenal. However, one study reported that depletion of neurogenesis by irradiation ameliorates the actions of antidepressants. This study indicated that mice irradiated at the level of the hippocampus failed to display anxiolytic responses to chronic antidepressant treatment in the novelty suppressed feeding task (NSF) (Santarelli *et al.*, 2003). This behavioral paradigm is one of the few available tests that mimic the situation in humans where only chronic but not acute administration of antidepressants exerts effects. Therefore, the authors conclude that antidepressants, at least in this task, require adult hippocampal neurogenesis. Recently, however, using a similar approach, the same group reported that environmental enriched housing conditions, known to increase neurogenesis, have similar anxiolytic effects in the NSF, but do not require hippocampal neurogenesis (Meshi *et al.*, 2006). It is noteworthy that hippocampal lesions have behavioral effects consistent with reductions, rather than increases, in anxiety, and rats with hippocampal lesions show reduced hyponeophagia, as tested in the NSF and therefore less anxiety. Interestingly, mice with increased survival and neurogenesis due to Bcl-2 overexpression (see above) show reduced anxiety-like behavior (Rondi-Reig and Mariani, 2002; Rondi-Reig *et al.*, 1997), whereas mice heterozygous to Bcl-2 (KO mice) display increased anxiety responses. However, the Bcl-2 overexpressing mice also show impaired performance in a hippocampal-dependent learning task (Rondi-Reig *et al.*, 2001), and therefore, clarification of the role of adult hippocampal neurogenesis on affective like behavior needs further research.

G. Cellular Plasticity in Other Brain Areas

Even though the hippocampus has been an excellent model for adult neuroplasticity, one must remember that other less-studied regions of the brain may be important targets for affective disorders such as BPD. However, the question of whether adult neurogenesis takes place in regions other than the hippocampus and the olfactory bulb, such as the cortex or the striatum, is somewhat controversial. Some researchers observed neurogenesis is in the neocortices of adult rats and macaques (Bernier *et al.*, 2002; Dayer *et al.*, 2005; Gould *et al.*, 1999, 2001) but not all (Kornack and Rakic, 2001; Koketsu *et al.*, 2003; Rakic, 2002). Others found neurogenesis in the adult striatum (Bedard *et al.*, 2006), while others found neocortical neurogenesis only after cortical injury (Gu *et al.*, 2000; Jiang *et al.*, 2001; Magavi *et al.*, 2000; Tonchev *et al.*, 2005) and still others have simply failed to find neurogenesis (Benraiss *et al.*, 2001; Pencea *et al.*, 2001; Teramoto *et al.*, 2003). Despite the conflicting data, the prevalence of adult born cells within the adult neocortex is unquestioned. The real question is what kind of cells are they? Proliferation is found in cortical regions (Arsenijevic *et al.*, 2001; Palmer *et al.*, 1999) and subcortical white matter. Most of these cells express the proteoglycan NG2 and the progenitor marker nestin but do not express GFAP (Dayer *et al.*, 2005; Goldman and Sim, 2005). Therefore, these cells are distinct from GFAP-expressing neural stem cells found in the SVZ or DG. NG2 has commonly been considered as a marker for oligodendrocytes precursors that reside in the subcortical white matter. However, these cells seem to be potent not only of generating myelinating oligodendrocytes, but also giving rise to other not as well-characterized cells. In the postnatal hippocampus, cells that express NG2 have displayed spontaneous synaptic currents and fire action potentials (Belachew *et al.*, 2003). *In vitro* these cells produce all three neural cell lineages (Nunes *et al.*, 2003). Therefore, NG2 positive cells might form a multipotent *in situ* progenitor pool in cortical and white matter regions of the adult CNS. The adult brain environment rather than the cells themselves seem to restrict them to a certain lineage (Goldman and Sim, 2005). Some studies report on NG2-positive progenitors with antidepressants. Repeated ECS treatments nearly doubles the number of proliferating cells in the frontal cortex of rodents and after 4 weeks ~40% of these cells have differentiated into oligodendrocytes, none of them being astrocytes or neurons (Madsen *et al.*, 2005). In addition, ECS induces proliferation of NG2-expressing cells in the amygdala where after a survival period of 3 weeks most cells were still NG2 positive but some had differentiated into oligodendrocytes (Wennstrom *et al.*, 2004). Similarly, chronic treatment with either the SSRI fluoxetine or the atypical antipsychotic olanzapine increased proliferation of these cells in the frontal cortex, but not in motor cortex or SVZ. This increase was not stable over a 4-week period, suggesting these cells might have a transient short-term function (Kodama *et al.*, 2004). At the present time, no data is available on the effects of mood stabilizers on cell proliferation or differentiation in cortical regions. However, the finding in humans that gray matter increases in volumes of patients on lithium therapy points toward a possible role of lithium on cortical cell proliferation.

Despite the lack of findings regarding newly generated astrocytes in the studies reviewed above, it is known that gliosis, the hypertrophy of astrocytes, can take place as a response to numerous forms of destructive events such as physical trauma, hypoxia, or hypoglycemia. Moreover, changes in astrocyte morphology have been observed following learning (Matsutani and Leon, 1993), environmental enrichment (Sirevaag and Greenough, 1991), and sensory deprivation (Hawrylak and Greenough, 1995). One recent study looking at the hippocampus of the adult tree shrew reported that psychosocial stress decreases the number and the volume of GFAP-positive astrocytes. Additionally, chronic antidepressant treatment prevents the stress-induced effects on cell number, but not on cell volume (Czeh *et al.*, 2006), whereas chronic lithium induces gliosis and modifies astrocyte morphology in the adult rat hippocampus (Rocha *et al.*, 1998).

The findings reviewed above give reason to contemplate dysfunctions of cell proliferation and neurogenesis as critical in the pathophysiology of BPD. However, the functional consequences of alterations of glio- or neurogenesis are not well understood. Therefore, future research is warranted to elucidate electrophysiological and behavioral effects of modifications in cellular plasticity. Transgenic animal models could help to specifically ablate proliferating cells in different areas of the adult brain (Imura *et al.*, 2003; Maeda *et al.*, 2006; Mathis *et al.*, 2000) and could therefore shed light on the role these events play in CNS function and behavior.

H. Neurite Growth, Synaptic Mechanisms, and Dendritic Remodeling

Another important structural correlate of neuroplasticity is the cellular morphology of neurons including their characteristic axonal and dendritic architecture. In neuronal cell cultures, growth and development of axons can be observed and quantified. Lithium and other mood stabilizers have significant effects on cellular morphology and neurite outgrowth (Table II). A lively debate is under way about which molecular mechanisms or pathways are underlying these effects, as outlined below (Fig. 2).

Therapeutic levels of lithium directly inhibit several key enzymes of the PI signaling pathway (York *et al.*, 1995). One of them, inositol-monophosphatase (IMPase), metabolizes inositol-1,4,5-trisphosphate (IP3) into myoinositol. IMPase is rate limiting for recycling IP3 back to phosphoinositide phosphate (PIP-2). Inhibition of this recycling leads to decreased availability of PIP-2. The PI signaling cascade starts with phospholipase C (PLC) activation by certain G proteins. PLC then catalyzes the hydrolysis of PIP-2 into two intracellular messenger molecules diacylglycerol (DAG) and IP3. DAG activates protein kinase C (PKC) that, among many other effects, activates myristoylated alanine-rich C kinase substrate (MARCKS). Lithium and VPA treatment both decrease levels of phosphorylated and total MARCKS (Chen *et al.*, 1994; Lenox *et al.*, 1992; Watson and Lenox, 1996) and heterozygous MARCKS

TABLE II
Mood Stabilizer Effects on Neurite Growth and Morphology[a]

Drug	Model	Findings	Proposed mechanism	References
In vivo				
Lithium	Rat (SD)	Prevention of stress-induced dendrite remodeling in CA3	Blockage of glutamate	Wood *et al.* (2004)
In vitro				
Lithium, VPA	DRG	Increase in growth cone spreading	PI signaling cascade	Williams *et al.* (2002)
Lithium	CGC	Increase in axonal spreading and branching, synapsin clustering	GSK-3	Lucas *et al.* (1998), Lucas and Salinas (1997)
Lithium	reHSC	Increase in axon number	GSK-3	Jiang *et al.* (2005)
Lithium, VPA	DRG	Increase in axonal branching, synapsin clustering	GSK-3	Hall *et al.* (2002)
VPA	SH-SY5Y	Increase in neurite growth	ERK pathway	Yuan *et al.* (2001)
VPA	CC	Increase in neurite growth	ERK pathway	Hao *et al.* (2004)
Lithium	RGCs	Axonal regeneration	Bcl-2	Huang *et al.* (2003)

[a]SD, Sprague-Dawley; DRG, rat dorsal root ganglia explants; CGC, cerebellar granule cell; CC, cerebral cortical culture; reHSC, rat embryonic hippocampal stem cells; SH-SY5Y, human neuroblastoma cells; RGCs, retinal ganglion cells; PI, phosphoinositol.

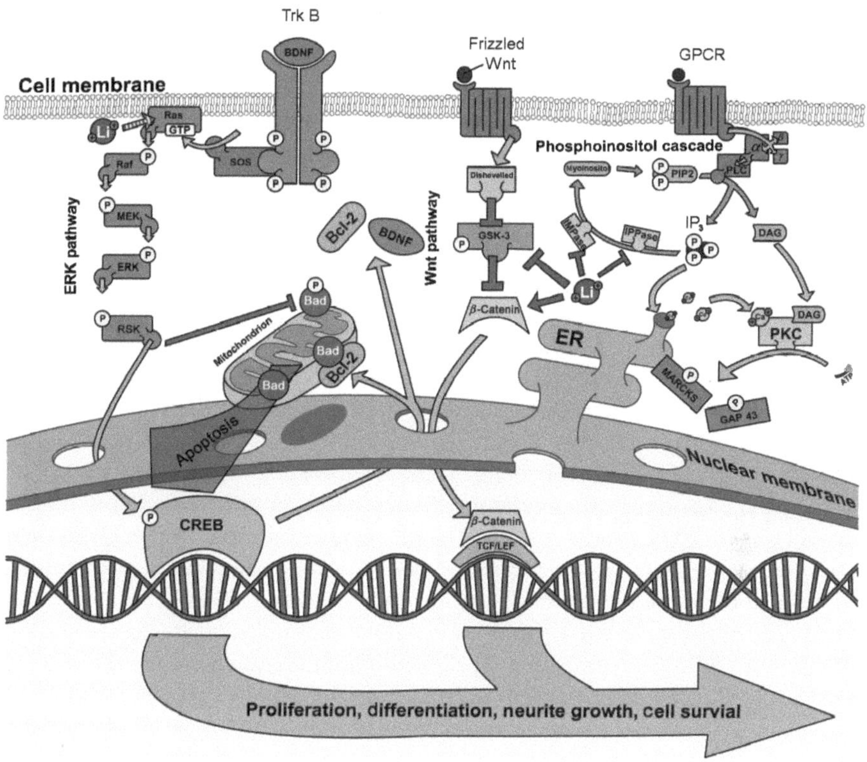

FIG. 2. Signaling cascades affected by mood stabilizers.

knockout mice display impaired LTP in hippocampus and spatial learning deficits (Hussain *et al.*, 2006; McNamara *et al.*, 2005). Lithium, VPA, and another mood stabilizer, carbamazepine, cause increased spreading of growth cones which are the dynamic actin-supported extensions of a developing axon seeking its synaptic target but notably can be prevented by exogenous myoinositol (Williams *et al.*, 2002). Stress, as well as psychostimulants, can trigger manic episodes in susceptible individuals and induce manic-like symptoms in animals via the activation of PKC. Stress-activated PKC leads to cognitive impairments related to prefrontal cortical function in rhesus monkeys and can be blocked by mood stabilizer treatment (Birnbaum *et al.*, 2004).

I. GSK-3 Inhibition and Wnt Pathway

Another important pathway controlling axonal remodeling, the Wnt pathway, is modulated by mood stabilizers through inhibition of glycogen synthase kinase 3 (GSK-3). In the Wnt signaling pathway, Wnt glycoproteins interact with the frizzled family of receptors to stimulate the disheveled-mediated inactivation of GSK-3. Inhibition of GSK-3 prevents β-catenin phosphorylation and thereby inhibits degradation of this transcription factor. This leads to increased transcription of β-catenin-dependent genes. Wnt proteins induce axonal spreading and branching in neuronal cell cultures and lithium mimics these effects (Lucas and Salinas, 1997; Lucas *et al.*, 1998). Also, inhibition of GSK-3 activity by lithium increases the number of axons in hippocampal neuron cultures (Jiang *et al.*, 2005) and both mood stabilizers, lithium and VPA, increase axon branching in cultures derived from dorsal root ganglia (Hall *et al.*, 2002). Wnt signaling also regulates synapsins, a family of proteins involved in neurotransmitter release, synaptogenesis, and axonal transport (Greengard *et al.*, 1993). Lithium as well as VPA increases the clustering of synapsin 1 in neuronal cell cultures and pontine explants which extend mossy fibers *in vitro* (Hall *et al.*, 2002; Lucas and Salinas, 1997) through a GSK-3-dependent mechanism. Inhibition of GSK-3 by pharmacological compounds attenuates amphetamine hyperactivity similar to lithium (Beaulieu *et al.*, 2004; Gould *et al.*, 2004). Interestingly, animals treated with these drugs, as well as heterozygous GSK-3 knockout mice, show antidepressant-like behavior in the forced swim test, a model frequently used to assess antidepressant actions (Beaulieu *et al.*, 2004; Gould *et al.*, 2004; Kaidanovich-Beilin *et al.*, 2004; O'Brien *et al.*, 2004). It seems that GSK-3 inhibition uniquely mimics these effects of lithium on behavior relevant to both mania and depression.

J. ERK Pathway

The ERK pathway mediates intracellular signaling of receptor tyrosine kinases (Trk). Neurotrophins such as nerve growth factor (NGF), BDNF, and NT-3/4 bind to specific Trks. Ligand binding to Trks causes intracellular cascade activation regulating downstream signaling molecules as well as gene regulatory proteins. This can cause an up-regulation of BDNF and the anti-apoptotic protein Bcl-2 (see above). The resulting changes in gene expression and protein activity lead to complex changes influencing neurite growth and remodeling among other neuroplasticity-related events (see cell proliferation in Section III.A) (Patapoutian and Reichardt, 2001; Poo, 2001). Chronic treatment of rodents with either lithium or VPA activates several members of the ERK pathway such as ERK, RSK1, Bad, and CREB (Chen and Manji, 2006; Ghribi *et al.*, 2003; Hao *et al.*, 2004). VPA increases the growth of neurites in human neuroblastoma SH-SY5Y cells (Yuan *et al.*, 2001) and primary cortical cells in culture which both depend on ERK pathway activation (Hao *et al.*, 2004). VPA improves axonal regeneration and recovery of motor functions after sciatic nerve axotomy in adult rats (Cui *et al.*, 2003) and lithium supports retinal ganglion cells (RGCs) axons regeneration after axotomy (Huang *et al.*, 2003). These effects seem to be due to increases of Bcl-2 because it was shown that depleting the expression of Bcl-2 in RGCs blocks the effects of lithium.

Animal behavior studies show that pharmacological inhibition of the ERK pathway results in increased activity (Einat *et al.*, 2003) and disruption of latent inhibition (Lewis *et al.*, 2004) that is similar to the effects of psychostimulants that may reflect increased distractibility. Depending on the brain area, activation of the ERK pathway by infusion of recombinant BDNF can lead to both antidepressant-like as well as pro-depressant-like effects. Infusions into the hippocampus result in behavior (measured 3 days later) similar to behavior induced by antidepressants (Shirayama *et al.*, 2002; Siuciak *et al.*, 1997). However, infusions into the ventral tegmental area-nucleus accumbens circuit, in which chronic stress increases BDNF expression, lead to depressive-like behavior (Eisch *et al.*, 2003). Taken together, these data imply a brain region-specific regulation of behavior by the activation of the ERK pathway.

Alterations of neurite morphology and function clearly play an important role in the CNS. However, most available data examining these mechanisms has so far been obtained only from *in vitro* cell culture models. More data on mood stabilizer effects on neurite morphology *in vivo* is clearly needed. However, one study did find that lithium was efficacious in preventing the structural remodeling of CA3 pyramidal neurons by chronic restraint stress (Wood *et al.*, 2004).

K. Cell Death

A dysfunction in cell proliferation and maturation in the adult brain is one possible explanation for the findings of cell loss and atrophy in postmortem brains of patients with affective disorders. However, the brain is highly sensitive to several kinds of neurotoxic events leading to cell death by apoptosis. Therefore, another important means to ameliorate the course of the disease could be augmentation of cellular resilience against deleterious events. Indeed, mood stabilizers prevent cell death in multiple preclinical *in vitro* and *in vivo* models. These neuroprotective properties *in vitro* have been found in several different cell culture systems. In cerebellar granule, cell cultures lithium-induced apoptosis in immature neurons, but promoted survival of mature neurons (D'Mello *et al.*, 1994). GABAergic neurons from cerebellum and cerebral cortex survived also longer in culture when treated with lithium (Volonte *et al.*, 1994).

Additionally, an important promoter of neurodegeneration in stroke, Huntington's disease, ALS, brain trauma, and possibly Parkinson's and Alzheimer's disease is glutamate-induced excitotoxicity (Yuan and Yankner, 2000). Too much glutamate release can be destructive and excite a neuron to death. Neurotoxicity follows as a response to overactivation of calcium-dependent enzymes and the generation of reactive oxygen species leading to apoptosis. Long-term lithium treatment protects cerebral granule cells in culture against glutamate-induced excitotoxicity (Nonaka *et al.*, 1998) and lithium as well as VPA reduce brain damage induced by transient focal cerebral ischemia in rats after middle cerebral artery occlusion (Nonaka and Chuang, 1998; Ren *et al.*, 2003, 2004). Besides decreasing the size of the infarct, mood stabilizer treatment also improves neurological deficits, including abnormal posture and hemiplegia. Lithium also protects against neurodegeneration in rodent models for Alzheimer's (De Ferrari *et al.*, 2003), Parkinson's (Youdim and Arraf, 2004), and Huntington's disease (Senatorov *et al.*, 2004; Wei *et al.*, 2001). β-Amyloid (Aβ), the main constituent of amyloid plaques in various neurological disorders, most prominently Alzheimer's disease, can induce cell death of neurons *in vitro* and *in vivo*. Aβ is produced by cleavage of amyloid precursor protein (APP). Lithium, as well as VPA, inhibits the production of Aβ in cell cultures models and in brains of APP-overproducing mice (De Ferrari *et al.*, 2003; Hoshi *et al.*, 2003; Phiel *et al.*, 2003; Su *et al.*, 2004). Another common pathological feature of many neurodegenerative disorders including Alzheimer's disease is neurofibrillary tangles. Abnormal phosphorylation of tau is one of the pathogenic mechanisms leading to tangle formation. Treatment of transgenic mice overexpressing mutant tau with lithium resulted in lower levels of tau phosphorylation and reduced levels of aggregated, insoluble tau (Noble *et al.*, 2005; Perez *et al.*, 2003). The effects of mood stabilizers on Aβ production and tau phosphorylation might be due to inhibition of GSK-3. Tau is a direct target of GSK-3, mediated by phosphorylation thus explains noted effects. Other important mechanisms by which lithium and VPA reduce neurodegeneration are the enhancements of anti-apoptoic factors such as Bcl-2 and neurotrophic factors such as BDNF and activation of the ERK pathway discussed above.

Cell death and atrophy causing dysregulation of the CNS functions are further likely candidates in the pathophysiology of BPD. More research is warranted to clarify the functional role of cell death and to categorize specific cells that are most at risk for cell death in the brains of BPD patients.

IV. Conclusions

Emerging results from a variety of clinical and preclinical experimental studies suggest an important role for neuroplasticity in the pathophysiology of BPD. Understanding the biological mechanisms underlying structural and functional abnormalities in the brains of BPD patients will provide us with greater opportunities to develop next generation therapeutics. Increasing evidence supports regional reductions in brain volume are most likely a consequence of impairments in neuroplasticity and cellular resilience. Thus, optimal long-term treatment of these severe illnesses will require therapeutic treatments that enhance neuroplasticity. Mood stabilizers, such as lithium and VPA, currently are the most effective available medications used in BPD, and many studies report remarkable effects on cellular plasticity and signaling cascades related to these medications. To understand the pathophysiology of BPD, more research is needed to elucidate the relevance certain impairments in neuroplasticity have on brain function and emotional behavior. Furthermore, discovering the biological targets of mood stabilizers is critical for the development of new and better therapeutics. The growing appreciation of the relevance of neuroplasticity to the pathophysiology of mood disorders has opened new doors to a better understanding of these conditions and holds great promise for the future development of novel therapeutics that could improve the lives of millions.

References

Adams, J. M., and Cory, S. (1998). The Bcl-2 protein family: Arbiters of cell survival. *Science* **281,** 1322–1326.

Alonso, R., Griebel, G., Pavone, G., Stemmelin, J., Le Fur, G., and Soubrie, P. (2004). Blockade of CRF(1) or V(1b) receptors reverses stress-induced suppression of neurogenesis in a mouse model of depression. *Mol. Psychiatry* **9,** 224, 278–286.

Altshuler, L. L., Bartzokis, G., Grieder, T., Curran, J., and Mintz, J. (1998). Amygdala enlargement in bipolar disorder and hippocampal reduction in schizophrenia: An MRI study demonstrating neuroanatomic specificity. *Arch. Gen. Psychiatry* **55,** 663–664.

Altshuler, L. L., Bartzokis, G., Grieder, T., Curran, J., Jimenez, T., Leight, K., Wilkins, J., Gerner, R., and Mintz, J. (2000). An MRI study of temporal lobe structures in men with bipolar disorder or schizophrenia. *Biol. Psychiatry* **48,** 147–162.

Arsenijevic, Y., Villemure, J. G., Brunet, J. F., Bloch, J. J., Deglon, N., Kostic, C., Zurn, A., and Aebischer, P. (2001). Isolation of multipotent neural precursors residing in the cortex of the adult human brain. *Exp. Neurol.* **170,** 48–62.

Banasr, M., Hery, M., Printemps, R., and Daszuta, A. (2004). Serotonin-induced increases in adult cell proliferation and neurogenesis are mediated through different and common 5-HT receptor subtypes in the dentate gyrus and the subventricular zone. *Neuropsychopharmacology* **29,** 450–460.

Baumann, B., Bielau, H., Krell, D., Agelink, M. W., Diekmann, S., Wurthmann, C., Trubner, K., Bernstein, H. G., Danos, P., and Bogerts, B. (2002). Circumscribed numerical deficit of dorsal raphe neurons in mood disorders. *Psychol. Med.* **32,** 93–103.

Baxter, L. R., Jr., Schwartz, J. M., Phelps, M. E., Mazziotta, J. C., Guze, B. H., Selin, C. E., Gerner, R. H., and Sumida, R. M. (1989). Reduction of prefrontal cortex glucose metabolism common to three types of depression. *Arch. Gen. Psychiatry* **46,** 243–250.

Beasley, C. L., Zhang, Z. J., Patten, I., and Reynolds, G. P. (2002). Selective deficits in prefrontal cortical GABAergic neurons in schizophrenia defined by the presence of calcium-binding proteins. *Biol. Psychiatry* **52,** 708–715.

Beaulieu, J. M., Sotnikova, T. D., Yao, W. D., Kockeritz, L., Woodgett, J. R., Gainetdinov, R. R., and Caron, M. G. (2004). Lithium antagonizes dopamine-dependent behaviors mediated by an AKT/glycogen synthase kinase 3 signaling cascade. *Proc. Natl. Acad. Sci. USA* **101,** 5099–5104.

Bedard, A., Gravel, C., and Parent, A. (2006). Chemical characterization of newly generated neurons in the striatum of adult primates. *Exp. Brain Res.* **170,** 501–512.

Belachew, S., Chittajallu, R., Aguirre, A. A., Yuan, X., Kirby, M., Anderson, S., and Gallo, V. (2003). Postnatal NG2 proteoglycan-expressing progenitor cells are intrinsically multipotent and generate functional neurons. *J. Cell Biol.* **161,** 169–186.

Benes, F. M., and Berretta, S. (2001). GABAergic interneurons: Implications for understanding schizophrenia and bipolar disorder. *Neuropsychopharmacology* **25,** 1–27.

Benes, F. M., Todtenkopf, M. S., and Kostoulakos, P. (2001a). GluR5,6,7 subunit immunoreactivity on apical pyramidal cell dendrites in hippocampus of schizophrenics and manic depressives. *Hippocampus* **11,** 482–491.

Benes, F. M., Vincent, S. L., and Todtenkopf, M. (2001b). The density of pyramidal and nonpyramidal neurons in anterior cingulate cortex of schizophrenic and bipolar subjects. *Biol. Psychiatry* **50,** 395–406.

Benraiss, A., Chmielnicki, E., Lerner, K., Roh, D., and Goldman, S. A. (2001). Adenoviral brain-derived neurotrophic factor induces both neostriatal and olfactory neuronal recruitment from endogenous progenitor cells in the adult forebrain. *J. Neurosci.* **21,** 6718–6731.

Bernier, P. J., Bedard, A., Vinet, J., Levesque, M., and Parent, A. (2002). Newly generated neurons in the amygdala and adjoining cortex of adult primates. *Proc. Natl. Acad. Sci. USA* **99,** 11464–11469.

Bertolino, A., Frye, M., Callicott, J. H., Mattay, V. S., Rakow, R., Shelton-Repella, J., Post, R., and Weinberger, D. R. (2003). Neuronal pathology in the hippocampal area of patients with bipolar disorder: A study with proton magnetic resonance spectroscopic imaging. *Biol. Psychiatry* **53,** 906–913.

Birnbaum, S. G., Yuan, P. X., Wang, M., Vijayraghavan, S., Bloom, A. K., Davis, D. J., Gobeske, K. T., Sweatt, J. D., Manji, H. K., and Arnsten, A. F. (2004). Protein kinase C overactivity impairs prefrontal cortical regulation of working memory. *Science* **306,** 882–884.

Blumberg, H. P., Stern, E., Ricketts, S., Martinez, D., de Asis, J., White, T., Epstein, J., Isenberg, N., McBride, P. A., Kemperman, I., Emmerich, S., Dhawan, V., *et al.* (1999). Rostral and orbital prefrontal cortex dysfunction in the manic state of bipolar disorder. *Am. J. Psychiatry* **156,** 1986–1988.

Blumberg, H. P., Stern, E., Martinez, D., Ricketts, S., de Asis, J., White, T., Epstein, J., McBride, P. A., Eidelberg, D., Kocsis, J. H., and Silbersweig, D. A. (2000). Increased anterior cingulate and caudate activity in bipolar mania. *Biol. Psychiatry* **48,** 1045–1152.

Botteron, K. N., Vannier, M. W., Geller, B., Todd, R. D., and Lee, B. C. (1995). Preliminary study of magnetic resonance imaging characteristics in 8- to 16-year-olds with mania. *J. Am. Acad. Child Adolesc. Psychiatry* **34,** 742–749.

Bouras, C., Kovari, E., Hof, P. R., Riederer, B. M., and Giannakopoulos, P. (2001). Anterior cingulate cortex pathology in schizophrenia and bipolar disorder. *Acta Neuropathol. (Berl.)* **102,** 373–379.

Bowden, C. L., Brugger, A. M., Swann, A. C., Calabrese, J. R., Janicak, P. G., Petty, F., Dilsaver, S. C., Davis, J. M., Rush, A. J., Small, J. G., *et al.* (1994). Efficacy of divalproex vs lithium and placebo in the treatment of mania. The depakote mania study group. *JAMA* **271,** 918–924.

Bowley, M. P., Drevets, W. C., Ongur, D., and Price, J. L. (2002). Low glial numbers in the amygdala in major depressive disorder. *Biol. Psychiatry* **52,** 404–412.

Brambilla, P., Harenski, K., Nicoletti, M., Mallinger, A. G., Frank, E., Kupfer, D. J., Keshavan, M. S., and Soares, J. C. (2001). Differential effects of age on brain gray matter in bipolar patients and healthy individuals. *Neuropsychobiology* **43,** 242–247.

Brambilla, P., Nicoletti, M. A., Harenski, K., Sassi, R. B., Mallinger, A. G., Frank, E., Kupfer, D. J., Keshavan, M. S., and Soares, J. C. (2002). Anatomical MRI study of subgenual prefrontal cortex in bipolar and unipolar subjects. *Neuropsychopharmacology* **27,** 792–799.

Brambilla, P., Harenski, K., Nicoletti, M., Sassi, R. B., Mallinger, A. G., Frank, E., Kupfer, D. J., Keshavan, M. S., and Soares, J. C. (2003). MRI investigation of temporal lobe structures in bipolar patients. *J. Psychiatr. Res.* **37,** 287–295.

Brezun, J. M., and Daszuta, A. (1999). Depletion in serotonin decreases neurogenesis in the dentate gyrus and the subventricular zone of adult rats. *Neuroscience* **89,** 999–1002.
Brezun, J. M., and Daszuta, A. (2000). Serotonin may stimulate granule cell proliferation in the adult hippocampus, as observed in rats grafted with foetal raphe neurons. *Eur. J. Neurosci.* **12,** 391–396.
Brown, G. W., and Harris, T. O. (1989). "Life Events and Illness." Guilford Press, New York.
Bruckheimer, E. M., Cho, S. H., Sarkiss, M., Herrmann, J., and McDonnell, T. J. (1998). The Bcl-2 gene family and apoptosis. *Adv. Biochem. Eng. Biotechnol.* **62,** 75–105.
Cameron, H. A., and Gould, E. (1994). Adult neurogenesis is regulated by adrenal steroids in the dentate gyrus. *Neuroscience* **61,** 203–209.
Cameron, H. A., and McKay, R. D. (1999). Restoring production of hippocampal neurons in old age. *Nat. Neurosci.* **2,** 894–897.
Cameron, H. A., Tanapat, P., and Gould, E. (1998). Adrenal steroids and N-methyl-D-aspartate receptor activation regulate neurogenesis in the dentate gyrus of adult rats through a common pathway. *Neuroscience* **82,** 349–354.
Cecil, K. M., DelBello, M. P., Morey, R., and Strakowski, S. M. (2002). Frontal lobe differences in bipolar disorder as determined by proton MR spectroscopy. *Bipolar Disord.* **4,** 357–365.
Chen, G., and Manji, H. K. (2006). The extracellular signal-regulated kinase pathway: An emerging promising target for mood stabilizers. *Curr. Opin. Psychiatry* **19,** 313–323.
Chen, G., Manji, H. K., Hawver, D. B., Wright, C. B., and Potter, W. Z. (1994). Chronic sodium valproate selectively decreases protein kinase C alpha and epsilon *in vitro*. *J. Neurochem.* **63,** 2361–2364.
Chen, G., Zeng, W. Z., Yuan, P. X., Huang, L. D., Jiang, Y. M., Zhao, Z. H., and Manji, H. K. (1999). The mood-stabilizing agents lithium and valproate robustly increase the levels of the neuroprotective protein bcl-2 in the CNS. *J. Neurochem.* **72,** 879–882.
Chen, G., Rajkowska, G., Du, F., Seraji-Bozorgzad, N., and Manji, H. K. (2000). Enhancement of hippocampal neurogenesis by lithium. *J. Neurochem.* **75,** 1729–1734.
Craddock, N., and Jones, I. (1999). Genetics of bipolar disorder. *J. Med. Genet.* **36,** 585–594.
Cui, S. S., Yang, C. P., Bowen, R. C., Bai, O., Li, X. M., Jiang, W., and Zhang, X. (2003). Valproic acid enhances axonal regeneration and recovery of motor function after sciatic nerve axotomy in adult rats. *Brain Res.* **975,** 229–236.
Czeh, B., Michaelis, T., Watanabe, T., Frahm, J., de Biurrun, G., van Kampen, M., Bartolomucci, A., and Fuchs, E. (2001). Stress-induced changes in cerebral metabolites, hippocampal volume, and cell proliferation are prevented by antidepressant treatment with tianeptine. *Proc. Natl. Acad. Sci. USA* **98,** 12796–12801.
Czeh, B., Welt, T., Fischer, A. K., Erhardt, A., Schmitt, W., Muller, M. B., Toschi, N., Fuchs, E., and Keck, M. E. (2002). Chronic psychosocial stress and concomitant repetitive transcranial magnetic stimulation: Effects on stress hormone levels and adult hippocampal neurogenesis. *Biol. Psychiatry* **52,** 1057–1065.
Czeh, B., Simon, M., Schmelting, B., Hiemke, C., and Fuchs, E. (2006). Astroglial plasticity in the hippocampus is affected by chronic psychosocial stress and concomitant fluoxetine treatment. *Neuropsychopharmacology* **31**(8), 1616–1626.
D'Mello, S. R., Anelli, R., and Calissano, P. (1994). Lithium induces apoptosis in immature cerebellar granule cells but promotes survival of mature neurons. *Exp. Cell Res.* **211,** 332–338.
Davidson, T. L., and Jarrard, L. E. (2004). The hippocampus and inhibitory learning: A 'Gray' area? *Neurosci. Biobehav. Rev.* **28,** 261–271.
Dayer, A. G., Cleaver, K. M., Abouantoun, T., and Cameron, H. A. (2005). New GABAergic interneurons in the adult neocortex and striatum are generated from different precursors. *J. Cell Biol.* **168,** 415–427.
De Ferrari, G. V., Chacon, M. A., Barria, M. I., Garrido, J. L., Godoy, J. A., Olivares, G., Reyes, A. E., Alvarez, A., Bronfman, M., and Inestrosa, N. C. (2003). Activation of Wnt signaling

rescues neurodegeneration and behavioral impairments induced by beta-amyloid fibrils. *Mol. Psychiatry* **8,** 195–208.

Deicken, R. F., Fein, G., and Weiner, M. W. (1995a). Abnormal frontal lobe phosphorous metabolism in bipolar disorder. *Am. J. Psychiatry* **152,** 915–918.

Deicken, R. F., Weiner, M. W., and Fein, G. (1995b). Decreased temporal lobe phosphomonoesters in bipolar disorder. *J. Affect. Disord.* **33,** 195–199.

Drevets, W. C., Price, J. L., Simpson, J. R., Jr., Todd, R. D., Reich, T., Vannier, M., and Raichle, M. E. (1997). Subgenual prefrontal cortex abnormalities in mood disorders. *Nature* **386,** 824–827.

Drevets, W. C., Price, J. L., Bardgett, M. E., Reich, T., Todd, R. D., and Raichle, M. E. (2002). Glucose metabolism in the amygdala in depression: Relationship to diagnostic subtype and plasma cortisol levels. *Pharmacol. Biochem. Behav.* **71,** 431–447.

Duman, R. S. (2002). Pathophysiology of depression: The concept of synaptic plasticity. *Eur. Psychiatry* **17**(Suppl. 3), 306–310.

Duman, R. S., and Monteggia, L. M. (2006). A neurotrophic model for stress-related mood disorders. *Biol. Psychiatry* **59**(12), 1116–1127.

Dunn, R. T., Kimbrell, T. A., Ketter, T. A., Frye, M. A., Willis, M. W., Luckenbaugh, D. A., and Post, R. M. (2002). Principal components of the beck depression inventory and regional cerebral metabolism in unipolar and bipolar depression. *Biol. Psychiatry* **51,** 387–399.

Eastwood, S. L., and Harrison, P. J. (2001). Synaptic pathology in the anterior cingulate cortex in schizophrenia and mood disorders. A review and a Western blot study of synaptophysin, GAP-43 and the complexins. *Brain Res. Bull.* **55,** 569–578.

Einat, H., Yuan, P., Gould, T. D., Li, J., Du, J., Zhang, L., Manji, H. K., and Chen, G. (2003). The role of the extracellular signal-regulated kinase signaling pathway in mood modulation. *J. Neurosci.* **23,** 7311–7316.

Eisch, A. J., and Harburg, G. C. (2006). Opiates, psychostimulants, and adult hippocampal neurogenesis: Insights for addiction and stem cell biology. *Hippocampus* **16,** 271–286.

Eisch, A. J., Bolanos, C. A., de Wit, J., Simonak, R. D., Pudiak, C. M., Barrot, M., Verhaagen, J., and Nestler, E. J. (2003). Brain-derived neurotrophic factor in the ventral midbrain-nucleus accumbens pathway: A role in depression. *Biol. Psychiatry* **54,** 994–1005.

Ellicott, A., Hammen, C., Gitlin, M., Brown, G., and Jamison, K. (1990). Life events and the course of bipolar disorder. *Am. J. Psychiatry* **147,** 1194–1198.

Fagiolini, A., Kupfer, D. J., Masalehdan, A., Scott, J. A., Houck, P. R., and Frank, E. (2005). Functional impairment in the remission phase of bipolar disorder. *Bipolar Disord.* **7,** 281–285.

Frangou, S., Raymont, V., and Bettany, D. (2002). The Maudsley bipolar disorder project. A survey of psychotropic prescribing patterns in bipolar I disorder. *Bipolar Disord.* **4,** 378–385.

Fukumoto, T., Morinobu, S., Okamoto, Y., Kagaya, A., and Yamawaki, S. (2001). Chronic lithium treatment increases the expression of brain-derived neurotrophic factor in the rat brain. *Psychopharmacology (Berl.)* **158,** 100–106.

Garcia, A., Steiner, B., Kronenberg, G., Bick-Sander, A., and Kempermann, G. (2004). Age-dependent expression of glucocorticoid- and mineralocorticoid receptors on neural precursor cell populations in the adult murine hippocampus. *Aging Cell* **3,** 363–371.

Ghaemi, S. N., and Goodwin, F. K. (1999). Use of atypical antipsychotic agents in bipolar and schizoaffective disorders: Review of the empirical literature. *J. Clin. Psychopharmacol.* **19,** 354–361.

Ghribi, O., Prammonjago, P., Herman, M. M., Spaulding, N. K., and Savory, J. (2003). Abeta (1–42)-induced JNK and ERK activation in rabbit hippocampus is differentially regulated by lithium but is not involved in the phosphorylation of tau. *Brain Res. Mol. Brain Res.* **119,** 201–206.

Gold, P. W., Chrousos, G., Kellner, C., Post, R., Roy, A., Augerinos, P., Schulte, H., Oldfield, E., and Loriaux, D. L. (1984). Psychiatric implications of basic and clinical studies with corticotropin-releasing factor. *Am. J. Psychiatry* **141,** 619–627.

Goldman, S. A., and Sim, F. (2005). Neural progenitor cells of the adult brain. *Novartis Found. Symp.* **265,** 66–80; discussion 82–97.

Gomez-Lira, G., Trillo, E., Ramirez, M., Asai, M., Sitges, M., and Gutierrez, R. (2002). The expression of GABA in mossy fiber synaptosomes coincides with the seizure-induced expression of GABAergic transmission in the mossy fiber synapse. *Exp. Neurol.* **177,** 276–283.

Gomez-Lira, G., Lamas, M., Romo-Parra, H., and Gutierrez, R. (2005). Programmed and induced phenotype of the hippocampal granule cells. *J. Neurosci.* **25,** 6939–6946.

Goodwin, F. K., and Jamison, K. R. (1990). "Manic-Depressive Illness." Oxford University Press, New York.

Goodwin, G. M., Cavanagh, J. T., Glabus, M. F., Kehoe, R. F., O'Carroll, R. E., and Ebmeier, K. P. (1997). Uptake of 99mTc-exametazime shown by single photon emission computed tomography before and after lithium withdrawal in bipolar patients: Associations with mania. *Br. J. Psychiatry* **170,** 426–430.

Gotz, R., Wiese, S., Takayama, S., Camarero, G. C., Rossoll, W., Schweizer, U., Troppmair, J., Jablonka, S., Holtmann, B., Reed, J. C., Rapp, U. R., and Sendtner, M. (2005). Bag1 is essential for differentiation and survival of hematopoietic and neuronal cells. *Nat. Neurosci.* **8,** 1169–1178.

Gould, E., Woolley, C. S., Cameron, H. A., Daniels, D. C., and McEwen, B. S. (1991). Adrenal steroids regulate postnatal development of the rat dentate gyrus: II. Effects of glucocorticoids and mineralocorticoids on cell birth. *J. Comp. Neurol.* **313,** 486–493.

Gould, E., Cameron, H. A., Daniels, D. C., Woolley, C. S., and McEwen, B. S. (1992). Adrenal hormones suppress cell division in the adult rat dentate gyrus. *J. Neurosci.* **12,** 3642–3650.

Gould, E., Tanapat, P., McEwen, B. S., Flugge, G., and Fuchs, E. (1998). Proliferation of granule cell precursors in the dentate gyrus of adult monkeys is diminished by stress. *Proc. Natl. Acad. Sci. USA* **95,** 3168–3171.

Gould, E., Reeves, A. J., Graziano, M. S., and Gross, C. G. (1999). Neurogenesis in the neocortex of adult primates. *Science* **286,** 548–552.

Gould, E., Vail, N., Wagers, M., and Gross, C. G. (2001). Adult-generated hippocampal and neocortical neurons in macaques have a transient existence. *Proc. Natl. Acad. Sci. USA* **98,** 10910–10917.

Gould, T. D., Einat, H., Bhat, R., and Manji, H. K. (2004). AR-A014418, a selective GSK-3 inhibitor, produces antidepressant-like effects in the forced swim test. *Int. J. Neuropsychopharmacol.* **7,** 387–390.

Grassi Zucconi, G., Cipriani, S., Balgkouranidou, I., and Scattoni, R. (2006). 'One night' sleep deprivation stimulates hippocampal neurogenesis. *Brain Res. Bull.* **69,** 375–381.

Greengard, P., Valtorta, F., Czernik, A. J., and Benfenati, F. (1993). Synaptic vesicle phosphoproteins and regulation of synaptic function. *Science* **259,** 780–785.

Grunhaus, L., Schreiber, S., Dolberg, O. T., Hirshman, S., and Dannon, P. N. (2002). Response to ECT in major depression: Are there differences between unipolar and bipolar depression? *Bipolar Disord.* **4**(Suppl. 1), 91–93.

Gu, W., Brannstrom, T., and Wester, P. (2000). Cortical neurogenesis in adult rats after reversible photothrombotic stroke. *J. Cereb. Blood Flow Metab.* **20,** 1166–1173.

Guzman-Marin, R., Suntsova, N., Methippara, M., Greiffenstein, R., Szymusiak, R., and McGinty, D. (2005). Sleep deprivation suppresses neurogenesis in the adult hippocampus of rats. *Eur. J. Neurosci.* **22,** 2111–2116.

Hairston, I. S., Little, M. T., Scanlon, M. D., Barakat, M. T., Palmer, T. D., Sapolsky, R. M., and Heller, H. C. (2005). Sleep restriction suppresses neurogenesis induced by hippocampus-dependent learning. *J. Neurophysiol.* **94,** 4224–4233.

Halim, N. D., Weickert, C. S., McClintock, B. W., Weinberger, D. R., and Lipska, B. K. (2004). Effects of chronic haloperidol and clozapine treatment on neurogenesis in the adult rat hippocampus. *Neuropsychopharmacology* **29,** 1063–1069.

Hall, A. C., Brennan, A., Goold, R. G., Cleverley, K., Lucas, F. R., Gordon-Weeks, P. R., and Salinas, P. C. (2002). Valproate regulates GSK-3-mediated axonal remodeling and synapsin I clustering in developing neurons. *Mol. Cell Neurosci.* **20,** 257–270.

Hao, Y., Creson, T., Zhang, L., Li, P., Du, F., Yuan, P., Gould, T. D., Manji, H. K., and Chen, G. (2004). Mood stabilizer valproate promotes ERK pathway-dependent cortical neuronal growth and neurogenesis. *J. Neurosci.* **24,** 6590–6599.

Hashimoto, R., Senatorov, V., Kanai, H., Leeds, P., and Chuang, D. M. (2003). Lithium stimulates progenitor proliferation in cultured brain neurons. *Neuroscience* **117,** 55–61.

Haskett, R. F. (1985). Diagnostic categorization of psychiatric disturbance in Cushing's syndrome. *Am. J. Psychiatry* **142,** 911–916.

Hawrylak, N., and Greenough, W. T. (1995). Monocular deprivation alters the morphology of glial fibrillary acidic protein-immunoreactive astrocytes in the rat visual cortex. *Brain Res.* **683,** 187–199.

Heckers, S., Stone, D., Walsh, J., Shick, J., Koul, P., and Benes, F. M. (2002). Differential hippocampal expression of glutamic acid decarboxylase 65 and 67 messenger RNA in bipolar disorder and schizophrenia. *Arch. Gen. Psychiatry* **59,** 521–529.

Heine, V. M., Maslam, S., Zareno, J., Joels, M., and Lucassen, P. J. (2004). Suppressed proliferation and apoptotic changes in the rat dentate gyrus after acute and chronic stress are reversible. *Eur. J. Neurosci.* **19,** 131–144.

Henn, F. A., and Vollmayr, B. (2004). Neurogenesis and depression: Etiology or epiphenomenon? *Biol. Psychiatry* **56,** 146–150.

Hlastala, S. A., Frank, E., Kowalski, J., Sherrill, J. T., Tu, X. M., Anderson, B., and Kupfer, D. J. (2000). Stressful life events, bipolar disorder, and the "kindling model." *J. Abnorm. Psychol.* **109,** 777–786.

Hoshi, M., Sato, M., Matsumoto, S., Noguchi, A., Yasutake, K., Yoshida, N., and Sato, K. (2003). Spherical aggregates of beta-amyloid (amylospheroid) show high neurotoxicity and activate tau protein kinase I/glycogen synthase kinase-3beta. *Proc. Natl. Acad. Sci. USA* **100,** 6370–6375.

Hsieh, J., Nakashima, K., Kuwabara, T., Mejia, E., and Gage, F. H. (2004). Histone deacetylase inhibition-mediated neuronal differentiation of multipotent adult neural progenitor cells. *Proc. Natl. Acad. Sci. USA* **101,** 16659–16664.

Huang, G. J., and Herbert, J. (2006). Stimulation of neurogenesis in the hippocampus of the adult rat by fluoxetine requires rhythmic change in corticosterone. *Biol. Psychiatry* **59,** 619–624.

Huang, X., Wu, D. Y., Chen, G., Manji, H., and Chen, D. F. (2003). Support of retinal ganglion cell survival and axon regeneration by lithium through a Bcl-2-dependent mechanism. *Invest. Ophthalmol. Vis. Sci.* **44,** 347–354.

Hudson, J. I., Hudson, M. S., Griffing, G. T., Melby, J. C., and Pope, H. G., Jr. (1987). Phenomenology and family history of affective disorder in Cushing's disease. *Am. J. Psychiatry* **144,** 951–953.

Hussain, R. J., Stumpo, D. J., Blackshear, P. J., Lenox, R. H., Abel, T., and McNamara, R. K. (2006). Myristoylated alanine rich C kinase substrate (MARCKS) heterozygous mutant mice exhibit deficits in hippocampal mossy fiber-CA3 long-term potentiation. *Hippocampus* **16,** 495–503.

Hyman, S. E., and Nestler, E. J. (1996). Initiation and adaptation: A paradigm for understanding psychotropic drug action. *Am. J. Psychiatry* **153,** 151–162.

Imura, T., Kornblum, H. I., and Sofroniew, M. V. (2003). The predominant neural stem cell isolated from postnatal and adult forebrain but not early embryonic forebrain expresses GFAP. *J. Neurosci.* **23,** 2824–2832.

Ito, H., Kawashima, R., Awata, S., Ono, S., Sato, K., Goto, R., Koyama, M., Sato, M., and Fukuda, H. (1996). Hypoperfusion in the limbic system and prefrontal cortex in depression: SPECT with anatomic standardization technique. *J. Nucl. Med.* **37,** 410–414.

Jiang, H., Guo, W., Liang, X., and Rao, Y. (2005). Both the establishment and the maintenance of neuronal polarity require active mechanisms: Critical roles of GSK-3beta and its upstream regulators. *Cell* **120,** 123–135.

Jiang, W., Gu, W., Brannstrom, T., Rosqvist, R., and Wester, P. (2001). Cortical neurogenesis in adult rats after transient middle cerebral artery occlusion. *Stroke* **32,** 1201–1207.

Johnson, S. L. (2005). Life events in bipolar disorder: Towards more specific models. *Clin. Psychol. Rev.* **25,** 1008–1027.

Johnson, S. L., Sandrow, D., Meyer, B., Winters, R., Miller, I., Solomon, D., and Keitner, G. (2000). Increases in manic symptoms after life events involving goal attainment. *J. Abnorm. Psychol.* **109,** 721–727.

Johnston-Wilson, N. L., Sims, C. D., Hofmann, J. P., Anderson, L., Shore, A. D., Torrey, E. F., and Yolken, R. H. (2000). Disease-specific alterations in frontal cortex brain proteins in schizophrenia, bipolar disorder, and major depressive disorder. The Stanley Neuropathology Consortium. *Mol. Psychiatry* **5,** 142–149.

Judd, L. L., and Akiskal, H. S. (2003). The prevalence and disability of bipolar spectrum disorders in the US population: Re-analysis of the ECA database taking into account subthreshold cases. *J. Affect. Disord.* **73,** 123–131.

Kaidanovich-Beilin, O., Milman, A., Weizman, A., Pick, C. G., and Eldar-Finkelman, H. (2004). Rapid antidepressive-like activity of specific glycogen synthase kinase-3 inhibitor and its effect on beta-catenin in mouse hippocampus. *Biol. Psychiatry* **55,** 781–784.

Kato, T., Shioiri, T., Murashita, J., Hamakawa, H., Inubushi, T., and Takahashi, S. (1994a). Phosphorus-31 magnetic resonance spectroscopy and ventricular enlargement in bipolar disorder. *Psychiatry Res.* **55,** 41–50.

Kato, T., Takahashi, S., Shioiri, T., Murashita, J., Hamakawa, H., and Inubushi, T. (1994b). Reduction of brain phosphocreatine in bipolar II disorder detected by phosphorus-31 magnetic resonance spectroscopy. *J. Affect. Disord.* **31,** 125–133.

Kato, T., Hamakawa, H., Shioiri, T., Murashita, J., Takahashi, Y., Takahashi, S., and Inubushi, T. (1996). Choline-containing compounds detected by proton magnetic resonance spectroscopy in the basal ganglia in bipolar disorder. *J. Psychiatry Neurosci.* **21,** 248–254.

Keck, P. E., Jr., and Manji, H. K. (2002). "Current and Emerging Treatments for Acute Mania and Long-Term Prophylaxis for Bipolar Disorder." Lippincott Williams & Wilkins, Philadelphia.

Keedwell, P. A., Andrew, C., Williams, S. C., Brammer, M. J., and Phillips, M. L. (2005). A double dissociation of ventromedial prefrontal cortical responses to sad and happy stimuli in depressed and healthy individuals. *Biol. Psychiatry* **58,** 495–503.

Kempermann, G., Jessberger, S., Steiner, B., and Kronenberg, G. (2004). Milestones of neuronal development in the adult hippocampus. *Trends Neurosci.* **27,** 447–452.

Kim, J. S., Chang, M. Y., Yu, I. T., Kim, J. H., Lee, S. H., Lee, Y. S., and Son, H. (2004). Lithium selectively increases neuronal differentiation of hippocampal neural progenitor cells both *in vitro* and *in vivo*. *J. Neurochem.* **89,** 324–336.

Kippin, T. E., Kapur, S., and van der Kooy, D. (2005). Dopamine specifically inhibits forebrain neural stem cell proliferation, suggesting a novel effect of antipsychotic drugs. *J. Neurosci.* **25,** 5815–5823.

Kodama, M., Fujioka, T., and Duman, R. S. (2004). Chronic olanzapine or fluoxetine administration increases cell proliferation in hippocampus and prefrontal cortex of adult rat. *Biol. Psychiatry* **56,** 570–580.

Koketsu, D., Mikami, A., Miyamoto, Y., and Hisatsune, T. (2003). Nonrenewal of neurons in the cerebral neocortex of adult macaque monkeys. *J. Neurosci.* **23,** 937–942.

Kornack, D. R., and Rakic, P. (2001). Cell proliferation without neurogenesis in adult primate neocortex. *Science* **294,** 2127–2130.

Kuhn, H. G., Biebl, M., Wilhelm, D., Li, M., Friedlander, R. M., and Winkler, J. (2005). Increased generation of granule cells in adult Bcl-2-overexpressing mice: A role for cell death during continued hippocampal neurogenesis. *Eur. J. Neurosci.* **22,** 1907–1915.

Kupfer, D. J. (2005). The increasing medical burden in bipolar disorder. *JAMA* **293,** 2528–2530.

Laeng, P., Pitts, R. L., Lemire, A. L., Drabik, C. E., Weiner, A., Tang, H., Thyagarajan, R., Mallon, B. S., and Altar, C. A. (2004). The mood stabilizer valproic acid stimulates GABA neurogenesis from rat forebrain stem cells. *J. Neurochem.* **91,** 238–251.

Lee, H. J., Kim, J. W., Yim, S. V., Kim, M. J., Kim, S. A., Kim, Y. J., Kim, C. J., and Chung, J. H. (2001). Fluoxetine enhances cell proliferation and prevents apoptosis in dentate gyrus of maternally separated rats. *Mol. Psychiatry* **6,** 610, 725–728.

Lemaire, V., Koehl, M., Le Moal, M., and Abrous, D. N. (2000). Prenatal stress produces learning deficits associated with an inhibition of neurogenesis in the hippocampus. *Proc. Natl. Acad. Sci. USA* **97,** 11032–11037.

Lemaire, V., Lamarque, S., Moal, M. L., Piazza, P. V., and Abrous, D. N. (2006). Postnatal stimulation of the pups counteracts prenatal stress-induced deficits in hippocampal neurogenesis. *Biol. Psychiatry* **59**(9), 786–792.

Lenox, R. H., Watson, D. G., Patel, J., and Ellis, J. (1992). Chronic lithium administration alters a prominent PKC substrate in rat hippocampus. *Brain Res.* **570,** 333–340.

Leuner, B., Gould, E., and Shors, T. J. (2006). Is there a link between adult neurogenesis and learning? *Hippocampus* **16,** 216–224.

Leverich, G. S., McElroy, S. L., Suppes, T., Keck, P. E., Jr., Denicoff, K. D., Nolen, W. A., Altshuler, L. L., Rush, A. J., Kupka, R., Frye, M. A., Autio, K. A., and Post, R. M. (2002). Early physical and sexual abuse associated with an adverse course of bipolar illness. *Biol. Psychiatry* **51,** 288–297.

Lewis, M. C., Davis, J. A., and Gould, T. J. (2004). Inhibition of mitogen-activated protein kinase-extracellular signal-regulated kinase disrupts latent inhibition of cued fear conditioning in C57BL/6 mice. *Behav. Neurosci.* **118,** 1444–1449.

Liu, S., Wang, J., Zhu, D., Fu, Y., Lukowiak, K., and Lu, Y. M. (2003). Generation of functional inhibitory neurons in the adult rat hippocampus. *J. Neurosci.* **23,** 732–736.

Lopez, J. F., Akil, H., and Watson, S. J. (1999). Neural circuits mediating stress. *Biol. Psychiatry* **46,** 1461–1471.

Lucas, F. R., and Salinas, P. C. (1997). WNT-7a induces axonal remodeling and increases synapsin I levels in cerebellar neurons. *Dev. Biol.* **192,** 31–44.

Lucas, F. R., Goold, R. G., Gordon-Weeks, P. R., and Salinas, P. C. (1998). Inhibition of GSK-3beta leading to the loss of phosphorylated MAP-1B is an early event in axonal remodelling induced by WNT-7a or lithium. *J. Cell Sci.* **111**(Pt. 10), 1351–1361.

Lyoo, I. K., Lee, H. K., Jung, J. H., Noam, G. G., and Renshaw, P. F. (2002). White matter hyperintensities on magnetic resonance imaging of the brain in children with psychiatric disorders. *Compr. Psychiatry* **43,** 361–368.

Macritchie, K., Geddes, J. R., Scott, J., Haslam, D., de Lima, M., and Goodwin, G. (2003). Valproate for acute mood episodes in bipolar disorder. *Cochrane Database Syst. Rev.,* CD004052.

Madsen, T. M., Treschow, A., Bengzon, J., Bolwig, T. G., Lindvall, O., and Tingstrom, A. (2000). Increased neurogenesis in a model of electroconvulsive therapy. *Biol. Psychiatry* **47,** 1043–1049.

Madsen, T. M., Yeh, D. D., Valentine, G. W., and Duman, R. S. (2005). Electroconvulsive seizure treatment increases cell proliferation in rat frontal cortex. *Neuropsychopharmacology* **30,** 27–34.

Maeda, M., Namikawa, K., Kobayashi, I., Ohba, N., Takahara, Y., Kadono, C., Tanaka, A., and Kiyama, H. (2006). Targeted gene therapy toward astrocytoma using a Cre/loxP-based adenovirus system. *Brain Res.* **1081,** 34–43.

Magavi, S. S., Leavitt, B. R., and Macklis, J. D. (2000). Induction of neurogenesis in the neocortex of adult mice. *Nature* **405,** 951–955.

Malberg, J. E., and Duman, R. S. (2003). Cell proliferation in adult hippocampus is decreased by inescapable stress: Reversal by fluoxetine treatment. *Neuropsychopharmacology* **28,** 1562–1571.

Malberg, J. E., Eisch, A. J., Nestler, E. J., and Duman, R. S. (2000). Chronic antidepressant treatment increases neurogenesis in adult rat hippocampus. *J. Neurosci.* **20,** 9104–9110.

Malkoff-Schwartz, S., Frank, E., Anderson, B., Sherrill, J. T., Siegel, L., Patterson, D., and Kupfer, D. J. (1998). Stressful life events and social rhythm disruption in the onset of manic and depressive bipolar episodes: A preliminary investigation. *Arch. Gen. Psychiatry* **55,** 702–707.

Malkoff-Schwartz, S., Frank, E., Anderson, B. P., Hlastala, S. A., Luther, J. F., Sherrill, J. T., Houck, P. R., and Kupfer, D. J. (2000). Social rhythm disruption and stressful life events in the onset of bipolar and unipolar episodes. *Psychol. Med.* **30,** 1005–1016.

Manev, H., Uz, T., Smalheiser, N. R., and Manev, R. (2001). Antidepressants alter cell proliferation in the adult brain *in vivo* and in neural cultures *in vitro*. *Eur. J. Pharmacol.* **411,** 67–70.

Manji, H. K., Moore, G. J., and Chen, G. (2000). Lithium up-regulates the cytoprotective protein Bcl-2 in the CNS *in vivo*: A role for neurotrophic and neuroprotective effects in manic depressive illness. *J. Clin. Psychiatry* **61**(Suppl. 9), 82–96.

Mathis, C., Hindelang, C., LeMeur, M., and Borrelli, E. (2000). A transgenic mouse model for inducible and reversible dysmyelination. *J. Neurosci.* **20,** 7698–7705.

Matsutani, S., and Leon, M. (1993). Elaboration of glial cell processes in the rat olfactory bulb associated with early learning. *Brain Res.* **613,** 317–320.

McNamara, R. K., Hussain, R. J., Simon, E. J., Stumpo, D. J., Blackshear, P. J., Abel, T., and Lenox, R. H. (2005). Effect of myristoylated alanine-rich C kinase substrate (MARCKS) overexpression on hippocampus-dependent learning and hippocampal synaptic plasticity in MARCKS transgenic mice. *Hippocampus* **15,** 675–683.

Meshi, D., Drew, M. R., Saxe, M., Ansorge, M. S., David, D., Santarelli, L., Malapani, C., Moore, H., and Hen, R. (2006). Hippocampal neurogenesis is not required for behavioral effects of environmental enrichment. *Nat. Neurosci.* **9**(6), 729–731.

Middleton, G., Pinon, L. G., Wyatt, S., and Davies, A. M. (1998). Bcl-2 accelerates the maturation of early sensory neurons. *J. Neurosci.* **18,** 3344–3350.

Migliorelli, R., Starkstein, S. E., Teson, A., de Quiros, G., Vazquez, S., Leiguarda, R., and Robinson, R. G. (1993). SPECT findings in patients with primary mania. *J. Neuropsychiatry Clin. Neurosci.* **5,** 379–383.

Mirescu, C., Peters, J. D., and Gould, E. (2004). Early life experience alters response of adult neurogenesis to stress. *Nat. Neurosci.* **7,** 841–846.

Moore, G. J., Bebchuk, J. M., Parrish, J. K., Faulk, M. W., Arfken, C. L., Strahl-Bevacqua, J., and Manji, H. K. (1999). Temporal dissociation between lithium-induced changes in frontal lobe myo-inositol and clinical response in manic-depressive illness. *Am. J. Psychiatry* **156,** 1902–1908.

Moore, C. M., Breeze, J. L., Gruber, S. A., Babb, S. M., Frederick, B. B., Villafuerte, R. A., Stoll, A. L., Hennen, J., Yurgelun-Todd, D. A., Cohen, B. M., and Renshaw, P. F. (2000a). Choline, myo-inositol and mood in bipolar disorder: A proton magnetic resonance spectroscopic imaging study of the anterior cingulate cortex. *Bipolar Disord.* **2,** 207–216.

Moore, G. J., Bebchuk, J. M., Hasanat, K., Chen, G., Seraji-Bozorgzad, N., Wilds, I. B., Faulk, M. W., Koch, S., Glitz, D. A., Jolkovsky, L., and Manji, H. K. (2000b). Lithium increases N-acetyl-aspartate in the human brain: *In vivo* evidence in support of bcl-2's neurotrophic effects? *Biol. Psychiatry* **48,** 1–8.

Moore, P. B., Shepherd, D. J., Eccleston, D., Macmillan, I. C., Goswami, U., McAllister, V. L., and Ferrier, I. N. (2001). Cerebral white matter lesions in bipolar affective disorder: Relationship to outcome. *Br. J. Psychiatry* **178,** 172–176.

Murray, C. J., and Lopez, A. D. (1997). Alternative projections of mortality and disability by cause 1990–2020: Global Burden of Disease Study. *Lancet* **349,** 1498–1504.

Nakagawa, S., Kim, J. E., Lee, R., Malberg, J. E., Chen, J., Steffen, C., Zhang, Y. J., Nestler, E. J., and Duman, R. S. (2002). Regulation of neurogenesis in adult mouse hippocampus by cAMP and the cAMP response element-binding protein. *J. Neurosci.* **22,** 3673–3682.

Nemeroff, C. B. (2003). Safety of available agents used to treat bipolar disorder: Focus on weight gain. *J. Clin. Psychiatry* **64,** 532–539.

Noble, W., Planel, E., Zehr, C., Olm, V., Meyerson, J., Suleman, F., Gaynor, K., Wang, L., LaFrancois, J., Feinstein, B., Burns, M., Krishnamurthy, P., *et al.* (2005). Inhibition of glycogen synthase kinase-3 by lithium correlates with reduced tauopathy and degeneration *in vivo*. *Proc. Natl. Acad. Sci. USA* **102,** 6990–6995.

Noga, J. T., Vladar, K., and Torrey, E. F. (2001). A volumetric magnetic resonance imaging study of monozygotic twins discordant for bipolar disorder. *Psychiatry Res.* **106,** 25–34.

Nonaka, S., and Chuang, D. M. (1998). Neuroprotective effects of chronic lithium on focal cerebral ischemia in rats. *Neuroreport* **9,** 2081–2084.

Nonaka, S., Katsube, N., and Chuang, D. M. (1998). Lithium protects rat cerebellar granule cells against apoptosis induced by anticonvulsants, phenytoin and carbamazepine. *J. Pharmacol. Exp. Ther.* **286,** 539–547.

Nunes, M. C., Roy, N. S., Keyoung, H. M., Goodman, R. R., McKhann, G., II, Jiang, L., Kang, J., Nedergaard, M., and Goldman, S. A. (2003). Identification and isolation of multipotential neural progenitor cells from the subcortical white matter of the adult human brain. *Nat. Med.* **9,** 439–447.

O'Brien, W. T., Harper, A. D., Jove, F., Woodgett, J. R., Maretto, S., Piccolo, S., and Klein, P. S. (2004). Glycogen synthase kinase-3beta haploinsufficiency mimics the behavioral and molecular effects of lithium. *J. Neurosci.* **24,** 6791–6798.

O'Connell, R. A., Van Heertum, R. L., Luck, D., Yudd, A. P., Cueva, J. E., Billick, S. B., Cordon, D. J., Gersh, R. J., and Masdeu, J. C. (1995). Single-photon emission computed tomography of the brain in acute mania and schizophrenia. *J. Neuroimaging* **5,** 101–104.

Ongur, D., Drevets, W. C., and Price, J. L. (1998). Glial reduction in the subgenual prefrontal cortex in mood disorders. *Proc. Natl. Acad. Sci. USA* **95,** 13290–13295.

Palmer, T. D., Markakis, E. A., Willhoite, A. R., Safar, F., and Gage, F. H. (1999). Fibroblast growth factor-2 activates a latent neurogenic program in neural stem cells from diverse regions of the adult CNS. *J. Neurosci.* **19,** 8487–8497.

Patapoutian, A., and Reichardt, L. F. (2001). Trk receptors: Mediators of neurotrophin action. *Curr. Opin. Neurobiol.* **11,** 272–280.

Pearlson, G. D., Barta, P. E., Powers, R. E., Menon, R. R., Richards, S. S., Aylward, E. H., Federman, E. B., Chase, G. A., Petty, R. G., and Tien, A. Y. (1997). Ziskind-Somerfeld Research Award 1996. Medial and superior temporal gyral volumes and cerebral asymmetry in schizophrenia versus bipolar disorder. *Biol. Psychiatry* **41,** 1–14.

Pencea, V., Bingaman, K. D., Wiegand, S. J., and Luskin, M. B. (2001). Infusion of brain-derived neurotrophic factor into the lateral ventricle of the adult rat leads to new neurons in the parenchyma of the striatum, septum, thalamus, and hypothalamus. *J. Neurosci.* **21,** 6706–6717.

Perez, M., Hernandez, F., Lim, F., Diaz-Nido, J., and Avila, J. (2003). Chronic lithium treatment decreases mutant tau protein aggregation in a transgenic mouse model. *J. Alzheimers Dis.* **5,** 301–308.

Pham, K., Nacher, J., Hof, P. R., and McEwen, B. S. (2003). Repeated restraint stress suppresses neurogenesis and induces biphasic PSA-NCAM expression in the adult rat dentate gyrus. *Eur. J. Neurosci.* **17,** 879–886.

Phiel, C. J., Wilson, C. A., Lee, V. M., and Klein, P. S. (2003). GSK-3alpha regulates production of Alzheimer's disease amyloid-beta peptides. *Nature* **423,** 435–439.

Phillips, M. L., Drevets, W. C., Rauch, S. L., and Lane, R. (2003). Neurobiology of emotion perception I: The neural basis of normal emotion perception. *Biol. Psychiatry* **54,** 504–514.

Pillai, J. J., Friedman, L., Stuve, T. A., Trinidad, S., Jesberger, J. A., Lewin, J. S., Findling, R. L., Swales, T. P., and Schulz, S. C. (2002). Increased presence of white matter hyperintensities in adolescent patients with bipolar disorder. *Psychiatry Res.* **114,** 51–56.

Poo, M. M. (2001). Neurotrophins as synaptic modulators. *Nat. Rev. Neurosci.* **2,** 24–32.

Poolsup, N., Li Wan Po, A., and de Oliveira, I. R. (2000). Systematic overview of lithium treatment in acute mania. *J. Clin. Pharm. Ther.* **25,** 139–156.

Pope, H. G., Jr., McElroy, S. L., Keck, P. E., Jr., and Hudson, J. I. (1991). Valproate in the treatment of acute mania. A placebo-controlled study. *Arch. Gen. Psychiatry* **48,** 62–68.

Post, R. M. (1992). Transduction of psychosocial stress into the neurobiology of recurrent affective disorder. *Am. J. Psychiatry* **149,** 999–1010.

Radley, J. J., and Jacobs, B. L. (2002). 5-HT1A receptor antagonist administration decreases cell proliferation in the dentate gyrus. *Brain Res.* **955,** 264–267.

Radley, J. J., and Jacobs, B. L. (2003). Pilocarpine-induced status epilepticus increases cell proliferation in the dentate gyrus of adult rats via a 5-HT1A receptor-dependent mechanism. *Brain Res.* **966,** 1–12.

Rajkowska, G., Halaris, A., and Selemon, L. D. (2001). Reductions in neuronal and glial density characterize the dorsolateral prefrontal cortex in bipolar disorder. *Biol. Psychiatry* **49,** 741–752.

Rakic, P. (2002). Neurogenesis in adult primate neocortex: An evaluation of the evidence. *Nat. Rev. Neurosci.* **3,** 65–71.

Ren, M., Senatorov, V. V., Chen, R. W., and Chuang, D. M. (2003). Postinsult treatment with lithium reduces brain damage and facilitates neurological recovery in a rat ischemia/reperfusion model. *Proc. Natl. Acad. Sci. USA* **100,** 6210–6215.

Ren, M., Leng, Y., Jeong, M., Leeds, P. R., and Chuang, D. M. (2004). Valproic acid reduces brain damage induced by transient focal cerebral ischemia in rats: Potential roles of histone deacetylase inhibition and heat shock protein induction. *J. Neurochem.* **89,** 1358–1367.

Robinson, L. J., and Ferrier, I. N. (2006). Evolution of cognitive impairment in bipolar disorder: A systematic review of cross-sectional evidence. *Bipolar Disord.* **8,** 103–116.

Rocha, E., Achaval, M., Santos, P., and Rodnight, R. (1998). Lithium treatment causes gliosis and modifies the morphology of hippocampal astrocytes in rats. *Neuroreport* **9,** 3971–3974.

Rondi-Reig, L., and Mariani, J. (2002). To die or not to die, does it change the function? Behavior of transgenic mice reveals a role for developmental cell death. *Brain Res. Bull.* **57,** 85–91.

Rondi-Reig, L., Lemaigre Dubreuil, Y., Martinou, J. C., Delhaye-Bouchaud, N., Caston, J., and Mariani, J. (1997). Fear decrease in transgenic mice overexpressing bcl-2 in neurons. *Neuroreport* **8,** 2429–2432.

Rondi-Reig, L., Lemaigre-Dubreuil, Y., Montecot, C., Muller, D., Martinou, J. C., Caston, J., and Mariani, J. (2001). Transgenic mice with neuronal overexpression of bcl-2 gene present navigation disabilities in a water task. *Neuroscience* **104,** 207–215.

Rosoklija, G., Toomayan, G., Ellis, S. P., Keilp, J., Mann, J. J., Latov, N., Hays, A. P., and Dwork, A. J. (2000). Structural abnormalities of subicular dendrites in subjects with schizophrenia and mood disorders: Preliminary findings. *Arch. Gen. Psychiatry* **57,** 349–356.

Rubin, E., Sackeim, H. A., Prohovnik, I., Moeller, J. R., Schnur, D. B., and Mukherjee, S. (1995). Regional cerebral blood flow in mood disorders: IV. Comparison of mania and depression. *Psychiatry Res.* **61,** 1–10.

Sairanen, M., Lucas, G., Ernfors, P., Castren, M., and Castren, E. (2005). Brain-derived neurotrophic factor and antidepressant drugs have different but coordinated effects on neuronal turnover, proliferation, and survival in the adult dentate gyrus. *J. Neurosci.* **25,** 1089–1094.

Santarelli, L., Saxe, M., Gross, C., Surget, A., Battaglia, F., Dulawa, S., Weisstaub, N., Lee, J., Duman, R., Arancio, O., Belzung, C., and Hen, R. (2003). Requirement of hippocampal neurogenesis for the behavioral effects of antidepressants. *Science* **301,** 805–809.

Sassi, R. B., Nicoletti, M., Brambilla, P., Harenski, K., Mallinger, A. G., Frank, E., Kupfer, D. J., Keshavan, M. S., and Soares, J. C. (2001). Decreased pituitary volume in patients with bipolar disorder. *Biol. Psychiatry* **50,** 271–280.

Schaaf, M. J., De Kloet, E. R., and Vreugdenhil, E. (2000). Corticosterone effects on BDNF expression in the hippocampus. Implications for memory formation. *Stress* **3,** 201–208.

Schlaepfer, T. E., Harris, G. J., Tien, A. Y., Peng, L. W., Lee, S., Federman, E. B., Chase, G. A., Barta, P. E., and Pearlson, G. D. (1994). Decreased regional cortical gray matter volume in schizophrenia. *Am. J. Psychiatry* **151,** 842–848.

Schmidt-Hieber, C., Jonas, P., and Bischofberger, J. (2004). Enhanced synaptic plasticity in newly generated granule cells of the adult hippocampus. *Nature* **429,** 184–187.

Senatorov, V. V., Ren, M., Kanai, H., Wei, H., and Chuang, D. M. (2004). Short-term lithium treatment promotes neuronal survival and proliferation in rat striatum infused with quinolinic acid, an excitotoxic model of Huntington's disease. *Mol. Psychiatry* **9,** 371–385.

Shirayama, Y., Chen, A. C., Nakagawa, S., Russell, D. S., and Duman, R. S. (2002). Brain-derived neurotrophic factor produces antidepressant effects in behavioral models of depression. *J. Neurosci.* **22,** 3251–3261.

Silverstone, P. H., Wu, R. H., O'Donnell, T., Ulrich, M., Asghar, S. J., and Hanstock, C. C. (2002). Chronic treatment with both lithium and sodium valproate may normalize phosphoinositol cycle activity in bipolar patients. *Hum. Psychopharmacol.* **17,** 321–327.

Sirevaag, A. M., and Greenough, W. T. (1991). Plasticity of GFAP-immunoreactive astrocyte size and number in visual cortex of rats reared in complex environments. *Brain Res.* **540,** 273–278.

Siuciak, J. A., Lewis, D. R., Wiegand, S. J., and Lindsay, R. M. (1997). Antidepressant-like effect of brain-derived neurotrophic factor (BDNF). *Pharmacol. Biochem. Behav.* **56,** 131–137.

Smoller, J. W., and Finn, C. T. (2003). Family, twin, and adoption studies of bipolar disorder. *Am. J. Med. Genet. C Semin. Med. Genet.* **123,** 48–58.

Snyder, J. S., Hong, N. S., McDonald, R. J., and Wojtowicz, J. M. (2005). A role for adult neurogenesis in spatial long-term memory. *Neuroscience* **130,** 843–852.

Soares, J. C., and Mann, J. J. (1997). The anatomy of mood disorders—review of structural neuroimaging studies. *Biol. Psychiatry* **41,** 86–106.

Son, H., Yu, I. T., Hwang, S. J., Kim, J. S., Lee, S. H., Lee, Y. S., and Kaang, B. K. (2003). Lithium enhances long-term potentiation independently of hippocampal neurogenesis in the rat dentate gyrus. *J. Neurochem.* **85,** 872–881.

Sonino, N., and Fava, G. A. (2001). Psychiatric disorders associated with Cushing's syndrome. Epidemiology, pathophysiology and treatment. *CNS Drugs* **15,** 361–373.

Starkman, M. N., Giordani, B., Gebarski, S. S., Berent, S., Schork, M. A., and Schteingart, D. E. (1999). Decrease in cortisol reverses human hippocampal atrophy following treatment of Cushing's disease. *Biol. Psychiatry* **46,** 1595–1602.

Strakowski, S. M., Wilson, D. R., Tohen, M., Woods, B. T., Douglass, A. W., and Stoll, A. L. (1993). Structural brain abnormalities in first-episode mania. *Biol. Psychiatry* **33,** 602–609.

Strakowski, S. M., DelBello, M. P., Sax, K. W., Zimmerman, M. E., Shear, P. K., Hawkins, J. M., and Larson, E. R. (1999). Brain magnetic resonance imaging of structural abnormalities in bipolar disorder. *Arch. Gen. Psychiatry* **56,** 254–260.

Su, Y., Ryder, J., Li, B., Wu, X., Fox, N., Solenberg, P., Brune, K., Paul, S., Zhou, Y., Liu, F., and Ni, B. (2004). Lithium, a common drug for bipolar disorder treatment, regulates amyloid-beta precursor protein processing. *Biochemistry* **43,** 6899–6908.

Suzuki, A., and Tsutomi, Y. (1998). Bcl-2 accelerates the neuronal differentiation: New evidence approaching to the biofunction of bcl-2 in the neuronal system. *Brain Res.* **801,** 59–66.

Swayze, V. W., II, Andreasen, N. C., Alliger, R. J., Yuh, W. T., and Ehrhardt, J. C. (1992). Subcortical and temporal structures in affective disorder and schizophrenia: A magnetic resonance imaging study. *Biol. Psychiatry* **31,** 221–240.

Tanapat, P., Galea, L. A., and Gould, E. (1998). Stress inhibits the proliferation of granule cell precursors in the developing dentate gyrus. *Int. J. Dev. Neurosci.* **16,** 235–239.

Tanapat, P., Hastings, N. B., Rydel, T. A., Galea, L. A., and Gould, E. (2001). Exposure to fox odor inhibits cell proliferation in the hippocampus of adult rats via an adrenal hormone-dependent mechanism. *J. Comp. Neurol.* **437,** 496–504.

Teramoto, T., Qiu, J., Plumier, J. C., and Moskowitz, M. A. (2003). EGF amplifies the replacement of parvalbumin-expressing striatal interneurons after ischemia. *J. Clin. Invest.* **111,** 1125–1132.

Tonchev, A. B., Yamashima, T., Sawamoto, K., and Okano, H. (2005). Enhanced proliferation of progenitor cells in the subventricular zone and limited neuronal production in the striatum and neocortex of adult macaque monkeys after global cerebral ischemia. *J. Neurosci. Res.* **81,** 776–788.

Van der Borght, K., Meerlo, P., Luiten, P. G., Eggen, B. J., and Van der Zee, E. A. (2005). Effects of active shock avoidance learning on hippocampal neurogenesis and plasma levels of corticosterone. *Behav. Brain Res.* **157,** 23–30.

van der Hart, M. G., Czeh, B., de Biurrun, G., Michaelis, T., Watanabe, T., Natt, O., Frahm, J., and Fuchs, E. (2002). Substance P receptor antagonist and clomipramine prevent stress-induced alterations in cerebral metabolites, cytogenesis in the dentate gyrus and hippocampal volume. *Mol. Psychiatry* **7,** 933–941.

van Praag, H., Schinder, A. F., Christie, B. R., Toni, N., Palmer, T. D., and Gage, F. H. (2002). Functional neurogenesis in the adult hippocampus. *Nature* **415,** 1030–1034.

Vieta, E., Gasto, C., Martinez de Osaba, M. J., Nieto, E., Canto, T. J., Otero, A., and Vallejo, J. (1997). Prediction of depressive relapse in remitted bipolar patients using corticotrophin-releasing hormone challenge test. *Acta Psychiatr. Scand.* **95,** 205–211.

Vieta, E., Martinez-De-Osaba, M. J., Colom, F., Martinez-Aran, A., Benabarre, A., and Gasto, C. (1999). Enhanced corticotropin response to corticotropin-releasing hormone as a predictor of mania in euthymic bipolar patients. *Psychol. Med.* **29,** 971–978.

Volonte, C., Ciotti, M. T., and Merlo, D. (1994). LiCl promotes survival of GABAergic neurons from cerebellum and cerebral cortex: LiCl induces survival of GABAergic neurons. *Neurosci. Lett.* **172,** 6–10.

Wakade, C. G., Mahadik, S. P., Waller, J. L., and Chiu, F. C. (2002). Atypical neuroleptics stimulate neurogenesis in adult rat brain. *J. Neurosci. Res.* **69,** 72–79.

Wang, H. D., Dunnavant, F. D., Jarman, T., and Deutch, A. Y. (2004). Effects of antipsychotic drugs on neurogenesis in the forebrain of the adult rat. *Neuropsychopharmacology* **29,** 1230–1238.

Watson, D. G., and Lenox, R. H. (1996). Chronic lithium-induced down-regulation of MARCKS in immortalized hippocampal cells: Potentiation by muscarinic receptor activation. *J. Neurochem.* **67,** 767–777.

Wei, H., Qin, Z. H., Senatorov, V. V., Wei, W., Wang, Y., Qian, Y., and Chuang, D. M. (2001). Lithium suppresses excitotoxicity-induced striatal lesions in a rat model of Huntington's disease. *Neuroscience* **106,** 603–612.

Weiner, I., Schiller, D., Gaisler-Salomon, I., Green, A., and Joel, D. (2003). A comparison of drug effects in latent inhibition and the forced swim test differentiates between the typical antipsychotic haloperidol, the atypical antipsychotics clozapine and olanzapine, and the antidepressants imipramine and paroxetine. *Behav. Pharmacol.* **14,** 215–222.

Wennstrom, M., Hellsten, J., and Tingstrom, A. (2004). Electroconvulsive seizures induce proliferation of NG2-expressing glial cells in adult rat amygdala. *Biol. Psychiatry* **55,** 464–471.

Westenbroek, C., Den Boer, J. A., Veenhuis, M., and Ter Horst, G. J. (2004). Chronic stress and social housing differentially affect neurogenesis in male and female rats. *Brain Res. Bull.* **64,** 303–308.

Williams, R. S., Cheng, L., Mudge, A. W., and Harwood, A. J. (2002). A common mechanism of action for three mood-stabilizing drugs. *Nature* **417,** 292–295.

Winsberg, M. E., Sachs, N., Tate, D. L., Adalsteinsson, E., Spielman, D., and Ketter, T. A. (2000). Decreased dorsolateral prefrontal N-acetyl aspartate in bipolar disorder. *Biol. Psychiatry* **47,** 475–481.

Wong, E. Y., and Herbert, J. (2006). Raised circulating corticosterone inhibits neuronal differentiation of progenitor cells in the adult hippocampus. *Neuroscience* **137,** 83–92.

Wood, G. E., Young, L. T., Reagan, L. P., Chen, B., and McEwen, B. S. (2004). Stress-induced structural remodeling in hippocampus: Prevention by lithium treatment. *Proc. Natl. Acad. Sci. USA* **101,** 3973–3978.

York, J. D., Ponder, J. W., and Majerus, P. W. (1995). Definition of a metal-dependent/Li(+)-inhibited phosphomonoesterase protein family based upon a conserved three-dimensional core structure. *Proc. Natl. Acad. Sci. USA* **92,** 5149–5153.

Youdim, M. B., and Arraf, Z. (2004). Prevention of MPTP (N-methyl-4-phenyl-1,2,3, 6-tetrahydropyridine) dopaminergic neurotoxicity in mice by chronic lithium: Involvements of Bcl-2 and Bax. *Neuropharmacology* **46,** 1130–1140.

Yu, I. T., Kim, J. S., Lee, S. H., Lee, Y. S., and Son, H. (2003). Chronic lithium enhances hippocampal long-term potentiation, but not neurogenesis, in the aged rat dentate gyrus. *Biochem. Biophys. Res. Commun.* **303,** 1193–1198.

Yuan, J., and Yankner, B. A. (2000). Apoptosis in the nervous system. *Nature* **407,** 802–809.

Yuan, P. X., Huang, L. D., Jiang, Y. M., Gutkind, J. S., Manji, H. K., and Chen, G. (2001). The mood stabilizer valproic acid activates mitogen-activated protein kinases and promotes neurite growth. *J. Biol. Chem.* **276,** 31674–31683.

Yurgelun-Todd, D. A., Gruber, S. A., Kanayama, G., Killgore, W. D., Baird, A. A., and Young, A. D. (2000). fMRI during affect discrimination in bipolar affective disorder. *Bipolar Disord.* **2,** 237–248.

Zhang, K. Z., Westberg, J. A., Holtta, E., and Andersson, L. C. (1996). BCL2 regulates neural differentiation. *Proc. Natl. Acad. Sci. USA* **93,** 4504–4508.

Zhou, D. F., Shen, Y. C., Shu, L. N., and Lo, H. C. (1987). Dexamethasone suppression test and urinary MHPG X SO4 determination in depressive disorders. *Biol. Psychiatry* **22,** 883–891.

Zipursky, R. B., Seeman, M. V., Bury, A., Langevin, R., Wortzman, G., and Katz, R. (1997). Deficits in gray matter volume are present in schizophrenia but not bipolar disorder. *Schizophr. Res.* **26,** 85–92.

NEUROREPLACEMENT, GROWTH FACTOR, AND SMALL MOLECULE NEUROTROPHIC APPROACHES FOR TREATING PARKINSON'S DISEASE

Michael J. O'Neill,* Marcus J. Messenger,* Viktor Lakics,* Tracey K. Murray,* Eric H. Karran,* Philip G. Szekeres,* Eric S. Nisenbaum,† and Kalpana M. Merchant†

*Eli Lilly and Co. Ltd., Lilly Research Centre, Erl Wood Manor, Windlesham Surrey GU20 6PH, United Kingdom

†Eli Lilly and Co. Ltd., Lilly Corporate Center, Indianapolis, Indiana 46285, USA

I. Introduction
A. Pathology and Clinical Manifestations
II. Etiology of PD
III. Current Treatment for PD
A. A Review of Current Medications
B. The Unmet Medical Need
IV. Rodent Models of PD Used to Evaluate Putative Neuroprotective Therapies
V. Approaches to Brain Repair in PD
A. Transplantation and Stem Cells
B. Growth Factors
C. Small Molecule Approaches
VI. Summary and Conclusions
References

Parkinson's disease (PD) is a movement disorder resulting from neurodegeneration of the basal ganglia. PD is usually diagnosed at ~55–60 years of age and affects ~1% of the population over 60 years. Primarily, dopaminergic projections whose cell bodies are located in the substantia nigra and terminals extend to the neostriatum are affected. This dopaminergic system degenerates slowly and displays a large compensatory capacity such that ~60% of cells are already lost when clinical motor symptoms first become evident. In addition to the classic triad of PD symptoms, resting tremor, muscular rigidity, and bradykinesia, abnormalities in postural reflexes, dementia, and depression are important comorbid conditions. Current therapies are aimed primarily at replacing dopamine with the dopamine precursor L-3,4-dihydroxyphenylalanine (levodopa or L-DOPA) or by the use of direct acting dopamine receptor agonists. Adjunctive treatments with monoamine oxidase (MAO) inhibitors, catechol-*O*-methyl transferase (COMT) inhibitors, and amantadine are also used. While providing reasonably effective symptomatic therapy in

INTERNATIONAL REVIEW OF
NEUROBIOLOGY, VOL. 77
DOI: 10.1016/S0074-7742(06)77006-4

0074-7742/07 $35.00

early stages of the disease, these agents fail to halt disease progression, resulting in significant worsening of the quality of life for PD patients 5–10 years after diagnosis. Therefore, in order to maintain an acceptable quality of life for patients with PD, therapies that provide not only symptomatic improvement but also slow or stop disease progression are greatly needed. In the present chapter, we will summarize briefly current understanding of the etiology of PD, clinical treatment options, and major unmet medical needs followed by specific discussion of various neurotrophic approaches for the disease. Under the neurotrophic approaches, the chapter will include a discussion on stem cell transplantation, biomolecules such as glial-derived neurotrophic factor (GDNF), and brain-derived neurotrophic factor (BDNF), as well as small molecule strategies targeting the orphan nuclear receptor, Nurr1. Finally, we will discuss a small molecule drug discovery approach undertaken at Eli Lilly, which targets the α-amino-3-hydroxy-5-methylisoxazole-4-propionic acid (AMPA) receptors in the brain. In particular, we have focused our efforts on a series (LY404187, LY450108, and LY503430) of systemically active allosteric AMPA receptor potentiators. Data indicates that these molecules increase BDNF levels in cortical and hippocampal primary neuronal cultures and provide functional and histological improvements against a severe 6-hydroxydopamine (6-OHDA) lesion of the nigrostriatal pathway. These improvements appear to be trophic in nature as the effect(s) was observed when administration of AMPA receptor modulators was initiated 1, 3, or 6 days after infusion of 6-OHDA into the substantia nigra. Studies to address the cellular mechanisms underlying the trophic effects of AMPA receptor potentiators are ongoing. However, small molecule agents that were capable of producing trophic effects have the potential to revolutionize treatment of PD.

I. Introduction

Parkinson's disease (PD) is an age-related, progressive neurodegenerative disorder that was characterized first in 1817 by James Parkinson in "An Essay on the Shaking Palsy" (Parkinson, 1817). The disease affects 1:1000 of the general population, which increases to 1:100 in people over 50 years and the prevalence continues to increase with age at least until about 90 years (O'Neill and Siemers, 2002; Zigmond and Strickler, 1989). The mean age of diagnosis of PD is 55 $\pm$ 11 years (Bezard *et al.*, 1998). The predominant clinical symptoms of PD include resting tremor, muscle rigidity, akinesia, and bradykinesia as well as postural instability and gait abnormalities. It is well recognized that these motor symptoms are caused by the progressive degeneration of the pigmented, dopamine-containing neurons of the substantia nigra pars compacta (SNc). Thus, currently available therapeutic agents primarily target reduction of symptom severity via drugs that directly or indirectly enhance dopaminergic neurotransmission. However, these agents do not affect the underlying neuropathology, have limited efficacy in late stages of the disease, and are encumbered by severe side effects. Thus, there remains a critical unmet medical need to treat PD, especially with strategies that address the neurodegeneration associated with PD. This chapter focuses on current investigative approaches aimed at providing trophic support to dopaminergic neurons. Recent advances in drug delivery technology, gene therapy, and small molecule therapeutics that may stimulate endogenous trophic factors make neurotrophic approaches an especially promising strategy to address the unmet medical needs of this devastating disorder. To establish appropriate context, we will begin with a brief description of the pathology, associated symptomatology, and the etiology of PD, followed by a section on current therapies and then discuss specific, investigative neurotrophic and neurorestorative approaches.

A. Pathology and Clinical Manifestations

The progressive loss of dopamine neurons in the SNc leads to dopamine depletion in the nigrostriatal pathway. Thus in 1960, Hornykiewicz and coworkers found that dopamine levels in the postmortem brains of PD patients were only about 10% of normal (reviewed by Hornykiewicz and Kish, 1987). As mentioned in the Introduction, the chronic depletion of dopamine underlies the characteristic motor symptoms of PD. Parkinson's sufferers can also exhibit immobile or rigid facial features (hypomimia), respiratory problems, fatigue, and psychiatric disturbances, including depression and cognitive decline (Lang and Lozano, 1998a,b). Interestingly, of all patients who exhibit parkinsonian symptoms, only 80% have idiopathic PD, while the remaining 20% have other forms of parkinsonism such as progressive supranuclear palsy, diffuse Lewy body disease, and olivopontocerebellar atrophy (Hingtgen and Siemers, 1998; Rajput and Rajput, 2002; Siemers, 1992). Indeed, PD could represent several pathological processes that converge in a common pathophysiology, resulting in similar clinical maifestations (Schapira, 1999).

Cell death also occurs in many other areas of the central nervous system (CNS), including the cortex, olfactory bulb, hypothalamus, and brain stem nuclei. Degeneration of cortical structures and, in particular, the nucleus basilis of Meynert leads to a reduction in ACh activity throughout the forebrain which, particularly in elderly sufferers, may be responsible for the cognitive defects such as dementia which is associated with late-stage PD (Ruberg *et al.*, 1982). Cell death in the hypothalamus and areas of the spinal cord has been implicated in autonomic defects in the disorder. Similarly, depletion of the noradrenergic locus coeruleus (which projects to the cortex and limbic structures) and serotonergic (5-hydroxytryptamine, 5-HT) nuclei, such as the dorsal raphe nucleus and amygdala, may account for the intellectual impairment and high prevalence of depression in PD (Agid *et al.,* 1989; Javoy-Agid *et al.,* 1984). However, although multiple neurotransmitter systems are affected in PD, degeneration of dopaminergic neurons in the SNc is the defining characteristic of the disorder.

In addition to the dopamine cell death, intracellular proteinaceous aggregates, termed Lewy bodies (LBs) and Lewy neurites, in the areas of neurodegeneration form the hallmark of PD pathology. First described by Lewy in 1912, these diffuse inclusions are found *postmortem* in the cytoplasm of surviving neurons (reviewed in Dunnett and Bjorklund, 1999; Holdorff, 2002). LBs appear as spherical intracellular aggregates, 5–25 μm in diameter, with a central core and lighter halo (Lang and Lozano, 1998a,b). LBs also appear in several neurodegenerative diseases related to PD, each with its own peculiar symptomatology and pathology, such as incidental Lewy body disease, multiple system atrophy, dementia with LBs, and Alzheimer's disease with LBs (Kotzbauer *et al.,* 2001; Salloway *et al.,* 2002). PD may be one of a number of LB diseases, although the relationship between the LB pathology of these disorders to the clinical features remains uncertain.

II. Etiology of PD

The etiology of dopamine cell death in PD remains unknown despite the intense interest it has received. Both environmental and genetic factors have been implicated with an emerging view being that most cases may result from an interaction between environmental insults and genetic susceptibility (Abou-Sleiman *et al.,* 2006; Betarbet *et al.,* 2006; Burke, 1998; Burke and Kholodilov, 1998; Dawson and Dawson, 2003; Farrer, 2006; Hardy *et al.,* 2003; Jenner and Olanow, 1998; Przedborski and Vila, 2003; Scott *et al.,* 1997; Singleton *et al.,* 2003; Tanner and Langston, 1990). A variety of cellular processes, such as oxidative stress, mitochondrial impairment, defects in ubiquitin-proteasomal activity as well as excitotoxicity have been hypothesized to contribute to the neurodegeneration associated with PD (Abou-Sleiman *et al.,* 2006; Beal, 1998, 2003; Jenner, 1998, 2001; Jenner and Olanow, 1996, 1998; McNaught *et al.,* 2001). Apoptosis, rather than necrosis, is ultimately associated with neurodegeneration in PD (Andersen, 2001; Tatton and Kish, 1997).

Among the exogenous environmental toxins, the heroin analogue, 1-methyl-4-phenyl-1,2,3,6-tetrahydropyridine (MPTP), rapidly induces a PD-like syndrome in humans (Langston *et al.,* 1983) and, hence, has been used to understand the mechanisms underlying dopaminergic cell loss. MPTP-induced nigral cell death results from its metabolism to *N*-methyl-4-phenylpyridinium ion (MPP^+), by monoamine oxidase (MAO)-B, in dopamine neurons (Przedborski *et al.,* 2001). MPP^+ in turn causes inhibition of mitochondrial complex I (Abou-Sleiman *et al.,* 2006; Jakowec and Petzinger, 2004; Przedborski *et al.,* 2000), thereby closely mimicking one aspect of the human pathology of PD. Because of these similarities, the MPTP-treated mouse and primate models of PD are now among the most frequently used to study the disorder (Jakowec and Petzinger, 2004; Miller *et al.*, 2005; Przedborski and Vila, 2003; Sundstrom *et al.*, 1990). Numerous industrial compounds and agrochemicals are also associated with an increased risk of PD (Semchuk *et al.*, 1992; Tanner and Langston, 1990; reviewed by Sherer *et al.*, 2002). For example, the pesticide rotenone, another mitochondrial complex I inhibitor, produces PD-like pathology when infused into rodents (Betarbet *et al.*, 2000). Postmortem, these rats not only have severe degeneration of the dopaminergic nigrostriatal tract but also exhibit LB-like inclusion bodies previously unseen in rodent models of PD.

Although PD is generally a sporadic disease, a small number of individuals show Mendelian inheritance of the disease (Farrer, 2006; Hardy *et al.*, 2003; Olanow and Tatton, 1999). Recent advances in molecular genetic studies of the familial PD cases have led to linkage of 11 genetic loci to the disease and identified pathogenic mutations in at least five genes, *SNCA, PRKN, DJ-1, PINK-1,* and *LRRK2* (Moore *et al.*, 2005). These genes respectively encode α-synuclein, parkin, DJ-1, PTEN-induced putative kinase-1 (PINK-1), and leucine-rich repeat kinase-2 (LRRK2, also termed dardarin). In addition, mutations and variants in several other genes (e.g., *MAPT*-encoding microtubule-associated protein tau) are associated with the clinical syndrome of parkinsonism (Ross and Farrer, 2005). Of these, both *SNCA* and *LRRK2* mutations show autosomal dominant inheritance in familian PD and the mutations appear to cause gain of function (Farrer, 2006; Gloeckner *et al.*, 2006; West *et al.*, 2005). A set of coding sequence mutations in *LRRK2* gene have received significant attention recently since they represent the most common cause of Mendelian inheritance of PD and appear to be associated also with some cases of sporadic PD (Farrer, 2006; Gosal *et al.*, 2005; Skipper *et al.*, 2005) and LB dementia (Ross *et al.*, 2006; Toft *et al.*, 2005). In contrast, mutations in parkin, DJ-1 and PINK-1 are recessive, include large deletions or truncations and appear to cause loss of function of these proteins. Not surprisingly, a highly active area of current research on PD etiology is focused on understanding whether the proteins identified through molecular genetic studies function in a common biochemical pathway leading to the disease. Indeed, the shared phenotype resulting from different genetic mutations lends support to the possibility that the distinct mutations may converge on to a discrete, unifying pathophysiological mechanism underlying PD, the identification of which could provide insights into the pathophysiology of sporadic PD and offer new therapeutic targets. Interestingly, recent biochemical studies on disease-causing mutations in α-synuclein, parkin, DJ-1, and PINK-1 continue to support a role of defects in cellular processes such as proteasome activity, mitochondrial function, and oxidative stress in dopaminergic neuronal loss associated with PD. However, the reasons underlying the relatively higher vulnerability of dopamine neurons to these defects or the contribution of protein misfolding and aggregation to progressive neurodegeneration remain speculative.

III. Current Treatment for PD

A. A Review of Current Medications

Based on the relationship between pathophysiology affecting the nigrostriatal dopaminergic system and clinical symptomatology, a number of approaches have been developed to restore the lost dopaminergic activity in the striatum (Dunnett and Bjorklund, 1999; Johnston and Brotchie 2004, 2006). Dopamine replacement therapies are the most effective pharmacological symptomatic treatment of PD. L-3,4-Dihydroxyphenylalanine (L-DOPA), a dopamine precursor, is considered the "gold standard" symptomatic treatment for PD. L-DOPA is the most common and effective therapeutic strategy used to restore dopaminergic neurotransmission in the basal ganglia and can dramatically alleviate parkinsonian symptoms (Marsden and Parkes, 1977). L-DOPA is coadministered with carbidopa or benserazide, peripheral dopamine decarboxylase inhibitors, which inhibit the conversion of L-DOPA to dopamine outside the brain and thus reduce unwanted peripheral side effects such as gastrointestinal upset. The MAO-B inhibitor selegiline, which prevents the breakdown of dopamine, can also be coadministered with L-DOPA or as a monotherapy and can reduce both the unwanted motor response fluctuations and dosage requirements for L-DOPA discussed below (Golbe *et al.*, 1988; Lew, 2005; Lieberman *et al.*, 1987; Tetrud and Langston, 1989). Another MAO-B inhibitor, rasagiline, has recently been launched, which may also have putative neuroprotective properties (Henchcliffe *et al.*, 2005; Rascol, 2005; Schapira *et al.*, 2005; Youdim *et al.*, 2005).

Dopamine receptor agonists can also mimic striatal dopamine receptor function to relieve symptoms of PD (Factor, 1999; Hurley and Jenner, 2006). These compounds activate the postsynaptic dopamine receptors on striatal GABAergic medium spiny neurons. Bromocriptine has a high affinity for D2 receptors and is a weak antagonist at D1 receptors. Pergolide has approximately equal affinity for both D2 and D1 receptors (Pezzoli *et al.*, 1994). Other dopamine agonists in clinical use include lisuride; ropinirole, a D2 receptor agonist; and pramipexole, a D3 receptor agonist (Goetz and Diederich, 1992; Shannon *et al.*, 1997; Tulloch, 1997). PD patients who are given either pramipexole, ropinirole, or pergolide rather than L-DOPA show a delay in the onset of motor fluctuations and these drugs can serve as an alternative treatment for patients in need of an increased dose of L-DOPA or whom exhibit severe motor fluctuations (Hundemer *et al.*, 2000; Rascol *et al.*, 2000).

Long-term use of L-DOPA and many dopamine receptor agonists can lead to dyskinetic and psychiatric side effects coupled with periods of so-called "on–off" motor response variations. The "on" periods occur when antiparkinsonian drugs are effective and "off" periods when there is no symptomatic response. At times when plasma and brain levels of L-DOPA are relatively high, patients can also experience dyskinesia, chorea, or dystonic movements. The efficacy of L-DOPA is also inversely proportional to the progression of neurodegeneration and thus becomes less effective as the course of PD continues. It was always considered that dopamine replacement therapies for PD, which produce significant improvement in motor function, appear to have little effect on the progression of neurodegeneration. Nevertheless, there are also several lines of evidence (Schapira, 2003) to suggest dopamine receptor agonists can have some form of protective effects *in vitro* (Presgraves *et al.*, 2004) and *in vivo* (Iravani *et al.*, 2006; Joyce *et al.*, 2004). Ropinirole and pramipexole have been evaluated in disease-modifying trials (Parkinson Study Group, 2002; Whone *et al.*, 2003) and while mild disease-modifying effects were reported, there is debate as to whether these effects were real. Both compounds were compared to L-DOPA and there has been debate as to whether L-DOPA might in fact increase the rate of decline (Fahn *et al.*, 2004, 2005). Another issue is that dopamine agonists may interfere with some of the imaging techniques used in the trials (Brooks, 2004; Brooks *et al.*, 2003) or there may be some disconnects between Unified Parkinson's Diease Rating Scores (UPDRS) and imaging endpoints. It may also be that dopamine agonists would only provide protection if administered very early in the course of the disease. For example, recent studies have reported that pretreatment, but not posttreatment, with pramipexole protects against MPTP toxicity in nonhuman primates (Iravani *et al.*, 2006).

COMT inhibitors such as tolcapone and entacapone have also been used to treat PD, particularly as an adjunct to L-DOPA therapy. These compounds increase the plasma half-life of L-DPOA, allowing more to cross the blood–brain barrier leading to a higher level of L-DOPA in the brain. In terms of L-DOPA-mediated motor fluctuations, COMT inhibitors consequently prolong "on" time and reduce "off" time. Tolcapone was the first COMT inhibitor to be used clinically but was associated with hepatic toxicity (Colosimo, 1999). However, entacapone, a newer COMT inhibitor, appears to reduce motor response fluctuations without any reported toxicity (Ruottinen and Rinne, 1996; Smith *et al.*, 2005). In the presence of monoamine inhibitors, the required dose of L-DOPA is markedly less since dopamine reuptake in the synaptic cleft is inhibited causing an increased concentration of active free dopamine.

Several surgical approaches have been tried as an alternate treatment of PD aimed at restoring normal pallidal-subthalamic-nigral neurotransmission in the basal ganglia (reviewed by Marsden *et al.*, 1997). Ablative surgery including pallidotomy or subthalamotomy or lesion to the ventral intermediate nucleus of the thalamus can significantly alleviate parkinsonian symptoms. Similarly, deep brain stimulation of the internal segment of the globus pallidus, subthalamic nucleus (STN), or thalamic ventral intermediate nuclei can also alleviate the motor symptoms of PD and may dramatically lessen dyskinesia and reduce response fluctuations. However, as with dopamine replacement therapies, surgical techniques do not slow disease progression and these neurosurgical techniques have other additional limitations (availability, cost, accessability, and the intrusive nature of the treatments).

B. The Unmet Medical Need

In summary, most current pharmacological treatments for PD increase dompamine levels or directly stimulate dopamine receptors in the brain. However, following chronic use "on–off" fluctuations and various adverse clinical symptoms develop (Marsden and Parkes, 1977; Nutt, 2002). Coadministration of adjunctive therapies such as MAO-B inhibitors and/or COMT inhibitors can delay and reduce these unwanted complications as can initiating treatment with doopamine agonists rather than L-DOPA. However, the main issue with current therapies is that they do not halt the advancing disease. Additional symptoms also start to develop that do not respond to dopamine replacement strategies. Among the most problematic of these are a cognitive decline and diminished postural reflexes. Surgical techniques can alleviate dyskinesia and improve fluctuations in motor response but do not slow PD progression and do not improve cognition, postural reflexes, or late-stage motor symptoms (Marsden *et al.,* 1997). Furthermore, these neurosurgical techniques have numerous limitations such as patient suitability, technique availability, cost, accessability, and the intrusive nature of the treatments, not to mention increased mortality risk.

Thus, a pharmacological treatment for PD, which can slow or halt the rate of neurodegeneration and improve the physiological function of the remaining neurons would be ideal. Postmortem studies of PD patients and imaging studies using positron emission tomography (PET) or single-photon emission computed tomography (SPECT) scans suggest that neuron loss in the SNc appears about 4–5 years before symptoms manifest (Brooks, 2004). The development of neuroprotective and neurorestorative drugs that could be used during the early stages of PD when the rate of cell loss appears to be more rapid would be of great value to retain and amplify the remaining dopamine activity in the striatum.

IV. Rodent Models of PD Used to Evaluate Putative Neuroprotective Therapies

A number of animal models have been developed to model the dopaminergic degeneration for use in understanding the disease and to aid in the discovery of improved symptomatic, neuroprotective, or neurorestorative agents. The direct application of neurotoxins such as 6-hydroxydopamine (6-OHDA) into either the cell bodies or terminal fields of the nigrostriatal pathway has been used to generate the neurodegeneration of this pathway, a process that is a hallmark of PD (Beal, 2001; Emborg, 2004; Zigmond and Strickler, 1989). The 6-OHDA model produces a discrete lesion that can be used to study neurotrophic agents. Systemic injection of MPTP produces a bilateral dopaminergic depletion in mice (Beal, 2001; Jakowec and Petzinger, 2004; Przedborski and Vila, 2003) and indeed in larger nonhuman primate species (Jakowec and Petzinger, 2004; Przedborski *et al.*, 2001). Various MPTP dosing protocols have been used in mice. These range from acute (4 injections over 24 hours) or subacute treatments, which do not represent the slow progressive nature of the disease to long-term treatments over weeks, that may have better validity (Fornai *et al.*, 2005). The MPTP model has good face validity because MPTP-induced parkinsonian symptoms in nonhuman primates resemble those of clinical, idiopathic PD in humans.

Acute 6-OHDA and MPTP (acutely or subchronically) produce relatively rapid dopaminergic damage but do not have cytoplasmic, LB-like inclusions. More recently, it has been reported that slow minipump infusion of rotenone (Betarbet *et al.*, 2000), MPTP (Fornai *et al.*, 2005), or periodic systemic injections of a proteasome inhibitor (McNaught *et al.*, 2004) produces a slow progressive degeneration with presence of inclusion bodies. Thus, these recent chronic infusion models do seem to offer advantages over the multiple systemic injection models in that a low dose of toxin is constantly stressing the system and this may be closer to the human condition. Although these models may be better suited for testing neuroprotective agents, they are relatively new and require further validation and characterization as well as assessment of "standard" neuroprotectants. In addition, a number of transgenic mouse (Maries *et al.*, 2003; Masliah *et al.*, 2000; Richfield *et al.*, 2002) models based on genetic etiology (α-synuclein overexpression, parkin knockout), which reproduce some of the PD pathology have been generated and characterized (reviewed by Fleming *et al.*, 2005).

V. Approaches to Brain Repair in PD

A. Transplantation and Stem Cells

PD represents an attractive target for transplantation strategies (Olanow *et al.*, 1997) because of the fairly localized degeneration of a specific cell type, and the existence of established toxin-based animal models, which provide a route to test the behavioral and histological outcome of the grafting procedures. Transplantation is usually performed by grafting dopaminergic (DAergic) neurons into the putamen, where these cells integrate into the patient's brain, and could provide additional benefits such as physiologically controlled release of dopamine and improved clinical efficacy and side effect profile over conventional pharmacological treatments, especially in the late stage of the disease.

Following the initial animal studies, transplantation of human fetal ventral mesencephalic (VM) tissue to the striatum of PD patients was performed first in 1987, and since then over 300 patients have received similar treatment in various clinical settings (Lindvall and Bjorklund, 2004). These studies served as a proof of principle to show that the grafted neurons can not only survive and form functional neuronal circuits in the patients' brain, but more importantly, the grafting procedure can result in a measurable clinical benefit. The extent of the clinical benefit varied between 10% and 40% on the UPDRS and in most cases the L-DOPA requirements could be reduced significantly (Brundin *et al.*, 2000; Hagell *et al.*, 1999; Hauser *et al.*, 1999). The changes in UPDRS have been more modest (18% vs 40%) in double-blinded placebo-controlled trials compared to the open-label ones, but were still significant (Freed *et al.*, 2001). Grafting provided a reasonably long-lasting benefit; usually lasting several years, in exceptional cases it was over 10 years (Piccini *et al.*, 1999). In some trials, the UPDRS scores returned to their preoperative values after 16–24 months, probably due to insufficient immunosuppression (Olanow *et al.*, 2003). It is also worth noting that after extensive studies, there has been no evidence suggesting that the grafted cells, once they got integrated, would be affected by the parkinsonian degeneration ongoing in the host, confirming that transplantation can be a disease-modifying treatment strategy for the long term.

However, the above studies also pointed out the limitations of the fetal VM tissue transplantation strategy. First of all, the need for three to six human fetuses per PD patient to be grafted made it clear that transplantation of fetal grafts cannot be a successful strategy for thousands of patients and also raised ethical concerns. The survival, differentiation, and integration of the grafted neurons and the extent of the clinical benefit showed great individual variation among patients, depending on a variety of different factors such as the age of patient, the disease stage, postmortem delay regarding the grafted tissue, surgical procedure, and solid versus dissociated grafts just to list a few. In addition, in several clinical trials, transplantation resulted in the development of graft-induced dyskinesias in a significant number of patients—15% according to Freed *et al.* (2001), while others reported up to 56% (Olanow *et al.*, 2003). Although their precise cause remains to be elucidated, the appearance of the dyskinesias indicates that grafts cannot completely recreate the same high-level organization of DAergic circuitry needed for the normal functioning of the striatal signaling.

Two of the major obstacles of fetal human VM tissue transplantation—lack of sufficient amounts of graft material and the requirements of standardization of the quality of the transplanted tissue—prompted several laboratories to find better tissue sources for grafting PD patients. Neuronal progenitor cells from human fetal VM tissue expanded and differentiated *in vitro* then transplanted to the PD patients' brain could help solve the above limitations. Although a few reports have described the generation of tyrosine hydroxylase (TH)-positive neurons from human mesencephalic neuronal progenitors, appropriate culture conditions, which would allow the extension and DAergic differentiation of these cells at scale have not yet been developed (Jin *et al.*, 2005; Riaz *et al.*, 2002; Wang *et al.*, 2004).

Multipotent adult stem cells found outside the CNS, such as bone marrow cells or stromal cells could easily be isolated from patients, alleviating problems with tissue rejection. While there are reports showing the appearance of certain neuronal and even DAergic markers on these cells after differentiation *in vitro* (Jiang *et al.*, 2002; Zhao *et al.*, 2002), there is only limited evidence for the development of *functional* neurons (Sigurjonsson *et al.*, 2005).

As opposed to neuronal progenitor cells and adult stem cells from outside the CNS, rodent embryonic stem (ES) cells can not only be produced and maintained at scale *in vitro*, but they can also be differentiated to functional DAergic neurons. Deacon *et al.* (1998) were the first to report that rodent ES cells can differentiate into DAergic neurons and can correct the phenotype in a mouse model of PD. McKay and colleagues developed detailed protocols to differentiate mouse ES cells into neurons *in vitro* by embryoid body formation combined with retinoic acid treatment, and showed that further addition of sonic hedgehog (shh) and fibroblast growth factor-8 can commit the progenitors toward DAergic differentiation (Lee *et al.*, 2000). These neurons resemble the specific A9 type DAergic neurons of the SNc, which are the most affected ones in PD, and they express all appropriate DAergic markers such as Nurr1, Lmx1b, Ptx3, Girk2, and others. Neurons generated by the above differentiation method demonstrated to be effective in an animal model of PD (Kim *et al.*, 2002). Alternative methods using coculture of mouse ES cells and stromal cells have also been developed, and successfully used to generate DAergic neurons, which after transplantation proved to be beneficial in animal models of PD (Barberi *et al.*, 2003).

Kawasaki *et al.* (2002) used primate ES cells to generate DAergic neurons by coculturing them with PA6 mouse stromal cells, and the grafted neurons attenuated the behavioral and histological consequences of MPTP administration in primates (Takagi *et al.*, 2005). Despite successful attempts with rodent and primate ES cells, DAergic differentiation and transplantation of human ES cells have resulted in limited success so far. Although the basic mechanisms of DAergic differentiation seem to be the same as for rodent and primate ES cells, human ES cells are more difficult to culture, and their survival is limited as a graft in animal models of PD. From transplantation studies with fetal VM tissue, it is estimated that at least 100,000 DAergic neurons need to be surviving in the graft in each human putamen to get a measurable clinical benefit [restoring dopamine (DA) uptake to 50% of normal] in PD (Hagell and Brundin, 2001). A few studies reported problems with DAergic differentiation of human ES cells using the established methods (Buytaert-Hoefen *et al.*, 2004; Thomson *et al.*, 1998); others were able to generate

DAergic neurons, these cells did not survive in sufficient numbers after transplantation and consequently provided little or no benefit in the animal models (Park *et al.*, 2005; Perrier *et al.*, 2004; Schulz *et al.*, 2004; Zeng *et al.*, 2004).

A possible solution to boost the survival and *in vitro* differentiation of ES cells is to genetically manipulate them to express factors relevant to DAergic lineage commitment, as shown in the case of mouse ES cells and Nurr1 (Chung *et al.*, 2002; Kim *et al.*, 2006) or overproduce anti-apoptotic factors like Bcl-xL (Shim *et al.*, 2004). In fact, Bcl-xL has been shown to significantly (1–2 orders of magnitude) increase the capacity for spontaneous DAergic differentiation of human neuronal stem cells *in vitro* and also *in vivo* in a rat PD model (Liste *et al.*, 2004). Similarly, expressing growth factors relevant to DAergic survival and differentiation, such as glial cell-derived neurotrophic factor (GDNF) or BDNF, could also be a very promising approach (Ostenfeld *et al.*, 2002). Genetic engineering can also provide potential solutions for a more precise control of DAergic differentiation, switching on transcription factors such as Nurr1 at the right time, using regulated gene expression systems such as the doxycycline-dependent tetracycline (TET) system (Sonntag *et al.*, 2004). In the future, conditionally expressed transgenes could help control the fate of the graft, switching the proliferation or growth factor production of transplanted cells off, or turning on apoptotic pathways to eliminate grafts causing unwanted side effects (Thomis *et al.*, 2001). More recently, the observation that the homeobox transcription factor, Lmx1a, crucially determines dopamine cell fate has provided further promise to this approach (Andersson *et al.*, 2006).

It has been shown recently that neurogenesis does occur in rodent, primate, and also human substantia nigra, confirming the presence of adult stem cells/neuronal progenitors potentially capable of repair (Lie *et al.*, 2002; Yoshimi *et al.*, 2005; Zhao *et al.*, 2003). This finding offers yet another neurotrophic strategy in PD, to trigger endogenous neuronal progenitor cells in order to enhance their proliferation and subsequent differentiation, which could eventually lead to the repair of the degenerating DAergic region. This approach would avoid many of the potential problems associated with the transplantation of ES-derived cells: potential tumor formation, rejection of the grafts, even the surgery itself. Although some of the pathways mediating enhanced progenitor proliferation and/or differentiation have been reported, finding small molecules capable of crossing the blood–brain barrier and inducing neurogenesis or sprouting specific for degenerating DAergic neurons represents a major therapeutic challenge in the future (for review see Longo *et al.*, 2006).

B. Growth Factors

In recent years, there has been extensive research into the role of neurotrophic factors as a possible neurorestorative treatment in PD (Bradford *et al.*, 1999).

In particular, these peptides can promote neuronal differentiation and growth during brain development. Exploiting the neurotrophic ability to induce axonal elongation, neurite outgrowth, and synaptogenesis in dopaminergic projections as well as to provide support to the surviving dopaminergic neuronal perikarya may modify disease progression while simultaneously providing symptomatic relief.

1. *Glial-Derived Neurotrophic Factor*

The neurotrophic ability of GDNF, a member of the transforming growth factor-β superfamily, as a potential treatment in PD has been extensively studied. GDNF appears to act preferentially on dopaminergic neurons. For instance, a single injection of GDNF in a rat produces a profound increase in striatal dopamine content (Hoffer *et al.*, 1994). Numerous studies report that the presence of GDNF facilitates the survival of dopamine neurons in culture (Hudson *et al.*, 1995; Lin *et al.*, 1993) and protects mesencephalic primary neuron cultures from 6-OHDA- or MPTP-mediated cell death (Hou *et al.*, 1996). These results have also been confirmed *in vivo* using rodent models of PD (Fox *et al.*, 2001). GDNF was effective at protecting and restoring dopaminergic neurons and terminals in both the 6-OHDA-lesioned rat and MPTP-treated mouse models (Kearns and Gash, 1995; Kearns *et al.*, 1997). Furthermore, postlesion infusion of GDNF was able to restore TH activity in the striatum and improve locomotion in 6-OHDA-treated rats (Hoffer *et al.*, 1994; Lapchak *et al.*, 1998; Tomac *et al.*, 1995). Thus, GDNF appears to be one of the major components of neurotrophic activity in the dopamine-depleted nigrostriatal tract (Nakajima *et al.*, 2001).

The restorative effect of GDNF has also been replicated in nonhuman primate models of PD (Gash *et al.*, 1996; Zhang *et al.*, 1997). The antiparkinsonian action of GDNF in MPTP-treated monkeys was demonstrated with behavioral assessment using a primate-adapted version of the UPDRS. Postmortem TH analysis showed repair of the dopaminergic nigrostriatal system in the GDNF-treated monkeys (Gash *et al.*, 1996; Tomac *et al.*, 1995). Neuroprotection was clearly demonstrated after intracerebroventricular (icv) infusion of recombinant GDNF into the lateral ventricles in the MPTP-treated monkey (Grondin *et al.*, 2002). Moreover, the treated monkeys needed significantly less L-DOPA and exhibited a reduced incidence and severity of treatment-associated dyskinesia (Miyoshi *et al.*, 1997).

Viral vectors have also been used to deliver and induce striatal GDNF expression in PD rodent models and nonhuman primates lesioned with MPTP (Bilang-Bleuel *et al.*, 1997; Choi-Lundberg *et al.*, 1997; reviewed by Kordower, 2003). These studies have been performed using replication-deficient adenovirus (Lapchak *et al.*, 1998) and lentivirus constructs expressing GDNF (Bjorklund *et al.*, 2000). Kordower *et al.* (2000) demonstrated that lentiviral vector delivery of GDNF into the striatum of healthy monkeys significantly increased local GDNF levels and raised dopamine activity as assessed by F^{18}–DOPA PET imaging. In the same study, GDNF expression induced by lentiviral administration was neuroprotective in primates treated with MPTP as measured using an adapted version of the UPDRS and postmortem tyrosine expression in the striatum and substantia nigra. Moreover, functional improvements in arm reaching were reported accompanied by a normalization in striatal F^{18}–DOPA uptake. These studies have also been mimicked in rodents using adenoviral vectors or microspheres to deliver the GDNF.

Several human PD trials of treatment with GDNF have also been undertaken (Kordower *et al.*, 1999; Nutt *et al.*, 2001). The method of GDNF administration has been the subject of much debate since, unfortunately, the peptide does not easily cross the blood–brain barrier and systemic administration in human causes deleterious side effects (Zurn *et al.*, 2001). icv administration of GDNF via an implanted catheter produced no improvements in parkinsonism as measured by UPDRS and neuropathological evidence. Furthermore, there were significant side effects, including loss of appetite, nausea, tingling, depression, hallucinations, and "inappropriate sexual conduct" (Kordower *et al.*, 1999). However, commentators have cast doubt on whether the concentration of GDNF was sufficient and that the duration of the trial was too short. The ability of GDNF to reach the target site following icv administration was also a subject of debate. Since patients receiving GDNF experienced quite serious side effects, it can be deduced that the delivery of GDNF to the lateral ventricles was centrally acting in proximal regions such as the hypothalamus but probably not in the striatum and substantia nigra located deeper.

More recently, Gill *et al.* performed a more successful PD clinical trial by infusing GDNF directly into the putamen via a catheter pump into five patients. After 12 months, there was a 48% improvement in activities of daily living scores in the UPDRS and a significant reduction in L-DOPA-induced dyskinesia (Gill *et al.*, 2003). Striatal dopamine uptake/activity was also increased in the putamen up to 18 months into the trial as assessed using functional imaging with F^{18}–DOPA. After 2 years on GDNF infusion, the patients had a 57% improvement in their off-medication motor and a 63% improvement in UPDRS (Patel *et al.*, 2005). Thus in these small studies, GDNF, when delivered locally in the basal ganglia but not when administered distally via the ventricles, appeared to be sufficient to restore dopamine nigrostriatal activity and improve UPDRS scores. This small trial has been followed up with a larger (Amgen-sponsored) randomized study (34 patients) in which recombinant GDNF was perfused into the putamen of PD sufferers for 6 months (Lang *et al.*, 2006). Dopaminergic activity in the striatum of subjects given GDNF was significantly increased as measured by F^{18}–DOPA uptake. Despite this, the UPDRS scores in those patients given GDNF improved, but were not significantly different from those receiving placebo.

There has been much debate about the difference in methodology and results between the small open-labeled studies (Gill *et al.*, 2003) and the larger double-blinded trials (Lang *et al.*, 2006). Some recent review articles have summarized the data with GDNF (Sherer *et al.*, 2006) and the challenges in the therapeutic use of growth factors (Kotzbauer and Holtzman, 2006). The lack of any clear efficacy in the double-blinded study (Lang *et al.*, 2006), together with unusual cerebellar toxicity in monkeys treated with high doses on GDNF (Sherer *et al.*, 2006) led Amgen to withdraw the treatment until further information was obtained. In addition, 10% of patients participating in the trial developed antibodies to human GDNF (Sherer *et al.*, 2006). There has been a flurry of public debates on the ethical implications of terminating the clinical development of GDNF by Amgen, which underscores the severe unmet medical need of PD patients (see editorial in *Lancet Neurology* 4, 787 (2005) and responses to the editorial in *Lancet Neurology* 5, 2000–2003 (2006) and Barker, 2006).

2. *Brain-Derived Neurotrophic Factor*

Brain-derived neurotrophic factor (BDNF) is a small, homodimeric member of the neurotrophin (NT) family of secreted growth factors that are vital for trophic support of neurons within both the peripheral nervous system and the CNS. BDNF signals through at least two different receptors—the higher affinity tyrosine kinase receptor B (TrkB) and the lower affinity p75 NT receptor ($p75^{NTR}$)—and is abundantly and widely expressed in the adult mammalian brain in both neuronal and glial cells (for a review of BDNF localization in the human brain see Murer *et al.*, 2001). Within the adult nigrostriatal pathway, BDNF is both synthesized by dopaminergic neurons (Seroogy *et al.*, 1994) and used by the same cells as a potent survival and growth factor, first demonstrated by Hyman *et al.* (1991). It is also important to remember that BDNF is subject to both retrograde and anterograde transport within the CNS (Mufson *et al.*, 1999; Siegel and Chauhan, 2000), so it may have effects at sites some distance from where it is synthesized.

Among the growth factors of potential therapeutic utility in PD, BDNF, and GDNF are the most studied, though a role for other factors such as the recently discovered mesencephalic astrocyte-derived neurotrophic factor (MANF; Petrova *et al.*, 2003) or other, as yet undiscovered factors, cannot be excluded. There is a large body of data showing that BDNF promotes the survival of dopaminergic cells in culture (Hyman *et al.*, 1991; Presgraves *et al.*, 2004; Spina *et al.*, 1992a,b). Several molecular pathways involved in this trophic effect of BDNF have been implicated, including activation of glutathione reductase and activation of P3K-AKT or MEK/mitogen-activated protein kinase kinase (MAPKK) pathways. Similarly, BDNF promotes the survival of rat fetal ventral mesencephalon tissue (Hoglinger *et al.*, 1998). Other studies have suggested that BDNF actually induces a dopaminergic phenotype, for example increased TH and dopamine production, in rat and human fetal cortical cells during development (Theofilopoulos *et al.*, 2001; Zhou *et al.*, 1998).

The trophic effects of BDNF seen *in vitro* translate into efficacy in *in vivo* models. Thus, BDNF can induce striatal neurogenesis in adult rats with 6-OHDA lesions (Mohapel *et al.*, 2005), while maturation of nigrostriatal dopaminergic neurons is diminished in BDNF-deficient mice (Baker *et al.*, 2005; Baquet *et al.*, 2005). Similarly, rats treated with antisense oligonucleotides to inhibit BDNF production show significant loss of nigral dopaminergic neurons and a phenotype that closely resembles classical animal models of PD (Porritt *et al.*, 2005).

Considering the above evidence implicating BDNF in the survival and development of dopaminergic neurons, it is not surprising that several groups have studied whether polymorphisms in the BDNF gene may be responsible for causing PD. The most frequently studied polymorphism is G196A (Val66Met), for which there are several case-control studies producing contradictory results. However, a recent meta-analysis of all studies to date did not support an association of this polymorphism with the risk of developing PD (Zintzaras and Hadjigeorgiou, 2005).

Irrespective of whether or not polymorphisms in the BDNF gene play a causal role in PD, there is growing evidence that reduced neurotrophic support is a significant factor in the pathogenesis of a number of neurodegenerative diseases, including PD. Many studies have shown that BDNF (and other NT) levels are reduced in human PD brains when compared to healthy controls (Chauhan *et al.*, 2001; Howells *et al.*, 2000; Imamura *et al.*, 2005; Mogi *et al.*, 1999; Murer *et al.*, 2001; Nagatsu *et al.*, 2000; Parain *et al.*, 1999). Similarly, the number of TrkB mRNA-containing neurons is dramatically reduced in the SNc and ventral tegmental area of PD patients compared with healthy controls (Benisty *et al.*, 1998). However, this study showed that the level of TrkB mRNA in the *remaining* neurons in diseased brains is unchanged. This is clearly an important point; both BDNF and TrkB are expressed in the very nigral dopaminergic neurons that are lost in PD. Thus, it is evident that the levels of these two proteins would be expected to decrease in the PD brain, complicating the interpretation of these studies and questioning whether the loss of BDNF/TrkB is a cause, or the result, of PD. Nevertheless, the fact that the remaining dopaminergic neurons in PD brain still express TrkB offers hope that one day therapy that increases BDNF levels may be effective in this disease. Indeed, one study has even suggested that glial BDNF expression is endogenously up-regulated in PD brain in response to signals released from failing nigral neurons (Knott *et al.*, 2002).

An interesting link between α-synuclein, the major component of LBs in PD, and BDNF was recently made by Kohno *et al.* (2004). It was suggested that α-synuclein increases BDNF production from glial cells while mutant forms of α-synuclein, such as A30P and A53T, known to be associated with early onset PD, did not. This loss of stimulation of BDNF production was therefore implicated in the onset of disease in patients with these mutant forms of α-synuclein.

Despite the wealth of data linking BDNF to the etiology of PD, there are still numerous issues pertaining to delivery and treatment regimens in humans that need to be addressed prior to any clinical application of BDNF itself. One of the greatest problems is achieving optimal delivery of the peptide to the human brain without using direct intracerebral injection, since NTs do not cross the blood–brain barrier. Recent novel approaches, such as conjugating BDNF to antibodies specifically engineered to cross the blood–brain barrier, may offer hope in this regard (Pardridge, 2002). Many pharmaceutical companies are taking a different approach, aiming to find small, brain-penetrant molecules that act upstream of BDNF to increase its expression levels in the human CNS. Indeed, expression of the BDNF gene can be altered by many physiological signals and regulation of BDNF levels and/or activity has been implicated in the biological actions of a vast array of small molecules and treatment regimens.

Several groups have shown that direct application of BDNF is efficacious in a variety of animal models of PD. The first such study showed that chronic infusion of BDNF above the substantia nigra protected against 6-OHDA-induced behavioral deficits in the rat (Altar *et al.,* 1994). Subsequently, it was shown that similar efficacy could be achieved using daily intrastriatal injections (Shults *et al.,* 1995). Details of efficacy in MPTP models of PD soon followed, both in Japanese monkeys (Tsukahara *et al.,* 1995) and in neonatal rats (Kirschner *et al.,* 1996). Isacson and coworkers confirmed that it is also possible to prevent neurotoxicity in the rodent MPTP model using implantation of fibroblasts that have been genetically engineered to secrete BDNF (Frim *et al.,* 1994; Galpern *et al.,* 1996).

In an effort to study the effects of long-term exogenous BDNF expression, Klein *et al.* (1999) employed somatic gene transfer to express the protein in the SNc for up to 9 months following a single injection of a recombinant BDNF adeno-associated virus in rats. This study showed that while exogenous expression of BDNF could prevent 6-OHDA-induced rotational behavior, it was without significant effect on nigrostriatal dopaminergic neuron survival, tempting the authors to speculate that the efficacy of BDNF may be the consequence of a modulatory influence on remaining dopaminergic neurons rather than as a result of any neuroprotective effect, at least in this particular model. Others, however, have found a clear protective effect of BDNF on neuronal survival, even at sites distant from the primary injury location in PD. For example, supranigral infusion of BDNF in rats given striatal-pallidal excitotoxic injections of ibotenic acid prevents the loss of nigral GABAergic neurons (Volpe *et al.,* 1998).

Few studies have directly compared the relative efficacy of BDNF with other growth factors, such as GDNF, in animal models of PD. This prompted

Sun *et al.* (2005) to compare the effects of viral expression of these two growth factors in the rodent 6-OHDA model of PD. The results showed that GDNF is significantly more effective in correcting behavioral deficits and protecting nigrostriatal neurons than expression of BDNF, at least in this particular model. In agreement with previous work by Klein *et al.* (1999), while BDNF was able to support behavioral correction, it failed to provide protection of nigrostriatal neurons. Interestingly, expression of both neurotrophic factors together was no more effective than GDNF alone, suggesting that there are not synergistic effects from expressing the two growth factors together.

C. Small Molecule Approaches

Although recent advances is gene therapy and drug delivery devices make biomolecule therapies approachable, several issues related to optimal and controllable drug delivery remain. Additionally, these approaches are invasive and require technical skills that are not yet widely available. It should be noted that small molecule agonists of growth factor receptors such as ret-A or TrkB have not met with success likely due to the size and complexity of the ligand-binding pockets of these receptors. A promising alternative approach relies on discovery of small molecules that stimulate endogenous trophic factors that support dopaminergic neurons.

1. *Nurr1 and shh Pathway*

Developmental studies have identified two key druggable factors required for dopamine neurons, shh (Tsuboi and Shults, 2002) and Nurr1 (Chu *et al.*, 2006; Jankovic *et al.*, 2005). Nurr1, a transcription factor belonging to the orphan nuclear receptor superfamily, is thought to play a critical role in the development and maintenance of the dopaminergic system. It is expressed in developing and adult ventral midbrain neurons and appears to regulate expression of phenotypic markers of dopamine neurons such as TH and dopamine transporter (DAT) AADC, VMAT2, and cRet in ES cells as well as primary neuronal cultures (Hermanson *et al.* 2003; Iwawaki *et al.* 2000; Jankovic *et al.*, 2005; Martinat *et al.*, 2006; Sacchetti *et al.*, 2001; Wallen *et al.*, 2001). Experimental studies in Nurr1 knockout mice indicate that Nurr1 haploinsufficiency leads to reduced levels of Nurr1 protein and dopamine and inability of the embryonic dopamine neurons to innervate striatal fields (Saucedo-Cardenas *et al.*, 1998; Zetterstrom *et al.*, 1997). Additionally, these mice are significantly more sensitive to MPTP than the wild-type littermates (Jankovic *et al.*, 2005; Le *et al.*, 1999). Nurr1 expression is reduced during the course of normal aging (Chu *et al.*, 2002) and this is especially evident in α-synuclein aggregate-containing neurons in autopsied brains from PD cases (Chu *et al.*, 2006). Furthermore, mutations in the gene encoding Nurr1, *NR4A2,* are associated with sporadic PD (Le *et al.*, 2003) and transcription of

NR4A2 is reduced by overexpression of α-synuclein in neuroblastoma cells (Baptista *et al.*, 2003). Although Nurr1 does not appear to have a classical ligand-binding domain, it can partner with other nuclear hormone receptors such as RXR to regulate target gene expression (Zetterstrom *et al.*, 1996). Taking advantage of this property, several groups have identified Nurr1-RXR heterodimer activators, which increase dopamine or TH levels *in vitro* and *in vivo* (Wallen-Mackenzie *et al.*, 2003; WO 2004/072050 A1, WO 2003/015780 A2). Although the small molecule activators of Nurr1-RXR heterodimer have not been tested *in vivo* in models of PD, validation of this therapeutic approach is seen in preliminary evidence of protection of dopamine neurons following viral-mediated transduction of Nurr1 expression in rats and monkeys (Apostolakis, 2001). However, it is critical to point out that potential liabilities of systemically administered Nurr1 activators remain unknown. This is an issue worthy of consideration for empirical testing since recent studies indicate a pleiotropic role of Nurr1 as a transcription regulator in a number of cell types (Hsu *et al.*, 2004; Martinez-Gonzalez and Badimon, 2005).

shh, a member of the hedgehog (hh) family of signaling molecules, is necessary for normal axial patterning and cellular differentiation in the developing CNS. shh has been reported to promote survival of cultured neural progenitor cells (Rafuse *et al.*, 2005) and induce a dopaminergic neuronal phenotype in the midbrain (Hynes *et al.*, 1995). Intrastriatal (Tsuboi and Shults, 2002) and supranigral (Dass *et al.*, 2002) admisitration of shh has been reported to provide protection against neurotoxin-induced toxicity in rats and marmosets. More recently, adenoviral vector delivery of shh has also been reported to protect dopaminergic neurons against 6-OHDA toxicity in rats (Dass *et al.*, 2005). It may also be possible to target shh pathways using small molecule approaches. For example, Curis Inc. has reported small molecule modulators of hh signaling and identified shh receptor agonists and antagonists (Frank-Kamenetsky *et al.*, 2002).

2. *AMPA Receptor Potentiators*

AMPA receptors are one class of ionotropic glutamate receptors that control fast synaptic transmission and play a role in neuroplasticity phenomena such as long-term potentiation (LTP). At Eli Lilly, we have explored allosteric agonists of AMPA receptors termed AMPA receptor potentiators (O'Neill *et al.*, 2004a) as neurotrophic agents in animal models of PD (Murray *et al.*, 2003; O'Neill *et al.*, 2004b, 2005). For the rest of the chapter, we will summarize the rationale and data supporting the use of biarylpropylsulfonamide class of AMPA receptor potentiators as neurorestorative agents for PD. All positive, allosteric modulators of AMPA receptors (i.e., AMPA receptor potentiators) lack direct agonist effects at the AMPA receptor but augment the effects of an agonist such as AMPA by blocking desensitization or deactivation of the channel (O'Neill *et al.*, 2004a; Quirk and Nisenbaum, 2002). However, these agents differ in their potency and efficacy to potentiate agonist effects as studied by Ca^{2+} flux using fluorescent imaging plate

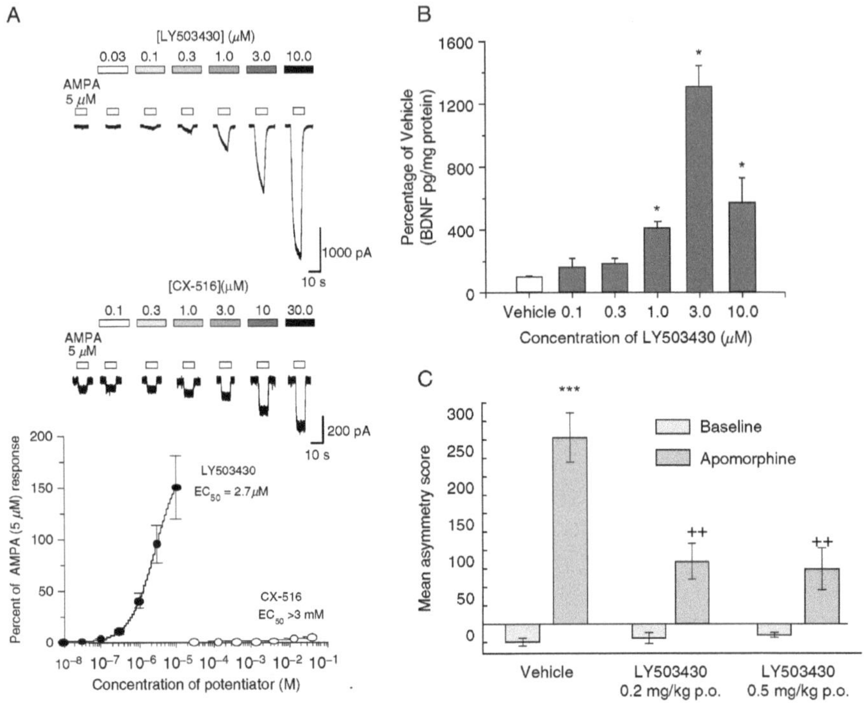

FIG. 1. Effects of LY503430 and CX-516 on native AMPA receptor activity *in vitro* (A), BDNF levels in cortical neurons (B), and functional improvements in 6-OHDA model (C). (A) The concentration-response profile for LY503430 (0.03–10.0 μM) potentiation was assessed by measuring the responses of acutely isolated substantia nigra dopamine neurons to 5-μM AMPA alone and in the presence of potentiator. The plots show the average degree of potentiation by LY503430 (0.03–10.0 μM) and CX-516 (0.3–3.0.0 mM) as a percentage of the 5-μM AMPA response for each concentration of compound tested in acutely isolated substantia nigra dopamine neurons. Data points in plots represent mean ± SEM. (B) The concentration-response profile for LY503430 (0.1–10.0 μM) at increasing BDNF expression in cortical neurons. (C) The effects of chronic treatment with LY503430 (0.2 and 0.5 mg/kg p.o. for 10 days starting 1 day after infusion of 6-OHDA into the nigra) on rotational behavior. Results indicate that both doses of LY503430 provided a significant correction of apomorphine-induced rotational asymmetry after unilateral infusion of 6-OHDA into the substantia nigra. Data are based on eight animals per group. $^{***}p < 0.001$ vs baseline rotations; $^{+}p < 0.05$, $^{++}p < 0.01$ vs vehicle treated animals.

reader (FLIPR) in cloned and native cells, patch clamp techniques (Miu *et al.*, 2001; Quirk and Nisenbaum, 2002), or *in vivo* using both electrophysiological and behavioral tests (Murray *et al.*, 2003; O'Neill *et al.*, 2004a,b; Vandergriff *et al.*, 2001). This is exemplified in the comparison of efficacy of CX-516 versus LY503430 at potentiating AMPA-mediated Ca^{2+} flux in native cells (Fig. 1A).

One of the downstream effects of AMPA receptor-mediated transmission appears to be induction of expression of the NT, BDNF *in vitro* (Hayashi *et al.*,

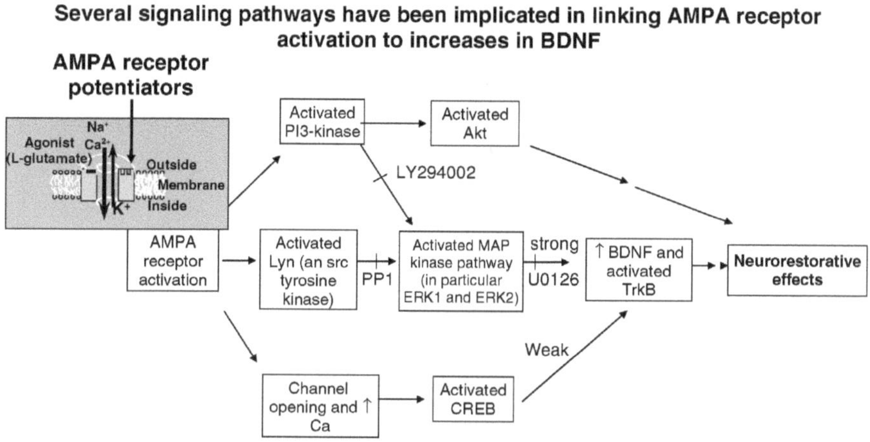

FIG. 2. An illustration of the pathways by which AMPA receptor modulation can increase the expression of genes and proteins from the BDNF family. The calcium-dependent pathway is shown on the bottom of the slide. Many of the biarylpropylsulfonamides (LY404187, LY503430) increase BDNF when given alone, suggesting that the Lyn tyrosine kinase pathway (center) may be predominant in some neuronal preparations [adapted from Hayashi *et al.* (1999) and Wu *et al.* (2004)]. (See Color Insert.)

1999; Zafra *et al.*, 1990; Fig. 1B). Mechanistic studies demonstrated that AMPA receptor-mediated increases in BDNF can be mediated via at least two distinct intracellular signaling pathways (Fig. 2). The first pathway involves increases in intracellular calcium concentration through voltage-gated L-type calcium channels activated as a consequence of the AMPA receptor-induced membrane depolarization. Increases in intracellular calcium lead to an increase in BDNF gene expression through calcium response elements on the BDNF promoter regions (Hardingham *et al.*, 1998; Tao *et al.*, 1998; Zafra *et al.*, 1990, 1992). The second pathway for BDNF production depends on activation of Lyn, a member of the src-family of protein tyrosine kinases, which can physically associate with AMPA receptor subunits. Lyn activates the MAPK pathway leading to increased BDNF expression (Hayashi *et al.*, 1999).

It has been demonstrated that biarylpropylsulfonamides can increase BDNF expression in both cerebellar granule, cortical and hippocampal neurons *in vitro* (Legutko *et al.*, 2001; Fig. 1B). The increases in BDNF observed with AMPA alone or in combination with a potentiator were blocked by L-type calcium channel blockers (nimodipine), while the increases produced by the potentiator alone were reduced by an MAPK inhibitor (PD 098059), suggesting that both pathways may contribute to the increased production in BDNF (Legutko *et al.*, 2001). It is also interesting that the biarylpropylsulfonamides alone can produce substantial BDNF increase in primary neuronal cultures (Fig. 1B). The data suggest that either there is sufficient glutamate in the media to induce the calcium

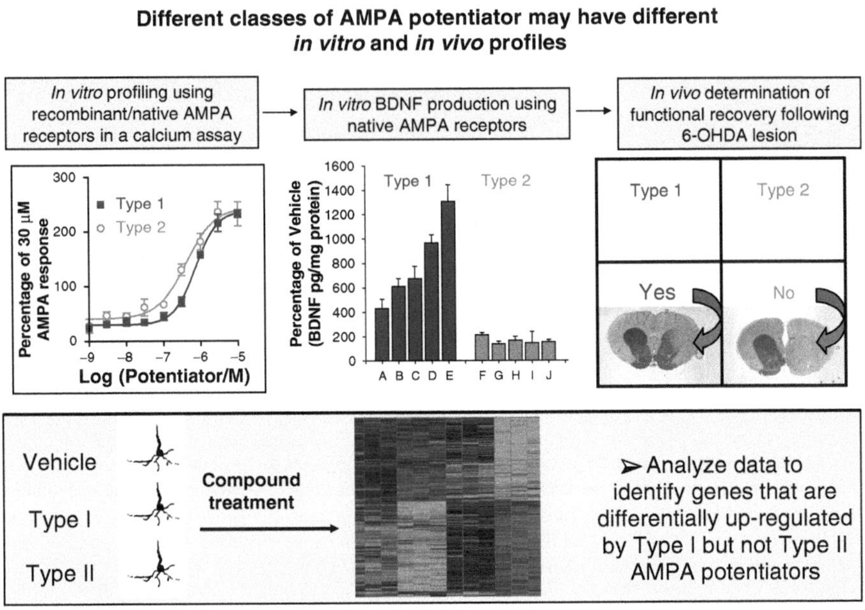

FIG. 3. Various AMPA potentiators have different *in vitro* and *in vivo* profiles. These differences could be used to help further our understanding into the mechanism of action of AMPA potentiators in the 6-OHDA model. The top panel shows some examples of AMPA potentiators that we have termed Type 1 and Type 2. Both classes potentiate AMPA responses in cloned and native cells, but only Type 1 increases BDNF levels in cortical neurons and shows efficacy in 6-OHDA model. The data demonstrate that various AMPA potentiatiors appear to have differential effects on downstream/secondary-signaling pathways. It is possible that by evaluating the genes expression profiles (bottom panel) of these potentiator classes, we may gain further insight into their neurotrophic actions. (See Color Insert.)

signaling or that the calcium-independent mechanisms are the predominant mechanism for inducing BDNF levels. We have also discovered a series of AMPA receptor potentiators that flux Ca^{2+} and potentiate AMPA responses at cloned and native AMPA receptors, but do not increase BDNF levels in cell cultures (Fig. 3). We are profiling these compounds together with compounds (such as LY404187, LY503430) that do increase BDNF to try to further characterize the mechanisms by which these agents increase BDNF (Fig. 3).

It has also been demonstrated that subchronic treatment with AMPA receptor potentiators increases BDNF mRNA (Lauterborn *et al.*, 2000; Mackowiak *et al.*, 2002) and protein (Mackowiak *et al.*, 2002) in the rat hippocampus. The *in vivo* induction in BDNF mRNA is dose and treatment paradigm dependent and shows specificity with respect to brain regions. In addition to their effects on BDNF levels, the biarylpropylsulfonamide class of AMPA potentiatior has also been demonstrated to increase cell proliferation (Bai *et al.*, 2003) in the rat brain.

TABLE I
THE EFFECTS OF A RANGE OF BIARYLPROPYLSULFONAMIDE AMPA RECEPTOR POTENTIATORS IN THE 6-OHDA MODEL

Compound	Rotation score	Ventral TH-I	Dorsal TH-I	ED_{50} (mg/kg)
LY404187	Active	Active	Active	<0.5
LY503403	Active	Active	Active	0.081
LY450108	Active	Active	Active	0.093
LY451395	Active	Active	Active	<0.5

Based on the observation that AMPA receptor potentiators increased BDNF expression and potentially neurogenesis, we hypothesized that AMPA receptor potentiators may have neurotrophic actions in PD models. The results of a series of studies indicated that both LY404187 and LY503430 protect against MPTP-induced neurotoxicity in mice and provide functional and histological improvement after infusion of 6-OHDA into the substantia nigra or striatum of rats (Murray *et al.*, 2003; O'Neill *et al.*, 2004b, 2005). The most robust functional and histological improvements were observed with subchronic AMPA potentiator treatment (10–14 days) initiated after a unilateral 6-OHDA lesion of the substantia nigra (Murray *et al.*, 2003, O'Neill *et al.*, 2005). Dose–response studies with a range of biarylpropylsulfonamides (LY404187, LY450108, LY451395, and LY503430) in the nigral 6-OHDA model were carried out (Table I). The results indicated that all compounds showed efficacy at 0.5 mg/kg and the ED_{50} for rotational behaviors ranged between 0.081 and 0.110 mg/kg p.o. A detailed review of the effects of LY503430 in various PD models has recently been published (O'Neill *et al.*, 2005).

Molecules were administered for 10 days with the first injection 24 hours after infusion of 6-OHDA into the substantia nigra. All molecules provided a significant correction of apomorphine-induced rotational asymmetry and improvements in both the dorsal (caudate) and ventral (accumbens) TH immunohistochemistry. The ED_{50} values listed in Table I are for rotational behavior ($n = 8$–10 per group).

Since the subchronic treatment of the AMPA receptor potentiators was initiated after 6-OHDA infusion, it appears that recruitment of some neuroplastic/neurotrophic pathway underlies the protective effects of these compounds in the rat model of dopaminergic neurodegeneration. The exact biochemical pathway(s) involved remains to be elucidated, but a brief summary of the key data to date is provided below.

1. Extensive profiling of LY503430 (0.5 mg/kg s.c.) indicates that acute administration does not affect either baseline rotational behavior or L-DOPA- or pergolide-stimulated rotations in 6-OHDA-lesioned rats (O'Neill *et al.*, 2005).

This observation is consistent with the lack of evidence that activation of AMPA receptors would directly alter dopamine release and implies that the AMPA receptor potentiator is unlikely to have acute symptomatic effects in PD. However, it cannot be ruled out that AMPA receptor potentiators may modulate other parts of the basal ganglia circuitry and this may not be evident from the experiments carried out using 6-OHDA.

2. The functional and protective actions of LY404187 and LY503430 appear to require some form of chronic dosing. Thus, 10–14 days dosing is robustly effective, but dosing for shorter periods (1 or 4 days) does not provide any functional or histological improvements. These data suggest that a repeated stimulation of the AMPA receptor is required to see the effect. The data also suggest that either a cascade of events is required for efficacy (activation of gene expression and/or downstream signaling pathways) or that some form of response to injury occurs over a 10- to 14-day period.

3. We have evaluated a number AMPA receptor poteniators in the nigral 6-OHDA that have similar potency at potentiating cloned and native AMPA receptors, but do not increase BDNF levels *in vitro*. These compounds failed to show efficacy in the 6-OHDA-lesioned model. This suggests that in addition to increasing calcium influx, the compounds that show efficacy in 6-OHDA have some additional downstream effects. BDNF expression *in vitro* appears to be a surrogate measure of efficacy *in vivo*.

4. The functional and protective actions of LY404187 and LY503430 appear to be associated with an increase in growth-associated protein-43 (GAP-43) expression in the striatum (O'Neill *et al.*, 2004b). It is well established that GAP-43 is involved in neuronal repair and sprouting and transgenic animals that overexpress GAP-43 exhibit increased sprouting and repair after exposure to lesions. It might be that AMPA potentiators such as LY503430 and related compounds may be responsible for initiating the biological cascade toward its production.

5. The preservation of striatal TH immunoreactivity was observed also when administration of AMPA receptor modulators was initiated 3 or 6 days after infusion of 6-OHDA into the substantia nigra (Murray *et al.*, 2003; O'Neill *et al.*, 2005). Importantly, the degree of TH immunoreactivity (preservation of terminals) was similar whether the drug was administered as cells start to die (1 day) or after severe dopamine depletion (80–90% loss of nigral cell bodies at 6 days after 6-OHDA infusion).

6. The functional and histological improvements were still present when animals were tested 28 days after "washing out" LY503430 (Fig. 17 in O'Neill *et al.*, 2005). One interpretation of these data is that LY503430 treatment for 10–14 days produced sprouting of the striatal terminals, which continue to remain intact even after the drug administration is stopped. Obviously, in this preclinical model, the neurotoxin (6-OHDA) was administrated once at the start

of the experiment, while in PD there is a progressive degeneration of the neurons, which will likely require long-term therapy with an AMPA receptor potentiator to maintain the trophic response.

7. Ongoing experiments to look at the effects of AMPA potentiators on gene expression *in vitro* and *in vivo* may help further elucidate the key signaling pathways leading to the large functional improvements after nigrostriatal lesions.

In summary, the data we have outlined above suggest that AMPA receptor potentiators may have both neuroprotective and neurotrophic actions in rodent models of PD. Thus, positive modulation of AMPA receptors may provide a means of both halting the progression and perhaps reversing the degeneration in PD. It should be noted that PD patients do not present with symptoms until 60–70% of the substantia nigra is damaged. So an agent that produced reinnervation and increased striatal terminals could in theory reverse the functional deficits. The molecular mechanism(s) of these effects are poorly elucidated but the involvement of a growth factor such as BDNF is possible. A number of ongoing studies (Fig. 3) aim to look at the types of mechanisms that could be contributing to these functional and histological improvements we have observed in the 6-OHDA model. The lesion models used in the studies to date do not fully mimic PD (e.g., do not have inclusion/LBs), but do produce large dopamine depletion in the relevant brain structures. It would be of interest to study these molecules in other models based on genetic etiology. Regardless of the mechanisms of nigral cell death in PD and whether Lewy bodies are contributing to the damage or helping protect the brain, it is likely that an agent that is able to produce reinnervation of the striatum would be of benefit.

VI. Summary and Conclusions

There has been a large amount of research into the mechanisms of cell death in PD, producing new and refining exisiting animal models of PD and evaluating methods of providing neurorepair after nigrostriatal degeneration. Many of the earlier transplantation and GDNF animal studies have now progressed into the clinical setting with mixed results. Fetal cell transplants provided some improvement in symptoms (UPDRS), but this was often associated with delayed side effects (dyskinesia or dystonia). GDNF has been administered in open trials and provided improved functional outcome and some improvement in F-dopa uptake in the caudate. However, this has yet to be replicated and debates over delivery methods (e.g., type of cannula, site of delivery) continue.

There are still many challenges to overcome, but in the last 3–4 years several new genes have been linked to familial PD and these developments coupled with newer models that show a "slower" progressive degeneration should allow new drug targets to be studied more robustly prior to progression to the clinic. Many other lines of research (i.e., evaluation of new trophic factors, understanding the delivery and integration of new cells into the brain, developing stem cell lines that can secrete trophic factors, and small molecule approaches to brain repair) are all progressing. Finally, the small molecule approaches described above may bear fruits and address the major unmet medical need of PD patients by arresting disease progression.

Acknowledgments

The authors would like to thank Deborah McCarty and Edward Siuda for collecting the BDNF and electrophysiology data in Figs. 1 and 3.

References

Abou-Sleiman, P. M., Muqit, M. K., and Wood, N. W. (2006). Expanding insights of mitochondrial dysfunction in Parkinson's disease. *Nat. Rev. Neurosci.* **7,** 207–219.

Agid, Y., Cervera, P., Hirsch, E., Javoy-Agid, F., Lehericy, S., Raisman, R., and Ruberg, M. (1989). Biochemistry of Parkinson's disease 28 years later: A critical review. *Mov. Disord.* **4**(Suppl. 1), S126–S144.

Altar, C. A., Boylan, C. B., Fritsche, M., Jones, B. E., Jackson, C., Wiegand, S. J., Lindsay, R. M., and Hyman, C. (1994). Efficacy of brain-derived neurotrophic factor and neurotrophin-3 on neurochemical and behavioral deficits associated with partial nigrostriatal dopamine lesions. *J. Neurochem.* **63**(3), 1021–1032.

Andersen, J. K. (2001). Does neuronal loss in Parkinson's disease involve programmed cell death? *Bioessays* **23,** 640–646.

Andersson, E., Tryggvason, U., Deng, Q., Frilling, S., Alekseenko, Z., Robert, B., Perlmann, T., and Erickson, J. (2006). Identification of intrinsic determinants of midbrain dopamine neurons. *Cell* **124,** 393–405.

Apostolakis, E. M. (2001). Society for Neuroscience abstract.

Bai, F., Bergeron, M., and Nelson, D. L. (2003). Chronic AMPA receptor potentiator (LY451646) treatment increases cell proliferation in adult rat hippocampus. *Neuropharmacology* **44,** 1013–1021.

Baker, S. A., Stanford, L. E., Brown, R. E., and Hagg, T. (2005). Maturation but not survival of dopaminergic nigrostriatal neurons is affected in developing and aging BDNF-deficient mice. *Brain Res.* **1039**(1–2), 177–188.

Baptista, M. J., O'Farrell, C., Daya, S., Ahmad, R., Miller, D. W., Hardy, J., Farrer, M. J., and Cookson, M. R. (2003). Co-ordinate transcriptional regulation of dopamine synthesis genes by α-synuclein in human neuroblastoma cell lines. *J. Neurochem.* **85,** 957–968.

Baquet, Z. C., Bickford, P. C., and Jones, K. R. (2005). Brain-derived neurotrophic factor is required for the establishment of the proper number of dopaminergic neurons in the substantia nigra pars compacta. *J. Neurosci.* **25**(26), 6251–6259.

Barberi, T., Klivenyi, P., Calingasan, N. Y., Lee, H., Kawamata, H., Loonam, K., Perrier, A. L., Bruses, J., Rubio, M. E., Topf, N., Tabar, V., Harrison, N. L., *et al.* (2003). Neural subtype specification of fertilization and nuclear transfer embryonic stem cells and application in parkinsonian mice. *Nat. Biotechnol.* **21,** 1200–1207.

Barker, R. (2006). Continuing trials of GDNF in Parkinson's disease. *Lancet Neurol.* **5,** 285–286.

Beal, M. F. (1998). Excitotoxicity and nitric oxide in Parkinson's disease pathogenesis. *Ann. Neurol.* **44**(3 Suppl. 1), S110–S114.

Beal, M. F. (2001). Experimental models of Parkinson's disease. *Nat. Rev. Neurosci.* **2**(5), 325–334.

Beal, M. F. (2003). Mitochondria, oxidative damage, and inflammation in Parkinson's disease. *Ann. NY Acad. Sci.* **991,** 120–131.

Benisty, S., Boissiere, F., Faucheux, B., Agid, Y., and Hirsch, E. C. (1998). trkB messenger RNA expression in normal human brain and in the substantia nigra of parkinsonian patients: An *in situ* hybridization study. *Neuroscience* **86**(3), 813–826.

Betarbet, R., Sherer, T. B., MacKenzie, G., Garcia-Osuna, M., Panov, A. V., and Greenamyre, J. T. (2000). Chronic systemic pesticide exposure reproduces features of Parkinson's disease. *Nat. Neurosci.* **3,** 1301–1306.

Betarbet, R., Canet-Aviles, R. M., Sherer, T. B., Mastroberardino, P. G., McLendon, C., Kim, J. H., Lund, S., Na, H. M., Taylor, G., Bence, N. F., Kopito, R., Seo, B. B., *et al.* (2006). Intersecting pathways to neurodegeneration in Parkinson's disease: Effects of the pesticide rotenone on DJ-1, alpha-synuclein, and the ubiquitin-proteasome system. *Neurobiol. Dis.* **22,** 404–420.

Bezard, E., Imbert, C., and Gross, C. E. (1998). Experimental models of Parkinson's disease: From static to the dynamic. *Rev. Neurosci.* **9,** 71–90.

Bilang-Bleuel, A., Revah, F., Colin, P., Locquet, I., Robert, J. J., Mallet, J., and Horellou, P. (1997). Intrastriatal injection of an adenoviral vector expressing glial-cell-line-derived neurotrophic factor prevents dopaminergic neuron degeneration and behavioral impairment in a rat model of Parkinson disease. *Proc. Natl. Acad. Sci. USA* **94**(16), 8818–8823.

Bjorklund, A., Kirik, D., Rosenblad, C., Georgievska, B., Lundberg, C., and Mandel, R. J. (2000). Towards a neuroprotective gene therapy for Parkinson's disease: Use of adenovirus, AAV and lentivirus vectors for gene transfer of GDNF to the nigrostriatal system in the rat Parkinson model. *Brain Res.* **886**(1–2), 82–98.

Bradford, H. F., Zhou, J., Pliego-Rivero, B., Stern, G. M., and Jauniaux, E. (1999). Neurotrophins in the pathogenesis and potential treatment of Parkinson's disease. *Adv. Neurol.* **80,** 19–25.

Brooks, D. J. (2004). Neuroimaging in Parkinson's disease. *NeuroRx* **1**(2), 243–254.

Brooks, D. J., Frey, K. A., Marek, K. L., Oakes, D., Paty, D., Prentice, R., Shults, C. W., and Stoessl, A. J. (2003). Assessment of neuroimaging techniques as biomarkers of the progression of Parkinson's disease. *Exp. Neurol.* **184**(Suppl. 1), S68–S79.

Brundin, P., Pogarell, O., Hagell, P., Piccini, P., Widner, H., Schrag, A., Kupsch, A., Crabb, L., Odin, P., Gustavii, B., Bjorklund, A., Brooks, D. J., *et al.* (2000). Bilateral caudate and putamen grafts of embryonic mesencephalic tissue treated with lazaroids in Parkinson's disease. *Brain* **123** (Pt. 7), 1380–1390.

Burke, R. E. (1998). Programmed cell death and Parkinson's disease. *Mov. Disord.* **13**(Suppl. 1), 17–23.

Burke, R. E., and Kholodilov, N. G. (1998). Programmed cell death: Does it play a role in Parkinson's disease? *Ann. Neurol.* **44,** S126–S133.

Buytaert-Hoefen, K. A., Alvarez, E., and Freed, C. R. (2004). Generation of tyrosine hydroxylase positive neurons from human embryonic stem cells after coculture with cellular substrates and exposure to GDNF. *Stem Cells* **22,** 669–674.

Chauhan, N. B., Siegel, G. J., and Lee, J. M. (2001). Depletion of glial cell line-derived neurotrophic factor in substantia nigra neurons of Parkinson's disease brain. *J. Chem. Neuroanat.* **21**(4), 277–288.

Choi-Lundberg, D. L., Lin, Q., Chang, Y. N., Chiang, Y. L., Hay, C. M., Mohajeri, H., Davidson, B. L., and Bohn, M. C. (1997). Dopaminergic neurons protected from degeneration by GDNF gene therapy. *Science* **275,** 838–841.

Chu, Y., Kompoliti, K., Cochran, E. J., Mufson, E. J., and Kordower, J. H. (2002). Age-related decreases in Nurr1 immunoreactivity in the human substantia nigra. *J. Comp. Neurol.* **450**(3), 203–214.

Chu, Y., Le, W., Kompoliti, K., Jankovic, J., Mufson, E. J., and Kordower, J. H. (2006). Nurr1 in Parkinson's disease and related disorders. *J. Comp. Neurol.* **494**(3), 495–514.

Chung, S., Sonntag, K. C., Andersson, T., Bjorklund, L. M., Park, J. J., Kim, D. W., Kang, U. J., Isacson, O., and Kim, K. S. (2002). Genetic engineering of mouse embryonic stem cells by Nurr1 enhances differentiation and maturation into dopaminergic neurons. *Eur. J. Neurosci.* **16,** 1829–1838.

Colosimo, C. (1999). The rise and fall of tolcapone. *J. Neurol.* **246,** 880–882.

Dass, B., Iravani, M. M., Jackson, M. J., Engber, T. M., Galdes, A., and Jenner, P. (2002). Behavioural and immunohistochemical changes following supranigral administration of sonic hedgehog in 1-methyl-4-phenyl-1,2,3,6-tetrahydropyridine-treated common marmosets. *Neuroscience* **114**(1), 99–109.

Dass, B., Iravani, M. M., Huang, C., Barsoum, J., Engber, T. M., Galdes, A., and Jenner, P. (2005). Sonic hedgehog delivered by an adeno-associated virus protects dopaminergic neurones against 6-OHDA toxicity in the rat. *J. Neural. Transm.* **112**(6), 763–778.

Dawson, T. M., and Dawson, V. L. (2003). Molecular pathways of neurodegeneration in Parkinson's disease. *Science* **302**(5646), 819–822.

Deacon, T., Dinsmore, J., Costantini, L. C., Ratliff, J., and Isacson, O. (1998). Blastula-stage stem cells can differentiate into dopaminergic and serotonergic neurons after transplantation. *Exp. Neurol.* **149,** 28–41.

Dunnett, S. B., and Bjorklund, A. (1999). Prospects for new restorative and neuroprotective treatments in Parkinson's disease. *Nature* **399,** A32–A39.

Emborg, M. E. (2004). Evaluation of animal models of Parkinson's disease for neuroprotective strategies. *J. Neurosci. Methods* **139**(2), 121–143.

Factor, S. A. (1999). Dopamine agonists. *Med. Clin. N. Amer.* **83,** 415–443.

Fahn, S., and the Parkinson Study Group (2005). Does levodopa slow or hasten the rate of progression of Parkinson's disease? *J. Neurol.* **252**(Suppl. 4), IV37–IV42.

Fahn, S., Oakes, D., Shoulson, I., Kieburtz, K., Rudolph, A., Lang, A., Olanow, C. W., Tanner, C., and Marek, K., and the Parkinson Study Group (2004). Levodopa and the progression of Parkinson's disease. *N. Engl. J. Med.* **351**(24), 2498–2508.

Farrer, M. J. (2006). Genetics of Parkinson disease: Paradigm shifts and future prospects. *Nat. Rev. Genet.* **7**(4), 306–318.

Fleming, S. M., Fernaught, P.-O., and Chesselet, M.-F. (2005). Genetic mouse models of parkinsonism: Strengths and limitations. *NeuroRx* **2,** 495–503.

Fornai, F., Schluter, O. M., Lenzi, P., Gesi, M., Ruffoli, R., Ferrucci, M., Lazzeri, G., Busceti, C. L., Pontarelli, F., Battaglia, G., Pellegrini, A., Nicoletti, F., *et al.* (2005). Parkinson-like syndrome induced by continuous MPTP infusion: Convergent roles of the ubiquitin-proteasome system and alpha-synuclein. *Proc. Natl. Acad. Sci. USA* **102**(9), 3413–3418.

Fox, C. M., Gash, D. M., Smoot, M. K., and Cass, W. A. (2001). Neuroprotective effects of GDNF against 6-OHDA in young and aged rats. *Brain Res.* **896,** 56–63.

Frank-Kamenetsky, M., Zhang, X. M., Bottega, S., Guicherit, O., Wichterle, H., Dudek, H., Bumcrot, D., Wang, F. Y., Jones, S., Shulok, J., Rubin, L. L., and Porter, J. A. (2002). Small-molecule modulators of Hedgehog signaling: Identification and characterization of Smoothened agonists and antagonists. *J. Biol.* **1**(2), 10.

Freed, C. R., Greene, P. E., Breeze, R. E., Tsai, W. Y., DuMouchel, W., Kao, R., Dillon, S., Winfield, H., Culver, S., Trojanowski, J. Q., Eidelberg, D., and Fahn, S. (2001). Transplantation of embryonic dopamine neurons for severe Parkinson's disease. *N. Engl. J. Med.* **344,** 710–719.

Frim, D. M., Uhler, T. A., Galpern, W. R., Beal, M. F., Breakefield, X. O., and Isacson, O. (1994). Implanted fibroblasts genetically engineered to produce brain-derived neurotrophic factor prevent 1-methyl-4-phenylpyridinium toxicity to dopaminergic neurons in the rat. *Proc. Natl. Acad. Sci. USA* **91**(11), 5104–5108.

Galpern, W. R., Frim, D. M., Tatter, S. B., Altar, C. A., Beal, M. F., and Isacson, O. (1996). Cell-mediated delivery of brain-derived neurotrophic factor enhances dopamine levels in an MPP+ rat model of substantia nigra degeneration. *Cell Transplant.* **5**(2), 225–232.

Gash, D. M., Zhang, Z., Ovadia, A., Cass, W. A., Yi, A., Simmerman, L., Russell, D., Martin, D., Lapchak, P. A., Collins, F., Hoffer, B. J., and Gerhardt, G. A. (1996). Functional recovery in parkinsonian monkeys treated with GDNF. *Nature* **380,** 252–255.

Gill, S. S., Patel, N. K.., Hotton, G. R., O'Sullivan, K., McCarter, R., Bunnage, M., Brooks, D. J., Svendsen, C. N., and Heywood, P. (2003). Direct brain infusion of glial cell line-derived neurotrophic factor in Parkinson's disease. *Nat. Med.* **9**(5), 589–595.

Gloeckner, C. J., Kinkl, N., Schumacher, A., Braun, R. J., O'Neill, E., Meitinger, T., Kolch, W., Prokisch, H., and Ueffing, M. (2006). The Parkinson disease causing LRRK2 mutation I2020T is associated with increased kinase activity. *Hum. Mol. Genet.* **15**(2), 223–232.

Goetz, C. G., and Diederich, N. J. (1992). Dopaminergic agonists in the treatment of Parkinson's disease. *Neurol. Clin.* **10**(2), 527–540.

Golbe, L. I., Lieberman, A. N., Muenter, M. D., Ahlskog, J. E., Gopinathan, G., Neophytides, A. N., Foo, S. H., and Duvoisin, R. C. (1988). Deprenyl in the treatment of symptom fluctuations in advanced Parkinson's disease. *Clin. Neuropharmacol.* **11,** 45–55.

Gosal, D., Ross, O. A., Wiley, J., Irvine, G. B., Johnston, J. A., Toft, M., Mata, I. F., Kachergus, J., Hulihan, M., Taylor, J. P., Lincoln, S. J., Farrer, M. J., *et al.* (2005). Clinical traits of LRRK2-associated Parkinson's disease in Ireland: A link between familial and idiopathic PD. *Parkinsonism Relat. Disord.* **11**(6), 349–352.

Grondin, R., Zhang, Z., Yi, A., Cass, W. A., Maswood, N., Andersen, A. H., Elsberry, D. D., Klein, M. C., Gerhardt, G. A., and Gash, D. M. (2002). Chronic, controlled GDNF infusion promotes structural and functional recovery in advanced parkinsonian monkeys. *Brain* **125**(Pt. 10), 2191–2201.

Hagell, P., and Brundin, P. (2001). Cell survival and clinical outcome following intrastriatal transplantation in Parkinson disease. *J. Neuropathol. Exp. Neurol.* **60,** 741–752.

Hagell, P., Schrag, A., Piccini, P., Jahanshahi, M., Brown, R., Rehncrona, S., Widner, H., Brundin, P., Rothwell, J. C., Odin, P., Wenning, G. K., Morrish, P., *et al.* (1999). Sequential bilateral transplantation in Parkinson's disease: Effects of the second graft. *Brain* **122**(Pt. 6), 1121–1132.

Hardingham, G. E., Cruzalegui, F. H., Chawla, S., and Bading, H. (1998). Mechanisms controlling gene expression by nuclear calcium signals. *Cell Calcium* **23,** 131–134.

Hardy, J., Cookson, M. R., and Singleton, A. (2003). Genes and parkinsonism. *Lancet Neurol.* **2**(4), 221–228.

Hauser, R. A., Freeman, T. B., Snow, B. J., Nauert, M., Gauger, L., Kordower, J. H., and Olanow, C. W. (1999). Long-term evaluation of bilateral fetal nigral transplantation in Parkinson disease. *Arch. Neurol.* **56,** 179–187.

Hayashi, T., Umemori, H., Mishina, M., and Yamamoto, T. (1999). The AMPA receptor interacts with and signals through the protein tyrosine kinase Lyn. *Nature* **397,** 72–76.

Henchcliffe, C., Schumacher, H. C., and Burgut, F. T. (2005). Recent advances in Parkinson's disease therapy: Use of monoamine oxidase inhibitors. *Expert Rev. Neurother.* **5**(6), 811–821.

Hermanson, E., Joseph, B., Castro, D., Lindqvist, E., Aarnisalo, P., Wallen, A., Benoit, G., Hengerer, B., Olson, L., and Perlmann, T. (2003). Nurr1 regulates dopamine synthesis and storage in MN9D dopamine cells. *Exp. Cell Res.* **288,** 324–334.

Hingtgen, C. M., and Siemers, E. (1998). The treatment of Parkinson's disease—current concepts and rationale. *Compr. Ther.* **24,** 560–566.

Hoffer, B. J., Hoffman, A., Bowenkamp, K., Huettl, P., Hudson, J., Martin, D., Lin, L. F., and Gerhardt, G. A. (1994). Glial cell line-derived neurotrophic factor reverses toxin-induced injury to midbrain dopaminergic neurons *in vivo*. *Neurosci. Lett.* **182**(1), 107–111.

Hoglinger, G. U., Sautter, J., Meyer, M., Spenger, C., Seiler, R. W., Oertel, W. H., and Widmer, H. R. (1998). Rat fetal ventral mesencephalon grown as solid tissue cultures: Influence of culture time and BDNF treatment on dopamine neuron survival and function. *Brain Res.* **813**(2), 313–322.

Holdorff, B. (2002). Friedrich Heinrich Lewy (1885–1950) and his work. *J. Hist. Neurosci.* **11**(1), 19–28.

Hornykiewicz, O., and Kish, S. J. (1987). Biochemical pathophysiology of Parkinson's disease. *Adv. Neurol.* **45,** 19–34.

Hou, J. G., Lin, L. F., and Mytilineou, C. (1996). Glial cell line-derived neurotrophic factor exerts neurotrophic effects on dopaminergic neurons *in vitro* and promotes their survival and regrowth after damage by 1-methyl-4-phenylpyridinium. *J. Neurochem.* **66**(1), 74–82.

Howells, D. W., Porritt, M. J., Wong, J. Y. F., Batchelor, P. E., Kalnins, R., Hughes, A. J., and Donnan, G. A. (2000). Reduced BDNF mRNA expression in the Parkinson's disease substantia nigra. *Exp. Neurol.* **166**(1), 127–135.

Hsu, H.-C., Zhou, T., and Mountz, J. D. (2004). Nur77 family of nuclear hormone receptors. *Curr. Drug Targets—Inflamm. Allergy* **3**(4), 413–423.

Hudson, J., Granholm, A. C., Gerhardt, G. A., Henry, M. A., Hoffman, A., Biddle, P., Leela, N. S., Mackerlova, L., Lile, J. D., Collins, F., and Hoffer, B. J. (1995). Glial cell line-derived neurotrophic factor augments midbrain dopaminergic circuits *in vivo*. *Brain Res. Bull.* **36**(5), 425–432.

Hundemer, H. P., Lledo, A., and van Laar, T. (2000). The safety of pergolide monotherapy in early-stage Parkinson's disease. One-year interim analysis of a 3-year double-blind, randomized study of pergolide versus levodopa. *Mov. Disord.* **15**(Suppl. 3), 115.

Hurley, M. J., and Jenner, P. (2006). What has been learnt from study of dopamine receptors in Parkinson's disease? *Pharmacol. Ther.* **111**(3), 715–728.

Hyman, C., Hofer, M., Barde, Y.-A., Juhasz, M., Yancopoulos, G. D., Squinto, S. P., and Lindsay, R. M. (1991). BDNF is a neurotrophic factor for dopaminergic neurons of the substantia nigra. *Nature* **350**(6315), 230–232.

Hynes, M., Porter, J. A., Chiang, C., Chang, D., Tessier-Lavigne, M., Beachy, P. A., and Rosenthal, A. (1995). Induction of dopaminergic neuron phenotype in the midbrain by sonic hedgehog protein. *Nat. Med.* **1**(11), 1184–1188.

Imamura, K., Hishikawa, N., Ono, K., Suzuki, H., Sawada, M., Nagatsu, T., Yoshida, M., and Hashizume, Y. (2005). Cytokine production of activated microglia and decrease in neurotrophic factors of neurons in the hippocampus of Lewy body disease brains. *Acta Neuropathol.* **109**(2), 141–150.

Iravani, M. M., Haddon, C. O., Cooper, J. M., Jenner, P., and Schapira, A. H. (2006). Pramipexole protects against MPTP toxicity in non-human primates. *J. Neurochem.* **96**(5), 1315–1321.

Iwawaki, T., Kohna, K., and Kobayashi, K. (2000). Identification of potential Nurr1 response element that activates the tyrosine hydroxylase gene promoter in cultured cells. *Biochem. Biophys. Res. Commun.* **274,** 590–595.

Jakowec, M. W., and Petzinger, G. M. (2004). 1-Methyl-4-phenyl-1,2,3,6-tetrahydropyridine-lesioned model of Parkinson's disease, with emphasis on mice and nonhuman primates. *Comp. Med.* **54,** 497–513.

Jankovic, J., Chen, S., and Le, W. D. (2005). The role of Nurr1 in the development of dopaminergic neurons and Parkinson's disease. *Prog. Neurobiol.* **77**(1–2), 128–138.

Javoy-Agid, F., Ruberg, M., Taquet, H., Bokobza, E., Agid, Y., Gaspar, P., Berger, B., N'Guyen-Legros, J., Alvarez, C., Gray, F., Escourolle, R., Scatton, B., *et al.* (1984). Biochemical neuropathology of Parkinson's disease. *Adv. Neurol.* **40,** 189–198.

Jenner, P. (1998). Oxidative mechanisms in nigral cell death in Parkinson's disease. *Mov. Dis.* **13,** 24–34.

Jenner, P. (2001). Parkinson's disease, pesticides and mitochondrial dysfunction. *Trends Neurosci.* **24,** 245–246.

Jenner, P., and Olanow, C. W. (1996). Oxidative stress and the pathogenesis of Parkinson's disease. *Neurology* **47**(6 Suppl. 3), S161–S170.

Jenner, P., and Olanow, C. W. (1998). Understanding cell death in Parkinson's disease. *Ann. Neurol.* **44,** S72–S84.

Jiang, Y., Jahagirdar, B. N., Reinhardt, R. L., Schwartz, R. E., Keene, C. D., Ortiz-Gonzalez, X. R., Reyes, M., Lenvik, T., Lund, T., Blackstad, M., Du, J., Aldrich, S., *et al.* (2002). Pluripotency of mesenchymal stem cells derived from adult marrow. *Nature* **418,** 41–49.

Jin, G., Tan, X., Tian, M., Qin, J., Zhu, H., Huang, Z., and Xu, H. (2005). The controlled differentiation of human neural stem cells into TH-immunoreactive (ir) neurons *in vitro*. *Neurosci. Lett.* **386,** 105–110.

Johnston, T. H., and Brotchie, J. M. (2004). Drugs in development for Parkinson's disease. *Curr. Opin. Investig. Drugs* **5**(7), 720–726.

Johnston, T. H., and Brotchie, J. M. (2006). Drugs in development for Parkinson's disease: An update. *Curr. Opin. Investig. Drugs* **7**(1), 25–32.

Joyce, J. N., Woolsey, C., Ryoo, H., Borwege, S., and Hagner, D. (2004). Low dose pramipexole is neuroprotective in the MPTP mouse model of Parkinson's disease, and downregulates the dopamine transporter via the D3 receptor. *BMC Biol.* **2,** 22.

Kawasaki, H., Suemori, H., Mizuseki, K., Watanabe, K., Urano, F., Ichinose, H., Haruta, M., Takahashi, M., Yoshikawa, K., Nishikawa, S., Nakatsuji, N., and Sasai, Y. (2002). Generation of dopaminergic neurons and pigmented epithelia from primate ES cells by stromal cell-derived inducing activity. *Proc. Natl. Acad. Sci. USA* **99,** 1580–1585.

Kearns, C. M., and Gash, D. M. (1995). GDNF protects nigral dopamine neurons against 6-hydroxydopamine *in vivo*. *Brain Res.* **672**(1–2), 104–111.

Kearns, C. M., Cass, W. A., Smoot, K., Kryscio, R., and Gash, D. M. (1997). GDNF protection against 6-OHDA: Time dependence and requirement for protein synthesis. *J. Neurosci.* **17**(18), 7111–7118.

Kim, D. W., Chung, S., Hwang, M., Ferree, A., Tsai, H. C., Park, J. J., Chung, S., Nam, T. S., Kang, U. J., Isacson, O., and Kim, K. S. (2006). Stromal cell-derived inducing activity, Nurr1 and signaling molecules synergistically induce dopaminergic neurons from mouse embryonic stem cells. *Stem Cells* **24**(3), 557–567.

Kim, J. H., Auerbach, J. M., Rodriguez-Gomez, J. A., Velasco, I., Gavin, D., Lumelsky, N., Lee, S. H., Nguyen, J., Sanchez-Pernaute, R., Bankiewicz, K., and McKay, R. (2002). Dopamine neurons derived from embryonic stem cells function in an animal model of Parkinson's disease. *Nature* **418,** 50–56.

Kirschner, P. B., Jenkins, B. G., Schulz, J. B., Finkelstein, S. P., Matthews, R. T., Rosen, B. R., and Beal, M. F. (1996). NGF, BDNF and NT-5, but not NT-3 protect against MPP + toxicity and oxidative stress in neonatal animals. *Brain Res.* **713**(1–2), 178–185.

Klein, R. L., Lewis, M. H., Muzyczka, N., and Meyer, E. M. (1999). Prevention of 6-hydroxydopamine-induced rotational behavior by BDNF somatic gene transfer. *Brain Res.* **847**(2), 314–320.

Knott, C., Stern, G., Kingsbury, A., Welcher, A. A., and Wilkin, G. P. (2002). Elevated glial brain-derived neurotrophic factor in Parkinson's diseased nigra. *Parkinsonism Relat. Disord.* **8**(5), 329–341.

Kohno, R., Sawada, H., Kawamoto, Y., Uemura, K., Shibasaki, H., and Shimohama, S. (2004). BDNF is induced by wild-type α-synuclein but not by the two mutants, A30P or A53T, in glioma cell line. *Biochem. Biophys. Res. Commun.* **318**(1), 113–118.

Kordower, J. H. (2003). *In vivo* gene delivery of glial cell line-derived neurotrophic factor for Parkinson's disease. *Ann. Neurol.* **53**(Suppl. 3), S120–S132; discussion S132–S134.

Kordower, J. H., Palfi, S., Chen, E., Ma, S. Y., Sendera, T., Cochran, E. J., Cochran, E. J., Mufson, E. J., Penn, R., Goetz, C. G., and Comella, C. D. (1999). Clinicopathological findings following intraventricular glial-derived neurotrophic factor treatment in a patient with Parkinson's disease. *Ann. Neurol.* **46,** 419–424.

Kordower, J. H., Emborg, M. E., Bloch, J., Ma, S. Y., Chu, Y., Leventhal, L., McBride, J., Chen, E. Y., Palfi, S., Roitberg, B. Z., Brown, W. D., Holden, J. E., *et al.* (2000). Neurodegeneration prevented by lentiviral vector delivery of GDNF in primate models of Parkinson's disease. *Science* **290,** 767–773.

Kotzbauer, P. T., and Holtzman, D. M. (2006). Expections and challenges in the therapeutic use of neurotrophic factors. *Ann. Neurol.* **59,** 444–447.

Kotzbauer, P. T., Trojanowsk, J. Q., and Lee, V. M. (2001). Lewy body pathology in Alzheimer's disease. *J. Mol. Neurosci.* **17**(2), 225–232.

Lang, A. E., and Lozano, A. M. (1998a). Parkinson's disease. First of two parts. *N. Engl. J. Med.* **339,** 1044–1053.

Lang, A. E., and Lozano, A. M. (1998b). Parkinson's disease. Second of two parts. *N. Engl. J. Med.* **339,** 1130–1143.

Lang, A. E., Gill, S., Patel, N. K., Lozano, A., Nutt, J. G., Penn, R., Brooks, D. J., Hotton, G., Moro, E., Heywood, P., Brodsky, M. A., Burchiel, K., *et al.* (2006). Randomized controlled trial of intraputamenal glial cell line-derived neurotrophic factor infusion in Parkinson disease. *Ann. Neurol.* **59**(3), 459–466.

Langston, J. W., Ballard, P., Tetrud, J. W., and Irwin, I. (1983). Chronic parkinsonism in humans due to a product of meperidine-analog synthesis. *Science* **219,** 979–980.

Lapchak, P. A., Araujo, D. M., Hilt, D. C., Jiao, S., Collin, F., Miyoshi, Y., Yi, A., Zhang, Z., and Gash, D. M. (1998). Topographical distribution of [125I]-glial cell line-derived neurotrophic factor in unlesioned and MPTP-lesioned rhesus monkey brain following a bolus intraventricular injection. *Brain Res.* **789,** 9–22.

Lauterborn, J. C., Lynch, G., Vanderklish, P., Arai, A., and Gall, C. M. (2000). Positive modulation of AMPA receptors increases neurotrophin expression by hippocampal and cortical neurons. *J. Neurosci.* **20**(1), 8–21.

Le, W., Conneely, O. M., He, Y., Jankovic, J., and Appel, S. H. (1999). Reduced Nurr1 expression increases the vulnerability of mesencephalic dopamine neurons to MPTP-induced injury. *J. Neurochem.* **73**(5), 2218–2221.

Le, W. D., Xu, P., Jankovic, J., Jiang, H., Appel, S. H., Smith, R. G., and Vassilatis, D. K. (2003). Mutations in NR4A2 associated with familial Parkinson disease. *Nat. Genet.* **33**(1), 85–89.

Lee, S. H., Lumelsky, N., Studer, L., Auerbach, J. M., and McKay, R. D. (2000). Efficient generation of midbrain and hindbrain neurons from mouse embryonic stem cells. *Nat. Biotechnol.* **18,** 675–679.

Legutko, B., Li, X., and Skolnick, P. (2001). Regulation of BDNF expression in primary neuron culture by LY392098, a novel AMPA receptor potentiator. *Neuropharmacology* **40**(8), 1019–1027.

Lew, M. F. (2005). Selegiline orally disintegrating tablets for the treatment of Parkinson's disease. *Expert Rev. Neurother.* **5**(6), 705–712.

Lie, D. C., Dziewczapolski, G., Willhoite, A. R., Kaspar, B. K., Shults, C. W., and Gage, F. H. (2002). The adult substantia nigra contains progenitor cells with neurogenic potential. *J. Neurosci.* **22,** 6639–6649.

Lieberman, A. N., Gopinathan, G., Neophytides, A., and Foo, S. H. (1987). Deprenyl versus placebo in Parkinson's disease. A double-blind study. *NY State J. Med.* **87,** 646–649.

Lin, L. F., Doherty, D. H., Lile, J. D., Bektesh, S., and Collins, F. (1993). GDNF: A glial cell line-derived neurotrophic factor for midbrain dopaminergic neurons. *Science* **260**(5111), 1130–1132.

Lindvall, O., and Bjorklund, A. (2004). Cell therapy in Parkinson's disease. *NeuroRx* **1,** 382–393.

Liste, I., Garcia-Garcia, E., and Martinez-Serrano, A. (2004). The generation of dopaminergic neurons by human neural stem cells is enhanced by Bcl-XL, both *in vitro* and *in vivo. J. Neurosci.* **24,** 10786–10795.

Longo, F. M., Yang, T., Xie, Y., and Massa, S. M. (2006). Small molecule approaches for promoting neurogenesis. *Curr. Alzheimer Res.* **3,** 5–10.

Mackowiak, M., O'Neill, M. J., Hicks, C. A., Bleakman, D., and Skolnick, P. (2002). An AMPA receptor potentiator modulates hippocampal expression of BDNF: An *in vivo* study. *Neuropharmacology* **43,** 1–10.

Maries, E., Dass, B., Collier, T. J., Kordower, J. H., and Steece-Collier, K. (2003). The role of alpha-synuclein in Parkinson's disease: Insights from animal models. *Nat. Rev. Neurosci.* **4**(9), 727–738.

Marsden, C. D., and Parkes, J. D. (1977). Success and problems of long-term levodopa therapy in Parkinson's disease. *Lancet* **1,** 345–349.

Marsden, C. D., Linazasoro, G., and Obeso, J. A. (1997). An introduction to the new surgery for Parkinson's disease. Past and present problems. *Adv. Neurol.* **74,** 143–147.

Martinat, C., Bacci, J. J., Leete, T., Kim, J., Vanti, W. B., Newman, A. H., Cha, J. H., Gether, U., Wang, H., and Abeliovich, A. (2006). Cooperative transcription activation by Nurr1 and Pitx3 induces embryonic stem cell maturation to the midbrain dopamine neuron phenotype. *Proc. Natl. Acad. Sci. USA* **103,** 2874–2879.

Martinez-Gonzalez, J., and Badimon, L. (2005). The NR4A subfamily of nuclear receptors: New early genes regulated by growth factors in vascular cells. *Cardiovasc. Res.* **65**(3), 609–618.

Masliah, E., Rockenstein, E., Veinbergs, I., Mallory, M., Hashimoto, M., Takeda, A., Sagara, Y., Sisk, A., and Mucke, L. (2000). Dopaminergic loss and inclusion body formation in alpha-synuclein mice: Implications for neurodegenerative disorders. *Science* **287**(5456), 1265–1269.

McNaught, K. S., Perl, D. P., Brownell, A. L., and Olanow, C. W. (2004). Systemic exposure to proteasome inhibitors causes a progressive model of Parkinson's disease. *Ann. Neurol.* **56**(1), 149–162.

McNaught, K. S. P., Olanow, C. W., Halliwell, B., Isacson, O., and Jenner, P. (2001). Failure of the ubiquitin-proteosome system in Parkinson's disease. *Nat. Rev. Neurosci.* **2,** 589–594.

Miller, R. M., Chen, L. L., Kiser, G. L., Giesler, T. L., Kaysser-Kranich, T. M., Palaniappan, C., and Federoff, H. J. (2005). Temporal evolution of mouse striatal gene expression following MPTP injury. *Neurobiol. Aging* **26,** 765–775.

Miu, P., Jarvie, K., Radhakrishnan, V., Gates, M., Ogden, A., Ornstein, P., Zarrinmayeh, H., Ho, K., Peters, D., Grabell, J., Gupta, A., Zimmerman, D., *et al.* (2001). Novel AMPA receptor potentiators LY392098 and LY404187: Effects on recombinant human AMPA receptors *in vitro. Neuropharmacology* **40,** 976–983.

Miyoshi, Y., Zhang, Z., Ovadia, A., Lapchak, P. A., Collins, F., Hilt, D., Lebel, C., Kryscio, R., and Gash, D. M. (1997). Glial cell line-derived neurotrophic factor-levodopa interactions and reduction of side effects in parkinsonian monkeys. *Ann. Neurol.* **42**(2), 208–214.

Mogi, M., Togari, A., Kondo, T., Mizuno, Y., Komure, O., Kuno, S., Ichinose, H., and Nagatsu, T. (1999). Brain-derived growth factor and nerve growth factor concentrations are decreased in the substantia nigra in Parkinson's disease. *Neurosci. Lett.* **270**(1), 45–48.

Mohapel, P., Frielingsdorf, H., Haggblad, J., Zachrisson, O., and Brundin, P. (2005). Platelet-derived growth factor (PDGF-BB) and brain-derived neurotrophic factor (BDNF) induce striatal neurogenesis in adult rats with 6-hydroxydopamine lesions. *Neuroscience* **132**(3), 767–776.

Moore, D. J., West, A. B., Dawson, V. L., and Dawson, T. M. (2005). Molecular pathophysiology of Parkinson's disease. *Annu. Rev. Neurosci.* **28,** 57–87.

Mufson, E. J., Kroin, J. S., Sendera, T. J., and Sobreviela, T. (1999). Distribution and retrograde transport of trophic factors in the central nervous system: Functional implications for the treatment of neurodegenerative diseases. *Prog. Neurobiol.* **57**(4), 451–484.

Murer, M. G., Yan, Q., and Raisman-Vozarin, R. (2001). Brain-derived neurotrophic factor in the control human brain, and in Alzheimer's disease and Parkinson's disease. *Prog. Neurobiol.* **63**(1), 71–124.

Murray, T. K., Whalley, K., Robinson, C. S., Ward, M. A., Hicks, C. A., Lodge, D., Vandergriff, J. L., Baumbarger, P., Siuda, E., Gates, M., Ogden, A. M., Skolnick, P., *et al.* (2003). LY503430, a novel a-amino-3-hydroxy-5-methylisoxazole-4-propionic acid receptor potentiator with functional, neuroprotective and neurotrophic effects in rodent models of Parkinson's disease. *J. Pharmacol. Exp. Ther.* **306**(2), 752–762.

Nagatsu, T., Mogi, M., Ichinose, H., and Togari, A. (2000). Changes in cytokines and neurotrophins in Parkinson's disease. *J. Neural Transm. Suppl.* **60,** 277–290.

Nakajima, K., Hida, H., Shimano, Y., Fujimoto, I., Hashitani, T., Kumazaki, M., Sakurai, T., and Nishino, H. (2001). GDNF is a major component of trophic activity in DA-depleted striatum for survival and neurite extension of DAergic neurons. *Brain Res.* **916,** 76–84.

Nutt, J. G. (2002). Motor fluctuations and dyskinesia. *In* "Parkinson's Disease. Diagnosis and Clinical Management" (S. A. Factor and W. J. Weiner, Eds.), pp. 445–453. Demos Medical Publishing, New York.

Nutt, J. G., Bronstein, J. M., and Carter, J. H. (2001). Intraventricular administration of GDNF in the treatment of Parkinson's disease. *Neurology* **56**(Suppl. 3), A375.

Olanow, C. W., and Tatton, W. G. (1999). Etiology and pathogenesis of Parkinson's disease. *Ann. Rev. Neurosci.* **22,** 123–144.

Olanow, C. W., Freeman, T. B., and Kordower, J. H. (1997). Neural transplantation as a therapy for Parkinson's disease. *Adv. Neurol.* **74,** 249–269.

Olanow, C. W., Goetz, C. G., Kordower, J. H., Stoessl, A. J., Sossi, V., Brin, M. F., Shannon, K. M., Nauert, G. M., Perl, D. P., Godbold, J., and Freeman, T. B. (2003). A double-blind controlled trial of bilateral fetal nigral transplantation in Parkinson's disease. *Ann. Neurol.* **54,** 403–414.

O'Neill, M. J., and Siemers, E. R. (2002). Pharmacological approaches to disease-modifying therapies in Parkinson's disease. *Expert Rev. Neurother.* **2**(6), 819–834.

O'Neill, M. J., Bleakman, D., Zimmermann, D. M., and Nisenbaum, E. S. (2004a). AMPA receptor potentiators for the treatment of CNS disorders. *Curr. Drug. Targets CNS Neurol. Disord.* **3,** 181–194.

O'Neill, M. J., Murray, T. K., Whalley, K., Ward, M. A., Hicks, C. A., Woodhouse, S., Osborne, D. J., and Skolnick, P. (2004b). Neurotrophic actions of the novel AMPA receptor potentiator, LY404187, in rodent models of Parkinson's disease. *Eur. J. Pharmacol.* **486**(2), 163–174.

O'Neill, M. J., Murray, T. K., Clay, M. P., Lindstrom, T., Yang, C. R., and Nisenbaum, E. S. (2005). LY503430: Pharmacology, pharmacokinetics, and effects in rodent models of Parkinson's disease. *CNS Drug Rev.* **11,** 77–96.

Ostenfeld, T., Tai, Y. T., Martin, P., Deglon, N., Aebischer, P., and Svendsen, C. N. (2002). Neurospheres modified to produce glial cell line-derived neurotrophic factor increase the survival of transplanted dopamine neurons. *J. Neurosci. Res.* **69,** 955–965.

Parain, K., Murer, M. G., Yan, Q., Faucheux, B., Agid, Y., Hirsch, E., and Vozari, R. (1999). Reduced expression of brain-derived neurotrophic factor protein in Parkinson's disease substantia nigra. *Neuroreport* **10**(3), 557–561.

Pardridge, W. M. (2002). Blood-brain barrier drug targeting enables neuroprotection in brain ischemia following delayed intravenous administration of neurotrophins. *Adv. Exp. Med. Biol.* **513,** 397–430.

Park, C. H., Minn, Y. K., Lee, J. Y., Choi, D. H., Chang, M. Y., Shim, J. W., Ko, J. Y., Koh, H. C., Kang, M. J., Kang, J. S., Rhie, D. J., Lee, Y. S., *et al.* (2005). *In vitro* and *in vivo* analyses of human embryonic stem cell-derived dopamine neurons. *J. Neurochem.* **92,** 1265–1276.

Parkinson, J. (1817). "An Essay on the Shaking Palsy." Sherwood, Nelly and Jones, London.

Parkinson Study Group (2002). Dopamine transporter brain imaging to assess the effects of pramipexole vs. levodopa on Parkinson's disease progression. *J. Am. Med. Assoc.* **287,** 1653–1661.

Patel, N. K., Bunnage, M., Plaha, P., Svendsen, C. N., Heywood, P., and Gill, S. S. (2005). Intraputamenal infusion of glial cell line-derived neurotrophic factor in PD: A two-year outcome study. *Ann. Neurol.* **57**(2), 298–302.

Perrier, A. L., Tabar, V., Barberi, T., Rubio, M. E., Bruses, J., Topf, N., Harrison, N. L., and Studer, L. (2004). Derivation of midbrain dopamine neurons from human embryonic stem cells. *Proc. Natl. Acad. Sci. USA* **101,** 12543–12548.

Petrova, P. S., Raibekas, A., Pevsner, J., Vigo, N., Anafi, M., Moore, M. K., Peaire, A. E., Shridhar, V., Smith, D. I., Kelly, J., Durocher, Y., and Commissiong, J. W. (2003). MANF: A new mesencephalic, astrocyte-derived neurotrophic factor with selectivity for dopaminergic neurons. *J. Mol. Neurosci.* **20**(2), 173–187.

Pezzoli, G., Martignoni, E., Pacchetti, C., Angeleri, V. A., Lamberti, P., Muratorio, A., Bonuccelli, U., De Mari, M., Foschi, N., Cossutta, E., Nicoletti, F., Giammona, F., *et al.* (1994). Pergolide compared with bromocriptine in Parkinson's disease: A multicenter, cross-over, conrolled study. *Mov. Disord.* **9,** 431–436.

Piccini, P., Brooks, D. J., Bjorklund, A., Gunn, R. N., Grasby, P. M., Rimoldi, O., Brundin, P., Hagell, P., Rehncrona, S., Widner, H., and Lindvall, O. (1999). Dopamine release from nigral transplants visualized *in vivo* in a Parkinson's patient. *Nat. Neurosci.* **2,** 1137–1140.

Porritt, M. J., Batchelor, P. E., and Howells, D. W. (2005). Inhibiting BDNF expression by antisense oligonucleotide infusion causes loss of nigral dopaminergic neurons. *Exp. Neurol.* **192**(1), 226–234.

Presgraves, S. P., Borwege, S., Millan, M. J., and Joyce, J. N. (2004). Involvement of dopamine D2/D3 receptors and BDNF in the neuroprotective effects of S32504 and pramipexole against 1-methyl-4-phenylpyridinium in terminally differentiated SH-SY5Y cells. *Exp. Neurol.* **190**(1), 157–170.

Przedborski, S., and Vila, M. (2003). The 1-methyl-4-phenyl-1,2,3,6-tetrahydropyridine mouse model: A tool to explore the pathogenesis of Parkinson's disease. *Ann. NY Acad. Sci.* **991,** 189–198.

Przedborski, S., Vila, M., Jackson-Lewis, V., and Dawson, T. M. (2000). Reply: A new look at the pathogenesis of Parkinson's disease. *Trends Pharmacol. Sci.* **21,** 165.

Przedborski, S., Jackson-Lewis, V., Naini, A. B., Jakowec, M., Petzinger, G., Miller, R., and Akram, M. (2001). The parkinsonian toxin 1-methyl-4-phenyl-1,2,3,6-tetrahydropyridine (MPTP): A technical review of its utility and safety. *J. Neurochem.* **76**(5), 1265–1274.

Quirk, J., and Nisenbaum, E. S. (2002). LY404187: A novel positive allosteric modulator of AMPA receptors. *CNS Drug Rev.* **8,** 255–282.

Rafuse, V. F., Soundararajan, P., Leopold, C., and Robertson, H. A. (2005). Neuroprotective properties of cultured neural progenitor cells are associated with the production of sonic hedgehog. *Neuroscience* **131**(4), 899–916.

Rajput, M. L., and Rajput, A. H. (2002). Epidemiology of parkinsonism. *In* "Parkinson's Disease. Diagnosis and Clinical Management" (S. A. Factor and W. J. Weiner, Eds.), pp. 31–40. Demos Medical Publishing, New York.

Rascol, O. (2005). Rasagiline in the pharmacotherapy of Parkinson's disease—a review. *Expert Opin. Pharmacother.* **6**(12), 2061–2075.

Rascol, O., Brooks, D. J., Korczyn, A. D., De Deyn, P. P., Clarke, C. E., and Lang, A. E., for the 056 Study Group (2000). A five-year study of the incidence of dyskinesia in patients with early Parkinson's disease who were treated with ropinirole or levodopa. *N. Engl. J. Med.* **342,** 1484–1491.

Riaz, S. S., Jauniaux, E., Stern, G. M., and Bradford, H. F. (2002). The controlled conversion of human neural progenitor cells derived from foetal ventral mesencephalon into dopaminergic neurons *in vitro*. *Brain Res. Dev. Brain Res.* **136,** 27–34.

Richfield, E. K., Thiruchelvam, M. J., Cory-Slechta, D. A., Wuertzer, C., Gainetdinov, R. R., Caron, M. G., Di Monte, D. A., and Federoff, H. J. (2002). Behavioral and neurochemical effects of wild-type and mutated human alpha-synuclein in transgenic mice. *Exp. Neurol.* **175**(1), 35–48.

Ross, O. A., and Farrer, M. J. (2005). Pathophysiology, pleiotrophy and paradigm shifts: Genetic lessons from Parkinson's disease. *Biochem. Soc. Trans.* **33**(Pt. 4), 586–590.

Ross, O. A., Toft, M., Whittle, A. J., Johnson, J. L., Papapetropoulos, S., Mash, D. C., Litvan, I., Gordon, M. F., Wszolek, Z. K., Farrer, M. J., and Dickson, D. W. (2006). Lrrk2 and Lewy body disease. *Ann. Neurol.* **59**(2), 388–393.

Ruberg, M., Ploska, A., Javoy-Agid, F., and Agid, Y. (1982). Muscarinic binding and choline acetyltransferase activity in parkinsonian subjects with reference to dementia. *Brain Res.* **232,** 129–139.

Ruottinen, H. M., and Rinne, U. K. (1996). Entacapone prolongs levodopa response in a one month double blind study in parkinsonian patients with levodopa related fluctuations. *J. Neurol. Neurosurg. Psychiatry* **60,** 36–40.

Sacchetti, P., Mitchell, T. R., Granneman, J. G., and Bannon, M. J. (2001). Nurr1 enhances transcription of the human dopamine transporter gene through a novel mechanism. *J. Neurochem.* **76,** 1565–1572.

Salloway, S., Correia, S., Peck, J., and Harrington, C. (2002). Dementia with Lewy bodies: A diagnostic and treatment challenge. *Med. Health RI* **85**(7), 207–209.

Saucedo-Cardenas, O., Quintana-Hau, J. D., Le, W. D., Smidt, M. P., Cox, J. J., De Mayo, F., Burbach, J. P., and Conneely, O. M. (1998). Nurr1 is essential for the induction of the dopaminergic phenotype and the survival of ventral mesencephalic late dopaminergic precursor neurons. *Proc. Natl. Acad. Sci. USA* **95**(7), 4013–4018.

Schapira, A., Bate, G., and Kirkpatrick, P. (2005). Rasagiline. *Nat. Rev. Drug Discov.* **4**(8), 625–626.

Schapira, A. H. (2003). Neuroprotection in PD—a role for dopamine agonists? *Neurology* **61**(6 Suppl. 3), S34–S42.

Schapira, A. H. V. (1999). Parkinson's disease. *Br. Med. J.* **318,** 311–314.

Schulz, T. C., Noggle, S. A., Palmarini, G. M., Weiler, D. A., Lyons, I. G., Pensa, K. A., Meedeniya, A. C., Davidson, B. P., Lambert, N. A., and Condie, B. G. (2004). Differentiation of human embryonic stem cells to dopaminergic neurons in serum-free suspension culture. *Stem Cells* **22,** 1218–1238.

Scott, W. K., Stajick, J. M., Yamaoka, L. H., Speer, M. C., Vance, J. M., Roses, A. D., and Pericak-Vance, M. A. (1997). Genetic complexity and Parkinson's disease. *Science* **277,** 387–388.

Semchuk, K. M., Love, E. J., and Lee, R. G. (1992). Parkinson's disease and exposure to agricultural work and pesticide chemicals. *Neurology* **42**(7), 1328–1335.

Seroogy, K. B., Lundgren, K. H., Tran, T. M. D., Guthrie, K. M., Isackson, P. J., and Gall, C. M. (1994). Dopaminergic neurons in rat ventral midbrain express brain-derived neurotrophic factor and neurotrophin-3 mRNAs. *J. Comp. Neurol.* **342**(3), 321–334.

Shannon, K. M., Bennett, J. P., and Friedman, J. H., for the Pramipexole Study Group (1997). Efficacy of pramipexole, a novel dopamine agonist, as monotherapy in mild to moderate Parkinson's disease. *Neurology* **49,** 724–728.

Sherer, T. B., Betarbet, R., and Greenamyre, J. T. (2002). Environment, mitochondria, and Parkinson's disease. *Neuroscientist* **8**(3), 192–197.

Sherer, T. B., Fiske, B. K., Svendsen, C. V., Lang, A. E., and Langston, J. W. (2006). Crossroads in GDNF therapy for Parkinson's disease. *Mov. Dis.* **21,** 136–141.

Shim, J. W., Koh, H. C., Chang, M. Y., Roh, E., Choi, C. Y., Oh, Y. J., Son, H., Lee, Y. S., Studer, L., and Lee, S. H. (2004). Enhanced *in vitro* midbrain dopamine neuron differentiation, dopaminergic function, neurite outgrowth, and 1-methyl-4-phenylpyridium resistance in mouse embryonic stem cells overexpressing Bcl-XL. *J. Neurosci.* **24,** 843–852.

Shults, C. W., Kimber, T., and Altar, C. A. (1995). BDNF attenuates the effects of intrastriatal injection of 6-hydroxydopamine. *Neuroreport* **6**(8), 1109–1112.

Siegel, G. J., and Chauhan, N. B. (2000). Neurotrophic factors in Alzheimer's and Parkinson's disease brain. *Brain Res. Rev.* **33**(2–3), 199–227.

Siemers, E. (1992). Recent progress in the treatment of Parkinson's disease. *Compr. Ther.* **18,** 20–24.

Sigurjonsson, O. E., Perreault, M. C., Egeland, T., and Glover, J. C. (2005). Adult human hematopoietic stem cells produce neurons efficiently in the regenerating chicken embryo spinal cord. *Proc. Natl. Acad. Sci. USA* **102,** 5227–5232.

Singleton, A. B., Farrer, M., Johnson, J., Singleton, A., Hague, S., Kachergus, J., Hulihan, M., Peuralinna, T., Dutra, A., Nussbaum, R., Lincoln, S., Crawley, A., *et al.* (2003). Alpha-synuclein locus triplication causes Parkinson's disease. *Science* **302**(5646), 841.

Skipper, L., Li, Y., Bonnard, C., Pavanni, R., Yih, Y., Chua, E., Sung, W. K., Tan, L., Wong, M. C., Tan, E. K., and Liu, J. (2005). Comprehensive evaluation of common genetic variation within LRRK2 reveals evidence for association with sporadic Parkinson's disease. *Hum. Mol. Genet.* **14** (23), 3549–3556.

Smith, L. A., Jackson, M. J., Al-Barghouthy, G., Rose, S., Kuoppamaki, M., Olanow, W., and Jenner, P. (2005). Multiple small doses of levodopa plus entacapone produce continuous dopaminergic stimulation and reduce dyskinesia induction in MPTP-treated drug-naive primates. *Mov. Disord.* **20**(3), 306–314.

Sonntag, K. C., Simantov, R., Kim, K. S., and Isacson, O. (2004). Temporally induced Nurr1 can induce a non-neuronal dopaminergic cell type in embryonic stem cell differentiation. *Eur. J. Neurosci.* **19,** 1141–1152.

Spina, M. B., Hyman, C., Squinto, S., and Lindsay, R. M. (1992a). Brain-derived neurotrophic factor protects dopaminergic cells from 6-hydroxydopamine toxicity. *Ann. NY Acad. Sci.* **648,** 348–350.

Spina, M. B., Squinto, S. P., Miller, J., Lindsay, R. M., and Hyman, C. (1992b). Brain-derived neurotrophic factor protects dopamine neurons against 6-hydroxydopamine and N-methyl-4-phenylpyridinium ion toxicity: Involvement of the glutathione system. *J. Neurochem.* **59**(1), 99–106.

Sun, M., Kong, L., Wang, X., Lu, X.-G., Gao, Q., and Geller, A. I. (2005). Comparison of the capability of GDNF, BDNF, or both, to protect nigrostriatal neurons in a rat model of Parkinson's disease. *Brain Res.* **1052**(2), 119–129.

Sundstrom, E., Fredriksson, A., and Archer, T. (1990). Chronic neurochemical and behavioral changes in MPTP-lesioned C57BL/6 mice: A model for Parkinson's disease. *Brain Res.* **528**(2), 181–188.

Takagi, Y., Takahashi, J., Saiki, H., Morizane, A., Hayashi, T., Kishi, Y., Fukuda, H., Okamoto, Y., Koyanagi, M., Ideguchi, M., Hayashi, H., Imazato, T., *et al.* (2005). Dopaminergic neurons generated from monkey embryonic stem cells function in a Parkinson primate model. *J. Clin. Invest.* **115,** 102–109.

Tanner, C. M., and Langston, J. W. (1990). Do environmental toxins cause Parkinson's disease? A critical review. *Neurology* **40,** 17–30.

Tao, X., Finkbeiner, S., Arnold, D. B., Shaywitz, A. J., and Greenberg, M. E. (1998). Ca^{2+} influx regulates BDNF transcription by a CREB family transcription factor-dependent mechanism. *Neuron* **20,** 709–726.

Tatton, N. A., and Kish, S. J. (1997). *In situ* detection of apoptotic nuclei in the substantia nigra compacta of 1-methyl-4-phenyl-1,2,3,6-tetrahydropyridine-treated mice using terminal deoxynucleotidyl transferase labelling and acridine orange staining. *Neuroscience* **77,** 1037–1048.

Tetrud, J., and Langston, W. (1989). The effect of deprenyl (selegiline) on the natural history of Parkinson's disease. *Science* **245,** 519–522.

Theofilopoulos, S., Goggi, J., Riaz, S. S., Jauniaux, E., Stern, G. M., and Bradford, H. F. (2001). Parallel induction of the formation of dopamine and its metabolites with induction of tyrosine hydroxylase expression in foetal rat and human cerebral cortical cells by brain-derived neurotrophic factor and glial-cell derived neurotrophic factor. *Brain Res. Dev. Brain Res.* **127**(2), 111–122.

Thomis, D. C., Marktel, S., Bonini, C., Traversari, C., Gilman, M., Bordignon, C., and Clackson, T. (2001). A Fas-based suicide switch in human T cells for the treatment of graft-versus-host disease. *Blood* **97,** 1249–1257.

Thomson, J. A., Itskovitz-Eldor, J., Shapiro, S. S., Waknitz, M. A., Swiergiel, J. J., Marshall, V. S., and Jones, J. M. (1998). Embryonic stem cell lines derived from human blastocysts. *Science* **282,** 1145–1147.

Toft, M., Mata, I. F., Kachergus, J. M., Ross, O. A., and Farrer, M. J. (2005). LRRK2 mutations and parkinsonism. *Lancet* **365**(9466), 1229–1230.

Tomac, A., Lindqvist, E., Lin, L. F. H., Ogren, S. O., Young, D., Hoffer, B. J., and Olson, L. (1995). Protection and repair of the nigrostriatal dopaminergic system by GDNF *in vivo*. *Nature* **373,** 335–339.

Tsuboi, K., and Shults, C. W. (2002). Intrastriatal injection of sonic hedgehog reduces behavioral impairment in a rat model of Parkinson's disease. *Exp. Neurol.* **173**(1), 95–104.

Tsukahara, T., Takeda, M., Shimohama, S., Ohara, O., Hashimoto, N., Bakay, R. A. E., and Redmond, D. E. (1995). Effects of brain-derived neurotrophic factor on 1-methyl-4-phenyl-1,2,3,6-tetrahydropyridine-induced parkinsonism in monkeys. *Neurosurgery* **37**(4), 733–741.

Tulloch, I. F. (1997). Pharmacologic profile of ropinirole: A nonergoline dopamine agonist. *Neurology* **49**(Suppl. 1), S58–S62.

Vandergriff, J., Huff, K., Bond, A., and Lodge, D. (2001). Potentiation of responses to AMPA on central neurones by LY392098 and LY404187 *in vivo*. *Neuropharmacology* **40**(8), 1003–1009.

Volpe, B. T., Wildmann, J., and Altars, C. A. (1998). Brain-derived neurotrophic factor prevents the loss of nigral neurons induced by excitotoxic striatal-pallidal lesions. *Neuroscience* **83**(3), 741–748.

Wallen, A. A., Castro, D. S., Zetterstrom, R. H., Karlen, M., Olson, L., Ericson, J., and Perlmann, T. (2001). Orphan nuclear receptor Nurr1 is essential for Ret expression in midbrain dopamine neurons and in the brain stem. *Mol. Cell Neurosci.* **18,** 649–653.

Wallen-Mackenzie, A., Mata de Urquiza, A., Petersson, S., Rodriguez, F. J., Friling, S., Wagner, J., Ordentlich, P., Lengqvist, J., Heyman, R. A., Arenas, E., and Perlmann, T. (2003). Nurr1-RXR heterodimers mediate RXR ligand-induced signaling in neuronal cells. *Genes Dev.* **17**(24), 3036–3047.

Wang, X., Li, X., Wang, K., Zhou, H., Xue, B., Li, L., and Wang, X. (2004). Forskolin cooperating with growth factor on generation of dopaminergic neurons from human fetal mesencephalic neural progenitor cells. *Neurosci. Lett.* **362,** 117–121.

West, A. B., Moore, D. J., Biskup, S., Bugayenko, A., Smith, W. W., Ross, C. A., Dawson, V. L., and Dawson, T. M. (2005). Parkinson's disease-associated mutations in leucine-rich repeat kinase 2 augment kinase activity. *Proc. Natl. Acad. Sci. USA* **102**(46), 16842–16847.

Whone, A. L., Watts, R. L., Stoessl, A. J., Davis, M., Reske, S., Nahmias, C., Lang, A. E., Rascol, O., Ribeiro, M. J., Remy, P., Poewe, W. H., Hauser, R. A., *et al.* (2003). Slower progression of Parkinson's disease with ropinirole versus levodopa: The REAL-PET study. *Ann. Neurol.* **54**(1), 93–101.

Wu, X., Zhu, D., Jiang, X., Okagaki, P., Mearow, K., Zhu, G., McCall, S., Banaudha, K., Lipsky, R. H., and Marini, A. M. (2004). AMPA protects cultured neurons against glutamate excitotoxicity through a phosphatidylinositol 3-kinase-dependent activation in extracellular signal-regulated kinase to upregulate BDNF gene expression. *J. Neurochem.* **90,** 807–818.

Yoshimi, K., Ren, Y. R., Seki, T., Yamada, M., Ooizumi, H., Onodera, M., Saito, Y., Murayama, S., Okano, H., Mizuno, Y., and Mochizuki, H. (2005). Possibility for neurogenesis in substantia nigra of parkinsonian brain. *Ann. Neurol.* **58,** 31–40.

Youdim, M. B., Maruyama, W., and Naoi, M. (2005). Neuropharmacological, neuroprotective and amyloid precursor processing properties of selective MAO-B inhibitor antiparkinsonian drug, rasagiline. *Drugs Today* **41**(6), 369–391.

Zafra, F., Hengerer, B., Leibrock, J., Thoenen, H., and Lindholm, D. (1990). Activity dependent regulation of BDNF and NGF mRNAs in the rat hippocampus is mediated by non-NMDA glutamate receptors. *EMBO J.* **9,** 3545–3550.

Zafra, F., Lindholm, D., Castren, E., Hartikka, J., and Thoenen, H. (1992). Regulation of brain-derived neurotrophic factor and nerve growth factor mRNA in primary cultures of hippocampal neurons and astrocytes. *J. Neurosci.* **12,** 4793–4799.

Zeng, X., Cai, J., Chen, J., Luo, Y., You, Z. B., Fotter, E., Wang, Y., Harvey, B., Miura, T., Backman, C., Chen, G. J., Rao, M. S., *et al.* (2004). Dopaminergic differentiation of human embryonic stem cells. *Stem Cells* **22,** 925–940.

Zetterstrom, R. H., Solomin, L., Mitsiadis, T., Olson, L., and Perlmann, T. (1996). Retinoid X receptor heterodimerization and developmental expression distinguish the orphan nuclear receptors NGFI-B, Nurr1, and Nor1. *Mol. Endocrinol.* **10**(12), 1656–1666.

Zetterstrom, R. H., Solomin, L., Jansson, L., Hoffer, B. J., Olson, L., and Perlmann, T. (1997). Dopamine neuron agenesis in Nurr1-deficient mice. *Science* **276**(5310), 248–250.

Zhang, Z., Miyoshi, Y., Lapchak, P. A., Collins, F., Hilt, D., Lebel, C., Kryscio, R., and Gash, D. M. (1997). Dose response to intraventricular glial cell line-derived neurotrophic factor administration in parkinsonian monkeys. *J. Pharmacol. Exp. Ther.* **282**(3), 1396–1401.

Zhao, L. R., Duan, W. M., Reyes, M., Keene, C. D., Verfaillie, C. M., and Low, W. C. (2002). Human bone marrow stem cells exhibit neural phenotypes and ameliorate neurological deficits after grafting into the ischemic brain of rats. *Exp. Neurol.* **174,** 11–20.

Zhao, M., Momma, S., Delfani, K., Carlen, M., Cassidy, R. M., Johansson, C. B., Brismar, H., Shupliakov, O., Frisen, J., and Janson, A. M. (2003). Evidence for neurogenesis in the adult mammalian substantia nigra. *Proc. Natl. Acad. Sci. USA* **100,** 7925–7930.

Zhou, J., Pliego-Rivero, B., Bradford, H. F., Stern, G. M., and Jauniaux, E. R. M. (1998). Induction of tyrosine hydroxylase gene expression in human foetal cerebral cortex. *Neurosci. Lett.* **252**(3), 215–217.

Zigmond, M. J., and Strickler, E. M. (1989). Animal models of parkinsonism using selective neurotoxins: Clinical and basic implications. *Int. Rev. Neurobiol.* **31,** 1–79.

Zintzaras, E., and Hadjigeorgiou, G. M. (2005). The role of G196A polymorphism in the brain-derived neurotrophic factor gene in the cause of Parkinson's disease: A meta-analysis. *J. Hum. Genet.* **50**(11), 560–566.

Zurn, A., Widmer, H., and Aebischer, P. (2001). Sustained delivery of GDNF: Towards a treatment for Parkinson's disease. *Brain Res. Rev.* **36,** 222–229.

USING *CAENORHABDITIS ELEGANS* MODELS OF NEURODEGENERATIVE DISEASE TO IDENTIFY NEUROPROTECTIVE STRATEGIES

Brian Kraemer*,† and Gerard D. Schellenberg*,†,‡,§

*Geriatric Research Education and Clinical Center, Veterans Affairs Puget Sound Health Care System, Seattle Division, Seattle, Washington 98108, USA
†Department of Medicine, Division of Gerontology and Geriatric Medicine University of Washington, Seattle, Washington 98195, USA
‡Department of Neurology, University of Washington, Seattle Washington 98195, USA
§Department of Pharmacology, University of Washington, Seattle Washington 98195, USA

I. Introduction
II. *C. elegans* Models of Neurodegenerative Disorders
  A. Necrotic Neurodegeneration
  B. Alzheimer's Disease and Tauopathy
  C. Parkinson's Disease
  D. Polyglutamine Disorders
  E. Amyotrophic Lateral Sclerosis
  F. Niemann-Pick Type C1 Disease
III. Identifying Candidate Modifier Genes of Neurodegeneration
  A. Genetic Modifiers of Necrotic Neurodegeneration
  B. Genetic Modifiers of Polyglutamine Aggregation
  C. Torsion Proteins and Neurodegeneration
IV. Molecular Inhibitors of Neurodegeneration
V. Neuroprotective Strategies
VI. Utility of HTS for Drugs in Worms
VII. Conclusions
  References

Human neurodegenerative diseases share common pathological features including protein aggregation, necrotic or apoptotic neurodegenerative changes, and loss of neurons. The nematode *Caenorhabditis elegans* has proven a useful model for both studying homologous human disease gene function and generating transgenic models of neurodegenerative disorders. Here we summarize progress in modeling aspects of human neurodegenerative diseases using *C. elegans*. Potential neuroprotective strategies arising from work with these models are also discussed.

INTERNATIONAL REVIEW OF
NEUROBIOLOGY, VOL. 77
DOI: 10.1016/S0074-7742(06)77007-6

I. Introduction

The use of animal models to study human disease has been a productive approach for gaining new insights into disease mechanisms. In the past decade, this approach has been extended to the study of neurodegenerative diseases using the worm *Caenorhabditis elegans*. This model organism offers a number of distinct experimental advantages for modeling human diseases, including small size, short generation time, rapid transgenics, robust classical genetics, accessible reverse genetics, a simple well-characterized nervous system, and well-studied behavior. For many disorders, the utility of worm models for disease remains unclear. However for other models, investigation into the function of worm genes has clearly illuminated the function of human disease genes. Models of both types will be reviewed here in the context of models for neurodegenerative disorders.

II. *C. elegans* Models of Neurodegenerative Disorders

A. Necrotic Neurodegeneration

Both programmed cell death and necrotic cell death occur in neurodegenerative disorders. While programmed cell death is necessary for a variety of developmental decisions and proper wiring of the developing nervous system, necrotic cell death only occurs after a genetic or environmental insult to the neuron that causes extreme cell stress. Necrotic cell death is characterized by swelling of the dying cell, mitochondrial dysfunction, rapid depletion of energy stores, membrane rupture, and spilling of cytoplasmic contents, resulting in an inflammatory response. The severity and type of cellular insult as well as the relative vulnerability of the cell contribute to a range of responses that includes aspects of either apoptosis pathways, necrotic pathways, or in some cases both (Artal-Sanz and Tavernarakis, 2005). Necrotic cell death can be induced by hypoxia, ischemia, stroke, trauma, Alzheimer's disease (AD), Parkinson's disease (PD), Huntington's disease (HD), and amyotrophic lateral sclerosis (ALS) (Artal-Sanz and Tavernarakis, 2005). Thus, understanding the molecular pathways involved in necrotic cell death is of therapeutic importance in a wide range of neurodegenerative disorders.

Chalfie and Wolinsky (1990) reported the first genetic model for neurodegeneration in *C. elegans* (see Table I for a summary of nontransgenic models of neurodegenerative disease). They isolated a single dominant allele of the *deg-1* gene that caused degeneration of the PVC interneurons. Death of the PVC cells causes worms to be touch insensitive. Other neurons also degenerate in *deg-1* mutant larvae, including IL1 sensory neurons, AVD interneurons, and the AVG interneuron, but these do not contribute to the touch insensitive phenotype. All of the cell deaths appear necrotic in nature with excessive swelling and vacuolation of degenerating neuron. *deg-1* encodes a non-voltage-gated ion channel whose abnormal activity leads to osmotic imbalance and subsequent neurodegeneration. Dominant mutations in a homologous gene, *mec-4,* also cause a neurodegenerative phenotype (Driscoll and Chalfie, 1991; Herman, 1987). *mec-4* encodes another non-voltage-gated ion channel of the DEG/ENaC family. Driscoll and Chalfie (1991) dissected the nature of the molecular defect in *mec-4* mutants using transgenic expression of many *mec-4* variants. The dominant *mec-4* mutations encoded T442A or V442A changes. Expression of T442A or V442A mutant MEC-4-encoding cDNAs in wild-type worms produced a Deg phenotype in transgenic animals. Likewise, expression of other MEC-4 changes that produced similar steric hindrance effects at position 442 (D442A, R442A, F442A, L442A, P442A) all recapitulated the Deg phenotype, while changes to nonsterically hindering amino acids (G442A, S442A, C442A) fail to cause a Deg phenotype.

Dominant mutations in *deg-3,* another ion channel gene, also cause a neurodegenerative phenotype (Treinin and Chalfie, 1995). *deg-3* encodes a nicotinic acetylcholine receptor α-subunit that causes touch cell degeneration similar to what is observed for *mec-4* and *deg-1* dominant mutants. The Deg-3 phenotype can be partially suppressed by treatment with nicotinic acetylcholine receptor antagonists.

A single genetic suppressor of *deg-1* and *mec-4* was identified by Xu *et al.* (2001). Loss-of-function (lf) mutations in *crt-1* ameliorate the toxic effects of *mec-4* and *deg-1* mutations, and partially suppresses expression of constitutively active heterotrimeric G protein α-subunit ($G\alpha_s$)-induced toxicity. *crt-1* encodes the worm homologue of calreticulin, a protein chaperone, and Ca^{2+}-binding protein localized to the endoplasmic reticulum (ER). CRT-1 participates in maintenance of intracellular Ca^{2+} stores. Reduction of calnexin expression also partially suppressed *mec-4*-induced cell death. *crt-1* suppresses *mec-4* neurotoxicity both by reducing MEC-4 protein levels and by altering neuronal Ca^{2+} balance.

Two similar transgenic models where expression of constitutively active heterotrimeric $G\alpha_s$ causes necrotic neurodegeneration have also been described (see Table II for a summary of transgenic models of neurodegenerative disease) (Berger *et al.,* 1998; Korswagen *et al.,* 1997). Korswagen *et al.* (1998) demonstrated that the worm heterotrimeric $G\alpha_s$ gene, *gsa-1,* is essential for viability. Mosaic loss of *gsa-1* caused uncoordinated locomotion (Unc) and egg laying defective (Egl) phenotypes consistent with GSA-1 regulating those behaviors. Transgenic overexpression of GSA-1 induced hyperactive locomotion and Egl. Korswagen *et al.* (1997) also employed a constitutively active mutant *gsa-1* transgene under control of a heat shock promoter that produced an inducible neurodegeneration phenotype similar to those described for *deg-1, deg-3,* and *mec-4.* Berger *et al.* (1998) describe a similar model where transgenic expression of the rat $G\alpha_s$ cDNA in worms using the *glr-1* glutamate-responsive cell-specific promoter causes

TABLE I

NONTRANSGENIC *C. ELEGANS* MODELS OF NEURODEGENERATIVE DISEASE

Disease[a]	Mutant gene	Pathology	Worm mutant(s)	Phenotype	Genetic modifiers	Molecular modifiers	Reference(s)
AD	PS1, PS2	Amyloid plaques, neurofibrillary tangles, neurodegeneration	*sel-12* or *hop-1*	Egg laying defective (Egl) suppresses *lin-12(d)*	*spr-1, spr-2, spr-3, spr-4, spr-5* suppress Egl defective phentoype		Eimer *et al.*, 2002; Jarriault and Greenwald, 2002; Lakowski *et al.*, 2003; Levitan and Greenwald, 1995; Levitan *et al.*, 1996; Li and Greenwald, 1996; Steiner *et al.*, 2001; Wen *et al.*, 2000
Necrotic cell death[b]	None	Necrotic neurodegeneration	*mec-4(d), deg-1,* or *deg-3*	Dominant neurodegeneration of PVC, IL1, AVD, and AVG neurons	*mec-6, crt-1* (lf) suppresses neurodegeneration		Chalfie, 1993; Chalfie and Wolinsky, 1990; Driscoll, 1992; Driscoll and Chalfie, 1991; Driscoll and Gerstbrein, 2003; Hall *et al.*, 1997; Hong and Driscoll, 1994; Huang and Chalfie, 1994; Syntichaki *et al.*, 2002; Treinin and Chalfie, 1995
NPC	NPC1	Accumulation of unesterified cholesterol causing neurodegeneration	*ncr-1;ncr-2*	Constitutive dauer larvae, reduced brood size, sensitive to cholesterol deprivation	DAF-9 overesxpression suppresses	Progesterone enhances, high cholesterol suppresses dauer formation	Li *et al.*, 2004; Sym *et al.*, 2000

PD	DJ-1	DAergic neuron loss	B0432.2 (RNAi)	Hypersensitivity to mitochondrial complex I inhibitors		TUDCA, probucol suppress	Ved *et al.*, 2005
PD	Parkin	DAergic neuron loss	*pdr-1*	Hypersensitive to proteotoxic stress, hypersensitive to mitochondrial complex I inhibitors	SNCA overexpression causes lethality	TUDCA, probucol suppress	Springer *et al.*, 2005; Ved *et al.*, 2005
PD	None[c]	DAergic neuron loss	None[c]	6-OHDA-induced DAergic neuron loss	*dat-1* required for neurotoxicity	Amphetamine, imipramine suppress	Nass *et al.*, 2002
SCA2	Ataxin2 (polyQ)	Degeneration of purkinje neurons	*atx-2*	Germ cell proliferation and germ line masculinization			Ciosk *et al.*, 2004

[a]Human diseases: AD, Alzheimer's disease; PD, Parkinson's disease; NPC, Niemann-Pick type C1 disease; SCA2, spinocerebellar ataxia type 2.

[b]Necrotic cell death is not a specific disease, but a more general description of a pathologic process brought on by stroke, ischemia, brain trauma, and so on. As such it has no specific associated mutations.

[c]The majority of PD is thought to be idiopathic; likewise, this model uses environmental toxicity to model DAergic neuron loss.

TABLE II

TRANSGENIC *C. ELEGANS* MODELS OF NEURODEGENERATIVE DISEASE

Disease[a]	Pathology	Worm transgene(s)	Transgene expression	Phenotype	Genetic modifiers	Molecular modifiers	Reference(s)
ALS	Motor neuron degeneration	*myo-3*::SOD1 hsp-16::SOD1	Body wall muscle or inducible	Paraquat hypersensitivity			Oeda *et al.*, 2001
AD	Amyloid plaques, neurofibrillary tangles, neurodegeneration	*unc-54::Aβ(1–42)*	Body wall muscle	Uncoordinated locomotion, paralysis, amyloid formation			Boyd-Kimball *et al.*, 2006; Butterfield *et al.*, 1999; Cottrell *et al.*, 2005; Drake *et al.*, 2003; Fay *et al.*, 1998; Fonte *et al.*, 2002; Gutierrez-Zepeda *et al.*, 2005; Hu *et al.*, 2000; Link, 1995, 2001, 2005; Link *et al.*, 1999, 2001, 2003, 2006; Yatin *et al.*, 1999
AD	Amyloid plaques, neurofibrillary tangles, neurodegeneration	*aex-3*::tau	All neurons	Unc, tau phosphorylation, tau aggregation, neurodegeneration			Kraemer *et al.*, 2003
AD		*mec-7*::tau	Touch neurons	Mec, tau phosphorylation, neurodegeneration	HSP70 homologue suppresses		Miyasaka *et al.*, 2005

HD	polyQ aggregation, neurodegeneration	*myo-3::polyQ-GFP*	Body wall muscle	Paralysis, age-dependent polyQ aggregation	HSP104, TOR-2 overexpression suppresses, 186 genes enhance torsionA overexpression suppresses		Caldwell *et al.*, 2003; Gidalevitz *et al.*, 2006; Morley *et al.*, 2002; Nollen *et al.*, 2004; Satyal *et al.*, 2000
HD	polyQ aggregation, neurodegeneration	*mec-3::polyQ-GFP*	Touch neurons	Mec, axon morphology abnormalities	SIR-2.1 overexpression suppresses defects	Resveratrol, fisetin suppress	Parker *et al.*, 2001, 2005
HD	polyQ aggregation, neurodegeneration	*osm-10::polyQ-GFP*	Body wall muscle	polyQ aggregation, ASH neuron degeneration, nose-touch defective	*pqe-1* enhances neurodegeneration		Faber *et al.*, 1999, 2002; Gurvitz *et al.*, 2000
PD	DAergic neuron loss in substantia nigra	*aex-3*::SNCA	All neurons	Unc, DAergic dendritic abnormalities			Lakso *et al.*, 2003
Necrotic cell death[b]	Necrotic neurodegeneration	*hsp-16::gsa-1*	Many neurons and other cells	Neurodegeneration, *gsa-1* (lf) is lethal	*sgs-1* suppresses neurodegeneration		Korswagen *et al.*, 1997, 1998
Necrotic cell death[b]	Necrotic neurodegeneration	*mec-7*::Gα_s	Touch neurons	Necrotic touch cell degeneration	*acy-1* suppresses		Berger *et al.*, 1998

[a]Human diseases: AD, Alzheimer's disease; ALS, Amyotrophic lateral sclerosis; HD, Huntington's disease; PD, Parkinson's disease.

[b]Necrotic cell death is not a specific disease, but a more general description of a pathologic process brought on by stroke, ischemia, brain trauma, and so on. As such no specific disease-causing mutations are known.

paralysis and neurodegeneration. This neurodegeneration is of a similar necrotic nature to those described above and likely occurs by a similar mechanism of osmotic stress as $G\alpha_s$ is thought to regulate ion channel activity among other cellular processes. Berger *et al.* (1998) also identified a genetic suppressor of the $G\alpha_s$-induced neurodegeneration, *acy-1*. Thus ACY-1, an adenylyl cyclase implicated in a wide variety of signaling pathways in other organisms, is required for $G\alpha_s$-induced neurodegeneration.

Work with *C. elegans* models of necrotic neuronal cell death suggests a couple of potential strategies to protect neurons. These models suggest that osmotic imbalance may be the ultimate cause of neurodegeneration. In all cases, these are dominant toxic gain-of-function mutations that can be ameliorated by blocking gene expression. Likewise, specific suppressor mutations can inhibit neurodegeneration of a given type. The question remains how specific some of these suppressors might be. For *crt-1*, suppression of both *deg-1* and $G\alpha_s$ suggests perhaps a common suppression of necrotic neurodegeneration. However, *crt-1* fails to suppress *deg-3*, indicating that its suppression may be specific to the degenerin family of ion channels (if acy-1 acts through degenerins). Also, whether *acy-1* mutations suppress a variety of neurodegenerative phenotypes or only those of overactive $G\alpha_s$ remains unknown.

B. Alzheimer's Disease and Tauopathy

AD is the most common aging-related neurodegenerative disorder and is characterized by progressive loss of memory and dementia. Two pathological hallmarks, amyloid plaques and neurofibrillary tangles (NFTs), define a positive diagnosis of AD. Human genetic evidence from familial early-onset AD and frontotemporal dementia with parkinsonism chromosome 17 type (FTDP-17) kindreds suggests that both lesions can play a causal role in neurodegeneration in addition to serving as markers for disease.

The amyloid deposits seen in AD consist primarily of extracellular aggregates of the $A\beta$ peptide which is derived from the proteolytic processing of the amyloid precursor protein (APP). Presenilin 1 (PS1) and PS2 are homologous proteases and form the core constituents of the γ-secretase, a complex involved in the cleavage of $A\beta$ peptide from the unprocessed APP molecule. Dominant mutations in APP, PS1, and PS2 can cause early-onset AD either by increasing the total amount of $A\beta$ produced or by altering the C-terminal cleavage site. In the latter case, a longer $A\beta$ molecule is formed ($A\beta$ 1–42) which aggregates more rapidly than the shorter form ($A\beta$ 1–39).

The other pathological hallmark of AD is the NFT which consists of insoluble filamentous aggregates of hyperphosphorylated tau protein. In the normal brain, tau binds to and stabilizes axon microtubules in neurons. Tau is alternatively spliced into six major isoforms expressed in brain; both exonic and intronic mutations that disrupt the normal splicing regulation of tau can cause FTDP-17. Likewise, some FTDP-17 mutations change the primary sequence of the tau protein and reduce its affinity for microtubules and/or increase propensity for self-aggregation. Abnormalities in tau protein function cause neurodegeneration in FTDP-17. Tau lesions from many FTDP-17 cases cannot be distinguished from those seen in AD (Sumi *et al.*, 1992), demonstrating the shared pathological features of the two disorders. The chief pathological feature that differentiates other tauopathies from AD is the lack of amyloid pathology and hippocampal involvement.

1. *Amyloid Pathology*

One transgenic model for amyloid pathology has been described (Link, 1995) and extensively characterized (Boyd-Kimball *et al.*, 2006; Drake *et al.*, 2003; Fay *et al.*, 1998; Fonte *et al.*, 2002; Link, 2001; Link *et al.*, 1999, 2001, 2003; Yatin *et al.*, 1999). In this model, an Aβ 1–42 peptide-encoding minigene is expressed as a fusion with a secretory signal sequence in the body wall muscle cells of the worm using the *unc-54* promoter. High levels of Aβ expression was observed in essentially all body wall muscle cells as well as coelomeocytes. Notably, Aβ expression caused a clear Unc phenotype characterized by progressive paralysis. Coelomocyte expression is consistent with Aβ peptide being scavenged from secreted muscle cell-derived Aβ. Thioflavin S, an amyloid plaque stain, labeled deposits of Aβ in transgenic animals indicative of authentic aggregation of amyloid in worm muscle. Coexpression of Aβ and transthyretin peptides reduced the number of these thioflavin S-positive deposits consistent with previous *in vitro* studies where transthyretin reduced Aβ aggregation (Link, 1995; Schwarzman *et al.*, 1994).

Fay *et al.* (1998) utilized the transgenic worm model for Aβ aggregation as an *in vivo* assay to analyze the propensity of a variety of Aβ variant peptides to aggregate. They observed that Aβ variants $^{L}17^{P}$ and $^{M}35^{C}$ both reduce amyloid formation *in vivo*, presumably due to disruption of the typical aggregated Aβ-pleated β-sheet structure by the introduced changes in Aβ primary sequence. Closer examination of amyloid deposits in transgenic worm body wall muscle using immunoelectron microscopy showed that worms share the characteristic fibrillar structure of authentic amyloid plaques from postmortem AD brains. Furthermore, worm amyloid deposits stained positive with X-34 (a Congo red derivative) even in living animals (Link *et al.*, 2001). Fonte *et al.* (2002) used coimmunoprecipitation to identify six proteins that associate with amyloid in worms. Among these Aβ-interacting proteins were a variety of chaperone proteins, including worm homologues of mammalian HSP70C (BiP), the cytoplasmic HSP70A, small glutamine-rich tetratricopeptide repeat-containing protein (SGT), and three αB crystallin homologues. Fonte *et al.* (2002) further demonstrated that expression of the normal Aβ 1–42 peptide induced a prominent paralysis phenotype and up-regulated expression of the six chaperones listed above, while an intramolecular Aβ dimer induced only weak paralysis and no chaperone expression.

Link *et al.* (2003) updated the previous model of amyloid pathology in *C. elegans* by utilizing a temperature inducible construct for Aβ expression that resulted in a strain giving approximately fivefold induction of Aβ 1–42 (Link *et al.,* 2003). Animals exhibited paralysis on induction of Aβ expression as previously reported for the constitutive expression of Aβ but no aggregated Aβ deposits occurred. Furthermore, while Aβ induction caused paralysis, it did not cause accumulations of aggregated amyloid, as in the constitutive model. This demonstrates in worms that Aβ pathology precedes accumulation of frank aggregated fibrillar amyloid deposits. This system allowed Link *et al.* (2003) to examine changes in mRNA abundance using microarray analysis. Sixty-seven genes were up-regulated and 240 genes down-regulated on Aβ induction. Drake *et al.* (2006) used the same inducible lines for Aβ expression to monitor the accumulation of oxidative damage in Aβ-expressing animals and found that Aβ induction caused an increase in protein carbonyl levels up to twofold above non-Aβ animals under identical conditions. Using a proteomic approach, Boyd-Kimball *et al.* (2006) further explored the role of oxidative damage as a potential driver of amyloid pathology by examining specific proteins oxidized in the induced Aβ worm model. The analysis of Aβ targets of oxidation revealed 16 specific proteins involved in energy metabolism (e.g., short- and medium-chain acyl-CoA dehydrogenases), proteasome function (e.g., proteasome α and β subunits), and scavenging of oxidants (glutathione S-transferase sigma).

While the expression of Aβ peptide in muscle is not a precise model of AD, it approximately models one of the pathological hallmarks of AD and does more accurately model another disorder called inclusion body myositis (IBM) (Link, 1995). Nonetheless, the pathological consequence of Aβ expression in worm muscle share similarities with the neurodegenerative pathology in AD, including fibrillar aggregation of Aβ, up-regulation of chaperon proteins, increased protein oxidation, and dysfunctional energy metabolism. Results from this model suggest that neuroprotective strategies should focus on the preaggregation state of Aβ rather than the prevention of aggregation, as cellular dysfunction occurs prior to amyloid fibril formation.

2. *Presenilin Function*

When initially cloned, nothing was known about the function of PS1 and PS2 as components of the γ-secretase. The identification of *sel-12* as a worm homologue of PS1 and PS2 provided an immediate arena for exploring presenilin function. To this end, Levitan and Greenwald (1995) observed that *sel-12* functions in the *lin-12/Notch* signaling pathways and shares sequence homology with human PS1. To probe topology of the predicted *sel-12* transmembrane domains *in vivo*, Li and Greenwald (1996) made SEL-12::lacZ gene fusions at various points along the length of the SEL-12 peptide sequence. Taking advantage of the relative inactivity of lacZ in the extracellular environment relative to cytoplasm, the authors were able to deduce the overall topology of SEL-12 and its membrane-spanning domains. Furthermore, the authors observed endoproteolytic cleavage of SEL-12, both observations borne out for PS1 in subsequent human studies. The functional similarity of PS1 and SEL-12 was demonstrated directly by rescuing the *sel-12* mutant phenotype by expressing a human PS1 cDNA under control of the *sel-12* promoter (Levitan *et al.*, 1996). The presence of AD causing point mutations in PS1 cDNAs reduced their ability to rescue the *sel-12* phenotype relative to the wild-type PS1 cDNA sequence, indicating that PS1 mutations cause a partial loss of function rather than a toxic gain of function. Li and Greenwald (1996) went on to identify another worm PS homologue in HOP-1. *hop-1* transgenic rescue experiments demonstrated that it can functionally replace SEL-12. Likewise, reducing *hop-1* function in the background of *sel-12* loss of function resulted in lethality, suggesting that *hop-1* and *sel-12* are both functionally redundant and essential for viability (as are PS1 and PS2 in mammals). Five genetic suppressors of *sel-12* have been identified, termed *spr-1* to *spr-5* genes. The *spr* genes all appear to act at the level of repressing transcription of sel-12 and/or downstream *lin-12/Notch* signaling pathway components (Eimer *et al.*, 2002; Springer *et al.*, 2005; Wen *et al.*, 2000; Yatin *et al.*, 1999). Thus while experiments in *C. elegans* with both worm and human homologues of the presenilins have not directly modeled human pathology, they have been extremely important for understanding the normal biological function of presenilins.

3. *Tau Pathology*

Two different transgenic *C. elegans* models for tau pathology have been described (Kraemer *et al.*, 2003; Miyasaka *et al.*, 2005). In both models, human tau expression causes neuronal dysfunction, resulting in progressive behavioral abnormalities. In one model, Kraemer *et al.* (2003) drove expression of the 1N4R isoform of human tau in all neurons using the *aex-3* promoter, while the other model of Miyasaka *et al.* (2005) employed the *mec-7* promoter to drive expression of human tau isoforms 0N4R and 0N3R in the six mechanosensory neurons. Likewise, Kraemer *et al.* (2003) used normal and FTDP-17 mutant tau sequences $^{V}337^{M}$ and $^{P}301^{L}$, while Miyasaka *et al.* (2005) used $^{P}301^{L}$ and $^{R}406^{W}$. Despite using different tau isoforms and FTDP-17 mutations, both models yielded similar, but complimentary findings in regards to the effects of neuronal tau expression.

Kraemer *et al.* (2003) observed that pan-neuronal expression of tau caused a pronounced Unc phenotype that worsened as animals aged. Furthermore, FTDP-17 mutations exacerbated the severity of the Unc phenotype. The lifespan of tau transgenic animals was reduced by up to 40%, with no significant difference in lifespan between normal and FTDP-17 mutant tau transgenic lines. Tau expression also caused impairment in presynaptic cholinergic neurotransmission. Tau transgenic animals were found to contain detergent insoluble tau, a pathological hallmark of AD. Insoluble tau did not begin to accumulate until late larval stages/adulthood and was most prominent in animals with $^{V}337^{M}$ mutant tau. Neuronal dysfunction preceded the appearance of aggregated tau, indicating that, at least in worms, the toxic species most likely is not large tau aggregates. Likewise, tau was phosphorylated at many of the sites observed in AD brain, including the phosphoepitopes for 12E8, AT8, AT270, CP13, PHF1, and pS422.

Kraemer *et al.* (2003) observed that tau was expressed in both the axons and cell bodies of virtually all neurons consistent with the expected *aex-3* promoter-driven expression pattern. Age-dependent neurodegeneration was directly observed in green fluorescent protein (GFP)-marked GABAergic neurons. Initially, young animals have structurally normal and intact GABAergic motoneurons, but as they age axons become disrupted, fill with large inclusions, and exhibit exaggerated varicosities. Animals also showed some GABAergic neuronal loss with age. Both normal and FTDP-17 tau lines were similar except near the end of life where FTDP-17 mutant lines exhibited a more severe level of axonal degeneration. Evidence of neurodegeneration was also evident in electron microscopy (EM) of aged worms. Degenerating neurons showed clearing of the axonal compartment, membranous infoldings, and amorphous tau-positive aggregates.

Consistent with the model described above, Miyasaka *et al.* (2005) observed that human tau expression in touch cells caused a progressive impairment in the touch response. As above, FTDP-17 mutations exacerbated the impairment in mechanosensation phenotype (Mec), while normal tau caused a modest defect relative to nontransgenic animals. Furthermore, the 0N3R tau isoform caused a modest defect in mechanosensation similar to that seen for the 0N4R isoform. Animals also exhibit morphological abnormalities, including tortuous processes, branching of the process, multiple processes, and meshwork-like neurites. These morphological abnormalities appear to be much worse in FTDP-17 mutant lines.

Miyasaka *et al.* (2005) observed that tau transgenic animals exhibited a loss-of-touch cell-specific tubulin as animals aged consistent with neuronal loss; loss of neurons appeared most severe in FTDP-17 mutant lines. As above, tau phosphorylated at the AT8, HT7, and PHF-1 epitopes accumulated in neurons, both in neurites and cell bodies of touch neurons. EM analysis confirmed the finding of abnormal or absent microtubules (MTs) in many touch cells. Miyasaka *et al.* (2005) tested the role of apoptotic pathways in the neurodegeneration phenotype caused by tau and found that neither *ced-3* nor *ced-4* loss-of-function mutants showed any effect on the tau phenotype. Heat shock of tau worms exacerbated the tau-induced Mec phenotype, while overexpression of HSP70 in the absence of heat shock partially ameliorated the Mec phenotype. Worm homologues of four known tau kinase genes (*gsk-3, cdk-5, jkk-1, mek-1*) were tested for their ability to modulate the tau-induced Mec phenotype. Only overexpression of *gsk-3* was shown to change the tau phenotype as it caused a mild enhancement of the tau-induced Mec phenotype.

The two models for tauopathy described above recapitulate many features of authentic human disease, including impaired behavior, abnormal neuronal morphology, decreased neurotransmission, accumulation of insoluble tau, phosphorylation of tau at pathological phosphoepitopes, axonal degeneration, and neuronal loss. Miyasaka *et al.* (2005) correlate functional impairment with fine structure morphological changes in neurites, while Kraemer *et al.* (2003) correlate functional impairment with accumulation of insoluble tau. Both of these changes are associated with progression of AD as dendritic pruning and NFT pathology increase as the disease progresses. Furthermore, the results of Miyasaka *et al.* (2005) suggest that the ultimate cause of neurodegeneration is of a necrotic rather than apoptotic nature. While tau becomes phosphorylated abundantly in both models, the role of phosphorylation in neuronal dysfunction and neurodegeneration remains unclear. The only directly neuroprotective strategy suggested by either model is the partial rescue of the tau-induced Mec phenotype by HSP70 overexpression. This result is consistent with HSP70 collaborating with carboxy terminal HSC70 interacting protein (CHIP) to mediate tau degradation as has been observed previously (Petrucelli *et al.*, 2004; Sahara *et al.*, 2005; Shimura *et al.*, 2004).

C. Parkinson's Disease

Another common progressive neurodegenerative disorder is PD. PD is a movement disorder characterized clinically by slowness of movement, rigidity, postural instability, and resting tremor. Selective degeneration of dopaminergic (DAergic) neurons in the substantia nigra causes these deficits in movement. Most cases of PD are sporadic with the underlying cause of DAergic neurons loss is unknown, but both rare autosomal dominant and recessive mutations can cause PD as well. Mutations in SNCA, *PRKN*, *PINK*, *DJ-1*, and *LRRK2* have all been associated with familial forms of PD. Like AD, PD exhibits distinctive pathological lesions in degenerating neurons. In PD, these spherical protein aggregates are called Lewy bodies (LB) and are composed primarily of fibrillar deposits of SNCA. However, some variants of PD exhibit DAergic neuronal loss in the absence of LB.

1. *DAergic Neurodegeneration*

Nass *et al.* (2002) developed a system whereby they monitored DAergic neurons for the effects of 6-hydroxydopamine (6-OHDA) exposure. 6-OHDA is a DAergic neuron-specific neurotoxin in a variety of model systems. In worms, acute exposure to 6-OHDA caused neurodegeneration of DAergic neurons. The DAergic neurons CEP and ADE proved to be the most vulnerable with rounding of neuronal soma and blebbing occurring within 2 hours of exposure to toxin. Dopamine transporter (DAT) activity is required for this neurodegeneration as worms exposed to DAT antagonists (imipramine or nisoxetine) were protected from 6-OHDA neurotoxicity. Furthermore, *dat-1* mutants were highly resistant to 6-OHDA. EM ultrastructural studies revealed advanced degenerative changes, including loss of CEP sensilla, shrunken dendritic endings, vacuolation, and nuclear condensation consistent with apoptotic cell death. However, genetic ablation of genes necessary for conventional apoptosis did not prevent 6-OHDA neurotoxicity. Likewise, no membranous whorls or cellular swelling indicative of necrotic neurodegeneration were evident. The authors suggest a novel apoptotic pathway may be responsible for 6-OHDA-induced neurodegeneration.

2. *SNCA Pathology*

Expression of human SNCA in worm neurons causes a variety of defects (Kuwahara *et al.,* 2006; Lakso *et al.,* 2003; Ved *et al.,* 2005). Lakso *et al.* (2003) demonstrated that pan-neuronal or motor neuron-targeted expression caused an Unc phenotype with reduced motility in liquid media. Expression in DAergic neurons alone did not cause an Unc phenotype. Both the wild-type and mutant SNCA transgenes caused defects in locomotion and induced neurodegenerative changes. Ved *et al.* (2005) describe a model where pan-neuronal expression of SNCA, knock down of the homologue of DJ-1 (B0432.2), or loss of the Parkin homologue (K08E3.7) cause an increase in vulnerability to mitochondrial complex I inhibitors, including rotenone, fenperoximate, pyridaben, and stigmatellin. Susceptibility was measured as survival relative to untreated and nontransgenic animals. Treatment with either drugs that activate mitochondrial complex II (D-β-hydroxybutyrate, DβHB) or drugs that inhibit apoptotic response (taurosodeoxycholic acid) protected worms from SNCA-induced toxicity and DJ-1 knockdown.

Kuwahara *et al.* (2006) report a model of α-synucleinopathy where overexpression of wild-type or mutant SNCA using the *dat-1* promoter causes degeneration of the neurites of CEP DAergic neurons. These animals also exhibit defects in food sensing and locomotion response to food. No LB pathology was detected by EM, but SNCA expression did cause significantly reduced dopamine content in transgenic animals. Furthermore, growth of SNCA transgenic worms on media supplemented with dopamine rescued the food-sensing defect, consistent with the clinical use of levodopa to treat some forms of PD.

3. parkin

The *C. elegans* homologue of *PRKN* is *pdr-1*. To study the genetic function of *PRKN* in worms a series of deletion mutants in *pdr-1* were generated (Springer *et al.,* 2005). *pdr-1* encodes the functional protein domains of *PRKN*, including a ubiquitin-like domain (UBD), unique parkin domain (UPD), two RING finger domains, and an in between RING finger domain. They observed PDR-1 expression in all embryonic cells and in both muscle and neuronal cells in larvae and adults. Springer *et al.* (2005) demonstrated a physical interaction between

PDR-1 and the *C. elegans* protein degradation machinery, as PDR-1 binds E2 enzymes (UBC-18, UBC-2, UBC-15) and an E4 ubiquitin ligase (CHN-1). Furthermore, PDR-1 is competent to self-ubiquitinylate in the presence of ubiquitin, E1, E3, and E4 enzymes. *pdr-1* mutants are homozygous viable, but exhibit mild hypersensitivity to ER stress. One particular in-frame deletion mutant lacking most of the UBL and UPD domains, *pdr-1(lg103)*, is strongly hypersensitive to ER stress. This hypersensitivity to ER stress is attributed to *pdr-1 (lg103)* encoding an aggregation-prone mutant version of PDR-1. Springer *et al.* (2005) also show overexpression of SNCA in a *pdr-1(lg103)* mutant background causes larval arrest and lethality, independent of the ER stress response. Since *pdr-1(lg103)*, an in-frame deletion, is cytotoxic, while simple loss-of-function alleles of *pdr-1* have no cytotoxic phenotype, the authors suggest *PRKN* mutations may be detrimental not because they eliminate parkin function, but because they cause parkin to misfold or aggregate.

The models for PD described here focus on the function of worm homologues of PD genes, toxic protein aggregation, and DAergic neuron degeneration. The ultimate causes of protein aggregation or neurodegeneration in these worm models appear diverse as either protein aggregation or chemical insult independently can cause DAergic neuron dysfunction and death. Similarly, the role of PD-related genes seems diverse because normal function of *DJ-1* and *PRKN* appears to protect worms from toxic stress while SNCA or misfolded parkin causes protein aggregation and cytotoxicity. These results suggest that the relationship among the growing list of PD-related genes will be complex. Perhaps future studies with the worm homologues of *PINK1* and/or *LLRK2* will shed light onto how all of these genes interrelate. Alternatively, mutations in the various PD-causing genes may act through divergent mechanisms with the similar endpoint of DAergic neuronal death.

D. Polyglutamine Disorders

The polyglutamine disorders are a constellation of diseases sharing pathological deposits of aggregated polyglutamine encoded by expansion of coding CAG trinucleotide repeats found in the exons of a variety of different genes that cause neurodegeneration. These disorders include HD, spinal and bulbar muscular atrophy (SBMA), and a number of spinocerebellar ataxias (SCAs). These disorders share a disparate set of symptoms caused by progressive neurodegeneration induced by polyglutamine peptides. While the underlying pathological mechanism of these disorders seems shared, the genetic causes are divergent (i.e., the CAG expansions are found in different genes). Pathologically, the expanded polyglutamine stretches appear to be causal for disease, as the length of polyglutamine expansion correlates with the severity of disease and rate of disease progression. Typically, onset is in mid-life, but extreme CAG expansions can cause more severe forms of disease including juvenile-onset forms. Furthermore, evidence from animal models indicates that a polyglutamine expansion placed in a variety of different genes is sufficient to cause neurodegeneration.

Faber *et al.* (1999) described the first model of polyglutamine (polyQ)-mediated neurotoxicity in worms using a 150 amino acid polyQ tract derived from a mutant human huntingtin (HTT) cDNA driven by the *osm-10* promoter. Expression of the Q150-HTT transgene in worm ASH sensory neurons caused mild late-onset structural defects in neurite structure or ciliary endings. Shorter polyQ2, polyQ23, or polyQ95 tracts failed to cause any measurable defect. Likewise, polyQ 150-expressing animals typically exhibited cytoplasmic aggregates of Q150-HTT by day 8, while shorter polyQ stretches fail to aggregate. Animals coexpressing polyQ and GFP exhibit an enhancement in ASH defects. Furthermore, worms coexpressing Q150 and *osm-10*::GFP also exhibit defects in the nose touch response consistent with ASH neuronal dysfunction. Expression of Q150 also caused over 10% of 8-day-old ASH neurons to undergo apoptosis. To further understand the mechanism of polyQ-induced neurotoxicity, Faber *et al.* (2002) conducted a screen for genes that normally protect neurons from polyQ toxicity. They identified a polyQ enhancer gene (*pqe-1*). In a *pqe-1* mutant background, polyQ150 causes both neuron morphological abnormality and neuronal death phenotypes in most ASH neurons. PQE-1 is a nuclear protein that contains a Q/P-rich domain that is sufficient to suppress polyQ150 toxicity when overexpressed.

Another model for polyQ disorders utilized yellow fluorescent protein (YFP)-Q82, or YFP-Q19 fusions expressed in body muscle cells via the *myo-3* promoter (Satyal *et al.*, 2000). This model resulted in cytoplasmic muscle cell inclusions of protein aggregates in Q82-expressing lines, but not in Q19-expressing lines. Also, Q82-expressing lines exhibit delayed larval development. The mobility of polyQ proteins under polyacrylamide gel electrophoresis were reduced indicative of protein aggregation. Furthermore, Q82-GFP induced a constitutive heat shock response as assessed by heat shock protein reporter constructs. Expression of HSP104, a molecular chaperone, both restored the larval growth defect and prevented Q82 aggregate formation. To explore the polyQ tract length necessary to aggregate, YFP-polyQ fusions of Q0, Q19, Q29, Q33, Q35, Q40, Q44, Q64, and Q82 were tested. PolyQ lengths of Q35 or less did not form aggregates, while those Q40 or longer did form aggregates (Morley *et al.*, 2002). Furthermore, as polyQ length increased so did the number of aggregates per worm and the severity of motility defects induced by polyQ. Likewise as age increased so did the number of polyQ aggregates and severity of the Unc phenotype. Extension of lifespan using the *age-1* mutant delays the accumulation of aggregates and onset of polyQ toxicity. Hsu *et al.* (2003) extend this finding by demonstrating that knockdown of *daf-16*, *hsf-1*, or DAF-16/HSF-1 regulatory targets exacerbate this aggregation phenotype. To explore the effects of polyQ on cellular capacity for protein folding, Gidalevitz *et al.* (2006) placed polyQ transgenes in the background of a variety of temperature sensitive (ts) mutations. Ts mutations typically introduce changes into a protein sequence that partially disrupt it structurally, resulting in a properly folded functional protein at the permissive temperature, but a misfolded nonfunctional protein at the nonpermissive temperature. Gidalevitz *et al.* (2006) demonstrate that when a variety of *C. elegans* ts mutations are placed in a polyQ-expressing background, they display mutant phenotypes at the normally permissive temperature. This indicates that in general proper protein folding is impaired in polyQ-expressing worms. Furthermore, ts mutations enhance the protein aggregation of polyQ transgenic animals.

A third type of polyQ aggregation model was devised by Parker *et al.* (2001) where polyQ19, polyQ88-GFP, or polyQ128-GFP constructs were expressed in touch neurons using the *mec-3* promoter. PolyQ88 and polyQ128 constructs caused touch receptor dysfunction as measured by mechanosensation assays with both being significantly impaired as compared to non-polyQ-expressing animals. Expression of polyQ-GFP fusions caused accumulation of aggregates to form in the axonal processes of PLM neurons. Likewise, the morphology of axonal processes in PLM neurons is abnormal in Q128 animals. Thus, polyQ aggregates correlated with severity of Mec phenotype and abnormal morphology of PLM neurons consistent with polyQ-inducing neuronal dysfunction via toxic polyQ protein aggregates. To identify worm proteins that interact with polyQ, Holbert *et al.* (2003) screened a yeast two-hybrid *C. elegans* cDNA library. They identified one gene, K08E3.3, that binds HTT in a polyQ sensitive manner. The human homologue of K08E3.3 is cdc42-interacting protein 4 (CIP4). They demonstrated that human CIP4 binds HTT both *in vitro* and *in vivo*. CIP4 is up-regulated in brain tissue from HD patients and expression of CIP4 induces cell death in cultured rat striatal neurons.

A model exploring the normal function of the worm homologue of ataxin-2, ATX-2, has been described (Ciosk *et al.*, 2004). Although the role of ataxin-2 in disease appears to be mediated through an expansion of CAG repeats encoding polyQ, its normal function remains unknown. In worms, knockdown of *atx-2* by RNAi causes a small gonad without oocytes. The ATX-2 protein is expressed in the cytoplasm of germ cells and interacts with poly(A)-binding protein. The Atx-2 phenotype is due to reduced GLP-1/Notch signaling. Furthermore, abnormal derepression of germ line mRNAs may be responsible for the Atx-2 phenotype as *rme-2* mRNA is expressed abnormally in *atx-2* RNAi-treated animals. Thus, Ciosk *et al.* (2004) suggest ATX-2's normal function is one of translational regulation and by analogy human ataxin-2 may function similarly.

The nature of polyQ toxicity has been explored in a number of diverse transgenic models where polyQ has been expressed in different sets of neurons or even muscle cells. The length of polyQ expansion determines the severity of cytotoxicity, protein aggregation, and general severity of phenotype both in these worm models and in human polyQ disorders. The underlying mechanism of toxicity remains unclear, but appears to involve misregulation of protein folding, at least in worm muscle. Sirtuin activation appears to be protective in neuronal models of polyQ toxicity, but the mechanism of this also remains unclear. Perhaps identification of suppressor genes for polyQ toxicity will provide more insight into the mechanisms of polyQ cytotoxicity.

E. Amyotrophic Lateral Sclerosis

ALS is an adult-onset neurodegenerative disorder with involvement of motor neurons of both the spinal cord and cerebral cortex. Like the other disorders described above, it exists in both familial and sporadic forms. Familial ALS has been shown to be caused by dominant mutations in the Cu/Zn superoxide dismutase 1 gene (SOD1). SOD1 catalyzes the conversion of O_2^- to O_2 and H_2O_2. Familial ALS patients also exhibit an increase in oxidative damage to macromolecules. Despite this, the mechanism whereby SOD1 mutations cause disease remains poorly understood.

To model SOD1-mediated toxicity, Oeda *et al.* (2001) expressed either normal or mutant human SOD1 cDNAs in *C. elegans* using either the *myo-3* muscle-specific promoter or the inducible *hsp16-2* promoter. Both wild-type and mutant transgenic SOD1 showed activity similar to that seen for native human SOD1 (Oeda *et al.*, 2001). When challenged with oxidative stress from paraquat treatment, only SOD1 mutant-expressing strains exhibited hypersensitivity relative to nontransgenic animals. Oxidative stress delayed the turnover of transgenic SOD1, but mutant SOD1 was degraded more rapidly than wild-type SOD1. Oxidative stress also caused aggregates containing both GFP and mutant SOD1 to form in body wall muscle cells, but wild-type SOD1 did not aggregate under identical treatment, suggesting that mutant SOD1 toxicity might involve protein aggregation. Thus, SOD1 mutations may induce neurotoxic protein aggregation as seen in AD, PD, and the polyQ disorders described above.

F. Niemann-Pick Type C1 Disease

Niemann-Pick type C1 disease (NPC) is caused by autosomal recessive mutations in the *NPC1* gene. It is a neurodegenerative condition with a wide range of clinical phenotypes and severity. Onset typically occurs in childhood, but may manifest anytime from birth to early adulthood. The disease typically exhibits liver, lung, and neurological involvement. Neurological symptoms may include loss of coordination, muscle weakness, and difficulty in swallowing. The most prominent cellular defect is one of lipid trafficking. Internalization and hydrolysis of low-density lipoprotein occurs normally, but further transport of unesterified cholesterol appears defective. This results in free cholesterol accumulation in the late endosomal and lysosomal compartments. The exact role played by NPC1 in mediating proper cholesterol transport remains poorly understood.

Sym *et al.* (2000) reported a genetic model of NPC1 function in *C. elegans.* Worms have two genes homologous to NPC1 named *ncr-1* and *ncr-2.* Sym *et al.* (2000) generated deletion mutants for both *ncr-1* and *ncr-2.* Since wild-type *C. elegans* cannot synthesize cholesterol, they require it exogenously in the growth media. Normal worms grown in the absence of cholesterol will survive and reproduce, but their progeny die as larvae from cholesterol starvation. *ncr-1* and *ncr-1;ncr-2* double-mutant animals are hypersensitive to cholesterol starvation as they usually fail to develop past the L3 larval stage and never produce progeny in the absence of cholesterol. *ncr-1* and *ncr-1;ncr-2* double-mutant animals also exhibit retarded development during embryogenesis and at each larval stage such that the time from embryo to adulthood is extended by about 24 hours relative to wild type. Egl is constitutive in *ncr-1* and *ncr-1;ncr-2* double-mutant animals. Dauer larvae are an alternative developmental pathway for third-stage larvae induced by stressful conditions. The *ncr-1;ncr-2* double-mutant animals constitutively form dauer larvae inappropriately, while either *ncr-1* or *ncr-2* mutations alone follow the normal nondauer developmental pathway under routine growth conditions.

Li *et al.* (2004) further explore the roles of *ncr-1* and *ncr-2* in *C. elegans* development. The *ncr-1*; *ncr-2* animals exhibit many of the pleiotropic phenotypes of dauer formation defective (*daf*) mutants, including partial dauer formation, abnormal vulval morphology, reduced brood size, reduced lifespan, gonad migration, and seam cell defects (Li *et al.*, 2004). While *ncr-1; ncr-2* animals are hypersensitive to cholesterol deficiency, media containing high cholesterol partially rescues the dauer and reduced lifespan phenotypes. Li *et al.* (2004) also demonstrated that progesterone treatment of *ncr-1* worms mimics cholesterol deprivation, indicating that progesterone inhibits cholesterol trafficking in worms, as in mammalian cells. *ncr-1* is expressed in many cell types, including intestine, somatic gonad, many neurons, XXX cells, muscle, and hypodermis. *ncr-2* is expressed in the gonadal sheath and XXX cells. XXX expression is consistent with the Daf phenotype as ablation of the XXX cells induces dauer formation. The *ncr-1; ncr-2* phenotype is suppressed by overexpression of DAF-9, while overexpression of NCR-1 or NCR-2 failed to suppress *daf-9*; indicating that *daf-9* is upstream of *ncr-1* and *ncr-2.* To determine the role of *ncr-1* and *ncr-2* in neuronal development, the morphology of ASER neurons in *ncr-1; ncr-2* animals was examined and found to be abnormal. This model of the human NPC1 disorder differs from the other models for neurodegeneration previously described in that no abnormal or aggregating proteins are present to cause the phenotypes observed. However, no frank neurodegeneration was described for this model despite the presence of changes in neuronal morphology.

III. Identifying Candidate Modifier Genes of Neurodegeneration

A. Genetic Modifiers of Necrotic Neurodegeneration

To identify genes required for necrotic neurodegeneration, Xu *et al.* (2001) employed a classical genetic screen to find genes that modify *mec-4(d)*-induced neurodegeneration (Syntichaki *et al.,* 2002). To this end, transgenic animals expressing an *unc-8::mec-4(d)* transgene in motorneurons were engineered, resulting in worms that exhibit extensive motoneuron degeneration and a strong Unc phenotype. They then screened for transgenic animals with suppression of the Unc phenotype. This screen yielded multiple alleles of a single genetic suppressor of both the Mec-4 and Deg-1 phentoypes. Loss-of-function mutations in *crt-1* ameliorate the toxic effects of *mec-4* and *deg-1* mutations in addition to partially suppressing $G\alpha_s$-induced toxicity. *crt-1* encodes the worm homologue of calreticulin, a protein chaperone, and Ca^{2+}-binding protein localized to the ER involved in maintaining intracellular Ca^{2+} stores. Reduction of calnexin expression also partially suppressed *mec-4*-induced cell death. *crt-1* suppresses *mec-4* neurotoxicity both by reducing MEC-4 protein levels and by altering neuronal Ca^{2+} balance. From this, Xu *et al.* (2001) hypothesize that mutant Mec-4 and other channels cause cellular stress signaling, which causes active release of Ca^{2+} stores from the ER via calreticulin leading to increased cytoplasmic Ca^{2+}, which in turn initiates necrotic cell death.

B. Genetic Modifiers of Polyglutamine Aggregation

To determine whether polyQ modifiers identified in other systems can modify polyQ aggregation in the worm model, *hsp-1 and hsf-1* were tested for their ability to exacerbate polyQ aggregate formation (Nollen *et al.,* 2004). Both were found to increase polyQ foci. To identify genes that normally prevent polyQ aggregation, Nollen *et al.* (2004) conducted a genome-wide reverse genetic screen using RNAi testing for genes whose knockdown caused an increase of discrete polyQ protein aggregate foci. Knockdown of 186 genes caused an enhancement of polyQ aggregation. The enhancer genes were distributed into five classes: genes involved in RNA metabolism, protein synthesis, protein folding, protein transport, and protein degradation. A subset of modifiers from each class was tested for the ability to form biochemically insoluble polyQ aggregates, with most causing an increase in detectable insoluble polyQ aggregates. The authors propose a model for how each class of modifiers increases polyQ aggregates. They hypothesize that the RNA processing and protein synthesis classes of modifiers increase production of misfolded proteins while the protein folding, protein transport, and protein degradation classes of enhancers decrease clearance of misfolded proteins.

C. Torsion Proteins and Neurodegeneration

Autosomal dominant mutations in the human *TOR1A* gene cause torsion dystonia, a disorder characterized by involuntary muscle contractions and twisting postures without detectable neuropathology or neurodegeneration. The worm genome contains two homologues of human torsion proteins called TOR-1 and TOR-2. Human torsions and TOR-1 and TOR-2 all belong to the AAA+ family of ATPases implicated in protein folding and turnover as well as a variety of other cellular processes. Caldwell *et al.* (2003) reported exploring worm TOR-2 protein function in the context of polyQ aggregation. They demonstrate that TOR-2 overexpression modestly decreases the level of polyQ proteins, while dramatically decreasing both the number and size of polyQ aggregates. A TOR-2(Δ368) deletion mutant that mimicked a mutation found in human torsion dystonia kindreds failed to modify polyQ aggregate size or abundance, but did slowdown the growth of polyQ aggregate-bearing worms. Normal TOR-2 overexpression strongly inhibited polyQ aggregation, while a TOR-2(Δ368) mutant mildly exacerbated polyQ aggregation. Both TOR-2 and ubiquitin localized to polyQ aggregates within muscle cells. However, TOR-2 is localized to the ER in the absence of polyQ proteins (Li *et al.*, 2004).

Cao *et al.* (2005) explored the role of torsion proteins in protecting worm neurons from cellular stress. They employed 6-OHDA to mediate DAergic neuron neurodegeneration (Nass *et al.*, 2001) and assayed the effect of TOR-2 overexpression as described above for polyQ toxicity. Both TOR-2 and human torsionA protein overexpression ameliorated 6-OHDA neurotoxicity, while mutants incapable of binding ATP gave only modest protection from 6-OHDA (Cao *et al.*, 2005). Note that nontransgenic worms express no endogenous TOR-2 in DAergic neurons. Dominant-negative mutants of torsionA reduced the ability of normal torsionA to protect from 6-OHDA toxicity. TOR-2 and torsionA also protect DAergic neurons against neurotoxicity caused by human SNCA overexpression. Thus, at least when overexpressed, torsion proteins can prevent neurotoxicity caused by molecular insult (6-OHDA), polyQ aggregation, or human SNCA.

IV. Molecular Inhibitors of Neurodegeneration

Parker *et al.* (2005) used their model of polyQ-mediated cytotoxicity to test compounds for the ability to prevent mutant polyQ cytotoxicity. They demonstrate that the Sir2-activating drugs resveratrol and fisetin partially rescue the Mec phenotype of polyQ128 transgenic worms. Likewise, overexpression of *sir-2.1* rescues the Mec phenotype, demonstrating that SIR-2.1 activation either pharmacologically or genetically is sufficient to ameliorate the effects of polyQ expression. This suppression of polyQ toxicity occurs in the absence of a change in polyQ protein expression. However, the number of axonal aggregates and dystrophic processes in polyQ19 and polyQ128 is down-regulated by SIR-2.1 activation to a similar extent, suggesting that neither is the cause of neuronal dysfunction. *daf-16* is the terminal regulator in the *daf-2*/insulin signaling pathway in worms in which mutations mediate dramatic extension in worm lifespan. Protection from polyQ toxicity by SIR-2.1 requires *daf-16* as mutants failed to respond to either SIR-2.1 overexpression or resveratrol treatment. The Sir2 activation findings were confirmed in a murine cell culture model of polyQ disorders where striatal neurons bearing an HTT-Q111 transgene were partially rescued by resveratrol treatment.

In a model for PD, Ved *et al.* (2005) report a variety of compounds that partially protect against the toxicity of mitochondrial complex I inhibition. They show that the antioxidant probucol significantly reduced rotenone toxicity in their model (Ved *et al.*, 2005). Likewise, treatment with compounds hypothesized to bypass rotenone toxicity by increasing substrates for mitochondrial complex II was partially successful in ameliorating rotenone toxicity. DβHB protected worms against rotenone toxicity at a concentration of 50 mM. Furthermore, treatment with a putative inhibitor of apoptosis, tauroursodeoxycholic acid (TUDCA) gave dose-dependent protection against rotenone-induced toxicity. Similarly, probucol protected worms from paraquat toxicity.

V. Neuroprotective Strategies

Work with animal models of disease has begun to suggest possibilities for intervening in neurodegeneration. In particular, the worm models for neurodegeneration outlined above have potential for rapidly testing a wide array of candidate therapeutic approaches. Of course, the limitation of these models is the potential lack of correspondence between the worm model and the authentic disease state in humans. For instance, the worm's lack of a mammalian-like inflammatory response suggests that any model will be a significant simplification for neurodegeneration disorders. However, given the relative dearth of knowledge about the pathogenic mechanism underlying many of these disorders, this may actually turn out to be an advantage for identifying root causes of neurodegeneration. Even if these simplified systems do fully recapitulate the neurodegenerative phenotypes of the authentic disease state, this drawback is offset by the experimental advantages and speed of worm models. The key to demonstrating the veracity of these worm models will be to extend the findings in worm models to other mammalian cell culture or animal models, or indeed to the genuine disease state itself. In some cases (Holbert *et al.*, 2003; Parker *et al.*, 2005; Springer *et al.*, 2005) findings have already been extended to mammalian systems, but in most they have not.

Work from these worm models suggests specific interventions that might warrant investigation in mammalian tissue culture or even animal model systems. For instance in *ncr-1;ncr-2* mutant worms, exposing worms to high cholesterol rescues the dauer defect. Since more cholesterol rescues worms from loss of *ncr* function, perhaps a high cholesterol diet would be therapeutic for loss of NPC1 function in mammals. This could be explored as a therapy in NPC1 knockout mice. Likewise, Ca^{2+} release from ER stores appears necessary for necrotic neuronal death in worms and is partially alleviated by Ca^{2+} chelation. This suggests drugs that act to restore Ca^{2+} homeostasis may be useful against necrotic cell death in stroke, ischemia, or trauma. Overexpression of torsion proteins has been shown in worms to prevent both polyQ and SNCA neurotoxicity. Possibly, molecular activators of AAA+ proteins should be considered as therapeutics for these aggregation diseases. Clearly, sirtuin inhibition in worms can rescue polyQ toxicity in neurons; the question is will they be of use in treating human polyQ disoders. Testing of these compounds in mouse models of polyQ disorders should be the next step. With well-tolerated compounds like resveratrol, drug trials in patients could be considered. A strategy that may have application in multiple aggregation disorders is one of boosting cellular resistance to proteotoxic stress. Activation of chaperones can boost resistance to polyQ in worms; it may be that careful small molecule manipulation of mammalian chaperones will likewise prove beneficial. Treating worms with DAergic neuron dysfunction using exogenous dopamine rescues the DAergic neurons, suggesting at least some therapies that work in humans (treating PD with levodopa) also work in worms. This observation should encourage the testing of successful worm therapies in mammalian systems.

VI. Utility of HTS for Drugs in Worms

One distinct advantage of *C. elegans* over other multicellular model organisms for neurodegeneration is capacity to grow them in standard 96-well microtiter plates. While, neurons can be cultured in this format, this is not in the context of the whole organism. Growth in a 96-well format allows application of high-throughput screening (HTS) technologies developed for drug discovery. These technologies were initially devised for use with *in vitro* enzymatic or even cellular assays, but in principle could also be adapted to the study of worms. The chief barrier to implementing this approach is one of image analysis. HTS methods need to be developed for data capture, measurement, data classification, and evaluation of worm behavior and/or morphology in solution. "Machine vision" has been developed and employed to address similar problems for plate phenotypes of *C. elegans* (Baek *et al.*, 2002; Feng *et al.*, 2004). However, for these methods to be applied to HTS, architecture compatible with standard HTS hardware and screening practice needs to be developed. At present any drug screen using live worms is limited by the number of wells humans can score in a day. Considering small molecule collections can contain several million compounds, the time and resources needed to exhaustively screen a large collection is prohibitive. Given an appropriately flexible system for real-time HTS imaging and image analysis, many of the worm models for neurodegenerative disease could potentially be used to screen large collections of compounds for small molecule inhibitors of neuronal dysfunction and/or neurodegeneration. However, without advances in HTS methodology, large-scale drug screens will not be possible and the utility of the worm disease models described above will be restricted.

VII. Conclusions

Thus far, worm models for human neurodegenerative disorders have yielded some valuable insights into the potential pathological mechanisms underlying a variety of diseases. From these models we can extract some ideas to test as candidate therapeutic approaches. Indeed, in some cases individual genes or molecules that suppress neurodegeneration of a given type have been identified. However, the potential of many of these models has yet to be fully developed. The robust forward and reverse genetics available in studies using *C. elegans* has only been employed for a fraction of the models reviewed here. The identification of more genes that suppress neurodegeneration will likely shed further insight into disease mechanisms and suggest additional neuroprotective strategies. The development of automated high-throughput methods for quantitatively screening worm phenotypes would allow testing of existing compound libraries used in drug discovery for compounds that ameliorate the disease phenotypes in these worm models. Such screens will be needed to realize the full potential of the disease models described here.

References

Artal-Sanz, M., and Tavernarakis, N. (2005). Proteolytic mechanisms in necrotic cell death and neurodegeneration. *FEBS Lett.* **579**(15), 3287–3296.

Baek, J. H., Cosman, P., Feng, Z., Silver, J., and Schafer, W. R. (2002). Using machine vision to analyze and classify *Caenorhabditis elegans* behavioral phenotypes quantitatively. *J. Neurosci. Methods* **118**(1), 9–21.

Berger, A. J., Hart, A. C., and Kaplan, J. M. (1998). G alphas-induced neurodegeneration in *Caenorhabditis elegans*. *J. Neurosci.* **18**(8), 2871–2880.

Boyd-Kimball, D., Poon, H. F., Lynn, B. C., Cai, J., Pierce, W. M., Jr., Klein, J. B., Ferguson, J., Link, C. D., and Butterfield, D. A. (2006). Proteomic identification of proteins specifically oxidized in *Caenorhabditis elegans* expressing human Abeta(1–42): Implications for Alzheimer's disease. *Neurobiol. Aging* **27**(9), 1239–1249.

Butterfield, D. A., Yatin, S. M., and Link, C. D. (1999). *In vitro* and *in vivo* protein oxidation induced by Alzheimer's disease amyloid beta-peptide (1–42). *Ann. NY Acad. Sci.* **893,** 265–268.

Caldwell, G. A., Cao, S., Sexton, E. G., Gelwix, C. C., Bevel, J. P., and Caldwell, K. A. (2003). Suppression of polyglutamine-induced protein aggregation in *Caenorhabditis elegans* by torsin proteins. *Hum. Mol. Genet.* **12**(3), 307–319.

Cao, S., Gelwix, C. C., Caldwell, K. A., and Caldwell, G. A. (2005). Torsin-mediated protection from cellular stress in the dopaminergic neurons of *Caenorhabditis elegans*. *J. Neurosci.* **25**(15), 3801–3812.

Chalfie, M. (1993). Touch receptor development and function in *Caenorhabditis elegans*. *J. Neurobiol.* **24**(10), 1433–1441.

Chalfie, M., and Wolinsky, E. (1990). The identification and suppression of inherited neurodegeneration in *Caenorhabditis elegans*. *Nature* **345**(6274), 410–416.

Ciosk, R., DePalma, M., and Priess, J. R. (2004). ATX-2, the *C. elegans* ortholog of ataxin 2, functions in translational regulation in the germline. *Development* **131**(19), 4831–4841.

Cottrell, B. A., Galvan, V., Banwait, S., Gorostiza, O., Lombardo, C. R., Williams, T., Schilling, B., Peel, A., Gibson, B., Koo, E. H., Link, C. D., and Bredesen, D. E. (2005). A pilot proteomic study of amyloid precursor interactors in Alzheimer's disease. *Ann. Neurol.* **58**(2), 277–289.

Drake, J., Link, C. D., and Butterfield, D. A. (2003). Oxidative stress precedes fibrillar deposition of Alzheimer's disease amyloid beta-peptide (1–42) in a transgenic *Caenorhabditis elegans* model. *Neurobiol. Aging* **24**(3), 415–420.

Driscoll, M. (1992). Molecular genetics of cell death in the nematode *Caenorhabditis elegans*. *J. Neurobiol.* **23**(9), 1327–1351.

Driscoll, M., and Chalfie, M. (1991). The mec-4 gene is a member of a family of *Caenorhabditis elegans* genes that can mutate to induce neuronal degeneration. *Nature* **349**(6310), 588–593.

Driscoll, M., and Gerstbrein, B. (2003). Dying for a cause: Invertebrate genetics takes on human neurodegeneration. *Nat. Rev. Genet.* **4**(3), 181–194.

Eimer, S., Lakowski, B., Donhauser, R., and Baumeister, R. (2002). Loss of Spr-5 bypasses the requirement for the *C. elegans* presenilin Sel-12 by derepressing Hop-1. *EMBO J.* **21**(21), 5787–5796.

Faber, P. W., Alter, J. R., MacDonald, M. E., and Hart, A. C. (1999). Polyglutamine-mediated dysfunction and apoptotic death of a *Caenorhabditis elegans* sensory neuron. *Proc. Natl. Acad. Sci. USA* **96**(1), 179–184.

Faber, P. W., Voisine, C., King, D. C., Bates, E. A., and Hart, A. C. (2002). Glutamine/proline-rich PQE-1 proteins protect *Caenorhabditis elegans* neurons from huntingtin polyglutamine neurotoxicity. *Proc. Natl. Acad. Sci. USA* **99**(6), 17131–17136.

Fay, D. S., Fluet, A., Johnson, C. J., and Link, C. D. (1998). *In vivo* aggregation of beta-amyloid peptide variants. *J. Neurochem.* **71**(4), 1616–1625.

Feng, Z., Cronin, C. J., Wittig, J. H., Jr., Sternberg, P. W., and Schafer, W. R. (2004). An imaging system for standardized quantitative analysis of *C. elegans* behavior. *BMC Bioinformatics* **5,** 115.

Fonte, V., Kapulkin, V., Taft, A., Fluet, A., Friedman, D., and Link, C. D. (2002). Interaction of intracellular beta amyloid peptide with chaperone proteins. *Proc. Natl. Acad. Sci. USA* **99**(14), 9439–9444.

Gidalevitz, T., Ben-Zvi, A., Ho, K. H., Brignull, H. R., and Morimoto, R. I. (2006). Progressive disruption of cellular protein folding in models of polyglutamine diseases. *Science* **311**(5766), 1471–1474.

Gurvitz, A., Langer, S., Piskacek, M., Hamilton, B., Ruis, H., and Hartig, A. (2000). Predicting the function and subcellular location of *Caenorhabditis elegans* proteins similar to *Saccharomyces cerevisiae* beta-oxidation enzymes. *Yeast* **17**(3), 188–200.

Gutierrez-Zepeda, A., Santell, R., Wu, Z., Brown, M., Wu, Y., Khan, I., Link, C. D., Zhao, B., and Luo, Y. (2005). Soy isoflavone glycitein protects against beta amyloid-induced toxicity and oxidative stress in transgenic *Caenorhabditis elegans*. *BMC Neurosci.* **6,** 54.

Hall, D. H., Gu, G., Garcia-Anoveros, J., Gong, L., Chalfie, M., and Driscoll, M. (1997). Neuropathology of degenerative cell death in *Caenorhabditis elegans*. *J. Neurosci.* **17**(3), 1033–1045.

Herman, R. K. (1987). Mosaic analysis of two genes that affect nervous system structure in *Caenorhabditis elegans*. *Genetics* **116**(3), 377–388.

Holbert, S., Dedeoglu, A., Humbert, S., Saudou, F., Ferrante, R. J., and Neri, C. (2003). Cdc42-interacting protein 4 binds to huntingtin: Neuropathologic and biological evidence for a role in Huntington's disease. *Proc. Natl. Acad. Sci. USA* **100**(5), 2712–2717.

Hong, K., and Driscoll, M. (1994). A transmembrane domain of the putative channel subunit MEC-4 influences mechanotransduction and neurodegeneration in *C. elegans*. *Nature* **367**(6462), 470–473.

Hsu, A. L., Murphy, C. T., and Kenyon, C. (2003). Regulation of aging and age-related disease by DAF-16 and heat-shock factor. *Science* **300**(5622), 1142–1145.

Hu, B., Finsinger, D., Peter, K., Guttenberg, Z., Barmann, M., Kessler, H., Escherich, A., Moroder, L., Bohm, J., Baumeister, W., Sui, S. F., and Sackmann, E. (2000). Intervesicle cross-linking with integrin alpha IIb beta 3 and cyclic-RGD-lipopeptide. A model of cell-adhesion processes. *Biochemistry* **39**(40), 12284–12294.

Huang, M., and Chalfie, M. (1994). Gene interactions affecting mechanosensory transduction in *Caenorhabditis elegans*. *Nature* **367**(6462), 467–470.

Jarriault, S., and Greenwald, I. (2002). Suppressors of the egg-laying defective phenotype of Sel-12 presenilin mutants implicate the CoREST corepressor complex in LIN-12/Notch signaling in *C. elegans*. *Genes Dev.* **16**(20), 2713–2728.

Korswagen, H. C., Park, J. H., Ohshima, Y., and Plasterk, R. H. (1997). An activating mutation in a *Caenorhabditis elegans* Gs protein induces neural degeneration. *Genes Dev.* **11**(12), 1493–1503.

Korswagen, H. C., van der Linden, A. M., and Plasterk, R. H. (1998). G protein hyperactivation of the *Caenorhabditis elegans* adenylyl cyclase SGS-1 induces neuronal degeneration. *EMBO J.* **17**(17), 5059–5065.

Kraemer, B. C., Zhang, B., Leverenz, J. B., Thomas, J. H., Trojanowski, J. Q., and Schellenberg, G. D. (2003). Neurodegeneration and defective neurotransmission in a *Caenorhabditis elegans* model of tauopathy. *Proc. Natl. Acad. Sci. USA* **100**(17), 9980–9985.

Kuwahara, T., Koyama, A., Gengyo-Ando, K., Masuda, M., Kowa, H., Tsunoda, M., Mitani, S., and Iwatsubo, T. (2006). Familial Parkinson mutant alpha-synuclein causes dopamine neuron dysfunction in transgenic *Caenorhabditis elegans*. *J. Biol. Chem.* **281**(1), 334–340.

Lakowski, B., Eimer, S., Gobel, C., Bottcher, A., Wagler, B., and Baumeister, R. (2003). Two suppressors of Sel-12 encode C2H2 zinc-finger proteins that regulate presenilin transcription in *Caenorhabditis elegans*. *Development* **130**(10), 2117–2128.

Lakso, M., Vartiainen, S., Moilanen, A. M., Sirvio, J., Thomas, J. H., Nass, R., Blakely, R. D., and Wong, G. (2003). Dopaminergic neuronal loss and motor deficits in *Caenorhabditis elegans* over-expressing human alpha-synuclein. *J. Neurochem.* **86**(1), 165–172.

Levitan, D., Doyle, T. G., Brousseau, D., Lee, M. K., Thinakaran, G., Slunt, H. H., Sisodia, S. S., and Greenwald, I. (1996). Assessment of normal and mutant human presenilin function in *Caenorhabditis elegans*. *Proc. Natl. Acad. Sci. USA* **93**(25), 14940–14944.

Levitan, D., and Greenwald, I. (1995). Facilitation of Lin-12-mediated signalling by Sel-12, a *Caenorhabditis elegans* S182 Alzheimer's disease gene. *Nature* **377**(6547), 351–354.

Li, X., and Greenwald, I. (1996). Membrane topology of the *C. Elegans* SEL-12 presenilin. *Neuron* **17** (5), 1015–1021.

Li, J., Brown, G., Ailion, M., Lee, S., and Thomas, J. H. (2004). NCR-1 and NCR-2, the *C. elegans* homologs of the human Niemann-Pick type C1 disease protein, function upstream of DAF-9 in the dauer formation pathways. *Development* **131**(22), 5741–5752.

Link, C. D. (1995). Expression of human beta-amyloid peptide in transgenic *Caenorhabditis elegans*. *Proc. Natl. Acad. Sci. USA* **92**(20), 9368–9372.

Link, C. D. (2001). Transgenic invertebrate models of age-associated neurodegenerative diseases. *Mech. Ageing Dev.* **122**(14), 1639–1649.

Link, C. D. (2005). Invertebrate models of Alzheimer's disease. *Genes Brain Behav.* **4**(3), 147–156.

Link, C. D., Cypser, J. R., Johnson, C. J., and Johnson, T. E. (1999). Direct observation of stress response in *Caenorhabditis elegans* using a reporter transgene. *Cell Stress Chaperones* **4**(4), 235–242.

Link, C. D., Fonte, V., Hiester, B., Yerg, J., Ferguson, J., Csontos, S., Silverman, M. A., and Stein, G. H. (2006). Conversion of green fluorescent protein into a toxic, aggregation-prone protein by C-terminal addition of a short peptide. *J. Biol. Chem.* **281**(3), 1808–1816.

Link, C. D., Johnson, C. J., Fonte, V., Paupard, M., Hall, D. H., Styren, S., Mathis, C. A., and Klunk, W. E. (2001). Visualization of fibrillar amyloid deposits in living, transgenic *Caenorhabditis elegans* animals using the sensitive amyloid dye, X-34. *Neurobiol. Aging* **22**(2), 217–226.

Link, C. D., Taft, A., Kapulkin, V., Duke, K., Kim, S., Fei, Q., Wood, D. E., and Sahagan, B. G. (2003). Gene expression analysis in a transgenic *Caenorhabditis elegans* Alzheimer's disease model. *Neurobiol. Aging* **24**(3), 397–413.

Miyasaka, T., Ding, Z., Gengyo-Ando, K., Oue, M., Yamaguchi, H., Mitani, S., and Ihara, Y. (2005). Progressive neurodegeneration in *C. elegans* model of tauopathy. *Neurobiol. Dis.* **20**(2), 372–383.

Morley, J. F., Brignull, H. R., Weyers, J. J., and Morimoto, R. I. (2002). The threshold for polyglutamine-expansion protein aggregation and cellular toxicity is dynamic and influenced by aging in *Caenorhabditis elegans*. *Proc. Natl. Acad. Sci. USA* **99**(16), 10417–10422.

Nass, R., Miller, D. M., and Blakely, R. D. (2001). *C. elegans*: A novel pharmacogenetic model to study Parkinson's disease. *Parkinsonism Relat. Disord.* **7**(3), 185–191.

Nass, R., Hall, D. H., Miller, D. M., 3rd, and Blakely, R. D. (2002). Neurotoxin-induced degeneration of dopamine neurons in *Caenorhabditis elegans*. *Proc. Natl. Acad. Sci. USA* **99**(5), 3264–3269.

Nollen, E. A., Garcia, S. M., van Haaften, G., Kim, S., Chavez, A., Morimoto, R. I., and Plasterk, R. H. (2004). Genome-wide RNA interference screen identifies previously undescribed regulators of polyglutamine aggregation. *Proc. Natl. Acad. Sci. USA* **101**(17), 6403–6408.

Oeda, T., Shimohama, S., Kitagawa, N., Kohno, R., Imura, T., Shibasaki, H., and Ishii, N. (2001). Oxidative stress causes abnormal accumulation of familial amyotrophic lateral sclerosis-related mutant SOD1 in transgenic *Caenorhabditis elegans*. *Hum. Mol. Genet.* **10**(19), 2013–2023.

Parker, J. A., Arango, M., Abderrahmane, S., Lambert, E., Tourette, C., Catoire, H., and Neri, C. (2005). Resveratrol rescues mutant polyglutamine cytotoxicity in nematode and mammalian neurons. *Nat. Genet.* **37**(4), 349–350.

Parker, J. A., Connolly, J. B., Wellington, C., Hayden, M., Dausset, J., and Neri, C. (2001). Expanded polyglutamines in *Caenorhabditis elegans* cause axonal abnormalities and severe dysfunction of PLM mechanosensory neurons without cell death. *Proc. Natl. Acad. Sci. USA* **98**(23), 13318–13323.

Petrucelli, L., Dickson, D., Kehoe, K., Taylor, J., Snyder, H., Grover, A., DeLucia, M., McGowan, E., Lewis, J., Prihar, G., Kim, J., Dillmann, W. H., *et al.* (2004). CHIP and Hsp70 regulate tau ubiquitination, degradation and aggregation. *Hum. Mol. Genet.* **13,** 703–714.

Sahara, N., Murayama, M., Mizoroki, T., Urushitani, M., Imai, Y., Takahashi, R., Murata, S., Tanaka, K., and Takashima, A. (2005). *In vivo* evidence of CHIP up-regulation attenuating tau aggregation. *J. Neurochem.* **94**(5), 1254–1263.

Satyal, S. H., Schmidt, E., Kitagawa, K., Sondheimer, N., Lindquist, S., Kramer, J. M., and Morimoto, R. I. (2000). Polyglutamine aggregates alter protein folding homeostasis in *Caenorhabditis elegans*. *Proc. Natl. Acad. Sci. USA* **97**(11), 5750–5755.

Schwarzman, A. L., Gregori, L., Vitek, M. P., Lyubski, S., Strittmatter, W. J., Enghilde, J. J., Bhasin, R., Silverman, J., Weisgraber, K. H., Coyle, P. K., *et al.* (1994). Transthyretin sequesters amyloid beta protein and prevents amyloid formation. *Proc. Natl. Acad. Sci. USA* **91**(18), 8368–8372.

Shimura, H., Schwartz, D., Gygi, S. P., and Kosik, K. S. (2004). CHIP-Hsc70 complex ubiquitinates phosphorylated tau and enhances cell survival. *J. Biol. Chem.* **279**(6), 4869–4876.

Springer, W., Hoppe, T., Schmidt, E., and Baumeister, R. (2005). A *Caenorhabditis elegans* Parkin mutant with altered solubility couples alpha-synuclein aggregation to proteotoxic stress. *Hum. Mol. Genet.* **14**(22), 3407–3423.

Steiner, H., Revesz, T., Neumann, M., Romig, H., Grim, M. G., Pesold, B., Kretzschmar, H. A., Hardy, J., Holton, J. L., Baumeister, R., Houlden, H., and Haass, C. (2001). A pathogenic presenilin-1 deletion causes abberrant Abeta 42 production in the absence of congophilic amyloid plaques. *J. Biol. Chem.* **276**(10), 7233–7239.

Sumi, S. M., Bird, T. D., Nochlin, D., and Raskind, M. A. (1992). Familial presenile dementia with psychosis associated with cortical neurofibrillary tangles and degeneration of the amygdala. *Neurology* **42**(1), 120–127.

Sym, M., Basson, M., and Johnson, C. (2000). A model for Niemann-Pick type C disease in the nematode *Caenorhabditis elegans*. *Curr. Biol.* **10**(9), 527–530.

Syntichaki, P., Xu, K., Driscoll, M., and Tavernarakis, N. (2002). Specific aspartyl and calpain proteases are required for neurodegeneration in *C. elegans*. *Nature* **419**(6910), 939–944.

Treinin, M., and Chalfie, M. (1995). A mutated acetylcholine receptor subunit causes neuronal degeneration in *C. elegans*. *Neuron* **14**(4), 871–877.

Ved, R., Saha, S., Westlund, B., Perier, C., Burnam, L., Sluder, A., Hoener, M., Rodrigues, C. M., Alfonso, A., Steer, C., Liu, L., Przedborski, S., *et al.* (2005). Similar patterns of mitochondrial vulnerability and rescue induced by genetic modification of alpha-synuclein, Parkin, and DJ-1 in *Caenorhabditis elegans*. *J. Biol. Chem.* **280**(52), 42655–42668.

Wen, C., Levitan, D., Li, X., and Greenwald, I. (2000). Spr-2, a suppressor of the egg-laying defect caused by loss of Sel-12 presenilin in *Caenorhabditis elegans*, is a member of the SET protein subfamily. *Proc. Natl. Acad. Sci. USA* **97**(26), 14524–14529.

Xu, K., Tavernarakis, N., and Driscoll, M. (2001). Necrotic cell death in *C. elegans* requires the function of calreticulin and regulators of Ca^{2+} release from the endoplasmic reticulum. *Neuron* **31,** 957–971.

Yatin, S. M., Varadarajan, S., Link, C. D., and Butterfield, D. A. (1999). *In vitro* and *in vivo* oxidative stress associated with Alzheimer's amyloid beta-peptide (1–42). *Neurobiol. Aging* **20**(3), 325–330; discussion 339–342.

NEUROPROTECTION AND ENHANCEMENT OF NEURITE OUTGROWTH WITH SMALL MOLECULAR WEIGHT COMPOUNDS FROM SCREENS OF CHEMICAL LIBRARIES

Donard S. Dwyer and Addie Dickson

Department of Psychiatry, Louisiana State University Health Sciences Center
Shreveport, Louisiana 71130, USA

I. Introduction
II. Brain Deficit Disorders: The Challenges
III. Treatment Objectives
  A. Neuroprotection and Neurite Outgrowth
  B. Enhancement of Glucose Metabolism
IV. Application to Neurodevelopmental Disorders: Schizophrenia
  A. Pathology, Etiology, and Treatment of Schizophrenia
  B. Initial Screening of Compound Libraries
  C. Role of Akt in Schizophrenia and the Actions of Antipsychotic Drugs
  D. Possible Role of G Proteins
  E. Brain Glucose Metabolism and Metabolic Flexibility
  F. Protection Against Glucose and Glutamine Deprivation
V. Screening and Identification of Novel Compounds for Drug Discovery
  A. Screening Assays
  B. Structure-Based Design
VI. Neurodevelopmental Disorders and Neurodegenerative Diseases
  A. Rett Syndrome
  B. Lissencephaly and Tuberous Sclerosis Complex
  C. Autism
  D. Alzheimer's Disease
VII. Future Directions
  References

Currently, few drugs are available to effectively restore neuronal function in neurodevelopmental disorders, such as autism, or to prevent the cell loss and functional decline associated with neurodegenerative diseases. To address this situation, new drugs need to be developed that promote neurite/dendrite outgrowth, optimize glucose metabolism, and enhance neuronal viability, or collectively produce what has been called neuroenhancement. This chapter will review efforts and strategies aimed at the discovery of novel neuroenhancing drugs. In addition, the term metabolic flexibility is introduced to describe how cells manage the dynamic utilization of different fuel sources for energy production. Initial studies of antipsychotic drugs revealed that olanzapine stimulated glucose metabolism,

INTERNATIONAL REVIEW OF
NEUROBIOLOGY, VOL. 77
DOI: 10.1016/S0074-7742(06)77008-8

0074-7742/07 $35.00

enhanced neurite outgrowth, and promoted cell survival *in vitro*. The chemical structure of olanzapine served as the template for structure-based screening of chemical libraries for additional compounds with these properties. Small molecules have been identified that are neuroprotective based, in part, on activation of the serine–threonine kinase, Akt (or protein kinase B), which regulates cell proliferation and glucose metabolism. In addition, the compounds enhance neurite outgrowth from the PC12 cell line, and promote metabolic flexibility as judged by protection of cells against glutamine and nutrient deprivation. Although the mechanism of action has not been fully characterized, the compounds appear to act through G proteins and downstream kinase signaling pathways. Ultimately, this research may lead to the development of novel drugs that produce neuroenhancement. These drugs may prove valuable in the treatment of neurodevelopmental disorders, including schizophrenia and autism, and neurodegenerative conditions such as Alzheimer's disease.

I. Introduction

Neurodevelopmental disorders and neurodegenerative diseases adversely affect a significant proportion of the world's population. For example, schizophrenia (a neurodevelopmental disorder) afflicts roughly 1% of the population (Lewis and Levitt, 2002; Mueser and McGurk, 2004), whereas Alzheimer's disease has an annual incidence rate of 7–8% in individuals over 65 (Ferri *et al.*, 2005). Nevertheless, few truly effective treatments are available to correct or ameliorate these serious conditions. Drugs that are available typically provide only modest improvement in those patients who show any positive response. This is not surprising because the drugs mainly treat the symptoms of the disease and not its underlying cause, for example failure of neuronal migration/synapse formation, or outright loss of neurons in particular brain regions. This situation calls for a dramatic reevaluation of our current therapeutic approaches and highlights the need for novel strategies for the treatment of neurodevelopmental and neurodegenerative disorders. For the purpose of this chapter, we will refer to these conditions collectively as brain deficit disorders. This chapter will summarize recent developments that suggest new approaches for the treatment of brain deficit disorders. The goal of these novel therapeutic strategies is to move beyond superficial treatment of symptoms by addressing fundamental deficits in neuronal function and survival. We will focus most of our attention on schizophrenia because the brain defects in this disease are representative of other neurodevelopmental disorders that are less well understood. Potential therapeutic targets will be discussed, and a plan will be outlined for discovering new drugs that provide neuroenhancement, that is, that enhance neurite outgrowth (function), optimize energy metabolism, and promote neuronal survival.

II. Brain Deficit Disorders: The Challenges

Various neurological and neuropsychiatric conditions can broadly be considered brain deficit disorders, including the neurodevelopmental disorders—schizophrenia, Rett syndrome, and autism—and the neurodegenerative diseases—Alzheimer's, Parkinson's, and Huntington's. Although these disorders have very different origins, they share certain pathological features that present great clinical challenges. Brain deficit disorders are characterized by *functional* loss of neurons, which may be quite localized (e.g., loss of cholinergic neurons in the basal forebrain in Alzheimer's disease) or highly diffuse (as in Rett syndrome). In addition, there is typically poor neuronal connectivity, including reductions in neurite arborization, dendrite output, and synaptic connections (Gonatas and Moss, 1975; LeBlanc, 2005; McGlashan and Hoffman, 2000; Zoghbi, 2003). The impoverished connectivity may result from a failure to establish proper neuronal geometry and communication during development, or from a loss of cells and processes over time as a function of aging or the disease process. Moreover, the brain deficit disorders all show signs of impaired energy metabolism in the CNS (Blass, 2002; Chugani *et al.*, 1999; Dwyer *et al.*, 2001; Heininger, 2000; Yoshikawa *et al.*, 1992). This impairment may be a primary deficit, or may arise secondarily from diminished trophic support or general neurotoxic effects.

Of course, there are also striking differences between these two major classes of brain disorder. In the neurodevelopmental conditions, the lesion or defect is usually manifested at an early stage of development. Even if clear-cut symptoms only emerge somewhat later in life (as in some schizophrenics), the underlying deficits are thought to arise in the developing brain before it achieves full maturity. Disruption of normal brain development may occur *in utero* or in relation to developmental milestones, for example, at the onset of speech in children or during the establishment of higher order circuitry involving the frontal cortex in adolescents. The lesion itself is viewed as a mostly static fixture of the brain, although the expression of resulting deficits may be dynamic (Neul and Zoghbi, 2004; Palmen *et al.*, 2004; Weinberger, 1987). By contrast, the neurodegenerative disorders typically emerge in adulthood with aging as a contributory factor. There is apparently normal brain function (and development) up until the time that symptoms are expressed. The lesion(s) is progressive with ongoing cellular loss or deterioration of functional activity of neurons (Bertram and Tanzi, 2005). The progressive nature of the disease presents a moving target for therapeutic intervention.

Brain deficit disorders may result when there has been a disturbance in: (1) neurogenesis, (2) neuronal migration and/or differentiation, (3) neurite/dendrite outgrowth, (4) functional stabilization of synaptic connectivity, and (5) neuronal survival. The first two processes do not contribute significantly to causation of the neurodegenerative diseases. The biological processes listed here are quite complex compared with more typical targets of therapeutic intervention, namely enzyme activity or receptor binding. Given this complexity, a superficial analysis might consider extensive restoration of brain function in these disorders an unlikely therapeutic prospect. However, there are hopeful signs on the horizon. Neurotrophic factors have shown some promise in early clinical studies (Allen and Dawbarn, 2006; Apfel, 2002; Freeman, 1999). For example, nerve growth factor (NGF) was recently found to improve cognitive function and measures of brain glucose metabolism in Alzheimer's patients who received implants of cells producing the growth factor (Tuszynski *et al.*, 2005). The results with NGF in the treatment of neuropathies have been mixed (Apfel, 2002; Freeman, 1999; Schifitto *et al.*, 2001), as have trials of brain-derived neurotrophic factor (BDNF) for amyotrophic lateral sclerosis (ALS) (Beck *et al.*, 2005; Wellmer *et al.*, 2001). However, achieving and maintaining therapeutic doses of the protein neurotrophic factors have been major issues. Interestingly, psychotropic drugs used to treat bipolar disorder, depression, and schizophrenia produce trophic-like effects as judged by imaging studies in patients (Lieberman *et al.*, 2005b; Molina *et al.*, 2004; Moore *et al.*, 2000), brain morphology studies in animals (Duman, 2002; Law *et al.*, 2004; Wakade *et al.*, 2002; Wang *et al.*, 2004), and observations of neuroprotection *in vitro* (Bai *et al.*, 2002; Dwyer *et al.*, 2003a; Wei *et al.*, 2003). These positive actions are unintended consequences and are not the primary reason for administration of the psychotropic drugs, which must be considered low-potency neuroenhancing agents at this point. Nevertheless, these findings encourage efforts to screen for compounds which produce neuroenhancement *in vitro*, and to develop high-potency analogues as part of a new drug discovery program for brain deficit disorders.

III. Treatment Objectives

What biological activities would be desirable in a new drug aimed at treating brain deficit disorders? As mentioned earlier, these conditions are characterized by functional loss of neurons (reduced cell viability), poor synaptic connections, and impaired energy metabolism. Therefore, a new efficacious drug would ideally provide trophic support for neurons, promote neurite/dendrite outgrowth and stabilization, and optimize glucose metabolism, thus enhancing neuronal function. Previously, we have used the term neuroenhancement to describe this novel treatment strategy (Dwyer *et al.*, 2003b). We argued that the antipsychotic drug, olanzapine, and the vitamin, thiamine, represent low-potency prototypes of a neuroenhancing agent. These agents enhance neuronal function, in part, by stimulating phosphorylation of the serine–threonine kinase, Akt, and up-regulating glucose transport.

A. Neuroprotection and Neurite Outgrowth

Neurotrophins are endogenous trophic factors that protect neurons from various insults and promote neuronal differentiation. NGF is an original member of the neurotrophin family with well-characterized actions on cultured neurons and in the brain (Levi-Montalcini and Angeletti, 1968; Sofroniew *et al.*, 2001). However, it does not pass the blood–brain barrier on its own, which makes delivery to the CNS problematic. Therefore, it would be advantageous to develop small molecule drugs that produce a similar spectrum of biological effects, but that easily reach the brain from systemic circulation. Several neuroprotective drugs have been developed for use in the treatment of ALS, Alzheimer's disease, and Parkinson's disease. Riluzole is approved for ALS and produces a modest extension of life span and an increase in quality of life measures (Bensimon and Doble, 2004; Gordon, 2005). It is presumed to act by inhibiting release of glutamate (McGeer and McGeer, 2005). The neuroprotective agent, xaliproden (Ruigt *et al.*, 1996), was recently evaluated in clinical trials for ALS and the results were generally disappointing (Meininger *et al.*, 2004). Sabeluzole improves memory function in some Alzheimer's patients, although its precise mechanism of action is not known (Feldman *et al.*, 2005; Mohr *et al.*, 1997). Finally, pramipexole, a dopamine agonist used in Parkinson's disease (Tolosa, 2003), was shown to have additional neuroprotective effects *in vitro* (Gu *et al.*, 2004). Therefore, the overall strategy of using small molecule drugs to produce neurotrophic effects has been validated even if the positive clinical benefits of this approach have been modest thus far.

In most brain deficit disorders, there is a simplification of neuronal processes (less branching and fewer neurites) and a reduction in dendrite output. Clearly, these pathological features will have an adverse impact on learning, memory, and cognitive function (Hickmott and Ethell, 2006). It is well established that learning and brain activity are associated with changes in the number and complexity of dendrites on neuronal cells (Greenough *et al.*, 1985; Tailby *et al.*, 2005). Furthermore, exposure to enriched environments increases dendrite output and stabilizes more complex arborization patterns in neurons (Kolb and Whishaw, 1998; Volkmar and Greenough, 1972). In brain deficit disorders, the relative paucity of neurites/dendrites may also contribute to the behavioral symptoms, which include cognitive impairment, alterations in mood, and diminished social interactions. Conversely, therapeutic strategies aimed at increasing neurite/dendrite outgrowth may correct some of these defects (Duman, 2002; Hickmott and Ethell, 2006; Rosoklija *et al.*, 2000). Over the past 15 years, a number of small molecule compounds have been reported to stimulate neurite outgrowth from neuronal cell lines and primary cultures of neurons, or to enhance outgrowth in concert with NGF or other growth factors (Dago *et al.*, 2002a,b; Nozawa *et al.*, 2002; Paul *et al.*, 1990; Riese *et al.*, 2004;

Sanjo *et al.*, 1998). Some of these compounds (e.g., MS-818 and NS 1231) showed protective effects in animal models of nerve damage (Jiang *et al.*, 1995) and ischemic injury (Dago *et al.*, 2002a), but none have progressed through clinical evaluation in patients. The lack of a precise understanding of how these compounds stimulate neurite outgrowth has hindered clinical drug development efforts.

B. Enhancement of Glucose Metabolism

Glucose is required for both neuroprotection (Delgado-Esteban *et al.*, 2000; Dwyer *et al.*, 1999a; Gupta *et al.*, 2001; Moley and Mueckler, 2000) and enhancement of neurite outgrowth (Cohen, 1959; Morelli *et al.*, 1986; Waki *et al.*, 2001). Glucose is needed for several important metabolic processes: (1) to generate energy (ATP) through glycolysis and oxidative phosphorylation; (2) to supply reducing equivalents nicotinamide adenine dinucleotide phosphate (NADPH) and precursor molecules via the pentose phosphate pathway (PPP) for the synthesis of lipids, sterols, and nucleotides; and (3) to serve as a precursor for the synthesis of various neurotransmitters (Ames, 2000; Baquer *et al.*, 1988). The role of glucose in neurite outgrowth and neuronal differentiation was first demonstrated by Cohen (1959) who showed that this sugar was required for the full expression of the response to NGF. Others confirmed that NGF stimulates a dramatic up-regulation of glucose utilization (Morelli *et al.*, 1986; Waki *et al.*, 2001). Conversely, withdrawal of NGF is accompanied by a rapid decline in glucose transport that precedes any loss of cell viability (Deckwerth and Johnson, 1993). As a general rule, growth factors up-regulate glucose transport and metabolism (Berridge and Tan, 1995; Vander Heiden *et al.*, 2001). While some of the increase in glucose uptake during differentiation of neurons is used to provide additional energy, a significant proportion is diverted for metabolism in the PPP (Angeletti *et al.*, 1964; Baquer *et al.*, 1988). This finding is consistent with the increase in nucleotide production (Morelli *et al.*, 1986) and lipid synthesis that accompanies neuronal differentiation (Levi-Montalcini and Angeletti, 1968). Moreover, increased glucose metabolism via the PPP is associated with renewal of neurons in the olfactory bulb (Ninfali *et al.*, 1999) and maintenance of synaptic functioning in neurons (Appel and Parrot, 1970; Biagiotti *et al.*, 2001; Hothersall *et al.*, 1982). The role of the PPP in synaptic function may reflect the requirement of NADPH for neurotransmitter synthesis, metabolism of biogenic amines via monoamine oxidase (MAO), and regulation of the local redox state, in particular reduction of protein sulfhydryls involved in signaling (Baquer *et al.*, 1988; Hothersall *et al.*, 1982).

Glucose that is needed to support neuronal differentiation and function enters cells through glucose transporter (GLUT) proteins residing in the plasma membrane (Maher *et al.*, 1994). The expression pattern and functional aspects of

GLUT expression in brain have been reviewed previously (Dwyer *et al.*, 2002b). Relevant to the present discussion, it is worth noting that the ability of cells to survive harmful injury is closely related to the capacity for glucose uptake and efficient utilization of glucose for energy (Berridge and Tan, 1995; Dwyer *et al.*, 1999a; Goossens *et al.*, 1996; Kan *et al.*, 1994; Moley and Mueckler, 2000). Thus, overexpression of GLUT1 in the brain with viral vectors protected rats against a variety of insults, including ischemia and seizures (Lawrence *et al.*, 1995, 1996). Similarly, neuronal precursor cells were protected against alcohol toxicity if they preferentially used glucose over glutamine to meet their energy needs (Dwyer *et al.*, 1999a). For some of the uses mentioned in this section, only glucose in its original form will suffice. For example, the PPP requires glucose as does the synthesis of nucleotides, and glycoproteins. On the other hand, various byproducts of glucose (e.g., lactate, pyruvate, and glutamine) can be consumed as fuel in addition to the sugar molecule. This is an important distinction as will be argued below.

During development, the mammalian brain shifts from a reliance on the use of ketone bodies for energy in the neonate to essentially exclusive use of glucose in the adult (Sokoloff, 1976). There is often a misconception about what this observation means. The strict reliance on glucose was demonstrated by the finding that glucose could not be replaced as a fuel by other potential substrates in the blood, including lactate and pyruvate (Sokoloff, 1976). However, once in the brain, glucose may be converted to other energy substrates that are readily consumed by neuronal cells. For example, the glucose–lactate shuttle refers to the uptake of glucose by astrocytes, metabolism to lactate, and export of lactate to neurons to support oxidative phosphorylation (Pellerin *et al.*, 1998). To a lesser extent, pyruvate may experience a similar fate. A second shuttle, involving glutamate and glutamine, provides additional substrates for the tricarboxylic acid (TCA) cycle (Hertz and Dienel, 2002). In this case, glutamate is released by neurons during synaptic transmission; it is taken up by astrocytes, which convert it to glutamine via glutamine synthetase. The glutamine is then exported to neurons where it can be reused as glutamate or enter the TCA (as α-ketoglutarate) for energy consumption. Thus, at a given moment a neuron might use glucose, lactate, pyruvate, and glutamine to supply its energy needs. During severe starvation, ketone bodies can be added to this list (Hawkins *et al.*, 1971). The use of lactate, pyruvate, and glutamine as alternative sources of energy is entirely consistent with the strict dependence of adult brain tissue on glucose, so long as these alternative fuels are derived from glucose that crossed the blood–brain barrier (Hertz and Dienel, 2002).

The nature of the regulation of energy metabolism is quite complex and specific. Metabolites generated along a certain pathway often provide feedback control over rate-limiting enzymes in that pathway. Moreover, the cyclical behavior of the TCA and PPP requires that entry of substrates into these cycles at some point along the pathway must be matched by removal of substrates at another point in the cycle. This is referred to as an anaplerotic sequence (Owen *et al.*, 2002). Taken together, these properties mean that conditions in the cell that are optimum for metabolism of glucose will be different from those that are optimum for the use of lactate or glutamine as energy substrates. Thus, if a neuron is to use an alternative fuel for energy, it must adjust its metabolic machinery accordingly. If more glutamine is being used in the TCA, more glucose may be shunted for other purposes besides oxidative phosphorylation, and so forth. In order for the neuron to continuously operate at its maximum level of efficiency and productivity, it needs to maintain a flexible posture with respect to the metabolism of different sources of energy. We refer to this as metabolic flexibility. Neurons with greater metabolic flexibility will sustain higher functional activity and remain viable over wider fluctuations in the availability of energy substrates. There is emerging evidence that drugs and small molecules can shift glucose and energy metabolism to match the supply of fuel (Dwyer *et al.*, 2003a). Enhancement of metabolic flexibility with small molecules is one goal of the drug discovery effort outlined here.

IV. Application to Neurodevelopmental Disorders: Schizophrenia

A. Pathology, Etiology, and Treatment of Schizophrenia

Collectively, the neurodevelopmental disorders reveal a glaring lack of treatment options for correcting or alleviating brain deficits wrought by abnormal development. We will use schizophrenia as an example of a neurodevelopmental disorder that is amenable to the application of compound screening and structure-based design for the discovery of novel therapeutic agents.

Schizophrenia is now widely viewed as a neurodevelopmental disorder with psychiatric symptoms (Feinberg, 1982; Lewis and Levitt, 2002; McGlashan and Hoffman, 2000; Weinberger, 1987). Various lines of evidence support this notion. In comparison to normal subjects, schizophrenics show increased ventricle size (decreased brain size) (McGlashan and Hoffman, 2000; Weinberger, 1987), altered cytoarchitecture in the prefrontal and entorhinal cortex (Arnold *et al.*, 1991; Benes *et al.*, 1986, 1991; Glantz and Lewis, 1997), axon/neurite simplification in postmortem brain sections (Benes *et al.*, 1986; McGlashan and Hoffman, 2000), migration deficits in the olfactory bulb (Arnold *et al.*, 2001; Moberg *et al.*, 1999), and loss of gray matter in childhood-onset schizophrenia (Thompson *et al.*, 2001). In addition, the incidence of schizophrenia is higher in cohorts of subjects who experienced famine *in utero* and in relation to season of birth, which may reflect detrimental effects of infections during pregnancy (Davis and Bracha, 1996; Hulshoff Pol *et al.*, 2000; Lewis and Levitt, 2002). Genetic studies recently identified an assortment of genes that may confer an increased risk of developing schizophrenia (Harrison and Weinberger, 2005; Millar *et al.*, 2005; Ujike and Nakata, 2003). The genes can roughly be classified as to whether they affect neuronal growth and survival (Akt1, calcineurin, and neuregulin), neurite outgrowth (DISC1 and NTF3), or G protein-coupled signaling (RGS4 and PDE4B). These findings are generally consistent with the hypothesis that schizophrenia may originate from growth factor deficiency (Moises *et al.*, 2002).

Currently, schizophrenia is treated with antipsychotic drugs that mainly antagonize the actions of dopamine and serotonin at their receptors (Meltzer *et al.*, 2003; Strange, 2001). The first-generation drugs (e.g., chlorpromazine, haloperidol, and fluphenazine) show higher affinity for dopamine than serotonin receptors, whereas the second-generation drugs (e.g., clozapine, olanzapine, and quetiapine) show the opposite profile. The second-generation drugs were touted as superior to the older drugs when they were first introduced into clinical practice. However, a large, long-term study revealed that, with the possible exception of olanzapine, there was no difference in the efficacy of first- and second-generation antipsychotics (Lieberman *et al.*, 2005a). In addition, this study found that nearly 75% of patients discontinued their medications over the course of the 18-month trial. Of the patients who stopped taking their drugs, nearly 35% cited lack of efficacy as the reason for discontinuing therapy. Many of the patients who are considered responsive show only a modest 20–30% reduction in symptoms according to rating scales, and their relapse rates remain high. These shortcomings of the current antipsychotic drugs should not be surprising because these medications were never designed to treat the underlying neurodevelopmental deficits (e.g., aberrant neurite outgrowth and poor synaptic connectivity) that give rise to schizophrenia. The drugs should be considered symptomatic treatments rather than true therapeutic agents.

Recent findings with olanzapine offer some hope for the future. Neuroimaging studies showed that olanzapine, but not haloperidol, prevented loss of gray matter in schizophrenic patients (Lieberman *et al.*, 2005b; Molina *et al.*, 2004). In addition, this second-generation antipsychotic produced positive effects on cognition (Keefe *et al.*, 2006) and on brain metabolic parameters detected by magnetic resonance spectroscopy (Renshaw *et al.*, 2003). In rats, olanzapine was reported to stimulate neurogenesis in the hippocampus (dentate gyrus) and prefrontal cortex (Kodama *et al.*, 2004; Wakade *et al.*, 2002; Wang *et al.*, 2004), and to increase dendritic markers and the output of dendritic spines (Law *et al.*, 2004). *In vitro*, olanzapine and certain second-generation antipsychotics protected neuronal cells against a variety of insults, including serum/nutrient deprivation and oxidative stress (Bai *et al.*, 2002; Dwyer *et al.*, 2003a; Wei *et al.*, 2003). In addition, olanzapine, quetiapine, and clozapine enhanced neurite outgrowth from PC12 cells in combination with NGF (Lu and Dwyer, 2005). Taken together, these findings provide support for the feasibility of using small molecule drugs to produce positive functional outcomes in schizophrenia. Of course, depending on the nature of the brain deficits and the timing of their expression (e.g., *in utero*), it may be impossible to fully restore neuronal function in a patient by the time psychosis is manifested. Nevertheless, it may be possible to correct some of the enduring defects and to optimize functional activity in affected neurons.

B. Initial Screening of Compound Libraries

With this rationale and the structure of olanzapine as a starting point, our group initiated screens of chemical databases to identify compounds that shared structural features with olanzapine and that would be good candidates for evaluation in phenotypic assays. The chemical structure of olanzapine is shown in Fig. 1 together with fluphenazine, a first-generation antipsychotic, which tends to antagonize the positive effects of olanzapine in our system. Automated searches and visual inspection of chemical databases provided an initial panel

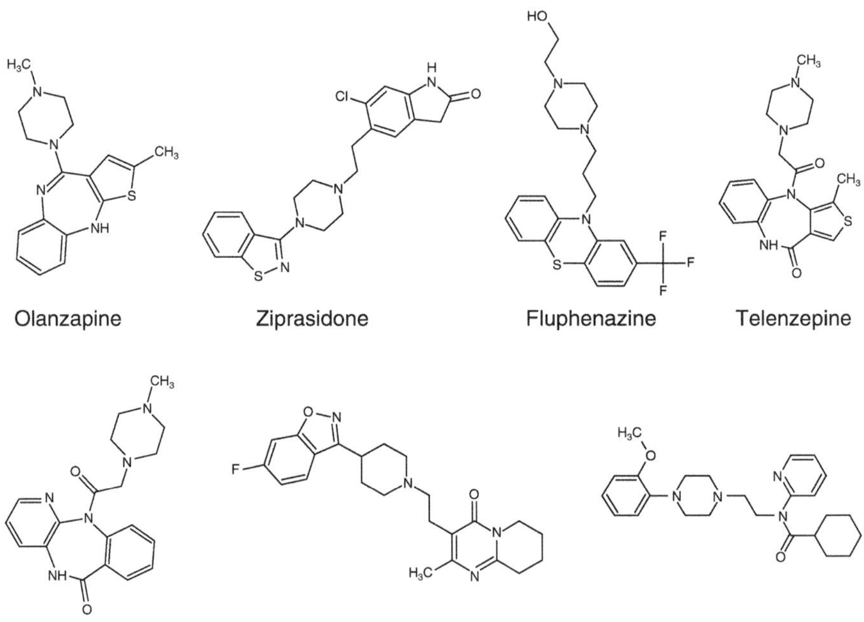

Fig. 1. Chemical structures of olanzapine and related molecules. Olanzapine, ziprasidone, and risperidone are second-generation antipsychotic drugs, whereas fluphenazine is a first-generation drug. Telenzepine and pirenzepine are selective M1 muscarinic receptor antagonists, and the negative compound, WAY-100635, is a 5-HT_{1A} serotonin receptor antagonist.

of compounds for biological screening. Because the positive effects of olanzapine on neuronal viability and neurite outgrowth could not be attributed to the established pharmacology of this drug (i.e., actions at serotonin or dopamine receptors), we could not use receptor-based assays in screening for analogues. Instead, we have used two phenotypic assays for screening: (1) protection against glutamine-free, low serum growth conditions and (2) enhancement of neurite outgrowth in the presence of suboptimum concentrations of NGF.

There are several advantages to phenotypic assays over highly selective screens based on binding to a particular target molecule. Compounds identified in phenotypic screens clearly produce the desired biological outcome, for example enhancement of cell viability or neurite outgrowth. These assays are unbiased, that is, they are not focused on a single receptor or enzyme target. Consequently, it is possible to turn up interesting leads that were unanticipated based on the known pharmacology of the system. With cell-based assays, cytotoxic compounds can be weeded out at an early stage of evaluation. The major disadvantage of phenotypic screens is that the precise mechanism of action is not known, that is, a compound could theoretically protect neuronal cells through a number of different pathways. Our strategy is to rely on phenotypic assays as an initial screen and to then confirm that positive compounds ("hits") from screening act through the relevant pathway (discussed in more detail in Section V.A).

For the initial studies presented here, PC12 cells were switched from normal Dulbecco's modified Eagle's medium (DMEM) with 10% horse (HS) and 5% fetal bovine serum (FBS) to glutamine-free DMEM with 1% HS and 0.5% FBS and were incubated for 3 days in the absence or presence of compounds from chemical libraries. Cell viability was measured by conversion of MTS dye reagent to formazan as described in detail elsewhere (Dwyer *et al.*, 1999a,b, 2003a). Control cells received solvent alone and compounds were tested over a range of concentrations typically from 5 to 160 μM. Under these conditions, olanzapine and quetiapine protect cells against serum and nutrient withdrawal. The data presented in Fig. 2 show that two compounds with a structural resemblance to olanzapine, pirenzepine, and telenzepine (Fig. 1) also provide neuroprotection in our system. Other structurally related compounds (including WAY-100635) were negative in the assay. Thus, we have identified additional neuroprotective compounds on the basis of structural similarity to olanzapine.

Our recent findings that olanzapine, quetiapine, and clozapine enhanced neurite outgrowth in PC12 cells encouraged us to evaluate two other second-generation drugs, ziprasidone and risperidone, for their effects on PC12 cell differentiation. For the studies reported here, we used a variant of the PC12 cell line derived in our laboratory. These cells were isolated on the basis of their greater adherence to plastic surfaces in the absence of any adhesion substrata such as poly-L-lysine. The variant, referred to as PC12.XL, rapidly responds to NGF and exhibits a fully differentiated phenotype in as little as 2–3 days after

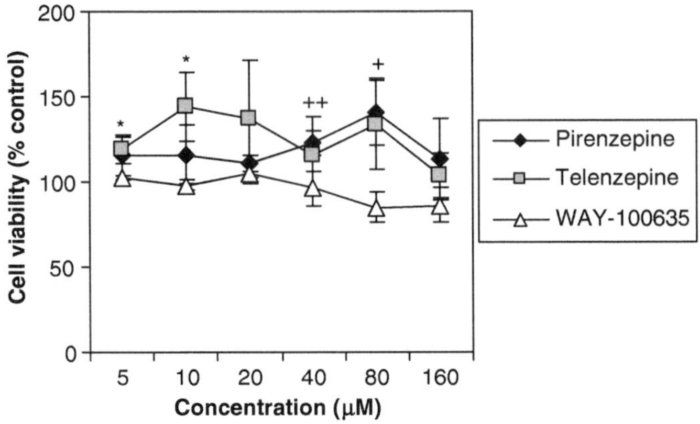

FIG. 2. Protection against low serum, glutamine-free growth conditions. PC12 cells were incubated for 3 days in solvent (DMSO; controls) or compounds at the concentrations indicated. Cell viability was measured and the data are expressed as the percentage of the control values. Significant differences from control are indicated by asterisks for telenzepine ($*p < 0.05$) or plus symbols for pirenzepine ($^{+}p < 0.05$, $^{++}p < 0.01$). WAY-100635 represents a structurally similar compound that showed no positive effects in the cell viability assay. Other muscarinic antagonists were negative in the cell viability assay (not shown).

addition of NGF. The effects of the drugs in the presence of suboptimum concentrations of NGF were quantified as described previously (Lu and Dwyer, 2005). We analyzed the number of cells (in six randomly chosen fields) that responded (i.e., sprouted neurites longer than one cell diameter), and the average length of the longest process extended. As seen in Fig. 3, ziprasidone produced significant enhancement of neurite outgrowth (both the number of responding cells and the average neurite length), whereas risperidone did not. Structurally, ziprasidone displays greater similarity to olanzapine than risperidone (Fig. 1), which suggests there might be a common mechanism of action with defined structural requirements. Furthermore, these data provide additional support for the notion that the second-generation drugs produce positive effects on neuronal growth and viability through mechanisms other than dopamine and serotonin receptors because the established pharmacology does not explain our results.

C. Role of Akt in Schizophrenia and the Actions of Antipsychotic Drugs

Akt [also known as protein kinase B (PKB)] is a serine–threonine kinase that is critical for the regulation of cell differentiation, growth, metbabolism, and survival (Hajduch *et al.*, 2001; Lawlor and Alessi, 2001; Whiteman *et al.*, 2002).

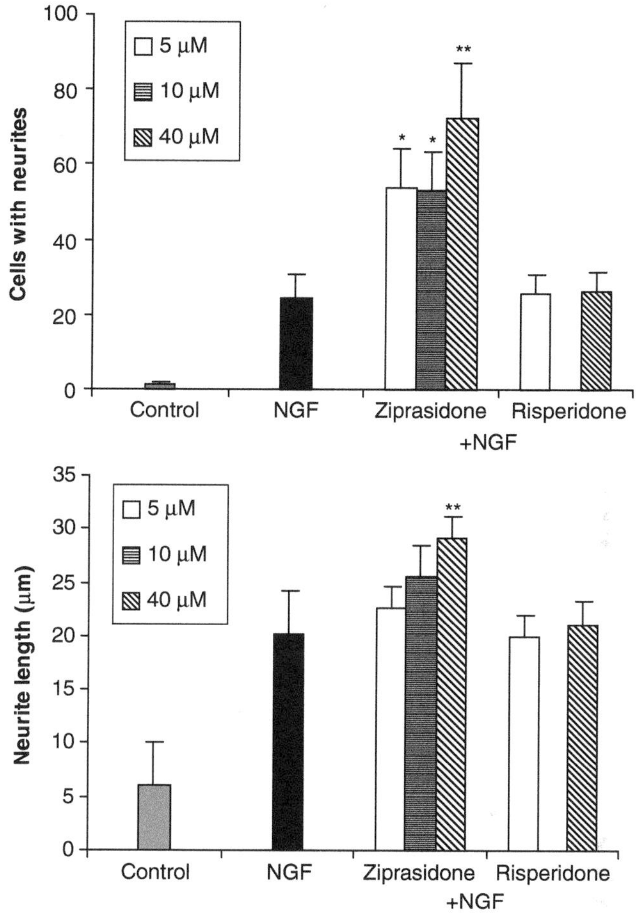

FIG. 3. Effect of ziprasidone and risperidone on neurite outgrowth in PC12 cells. PC12.XL cells were grown in the presence of solvent alone (control), suboptimum concentrations of NGF (20 ng/ml) (NGF), or NGF plus ziprasidone or risperidone at the concentrations indicated for 3 days. Photographs were taken and neurite outgrowth was quantified as described previously (Lu and Dwyer, 2005). Significant differences from the control group are indicated by asterisks: $*p < 0.05$, $**p < 0.01$.

The Akt pathway comprises a complex array of signaling molecules. Upstream of Akt is phosphatidylinositol 3-kinase (PI3K) (Vanhaesebroeck *et al.*, 2001), and important downstream targets include glycogen synthase kinase 3β (GSK3β), the tuberous sclerosis complex protein (TSC2), forkhead transcription factors, Raf1, Bad, and Mdm2, to name a few (Brazil *et al.*, 2002; Lawlor and Alessi, 2001; Plas and Thompson, 2005). This pathway coordinates a variety of cellular processes, including nutrient (glucose and amino acids) transport and protein synthesis, to foster cell growth and cell division. The fact that Akt stimulates both cell division and differentiation appears paradoxical at first. However, cellular differentiation typically requires one last round of cell division in order to initiate specific genetic programs. Thus, Akt may function in both capacities.

For the present discussion, the role of Akt in the regulation of glucose metabolism and cell size is particularly relevant. Akt is best known for its involvement in insulin signaling, specifically up-regulation of GLUTs at the cell surface in insulin-responsive tissues (Hajduch *et al.*, 2001; Lawlor and Alessi, 2001; Whiteman *et al.*, 2002). This produces an increase in glucose disposal and a reduction in blood glucose levels. However, Akt also regulates several other aspects of glucose metabolism. Akt signaling increases association of hexokinase with mitochondria as part of a protective, anti-apoptotic response (Gottlob *et al.*, 2001). In addition, it was recently reported that Akt orchestrates conversion of cellular energy metabolism from oxidative phosphorylation to aerobic glycolysis (Plas and Thompson, 2005). Moreover, Akt stimulates expression of glucose-6-phosphate dehydrogenase, the rate-limiting step in the PPP (Wagle *et al.*, 1998). Recent findings with Akt knockout mice confirm that this signaling pathway exerts significant control over glucose metabolism. In particular, mice lacking the Akt2 isoform are insulin-resistant and develop a type 2 diabetes-like syndrome (Cho *et al.*, 2001a). There are disturbances in glucose tolerance and hepatic glucose output in these mice. By contrast, Akt1 knockout mice maintain normal glucose tolerance and insulin responsiveness, but are significantly smaller in size than wild-type counterparts (Chen *et al.*, 2001; Cho *et al.*, 2001b). Interestingly, Akt3 null mice have normal glucose metabolism and overall body size; however, their brains are selectively decreased in size by 20% compared to controls (Easton *et al.*, 2005). Brain cells are smaller in the Akt3 knockout mice and are fewer in number. The TSC1/2-mTOR pathway appears to be involved in these brain-selective effects.

Previous studies by our group revealed that Akt plays a significant role in the neuroprotective effects of olanzapine (Lu *et al.*, 2004) and in neurite outgrowth elicited by the combination of NGF and second-generation antipsychotic (Lu and Dwyer, 2005). In early 2003, we (Dwyer *et al.*, 2003b) proposed that Akt may be a fruitful target for drug discovery in schizophrenia. We suggested that activation of Akt with therapeutic agents may enhance glucose metabolism in cells thus providing neuroprotection and maximizing neuronal function. Subsequently, Emamian *et al.* (2004) identified Akt as a risk factor for schizophrenia. They found decreased levels of Akt1 in the brains and peripheral blood lymphocytes of schizophrenic patients in comparison to control subjects. Moreover, they identified a particular genetic haplotype consisting of single nucleotide polymorphisms (SNPs) that was associated with schizophrenia (about a twofold increase of the haplotype in schizophrenics). We reported that olanzapine, quetiapine, and clozapine stimulated Akt phosphorylation (Lu and Dwyer, 2005; Lu *et al.*, 2004) and that inhibition of Akt with a selective inhibitor abolished the neuroprotective effects of olanzapine. Similarly, the positive effects of certain second-generation antipsychotics on neurite outgrowth were blocked by inhibition of PI3K (Lu and Dwyer, 2005). Thus, the PI3K-Akt signaling pathway appears to be highly relevant for both the neuroenhancing effects of antipsychotic drugs and possibly the pathology of schizophrenia.

On the basis of these findings with Akt, we evaluated "hits" from screening for their ability to stimulate Akt phosphorylation to ensure that the compounds worked through the same molecular mechanism. This information is important for structure-activity relationship (SAR) analysis in support of structure-based drug design (Section V.B). For these studies, PC12 cells were incubated overnight in serum-free medium to reduce background (growth factor-induced) levels of phosphorylated Akt (pAkt). The cells were grown in 96-well tissue culture plates treated with 100 μg/ml of poly-L-lysine to promote tight adherence. Phosphorylated and total Akt were then detected with specific antisera in the fast-activated cell-based ELISA (FACE) from Active Motif (Carlsbad, CA). Drugs or compounds were added for 20–30 min prior to rapid and direct fixation of cells in the wells with an 8% formaldehyde solution. The results of the FACE assays for a select group of compounds are shown in Fig. 4. Several of the compounds are referred to here by the numbers assigned during screening due to proprietary nature of the structural information at this point. The main positive control for this assay was NGF, which elicited a six- to tenfold induction of pAkt. Olanzapine and quetiapine stimulated Akt phosphorylation as demonstrated in previous reports based on Western blot analysis or ELISAs on cell extracts (Lu and Dwyer, 2005; Lu *et al.*, 2004). The neuroprotective compounds, LSU-14, LSU-31, LSU-36, and LSU-40, also produced activation of Akt in the FACE assay. Therefore, compounds selected for testing on the basis of their structural similarities to olanzapine proved to be neuroprotective and to activate Akt similarly to the prototype drug.

The neuroprotective drugs, olanzapine and quetiapine, promoted neurite outgrowth from PC12 cells in combination with NGF. Therefore, we examined whether the hits from neuroprotection screens enhanced neurite outgrowth in our system. For these experiments, PC12.XL cells were employed. As seen in Table I, several compounds from the screen enhanced neurite outgrowth in the presence of suboptimum concentrations of NGF as anticipated. Although the precise mechanism of action is not known, we showed previously that pertussis toxin (a selective modulator of $G_{i/o}$ proteins) and inhibitors of PI3K abolished the effects of second-generation antipsychotics on neurite outgrowth. Therefore, a signaling pathway involving G proteins, PI3K, and Akt may contribute to the positive effects of the drugs in our phenotypic assays. The results are very encouraging thus far because they suggest that small molecular weight compounds

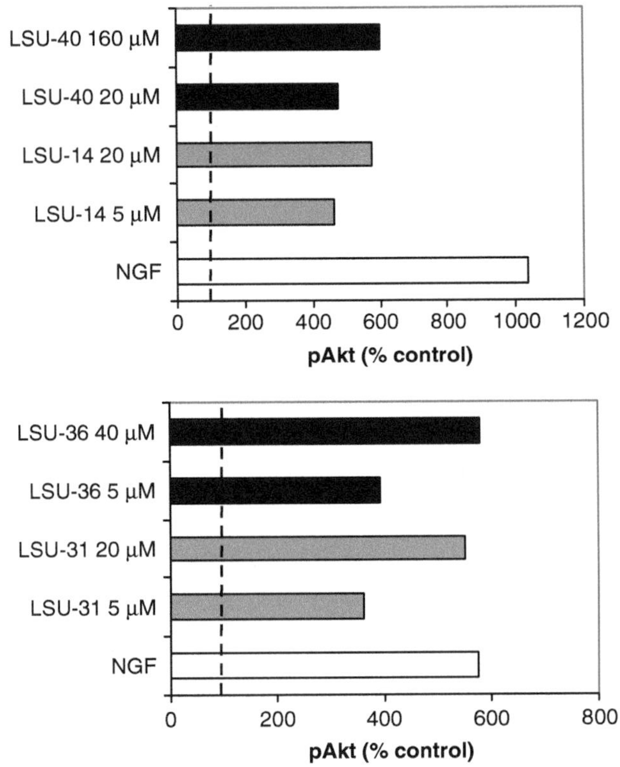

Fig. 4. Stimulation of Akt phosphorylation by compounds from screening. PC12.XL cells were cultured for 20 hours in serum-free medium in wells treated with poly-L-lysine. Compounds or NGFs (100 ng/ml) were added for 20 min and the plates were rapidly fixed by exchanging medium with 8% formaldehyde solution. The FACE assay was performed using antibodies specific for the phosphorylated form of Akt (Ser473). The values from the assay were normalized to total cell numbers derived from crystal violet staining of the wells and the data are expressed as the percentage of control cells. The dotted line represents the control values. The assays were performed in duplicate.

can effectively protect neuronal cells from harmful insults and enhance neuronal functionality by promoting expression of neurites.

D. Possible Role of G Proteins

How do the drugs/compounds activate PI3K-Akt? Published data from our group suggest that at some level G proteins may be involved in the overall signaling pathway (Lu and Dwyer, 2005; Lu *et al.*, 2004). A role for G proteins in the neuroenhancement responses is quite intriguing in view of the fact that two

TABLE I
EFFECTS OF COMPOUNDS FROM SCREENING ON NEURITE OUTGROWTH[a]

		Number of cells with neurites (±S.D.)	Average length of process (μm) (±S.D.)
Control	NGF	5.8 (2.5)	29.8 (5.3)
LSU-14	5 μM	14.0 (4.3)**	31.8 (4.0)
	80 μM	13.7 (4.8)**	33.4 (7.8)
Control	NGF	17.7 (3.7)	28.9 (5.1)
LSU-36	80 μM	29.5 (6.5)**	28.7 (1.9)
Control	NGF	19.0 (6.5)	31.3 (3.7)
LSU-145	80 μM	38.2 (7.4)**	37.0 (1.7)*
	160 μM	46.8 (13.9)*	40.5 (2.9)**

[a]PC12.XL cells were incubated with NGF (20 ng/ml) or NGF plus compounds from screening at the concentrations indicated. After 3 days, neurite outgrowth was quantified as described previously (Lu and Dwyer, 2005). The data represent averages (± the standard deviation, S.D.) from at least six randomly chosen fields. Significant differences from control are indicated by asterisks: *$p < 0.05$, **$p < 0.01$.

genetic risk factors for schizophrenia code for G protein-related molecules, namely RGS4 (Chowdari *et al.*, 2002; Mirnics *et al.*, 2001) and PDE4B (Millar *et al.*, 2005). RGS4 is a direct regulator of G protein signaling and binds to $G_{i/o}$ and $G_q\alpha$ subunits (Hollinger and Hepler, 2002). PDE4B is a phosphodiesterase that hydrolyzes cAMP, a product of G protein signaling. This enzyme interacts with another protein implicated as a risk factor for schizophrenia, namely disrupted in schizophrenia 1 (DISC1) (Harrison and Weinberger, 2005; Millar *et al.*, 2000). Knockout mice which lack various $G_i\alpha$ subunits develop normally, although $G_i\alpha_2$ knockout animals develop a type 2 diabetes-like condition (Moxham and Malbon, 1996) and reduced growth in certain tissues (Moxham *et al.*, 1993). Transgenic mice deficient in $G_o\alpha$ subunits showed no gross abnormalities in brain tissue, although they did experience higher postnatal mortality rates and exhibited neurological disturbances, including seizures (Offermanns, 1999). The $G_{i/o}$ proteins implicated in our studies on the actions of the antipsychotic drugs are clearly involved in the maintenance of optimum brain function. In terms of the PI3K-Akt signaling pathway, it is well established that $G\beta\gamma$ subunits directly activate PI3K on dissociation from the $G\alpha$ subunit (Clapham and Neer, 1997; Morris *et al.*, 1995; Zhang *et al.*, 1995). These findings reveal how G protein signaling might mediate the downstream activation of Akt.

The question remains as to how the drugs/compounds activate G protein signaling. Several possibilities can be entertained. (1) Theoretically, the drugs and compounds could bind to G protein-coupled receptors (GPCRs) to activate G proteins. The antipsychotic drugs bind to several different GPCRs; however, they typically antagonize actions of neurotransmitters at these receptors.

Furthermore, we found no evidence that a particular GPCR adequately explained the effects of these drugs in our system. (2) The drugs/compounds may directly activate G proteins by binding at critical G protein regulatory interfaces. A number of such sites are known and include the GPCR-G protein interface, the RGS-G protein site, and other protein–protein interaction sites (Ghosh *et al.*, 2003; Higashijima *et al.*, 1990; Kimple *et al.*, 2001; Taylor *et al.*, 1994). Helical peptides, such as mastoparan and melittin, directly activate G proteins through interactions with the Gα subunit (Higashijima and Ross, 1991). Moreover, small molecular weight drugs and compounds have been reported to bind to G proteins and cause activation of G protein signaling (Burde *et al.*, 1996; Hageluken *et al.*, 1994; Odagaki *et al.*, 1998). Therefore, direct actions of olanzapine and other similar molecules on G proteins must be considered a possibility. (3) The drugs/compounds may indirectly activate G proteins through transactivation. Binding of growth factors to receptor tyrosine kinases (RTKs) is known to produce secondary activation of G proteins, this process is called transactivation (Waters *et al.*, 2004). Perhaps, olanzapine and the hits from screening primarily stimulate activation of RTKs and indirectly activate G protein signaling. Either possibility (2) or (3) might explain the unusual pharmacology of the response to olanzapine and quetiapine, including the requirement for much higher concentrations of drug for neuroprotection (40–80 μM) than for binding to dopamine and serotonin receptors (in the nanomolar range).

E. Brain Glucose Metabolism and Metabolic Flexibility

This section will examine evidence for impaired brain glucose metabolism in schizophrenia and present arguments for maximizing metabolic flexibility as part of an overall therapeutic strategy. In fact, several groups have suggested that a disturbance in glucose/energy metabolism contributes to the pathology in psychotic illnesses (Ben-Shachar, 2002; Blass, 2002; Dwyer *et al.*, 2001). Fascinating connections between schizophrenia (psychoses more broadly speaking) and aberrant glucose metabolism have been noted over the years dating back as far as the late 1800s. Maudsley observed that diabetes occurred more frequently in families with a history of insanity than in normal individuals (as cited by Mukherjee *et al.*, 1989). In the 1930s, insulin shock therapy was used to treat severe psychoses (Sakel, 1938). It was found that psychotic patients required a much higher dose of insulin to induce a coma than nonpsychotic subjects, which indicated a degree of insulin resistance in the psychotic population. Some studies suggested an increased incidence of insulin resistance and diabetes in unmedicated schizophrenic patients (Braceland *et al.*, 1945; Freeman *et al.*, 1943; Waitzkin, 1966), although more recent studies failed to reach this same conclusion (Arranz *et al.*, 2004). A number of groups using brain-imaging techniques reported reduced glucose utilization in different brain regions, especially the frontal cortex of schizophrenic patients (Buchsbaum *et al.*, 1990; Ingvar and Franzen, 1974; Weinberger *et al.*, 1986; Wolkin *et al.*, 1985). This decrease in regional glucose metabolism was referred to as hypofrontality (Ingvar and Franzen, 1974). There are reports of normalization of glucose metabolism in some brain regions with successful treatment of schizophrenia/psychosis, although this remains controversial (Bartlett *et al.*, 1998; Buchsbaum *et al.*, 1992). Blass (2002) and others (Mauer *et al.*, 2001) observed decreased mitochondrial activity (specifically oxidative phosphorylation) in schizophrenic brains as compared with controls. Ben-Shachar (2002) proposed that schizophrenia arises from mitochondrial dysfunction that is modulated by dopaminergic activity. It is possible that mitochondrial deficits explain the hypofrontality noted in imaging studies of schizophrenics. In a related vein, Middleton *et al.* (2002) performed cDNA array profiling of the prefrontal cortex of postmortem brain tissue from schizophrenic patients in comparison to matched control subjects. They found significant reductions in key metabolic genes from the mitochondrial malate shuttle system and the TCA.

Additional connections between schizophrenia and brain glucose metabolism (as compared to overall energy metabolism) are illustrated by several findings. Thiamine is a critical cofactor in pyruvate dehydrogenase, an important enzyme of the glycolytic pathway (Cooper and Pincus, 1979). Severe thiamine deficiency is associated with Korsakoff's psychosis (Feinberg, 1980). Moreover, chronically ill psychotic patients were successfully treated with thiamine supplementation in the 1970s (Sacks *et al.*, 1989). Furthermore, it was reported that administration of glucose to schizophrenic patients produced significant improvements in memory and cognition (Newcomer *et al.*, 1999). Taken together, these findings suggest that suboptimum brain glucose metabolism in schizophrenia can be partially corrected by therapeutic intervention.

Antipsychotic drugs produce complex effects related to glucose metabolism in neuronal cells (Dwyer *et al.*, 1999b, 2002a, 2003a). Many antipsychotics, both first and second generation, inhibit glucose transport into neurons and other cell types (Ardizzone *et al.*, 2001; Dwyer *et al.*, 1999b). They appear to act by directly blocking GLUT proteins at an intracellular site (Ardizzone *et al.*, 2001; Dwyer *et al.*, 2002a). Following 24-hour incubation with the drugs, there is a significant increase in the cellular expression of GLUT1 and GLUT3 proteins, perhaps as an adaptation to extended glucose deprivation (Dwyer *et al.*, 1999b, 2003a). In the case of olanzapine, but not fluphenazine or clozapine, 6- to 12-hour incubation with drug is accompanied by an increase in GLUT expression at the cell surface and an increase in glucose uptake as a result of this presumed adaptive response (Dwyer *et al.*, 2003a). In mice, the antipsychotic drugs induce acute hyperglycemia in close correspondence to their ability to inhibit glucose transport in cell lines (Dwyer and Donohoe, 2003). This is consistent with observations by various groups that antipsychotic medications, especially second-generation drugs, produce alterations in glucose tolerance in patients, including diabetes in some (Henderson and Ettinger, 2002; Newcomer *et al.*, 2002). Thus, at various levels there are functional connections between glucose metabolism, schizophrenia, and the drugs used to treat this disorder.

The concept of metabolic flexibility introduced in Section III.B grew out of two sets of observations. First, neuronal cells at any given moment may use glucose, lactate, glutamine, and pyruvate for energy. Depending on which fuel source is dominant, cellular metabolic pathways must be adjusted to maintain a balance in the supply of energy substrates and the disposal of metabolic by-products, and to provide sufficient ATP where it is needed. Second, recent work from our group demonstrates that the cellular response to glucose versus glutamine deprivation is quite distinct. In other words, the optimum conditions for consumption of glucose as the primary energy source are very different from optimum conditions for utilization of alternative TCA cycle components, namely glutamine. Figure 5 depicts a minimal energy circuit in neuronal cells that reflects utilization of several different sources of energy: glucose, lactate, and glutamine. Glucose is also needed for other purposes, including the biosynthesis of neurotransmitters, nucleotides, and fatty acids. When glucose is diverted for biosynthetic purposes, alternative energy sources, such as glutamine and lactate, must be tapped for oxidative phosphorylation. Some have proposed that lactate is the major energy source for neurons and is provided by astrocytes as part of a lactate shuttle (Pellerin *et al.*, 1998). However, lactate will only be produced in significant concentrations when oxygen is limiting or when the rate of glycolysis surpasses the rate of entry of substrates into the TCA cycle. If lactate was used as the primary energy source of neurons, the accumulation of glucose-6-phosphate would lead to inhibition of hexokinase and disruption of ATP efflux from the mitochondria. Consequently, there must be coordinated regulation of the use of different energy substrates in the cell.

Glutamine from astrocyctes clearly provides a supplemental source of fuel for neurons (Hertz and Dienel, 2002). However, to enter the TCA cycle, glutamine must be converted to α-ketoglutarate. Under normal conditions, this is accomplished via a transamination reaction coupled to the conversion of pyruvate to alanine (Moreadith and Lehninger, 1984). Therefore, the use of glutamine as an energy source normally requires an adequate supply of pyruvate, which could be derived from glucose (via glycolysis) or lactate. In glucose deficiency and in synaptosomes, deamination of glutamate by glutamate dehydrogenase plays a greater role in the enzymatic production of α-ketoglutarate (McKenna *et al.*, 2000). Depending on which route is used, there may be increased production of ammonia and shifts in metabolites and oxidation–reduction states that may have adverse consequences for the cell (Honegger *et al.*, 2002). It is important to note that cells can switch between the various energy sources in relation to

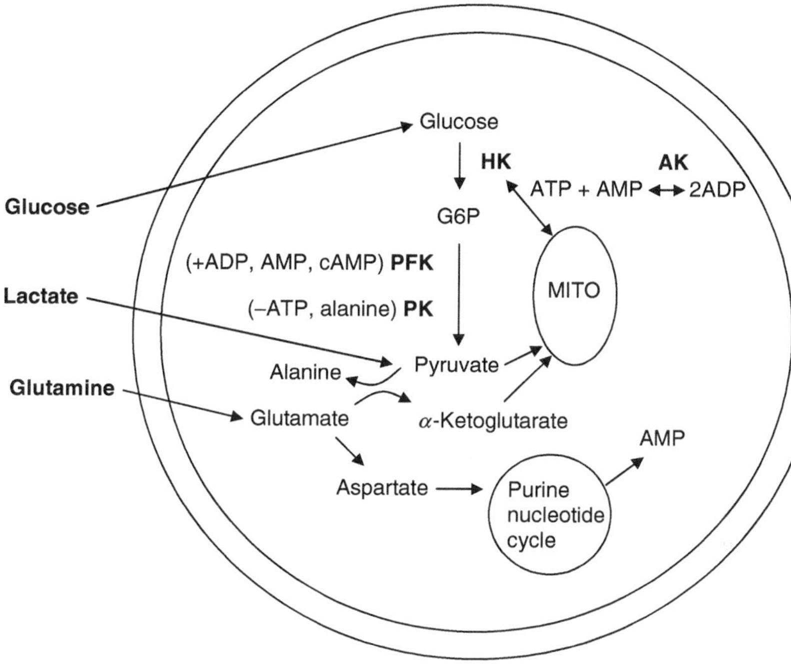

FIG. 5. Interconnections between major metabolic pathways in neurons. Several potential energy substrates are listed here together with partial pathways for their metabolism and the metabolites produced for oxidative phosphorylation in the mitochondria (MITO). Glucose is phosphorylated by hexokinase (HK) to produce glucose-6-phosphate (G6P), which may undergo glycolysis or metabolism in the PPP (not shown). Pyruvate derived from glycolysis (or lactate) may enter the TCA after conversion to acetyl-CoA or may undergo transamination (to alanine) coupled to glutamate conversion to α-ketoglutarate. Alternatively, glutamate may be used (after conversion to aspartate) in the purine nucleotide cycle. This cycle links energy utilization and the availability of substrates to the production of the adenine nucleotide AMP, which indirectly affects the reserves of ATP and ADP. Finally, elements of the phosphotransfer system, adenylate kinase (AK), phosphofructokinase (PFK), and pyruvate kinase (PK) are shown here. Nucleotidase, adenosine kinase, and phosphodiesterase (not shown) also contribute to the relative ratios of energy currency (ATP, ADP, and so on) in the cell.

the supplies available. Nevertheless, certain metabolic states (e.g., greater relative flux of glucose directly into the TCA) may support higher functional activity in neurons in comparison to other conditions (e.g., increased utilization of lactate for energy).

In order to achieve effective coordination of energy metabolism in the cell, there must be continuous renewal of various adenine nucleotide pools via phosphotransfer reactions (Fig. 5). The relative ratio of ATP to ADP and AMP determines the energy status and regulates metabolic enzymes, signaling pathways, and functional activity of cells (Ataullakhanov and Vitvitsky, 2002;

Brown, 1992). In addition, ATP (and the other adenine nucleotides for that matter) must be shuttled to their appropriate locations in the cell. Since diffusion is a very inefficient system for allocating ATP, ADP, and AMP resources, phosphotransfer systems have evolved for more precise delivery in cells (Dzeja *et al.*, 1998) and for buffering (Ames, 2000). ATP:ADP/AMP ratios and phosphotransfer pathways are maintained in neurons by adenylate kinase, nucleoside diphosphate kinase (NDPK), pyruvate kinase, phosphofructokinase (PFK), and to a lesser extent by creatine kinase (Ames, 2000; Dzeja *et al.*, 1998; Rabinovitz, 1992). Adenylate kinase is up-regulated during neuronal differentiation and protein levels increase in the brain in relation to synaptic activity (Inouye *et al.*, 1998, 1999). Studies have implicated NDPK in neuronal growth and differentiation (Ishijima *et al.*, 1999). NDPK is also involved in the maintenance of synaptic activity (Krishnan *et al.*, 2001). Thus, the biological activities impacted by the phosphotransfer system in neurons coincide with the cellular processes that we hope to affect through neuroenhancement. Data presented in the next section provide initial support for the possibility that small molecular weight compounds can be used to optimize metabolic flexibility via the mechanisms outlined in Fig. 5.

F. Protection Against Glucose and Glutamine Deprivation

When cultured cell lines are deprived of glucose, they typically undergo cell death over a 48- to 72-hour time period. Cells grown in glucose-free medium slow their rate of cell division and some cells die within 72 hours, depending on the cell line and its metabolic properties. In parallel with our phenotypic screens, we evaluated certain compounds for neuroprotection based on previous work. For example, it is known that cAMP protects against growth factor withdrawal and ischemic conditions (Boniece and Wagner, 1993). Therefore, cAMP, in the form of dibutyryl cAMP and similar analogues, was tested for its ability to protect PC12 cells against glucose deprivation. As seen in Table II, dibutyryl cAMP and to a much lesser degree, dioctanoyl cAMP, protected cells against total glucose deprivation. It was surprising to learn that these effects did not appear to be mediated via activation of PKA. This conclusion was based on the fact that PKA inhibitors (H-89 and KT5720) failed to abolish the response to dibutyryl cAMP. Furthermore, a chlorophenylthio analogue of cAMP (CPT-cAMP) did not protect against glucose deprivation, although this compound activates PKA. cAMP activates additional signaling proteins besides PKA, including Epac (exchange protein directly activated by cAMP) (de Rooij *et al.*, 1998), which is also known as cAMP-regulated guanine nucleotide exchange factor (cAMP-GEF) (Kawasaki *et al.*, 1998). CPT-cAMP activates Epac yet it did not protect against glucose deprivation like dibutyryl cAMP, and a more selective Epac activator, a methylated form of CPT (CPT-Me-cAMP) was also negative in the protection assay (Table II).

TABLE II

SMALL MOLECULE PROTECTION AGAINST GLUCOSE DEPRIVATION[a]

Compound	Concentration (μM)	Inhibitor	Cell viability (percentage of control ± S.E.)
Dibutyryl cAMP	62.5	–	146.6 ± 12.5
	250	–	197.5 ± 17.1
	500	–	246.8 ± 29.0
	250	H-89 (500 nM)	264.0 ± 29.4
	500	H-89 (500 nM)	320.6 ± 92.4
	250	KT5720 (500 nM)	211.1 ± 12.8
	500	KT5720 (500 nM)	236.8 ± 43.1
	250	Dipyridamole (80 μM)	137.6 ± 17.9
	500	Dipyridamole (80 μM)	115.6 ± 11.5
Dioctanoyl cAMP	62.5	–	130.1 ± 12.9
	250	–	172.5 ± 15.9
	500	–	122.0 ± 13.0
CPT-cAMP	15	–	79.3 ± 4.8
	62.5	–	85.7 ± 6.0
	125	–	89.5 ± 4.6
CPT-Me-cAMP	10	–	99.6 ± 22.6
	40	–	107.7 ± 8.9
	80	–	119.0 ± 13.8
AICAR	125	–	215.5 ± 27.8
	250	–	218.8 ± 30.9
	500	–	237.0 ± 24.1
	250	5-Iodotubercidin (500 nM)	103.8 ± 4.9
	500	5-Iodotubercidin (500 nM)	101.7 ± 4.6
	250	Dipyridamole (80 μM)	135.5 ± 17.5
	500	Dipyridamole (80 μM)	145.0 ± 24.1

[a]PC12 cells were incubated in glucose-free medium with normal concentrations of serum (10% horse serum, 5% fetal bovine serum) that had been dialyzed to remove glucose. The various compounds listed here were added alone or in combination at the concentrations indicated. Cell viability was measured after 2–3 days and was compared to control conditions (solvent alone). The averages from at least three separate experiments are shown here.

Although it is not widely appreciated, cAMP, including the dibutyryl form, can be converted to AMP in cell cultures (Braumann *et al.*, 1986). An increase in cellular AMP will shift the ATP:AMP ratio and potentially activate 5′-AMP kinase (AMPK) (Carling, 2004; Hardie *et al.*, 2003; Rutter *et al.*, 2003). AMPK plays an essential role in the regulation of energy metabolism through the coordination of various biological activities. The overall objective of the processes regulated by AMPK is to increase the level of ATP in the cell by generating more ATP through glycolysis and fatty acid oxidation, and by conserving ATP through inhibition of anabolic processes and ion flux (Carling, 2004; Hardie *et al.*, 2003;

Rutter *et al.*, 2003). Therefore, we evaluated whether an established activator of AMPK, 5-aminoimidazole-4-carboxamide ribonucleoside (AICAR), mimicked the protective effects of cAMP against glucose deprivation. AICAR enters cells via adenosine transporters and is rapidly phosphorylated by adenosine kinase to produce ZMP, an analogue of AMP. In our system, AICAR effectively mimicked the actions of dibutyryl cAMP in the glucose deprivation assay, and its protective effects were abolished by inhibition of the adenosine transporter with dipyridamole or adenosine kinase with iodotubercidin (Table II). Therefore, we propose that the PKA-independent protection against glucose deprivation rendered by cAMP is due to activation of AMPK. AMPK-mediated protection of neuronal cells against glucose deprivation is consistent with the data of Culmsee *et al.* (2001). Studies that are underway will examine possible molecular mechanisms in greater detail.

cAMP also protects neuronal cells against glutamine deprivation. For these studies, PC12 cells were grown in glutamine-free medium in the absence or presence of dibutyryl cAMP and other compounds as before. As seen in Table III, dibutyryl cAMP protected cells against glutamine withdrawal and this effect was not diminished by inhibition of PKA with H-89 and KT5720. AICAR did not protect the cells against glutamine deprivation (Table III), which argues against a role for AMPK. Furthermore, the selective activator of Epac, CPT-Me-cAMP, failed to produce significant protection in the assay. Therefore, it appears that cAMP protects cells against glucose and glutamine deprivation through two distinct pathways. Protection against glucose deprivation is mediated via AMPK, whereas protection against glutamine deprivation is not mediated through any of the known signaling pathways involving cAMP.

The data presented here suggest that cellular levels of AMP-related molecules determine the direction of metabolic signaling, and the tone of phosphotransfer pathways, which include adenylate kinase and PFK. This conclusion emerged from our discovery that addition of adenosine kinase inhibitors, iodotubercidin, and 5-amino-5′-deoxyadenosine (AMDA) provided neuroprotection against glutamine deprivation (Table III), but not glucose deprivation. Phosphotransfer systems in the cell are involved in optimization of the production and utilization of high energy phosphates such as ATP and creatine-phosphate. Our data suggest that adenosine kinase plays an underappreciated role in maintaining oxidative phosphorylation when cells use significant amounts of glutamine for energy (Bontemps *et al.*, 1983). Glutamine metabolism is tightly linked to the function of the purine nucleotide cycle (Kovacevic and McGivan, 1983). This cycle regulates the levels of adenine nucleotides, generates ammonia, provides intermediates for the TCA, supplies certain amino acids for energy, and regulates PFK and glycolysis (Lowenstein, 1972). When cells are grown in glutamine-free medium, the addition of cAMP may modulate the operation of

TABLE III
SMALL MOLECULE PROTECTION AGAINST GLUTAMINE DEPRIVATION[a]

Compound	Concentration (μM)	Inhibitor	Cell viability (percentage of control ± S.E.)
Dibutyryl cAMP	62.5	–	158.4 ± 21.7
	250	–	219.6 ± 16.2
	500	–	230.6 ± 23.2
	250	H-89 (500 nM)	216.0 ± 30.6
	500	H-89 (500 nM)	257.1 ± 52.4
	250	KT5270 (500 nM)	188.0 ± 17.7
	500	KT5270 (500 nM)	196.0 ± 24.8
CPT-cAMP	12.5	–	212.2 ± 20.4
	50	–	295.0 ± 49.0
	100	–	275.0 ± 43.1
CPT-Me-cAMP	25	–	98.7 ± 4.4
	100	–	93.0 ± 4.0
	200	–	93.5 ± 1.5
AICAR	125	–	110.0 ± 8.8
	250	–	95.0 ± 6.5
	500	–	111.6 ± 16.4
5-Iodotubercidin	0.06	–	281.0 ± 64.6
	0.25	–	290.8 ± 64.1
	1.0	–	204.0 ± 63.3
5-Amino-5-deoxyadenosine	5.0	–	195.0 ± 52.8
	20	–	212.3 ± 31.4
	80	–	198.0 ± 55.1

[a]PC12 cells were grown in low serum, glutamine-free medium for 3 days with the compounds listed here and were assayed for cell viability as described above.

the purine nucleotide cycle and shunt pyruvate derived from glucose metabolism into the TCA to replace the glutamine (α-ketoglutarate) that was previously being used for energy and adenylate turnover. In addition, the cAMP may activate PFK to enhance the coupling between glycolysis and oxidative phosphorylation. The purine nucleotide cycle is active in the brain, and deficiencies in enzymes from this cycle are associated with autism, seizures, and psychomotor retardation (Van den Berghe *et al.*, 1992). Taken together, the data suggest a complex role for cAMP and adenine nucleotides in the coordinated regulation of neuronal energy metabolism and survival (see the chapter by Hannila and colleagues for more discussion of cAMP and nerve regeneration). The fact that small molecule agents can be used to regulate energy metabolism in neuronal cells provides additional support for the feasibility of the drug discovery efforts outlined in this chapter.

V. Screening and Identification of Novel Compounds for Drug Discovery

This section will delineate a general strategy for new drug discovery aimed at neuroprotection and promotion of neurite outgrowth. We will include discussion of issues related to structure-based drug design.

A. Screening Assays

There are several approaches to screening chemical or combinatorial libraries for active compounds that might constitute leads for drug discovery/optimization. In many cases, a purified receptor or drug target is incorporated into a binding assay that is amenable for high-throughput screening (HTS). "Hits" consist of compounds that bind to the target; they may be agonists or antagonists depending on the binding mode. Alternatively, the screening assay may be more complex and involve a genetic readout. The expression of a readily detectable gene product is linked to upstream regulatory processes, and a hit in this case drives expression of the gene product. So-called phenotypic screens represent a higher order of assay complexity. In this case, hits are identified in a biological assay such as cell survival or cell differentiation, which can easily be scored either manually or automatically. Advances in robotics and assay methodology have enabled HTS in phenotypic assays with biological readouts.

The work described here is based on phenotypic screens, specifically protection against serum and glutamine withdrawal, and enhancement of neurite outgrowth. There are numerous advantages to phenotypic assays. We have attempted to overcome the disadvantages of these assays by setting strict criteria for compounds considered to be *bona fide* hits. For inclusion in the SAR analysis, a compound must: (1) provide neuroprotection in the cell viability assay, (2) enhance neurite outgrowth, (3) activate Akt, and (4) induce these activities in a pertussis toxin-sensitive manner. In this way, we increase the likelihood that the hits are working through the same molecular mechanism.

B. Structure-Based Design

Under ideal circumstances, structure-based drug design refers to the discovery of lead compounds and drug candidates based on their interaction with a protein target whose X-ray crystal structure is already established (Klebe, 2000; Laird and Blake, 2004). Moreover, the availability of cocrystals of compounds bound at the active (or receptor) site of the target protein greatly enhances such efforts. Of course, this high-level approach requires intimate knowledge about the structure of the target molecule. However, there are less exacting approaches to structure-based design that are more widely applicable. In this case, the structure of an initial prototype or lead molecule is used as the basis for identification of active compounds with similar structures through screening (Verdonk and Hartshorn, 2004). A general outline of this approach is discussed here.

Olanzapine consists of a thienobenzodiazepine core with an attached piperazinyl ring. It represents the prototype for our neuroenhancement strategy (Dwyer *et al.*, 2003b). Commercially available chemical libraries were searched for compounds that resembled substructures in the olanzapine molecule, for instance, compounds that included piperazinyl rings or thieno groups, and so on. Substructure analogues were screened in the phenotypic assays and some of the hits were described in Sections IV.B and C. SAR analysis of the hits in comparison to olanzapine led to the formulation of an initial pharmacophore hypothesis to relate chemical structure to functional activity in our battery of assays, including activation of Akt. Further searches of chemical libraries identified additional compounds for evaluation and yielded several more hits. Structural information from the second round of screening allowed us to refine the pharmacophore hypothesis. Thus, structure-based discovery entails iterative rounds of screening, revision of the pharmacophore structure, additional searches of chemical libraries for compounds with different substituent groups, evaluation of the new compounds, and so on. The goal is to identify the most potent substructures and if possible combinations of substructures that produce the desired biological activity as a starting point for *ab initio* design of novel lead compounds.

The design phase relies on two sets of information as a guide. (1) Structural information from the collection of *bone fide* hits is used to define substructures and substituent groups that are critical for biological activity. (2) Even in the absence of any information about the ultimate drug target, it is possible to incorporate general ideas about the receptor site into structure-based design. Thus, the pharmacophore can be viewed as an inverse image of the receptor. Loss or gain of biological activity associated with changes at a particular atom identifies that region as critical for activity. The chemical nature of key substituents also provides useful information. A crucial amide group in the pharmacophore may imply that the receptor has a hydrogen-bond acceptor (e.g., a carbonyl group) in the binding site, whereas a requirement for a ketone group in the active compound might suggest that a hydrogen-bond donor complements this group in the target protein. Similarly, bulky aromatic rings in the pharmacophore might be accommodated by a hydrophobic groove in the receptor or by aromatic amino acid side chains at the interface of secondary structural elements in proteins. We have previously used this inverse-image approach to characterize alcohol-binding sites on proteins (Dwyer and Bradley, 2000).

In structure-based screening and SAR analysis, initial similarities between compounds are typically based on two-dimensional (2-D) chemical structures. This information can be sufficient for the identification of analogues that produce the same biological activity as the prototype drug or compound. However, the design of novel compounds for synthesis is greatly enhanced by analysis of the 3-D geometry of active compounds in the collection. The 3-D structure of the active hit may even provide insight into the nature of the receptor or target molecule. In our research program, chemical structures of the hits from screening were created with the Builder module of the Insight II software from Accelrys (San Diego, CA). The structures were then subjected to energy minimization to convergence in order to obtain a likely 3-D conformation. The minimized 3-D structures of the hits were then compared to identify common structural motifs, which included orientation of key atoms in space and distances between active centers. In some cases, we performed further minimization of structures by quantum mechanics calculations to explore other possible conformations of the molecules.

Our 3-D structural analysis is incomplete; however, some trends are emerging from the comparisons. A piperazinyl ring is a common feature among the hits. In addition, the linkage of two substructures in tandem tends to increase the relative potencies of the compounds. Molecular-modeling data suggest that the compounds mimic an α-helical structure, which would be consistent with binding at a protein–protein interface. Additional studies will be needed to further refine the pharmacophore hypothesis and to learn more about the effects of substitution at particular atoms of interest.

To summarize thus far, we used the structure of olanzapine as a template to search for similar compounds in chemical libraries. More than 150 compounds were screened in phenotypic assays and 15 hits were identified. Most of the compounds appeared to produce the same spectrum of positive effects as the prototype drug. This included enhancement of cell viability, promotion of neurite outgrowth, and activation of Akt. The biological actions of the drugs/compounds appeared to be mediated through G proteins and involve optimization of glucose metabolism. The starting point for these discovery efforts was an antipsychotic medication. Drugs that emerge from this research may similarly be useful for the treatment of schizophrenia; however, they may also provide effective treatment for other brain deficit disorders.

VI. Neurodevelopmental Disorders and Neurodegenerative Diseases

As mentioned in Section I, few drugs are available for effective treatment of most brain deficit disorders. This includes devastating neurodevelopmental disorders, such as Rett syndrome, lissencephaly, and autism, and neurodegenerative diseases, such as Alzheimer's. The neurodevelopmental disorders present a particularly difficult challenge for treatment because much of the brain pathology occurs during early development *in utero* (Percy, 2002). Nevertheless, a therapeutic strategy aimed at neuroenhancement may provide a starting point for more effective treatment.

A. Rett Syndrome

Rett syndrome mainly affects girls and is characterized by developmental regression and mental retardation (Neul and Zoghbi, 2004). Stereotypic movements and autistic features are often present. The disorder in its purest form is caused by mutations in the methyl-CpG-binding protein, MeCP2, which is involved in the regulation of gene expression through direct interaction with DNA and other transcriptional regulators (Neul and Zoghbi, 2004). It was recently shown that MeCP2 regulates the expression of BDNF (Chen *et al.*, 2003), which serves as a neurotrophic factor for various neuronal populations, including dopaminergic neurons (Binder and Scharfman, 2004; Hyman *et al.*, 1991). BDNF has been implicated in affective disorders and schizophrenia (Angelucci *et al.*, 2005). The brains of girls with Rett syndrome are unusually small, there are signs of atrophy, neuronal cell bodies are also smaller in size, and dendrites and axons are abnormal (Neul and Zoghbi, 2004). This is somewhat reminiscent of the defects observed in Akt3 knockout mice (Easton *et al.*, 2005). Therefore, a drug that activates Akt may increase cell size and by enhancing neurite outgrowth promote the expression of dendrites and the stabilization of synaptic connections. By optimizing glucose metabolism in neurons, the ideal drug may also correct the metabolic abnormality in brain tissue that has recently been characterized in MeCP2 knockout mice (Saywell *et al.*, 2006) and in imaging studies of patients (Yoshikawa *et al.*, 1992).

B. Lissencephaly and Tuberous Sclerosis Complex

Lissencephaly is a more profound disturbance in brain development and offers less hope for effective intervention. However, a better understanding of this disorder may provide insight into other more treatable neurodevelopmental conditions. Lissencephaly and the related Miller–Dieker syndrome arise from mutations in the *Lis1* gene on chromosome 17 (Dobyns *et al.*, 1993; Lo Nigro *et al.*, 1997; Reiner *et al.*, 1993). Brain development, especially of the cortex, is radically altered by functional loss of the *Lis1* gene product, which exhibits a clear dose–response effect (Reiner, 2000). *Lis1* participates in various protein–protein interactions that ultimately regulate microtubule/dynein function and neuronal proliferation and migration (Feng *et al.*, 2000; Niethammer *et al.*, 2000). One of its binding partners, disrupted in schizophrenia 1 (DISC1), has been reported to be a genetic factor associated with increased risk for schizophrenia (Brandon *et al.*, 2004). Recently, *Lis1* was shown to indirectly regulate the activity of Ras-related (Rho-family) GTPases (Cdc42 and RhoA) by interaction with IQGAP1 (Kholmanskikh *et al.*, 2006). The Rho-family of GTPase proteins are involved in the control of cell motility and neurite outgrowth (Kaibuchi *et al.*, 1999).

It is encouraging to note that the prototype drug for our discovery efforts, olanzapine, enhances neurite outgrowth from PC12 cells.

Another genetic disorder with less severe developmental consequences is tuberous sclerosis complex (Astrinidis and Henske, 2005). Similar to lissencephaly, small Ras-related GTPase proteins are indirectly affected by mutations in the gene implicated as a causative factor, namely tuberin or TSC2. Tuberous sclerosis complex is characterized by benign tumors of many tissues (including brain), mental retardation, seizures, and autistic symptoms in some cases. TSC2 is a GTPase activating protein (GAP) that down-modulates Rheb (Ras homologue enriched in brain), which regulates signaling via mTOR (involved in growth) and B-Raf (involved in differentiation) (Astrinidis and Henske, 2005; Inoki *et al.*, 2003). Tuberin (TSC2) is expressed in PC12 cells and plays a role in the neuronal differentiation induced by NGF (Wu and Wong, 2005). This response is mediated in part via $G_{i/o}$ proteins. TSC2 is downstream of both AMPK and Akt and provides control of metabolic processes aimed at protecting cells against energy deprivation (Inoki *et al.*, 2003). Thus, drugs which modulate $G_{i/o}$ proteins and Akt, and protect against glutamine and/or glucose deprivation, may correct some of the defects that arise from the lack of TSC2.

C. Autism

Autism and autistic spectrum disorders might be amenable to the general therapeutic strategy proposed here. In autism, there is defective maturation of the frontal cortex and widespread, but often subtle, neuropathological features in other brain regions. This includes a higher density of neurons, but of smaller size, in limbic areas, and decreased numbers of neurons in the cerebellum and cerebral cortex (Palmen *et al.*, 2004). It has been reported that brain energy metabolism is disturbed in autistic patients (Zilbovicius *et al.*, 1995) and the response to insulin is abnormal (Maher *et al.*, 1975). There may be a decrease in the glutamate–glutamine cycle between astrocytes and neurons that reflects an overreliance on glucose (and possibly lactate) for energy (Chugani *et al.*, 1999; Rumsey *et al.*, 1985). A decrease in the use of glutamine for energy in the brain may tilt the balance toward higher levels of the excitatory neurotransmitter, glutamate. This could contribute to the higher incidence of epilepsy in autistic children. Although it may not be possible to restore normal cell numbers and neuronal organization in the brains of autistic patients with medications, a drug that enhances the functional activity and metabolic flexibility of the existing neurons may nevertheless provide tangible clinical benefits.

D. Alzheimer's Disease

Alzheimer's disease is an example of a brain deficit disorder that should be a good candidate for treatment by neuroenhancement. There is a progressive loss of neurons due to the combined accumulation of toxic species and loss of trophic support (Heininger, 2000; Hock *et al.*, 2000). Impairments in Akt function in Alzheimer's disease have been reported (Griffin *et al.*, 2005). Numerous metabolic disturbances have been described, including altered energy homeostasis, a reduction in GLUTs in various brain regions (Kalaria and Harik, 1989; Simpson *et al.*, 1994), and deficiencies in α-ketoglutarate dehydrogenase (Blass, 2002). Approved therapeutic strategies are mainly aimed at preservation of cholinergic function in the brain by inhibition of acetylcholinesterase. However, recent clinical trials have focused on neuroprotective drugs or administration of trophic factors (Allen and Dawbarn, 2006; Tuszynski *et al.*, 2005). Therefore, a drug that produces neuroenhancement would be a prime candidate for clinical testing is Alzheimer's disease. This is not to say that a single drug will be effective in the treatment of all the brain deficit disorders discussed here. Rather, the hope is to develop an assortment of innovative new drugs that are neuroprotective, enhance neurite/dendrite outgrowth and stabilization, and promote metabolic flexibility. Specific neuroenhancing drugs can then be tailored to maximize functional activity in those pathways, circuits, or cells that are mainly affected in a particular brain deficit disorder.

VII. Future Directions

In order to succeed with this novel therapeutic strategy of designing neuroenhancing drugs, significant advances will be necessary on several fronts. First, we need to gain a better understanding of the molecular mechanisms involved in neuroprotection and neurite outgrowth induced by the drugs/compounds. Initial studies implicate $G_{i/o}$ proteins and PI3K-Akt signaling pathways, but the intervening steps remain a mystery. The emerging concept of metabolic flexibility needs to be fleshed out and pharmacological approaches to the modulation of metabolic processes must be developed further. It is now clear that neurons are capable of using multiple sources of energy at a given moment in time. Is the use of a particular energy substrate better suited for particular functional states of the neuron (e.g., high vs low activity) or for maximum efficiency of astrocyte–neuron collaboration? At a fundamental level, we need to determine the extent to which it is possible to reverse neurodevelopmental insults. Other chapters in this volume have dealt with the issue of ongoing neurogenesis in the brain, which represents one possible avenue for restoring functional activity. If a drug stimulates neurite outgrowth or promotes stabilization of dendrite expression, will this enhance communication between neurons affected by the neurodevelopmental insult?

Thus far, our work has focused mainly on the role of Akt, ERK, and G proteins in neuroprotection. However, other targets appear to be promising. AMPK, TORC2, and the purine nucleotide cycle play critical roles in the regulation of energy metabolism and may represent worthwhile drug targets depending on the goal of therapy (e.g., protection against ischemic injury vs boosting suboptimum energy performance). *Lis1*, DISC1, and small GTPases (such as Rho, Rheb, and G_o) may represent attractive targets for the enhancement of neurite outgrowth and synaptic stabilization. The application of structure-based design to the discovery of drugs that interact with these proteins will be facilitated by the availability of 3-D structures of the potential drug targets and by advances in virtual screening of chemical libraries. Finally, the strategy outlined here offers a path for moving away from symptomatic therapies, which fail to address underlying deficits, and toward new drugs, which aim to maximize neuronal function and enhance metabolic flexibility in brain deficit disorders.

Acknowledgments

The authors are grateful for research support from the National Institutes of Health (grant #MH68385).

References

Allen, S. J., and Dawbarn, D. (2006). Clinical relevance of the neurotrophins and their receptors. *Clin. Sci.* **110,** 175–191.

Ames, A., III. (2000). CNS energy metabolism as related to function. *Brain Res. Brain Res. Rev.* **34,** 42–68.

Angeletti, P. U., Liuzzi, A., Levi-Montalcini, R., and Attardi, D. G. (1964). Effect of a nerve growth factor on glucose metabolism by sympathetic and sensory nerve cells. *Biochim. Biophys. Acta* **90,** 445–450.

Angelucci, F., Brene, S., and Mathe, A. A. (2005). BDNF in schizophrenia, depression and corresponding animal models. *Mol. Psychiatry* **10,** 345–352.

Apfel, S. C. (2002). Nerve growth factor for the treatment of diabetic neuropathy: What went wrong, what went right, and what does the future hold? *Int. Rev. Neurobiol.* **50,** 393–413.

Appel, S. H., and Parrot, B. L. (1970). Hexose monophosphate pathway in synapses. *J. Neurochem.* **17,** 1619–1625.

Ardizzone, T. D., Bradley, R. J., Freeman, A. M., III, and Dwyer, D. S. (2001). Inhibition of glucose transport in PC12 cells by the atypical antipsychotic drugs risperidone and clozapine, and structural analogs of clozapine. *Brain Res.* **923,** 87–90.

Arnold, S. E., Hyman, B. T., Van Hoesen, G. W., and Damasio, A. R. (1991). Some cytoarchitectural abnormalities of the entorhinal cortex in schizophrenia. *Arch. Gen. Psychiatry* **48,** 625–632.

Arnold, S. E., Han, L.-Y., Moberg, P. J., Turetsky, B. I., Gur, R. E., Trojanowski, J. Q., and Hahn, C.-G. (2001). Dysregulation of olfactory receptor neuron lineage in schizophrenia. *Arch. Gen. Psychiatry* **58,** 829–835.

Arranz, B., Rosel, P., Ramirez, N., Duenas, R., Fernandez, P., Sanchez, J. M., Navarro, M. A., and San, L. (2004). Insulin resistance and increased leptin concentrations in noncompliant schizophrenia patients but not in antipsychotic-naive first-episode schizophrenia patients. *J. Clin. Psychiatry* **65,** 1335–1342.

Astrinidis, A., and Henske, E. R. (2005). Tuberous sclerosis complex: Linking growth and energy signaling pathways with human disease. *Oncogene* **24,** 7475–7481.

Ataullakhanov, F. I., and Vitvitsky, V. M. (2002). What determines the intracellular ATP concentration. *Biosci. Rep.* **22,** 501–511.

Bai, O., Wei, Z., Lu, W., Bowen, R., Keegan, D., and Li, X.-M. (2002). Protective effects of atypical antipsychotic drugs on PC12 cells after serum withdrawal. *J. Neurosci. Res.* **69,** 278–283.

Baquer, N. Z., Hothersall, J. S., and McClean, P. (1988). Function and regulation of the pentose phosphate pathway in brain. *Curr. Top. Cell Regul.* **29,** 265–289.

Bartlett, E. J., Brodie, J. D., Simkowitz, P., Schloesser, R., Dewey, S. L., Lindenmayer, J.-P., Rusinek, H., Wolkin, A., Cancro, R., and Schiffer, W. (1998). Effect of a haloperidol challenge on regional brain metabolism in neuroleptic-responsive and nonresponsive schizophrenic patients. *Am. J. Psychiatry* **155,** 337–343.

Beck, M., Flachenecker, P., Magnus, T., Giess, R., Reiners, K., Toyka, K. V., and Naumann, M. (2005). Autonomic dysfunction in ALS: A preliminary study on the effects of intrathecal BDNF. *Amyotroph. Lateral Scler. Other Motor Neuron Disord.* **6,** 100–103.

Benes, F. M., Davidson, J., and Bird, E. D. (1986). Quantitative cytoarchitectural studies of the cerebral cortex of schizophrenics. *Arch. Gen. Psychiatry* **43,** 31–35.

Benes, F. M., McSparren, J., Bird, E. D., SanGiovanni, J. P., and Vincent, S. L. (1991). Deficits in small interneurons in prefrontal and cingulated cortices of schizophrenic and schizoaffective patients. *Arch. Gen. Psychiatry* **48,** 996–1001.

Ben-Shachar, D. (2002). Mitochondrial dysfunction in schizophrenia: A possible linkage to dopamine. *J. Neurochem.* **83,** 1241–1251.

Bensimon, G., and Doble, A. (2004). The tolerability of riluzole in the treatment of patients with amyotrophic lateral sclerosis. *Expert Opin. Drug Saf.* **3,** 525–534.

Berridge, M. V., and Tan, A. S. (1995). Interleukin-3 facilitates glucose transport in a myeloid cell line by regulating the affinity of the glucose transporter for glucose: Involvement of protein phosphorylation in transporter activation. *Biochem. J.* **305,** 843–851.

Bertram, L., and Tanzi, R. E. (2005). The genetic epidemiology of neurodegenerative disease. *J. Clin. Invest.* **115,** 1449–1457.

Biagiotti, E., Guidi, L., Capellacci, S., Ambrogini, P., Papa, S., Del Grande, P., and Ninfali, P. (2001). Glucose-6-phosphate dehydrogenase supports the functioning of the synapses in rat cerebellar cortex. *Brain Res.* **911,** 152–157.

Binder, D. K., and Scharfman, H. E. (2004). Brain-derived neurotrophic factor. *Growth Factors* **22,** 123–131.

Blass, J. P. (2002). Glucose/mitochondria in neurological conditions. *Int. Rev. Neurobiol.* **51,** 325–376.

Boniece, I. R., and Wagner, J. A. (1993). Growth factors protect PC12 cells against ischemia by a mechanism that is independent of PKA, PKC, and protein synthesis. *J. Neurosci.* **13,** 4220–4228.

Bontemps, F., Van den Berghe, G., and Hers, H.-G. (1983). Evidence for a substrate cycle between AMP and adenosine in isolated hepatocytes. *Proc. Natl. Acad. Sci. USA* **80,** 2829–2833.

Braceland, F. J., Meduna, L. J., and Vaichulis, J. A. (1945). Delayed action of insulin in schizophrenia. *Am. J. Psychiatry* **102,** 108–110.

Brandon, N. J., Handford, E. J., Schurov, I., Rain, J.-C., Pelling, M., Duran-Jimeniz, B., Camargo, L. M., Oliver, K. R., Beher, D., Shearman, M. S., and Whiting, P. J. (2004). Disrupted in schizophrenia 1 and Nudel form a neurodevelopmentally regulated protein complex: Implications for schizophrenia and other neurological disorders. *Mol. Cell. Neurosci.* **25,** 42–55.

Braumann, T., Jastorff, B., and Richter-Landsberg, C. (1986). Fate of cyclic nucleotides in PC12 cell cultures: Uptake, metabolism, and effects of metabolites on nerve growth factor-induced neurite outgrowth. *J. Neurochem.* **47,** 912–919.

Brazil, D. P., Park, J., and Hemmings, B. A. (2002). PKB binding proteins: Getting in on the Akt. *Cell* **111,** 293–303.

Brown, G. C. (1992). Control of respiration and ATP synthesis in mammalian mitochondria and cells. *Biochem. J.* **284,** 1–13.

Buchsbaum, M. S., Nuechterlein, K. H., Haier, R. J., Wu, J., Sicotte, N., Hazlett, E., Asarnow, R., Potkin, S., and Guich, S. (1990). Glucose metabolic rate in normals and schizophrenics during the continuous performance test assessed by positron emission tomography. *Br. J. Psychiatry* **156,** 216–227.

Buchsbaum, M. S., Potkin, S. G., Marshall, J. F., Lottenberg, S., Teng, C., Heh, C. W., Tafalla, R., Reynolds, C., Abel, L., Plon, L., and Bunney, W. E., Jr. (1992). Effects of clozapine and thiothixene on glucose metabolic rate in schizophrenia. *Neuropsychopharmacology* **6,** 155–163.

Burde, R., Dippel, E., and Seifert, R. (1996). Receptor-independent G protein activation may account for the stimulatory effects of first-generation H1-receptor antagonists in HL-60 cells, basophils, and mast cells. *Biochem. Pharmacol.* **51,** 125–131.

Carling, D. (2004). The AMP-activated protein kinase cascade—a unifying system for energy control. *Trends Biochem. Sci.* **29,** 18–24.

Chen, W. G., Chang, Q., Lin, Y., Meissner, A., West, A. E., Griffith, E. C., Jaenisch, R., and Greenberg, M. E. (2003). Derepression of BDNF transcription involves calcium-dependent phosphorylation of MeCP2. *Science* **302,** 885–889.

Chen, W. S., Xu, P.-Z., Gottlob, K., Chen, M.-L., Sokol, K., Shiyanova, T., Roninson, I., Weng, W., Suzuki, R., Tobe, K., Kadowaki, T., and Hay, N. (2001). Growth retardation and increased apoptosis in mice with homozygous disruption of the *akt1* gene. *Genes Dev.* **15,** 2203–2208.

Cho, H., Mu, J., Kim, J. K., Thorvaldsen, J. L., Chu, Q., Crenshaw, E. B., III, Kaestner, K. H., Bartolomei, M. S., Shulman, G. I., and Birnbaum, M. J. (2001a). Insulin resistance and a diabetes-like syndrome in mice lacking the protein kinase Akt2 (PKBβ). *Science* **292,** 1728–1731.

Cho, H., Thorvaldsen, J. L., Chu, Q., Feng, F., and Birnbaum, M. J. (2001b). Akt1/PKBα is required for normal growth but dispensable for maintenance of glucose homeostasis in mice. *J. Biol. Chem.* **276,** 38349–38352.

Chowdari, K. V., Mirnics, K., Semwal, P., Wood, J., Lawrence, E., Bhatia, T., Deshpande, S. N. T., Ferrell, R. E., Middleton, F. A., Devlin, B., Levitt, P., Lewis, D. A., *et al.* (2002). Association and linkage analyses of RGS4 polymorphisms in schizophrenia. *Hum. Mol. Genet.* **11,** 1373–1380.

Chugani, D. C., Sundram, B. S., Behen, M., Lee, M. L., and Moore, G. J. (1999). Evidence of altered energy metabolism in autistic children. *Prog. Neuropsychopharmacol. Biol. Psychiatry* **23,** 635–641.

Clapham, D. E., and Neer, E. J. (1997). G protein $\beta\gamma$ subunits. *Annu. Rev. Pharmacol. Toxicol.* **37,** 167–203.

Cohen, S. (1959). Purification and metabolic effects of a nerve growth-promoting protein from snake venom. *J. Biol. Chem.* **234,** 1129–1137.

Cooper, J. R., and Pincus, J. H. (1979). The role of thiamine in the nervous system. *Neurochem. Res.* **4,** 223–239.

Culmsee, C., Monnig, J., Kemp, B. E., and Mattson, M. P. (2001). AMP-activated protein kinase is highly expressed in neurons in the developing rat brain and promotes neuronal survival following glucose deprivation. *J. Mol. Neurosci.* **17,** 45–58.

Dago, L., Bonde, C., Peters, D., Moller, A., Bomholt, S. F., Hartz, J. B., Meyer, M., Drejer, J., and Gronborg, M. (2002a). NS 1231, a novel compound with neurotrophic-like effects *in vitro* and *in vivo*. *J. Neurochem.* **81,** 17–24.

Dago, L., Peters, D., Meyer, M., Hartz, B., Kruse, V., Drejer, J., and Gronborg, M. (2002b). NS-417, a novel compound with neurotrophic-like effects. *Neurochem. Res.* **27,** 107–111.

Davis, J. O., and Bracha, H. S. (1996). Famine and schizophrenia: First trimester malnutrition or second trimester beriberi? *Biol. Psychiatry* **40,** 1–3.

Deckwerth, T. L., and Johnson, E. M., Jr. (1993). Temporal analysis of events associated with programmed cell death (apoptosis) of sympathetic neurons deprived of nerve growth factor. *J. Cell Biol.* **123,** 1207–1222.

Delgado-Esteban, M., Almeida, A., and Bolanos, J. P. (2000). D-Glucose prevents glutathione oxidation and mitochondrial damage after glutamate receptor stimulation in rat cortical primary neurons. *J. Neurochem.* **75,** 1618–1624.

de Rooij, J., Zwartkruis, F. J., Verhaijen, M. H., Cool, R. H., Nijman, S. M. B., Wittinghofer, A., and Bos, J. L. (1998). Epac is a Rap1 guanine-nucleotide-exchange factor directly activated by cyclic AMP. *Nature* **396,** 474–477.

Dobyns, W. B., Reiner, O., Carrozzo, R., and Ledbetter, D. H. (1993). Lissencephaly: A human brain malformation associated with deletion of the *LIS1* gene located at chromosome 17p13. *JAMA* **270,** 2838–2842.

Duman, R. S. (2002). Synaptic plasticity and mood disorders. *Mol. Psychiatry* **7**(Suppl. 1), S29–S34.

Dwyer, D. S., and Bradley, R. J. (2000). Chemical properties of alcohols and their protein binding sites. *Cell. Mol. Life Sci.* **57,** 265–275.

Dwyer, D. S., and Donohoe, D. (2003). Induction of hyperglycemia in mice by atypical antipsychotic drugs. *Pharmacol. Biochem. Behav.* **75,** 255–261.

Dwyer, D. S., Liu, Y., and Bradley, R. J. (1999a). Ethanol-sensitive variant of the PC12 neuronal cell line: Sensitivity to alcohol is associated with increased cell adhesion and decreased glucose accumulation. *J. Cell. Physiol.* **178,** 93–101.

Dwyer, D. S., Pinkofsky, H. B., Liu, Y., and Bradley, R. J. (1999b). Antipsychotic drugs affect glucose uptake and the expression of glucose transporters in PC12 cells. *Prog. Neuropsychopharmacol. Biol. Psychiatry* **23,** 69–80.

Dwyer, D. S., Bradley, R. J., Kablinger, A. S., and Freeman, A. M., III. (2001). Glucose metabolism in relation to schizophrenia and antipsychotic drug treatment. *Ann. Clin. Psychiatry* **13,** 103–113.

Dwyer, D. S., Ardizzone, T. D., and Bradley, R. J. (2002a). Psychoactive drugs affect glucose transport and the regulation of glucose metabolism. *Int. Rev. Neurobiol.* **51,** 503–530.

Dwyer, D. S., Vannucci, S. J., and Simpson, I. A. (2002b). Expression, regulation, and functional role of glucose transporters (GLUTs) in brain. *Int. Rev. Neurobiol.* **51,** 159–188.

Dwyer, D. S., Lu, X.-H., and Bradley, R. J. (2003a). Cytotoxicity of conventional and atypical antipsychotic drugs in relation to glucose metabolism. *Brain Res.* **971,** 31–39.

Dwyer, D. S., Lu, X.-H., and Freeman, A. M., III. (2003b). Neuronal glucose metabolism and schizophrenia: Therapeutic prospects? *Expert Rev. Neurother.* **3,** 29–40.

Dzeja, P. P., Zeleznikar, R. J., and Goldberg, N. D. (1998). Adenylate kinase: Kinetic behavior in intact cells indicates that it is integral to multiple cellular processes. *Mol. Cell. Biochem.* **184,** 169–182.

Easton, R. M., Cho, H., Roovers, K., Shineman, D. W., Mizrahi, M., Forman, M. S., Lee, V. M., Szabolcs, M., de Jong, R., Oltersdorf, T., Ludwig, T., Efstratiadis, A., *et al.* (2005). Role for Akt3/Protein kinase Bγ in attainment of normal brain size. *Mol. Cell. Biol.* **25,** 1869–1878.

Emamian, E. S., Hall, D., Birnbaum, M. J., Karayiorgou, M., and Gogos, J. A. (2004). Convergent evidence for impaired AKT1-GSK3beta signaling in schizophrenia. *Nat. Genet.* **36,** 131–137.

Feinberg, I. (1982). Schizophrenia: Caused by a fault in programmed synaptic elimination during adolescence? *J. Psychiatr. Res.* **17,** 319–334.

Feinberg, J. F. (1980). The Wernicke-Korsakoff syndrome. *Am. Fam. Physician* **22,** 129–133.

Feldman, H. H., Van Baelen, B., Kavanagh, S. M., and Torfs, K. E. (2005). Cognition, function, and caregiving time patterns in patients with mild-to-moderate Alzheimer disease: A 12-month analysis. *Alzheimer Dis. Assoc. Disord.* **19,** 29–36.

Feng, Y., Olson, E. C., Stukenberg, P. T., Flanagan, L. A., Kirschner, M. W., and Walsh, C. A. (2000). LIS1 regulates CNS lamination by interacting with mNudE, a central component of the centrosome. *Neuron* **28,** 665–679.

Ferri, C. P., Prince, M., Brayne, C., Brodaty, H., Fratiglioni, L., Ganguli, M., Hall, K., Hasegawa, K., Hendrie, H., Huang, Y., Jorm, A., Mathers, C., *et al.* (2005). Global prevalence of dementia: A Delphi consensus study. *Lancet* **366,** 2112–2117.

Freeman, H., Looney, J. M., Hoskins, R. G., and Dyer, C. G. (1943). Results of insulin and epinephrine tolerance tests in schizophrenic patients and in normal subjects. *Arch. Neurol. Psychiatry* **49,** 195–203.

Freeman, R. (1999). Human studies of recombinant human nerve growth factor and diabetic peripheral neuropathy. *Eur. Neurol.* **41**(Suppl. 1), 20–26.

Ghosh, M., Peterson, Y. K., Lanier, S. M., and Smrcka, A. V. (2003). Receptor- and nucleotide exchange-independent mechanisms for promoting G protein subunit dissociation. *J. Biol. Chem.* **278,** 34747–34750.

Glantz, L. A., and Lewis, D. A. (1997). Reduction of synaptophysin immunoreactivity in the prefrontal cortex of subjects with schizophrenia. *Arch. Gen. Psychiatry* **54,** 943–952.

Gonatas, N. K., and Moss, A. (1975). Pathologic axons and synapses in human neuropsychiatric disorders. *Hum. Pathol.* **6,** 571–581.

Gottlob, K., Majewski, N., Kennedy, S., Kandel, E., Robey, R. B., and Hay, N. (2001). Inhibition of early apoptotic events by Akt/PKB is dependent on the first committed step of glycolysis and mitochondrial hexokinase. *Genes Dev.* **15,** 1406–1418.

Goossens, V., Grooten, J., and Fiers, W. (1996). The oxidative metabolism of glutamine. *J. Biol. Chem.* **271,** 192–196.

Gordon, P. H. (2005). Advances in clinical trials for amyotrophic lateral sclerosis. *Curr. Neurol. Neurosci. Rep.* **5,** 48–54.

Greenough, W. T., Larson, J. R., and Withers, G. S. (1985). Effects of unilateral and bilateral training in a reaching task on dendritic branching of neurons in the rat motor-sensory forelimb cortex. *Behav. Neural Biol.* **44,** 301–314.

Griffin, R. J., Moloney, A., Kelliher, M., Johnston, J. A., Ravid, R., Dockery, P., O'Connor, R., and O'Neill, C. (2005). Activation of Akt/PKB, increased phosphorylation of Akt substrates and loss and altered distribution of Akt and PTEN are features of Alzheimer's disease pathology. *J. Neurochem.* **93,** 105–117.

Gu, M., Irvani, M., Cooper, J. M., King, D., Jenner, P., and Schapira, A. H. (2004). Pramipexole protects against apoptotic cell death by non-dopaminergic neurons. *J. Neurochem.* **91,** 1075–1081.

Gupta, A., Ho, D. Y., Brooke, S., Franklin, L., Roy, M., McLaughlin, J., Fink, S. L., and Sapolsky, R. M. (2001). Neuroprotective effects of an adenoviral vector expressing the glucose transporter: A detailed description of the mediating cellular events. *Brain Res.* **908,** 49–57.

Hageluken, A., Grunbaum, L., Nurnberg, B., Harhammer, R., Schunack, W., and Seifert, R. (1994). Lipophilic β-adrenoceptor antagonists and local anesthetics are effective direct activators of G-proteins. *Biochem. Pharmacol.* **47,** 1789–1795.

Hajduch, E., Litherland, G. J., and Hundal, H. S. (2001). Protein kinase B (PKB/Akt)—a key regulator of glucose transport? *FEBS Lett.* **492,** 199–203.

Hardie, D. G., Scott, J. W., Pan, D. A., and Hudson, E. R. (2003). Management of cellular energy by the AMP-activated protein kinase system. *FEBS Lett.* **546,** 113–120.

Harrison, P. J., and Weinberger, D. R. (2005). Schizophrenia genes, gene expression, and neuropathology: On the matter of their convergence. *Mol. Psychiatry* **10,** 40–68.

Hawkins, R. A., Williamson, D. H., and Krebs, H. A. (1971). Ketone-body utilization by adult and suckling rat brain *in vivo*. *Biochem. J.* **122,** 13–18.

Heininger, K. (2000). A unifying hypothesis of Alzheimer's disease. IV. Causation and sequence of events. *Rev. Neurosci.* **11,** 213–328.

Henderson, D. C., and Ettinger, E. R. (2002). Schizophrenia and diabetes. *Int. Rev. Neurobiol.* **51,** 481–501.

Hertz, L., and Dienel, G. A. (2002). Energy metabolism in the brain. *Int. Rev. Neurobiol.* **51,** 1–102.

Hickmott, P. W., and Ethell, I. M. (2006). Dendritic plasticity in the adult neocortex. *Neuroscientist* **12,** 16–28.

Higashijima, T., and Ross, E. M. (1991). Mapping of the mastoparan-binding site of G proteins. *J. Biol. Chem.* **266,** 12655–12661.

Higashijima, T., Burnier, J., and Ross, E. M. (1990). Regulation of G_i and G_o by mastoparan, related amphiphilic peptides, and hydrophobic amines. *J. Biol. Chem.* **265,** 14176–14186.

Hock, C., Heese, K., Hulette, C., Rosenberg, C., and Otten, U. (2000). Region-specific neurotrophin imbalances in Alzheimer disease: Decreased levels of brain-derived neurotrophic factor and increased levels of nerve growth factor in hippocampus and cortical areas. *Arch. Neurol.* **57,** 846–851.

Hollinger, S., and Hepler, J. R. (2002). Cellular regulation of RGS proteins: Modulators and integrators of G protein signaling. *Pharmacol. Rev.* **54,** 527–559.

Honegger, P., Braissant, O., Henry, H., Boulat, O., Bachmann, C., Zurich, M.-G., and Pardo, B. (2002). Alteration of amino acid metabolism in neuronal aggregate cultures exposed to hypoglycaemic conditions. *J. Neurochem.* **81,** 1141–1151.

Hothersall, J. S., Greenbaum, A. L., and McLean, P. (1982). The functional significance of the pentose phosphate pathway in synaptosomes: Protection against peroxidative damage by catecholamines and oxidants. *J. Neurochem.* **39,** 1325–1332.

Hulshoff Pol, H. E., Hoek, H. W., Susser, E., Brown, A. S., Dingemans, A., Schnack, H. G., van Haren, N. E., Pereira Ramos, L. M., Gispen-de Wied, C. C., and Kahn, R. S. (Pol 2000). Prenatal exposure to famine and brain morphology in schizophrenia. *Am. J. Psychiatry* **157,** 1170–1172.

Hyman, C., Hofer, M., Barde, Y. A., Juhasz, M., Yancopoulos, G. D., Squinto, S. P., and Lindsay, R. M. (1991). BDNF is a neurotrophic factor for dopaminergic neurons of the substantia nigra. *Nature* **350,** 230–232.

Ingvar, D. H., and Franzen, G. (1974). Abnormalities of cerebral blood flow distribution in patients with chronic schizophrenia. *Acta Psychiat. Scand.* **50,** 425–462.

Inoki, K., Zhu, T., and Guan, K.-L. (2003). TSC2 mediates cellular energy response to control cell growth and survival. *Cell* **115,** 577–590.

Inouye, S., Seo, M., Yamada, Y., and Nakazawa, A. (1998). Increase of adenylate kinase isozyme 1 protein during neuronal differentiation in mouse embryonal carcinoma P19 cells and in rat brain primary cultured cells. *J. Neurochem.* **71,** 125–133.

Inouye, S., Yamada, Y., Miura, K., Suzuki, H., Kawata, K., Shinoda, K., and Nakazawa, A. (1999). Distribution and developmental changes of adenylate kinase isozymes in the rat brain: Localization of adenylate kinase 1 in the olfactory bulb. *Biochem. Biophys. Res. Commun.* **254,** 618–622.

Ishijima, Y., Shimada, N., Fukuda, M., Miyazaki, H., Orlov, N. Y., Orlova, T. G., Yamada, T., and Kimura, N. (1999). Overexpression of nucleoside diphosphate kinases induces neurite outgrowth and their substitution to inactive forms leads to suppression of nerve growth factor- and dibutyryl cyclic AMP-induced effects in PC12D cells. *FEBS Lett.* **445,** 155–159.

Jiang, X. M., Ohnishi, A., Yamamoto, T., Murai, Y., Awaya, A., and Ikeda, M. (1995). The effect of MS-818, a pyrimidine compound, on the regeneration of peripheral nerve fibers of mice after a crush injury. *Acta Neuropathol.* **90,** 130–134.

Kaibuchi, K., Kuroda, S., and Amano, M. (1999). Regulation of the cytoskeleton and cell adhesion by the Rho family GTPases in mammalian cells. *Annu. Rev. Biochem.* **68,** 459–486.

Kalaria, R. N., and Harik, S. I. (1989). Reduced glucose transport at the blood-brain barrier and in cerebral cortex in Alzheimer disease. *J. Neurochem.* **53,** 1083–1088.

Kan, O., Baldwin, S., and Whetton, A. (1994). Apoptosis is regulated by the rate of glucose transport in an interleukin-3-dependent cell line. *J. Exp. Med.* **180,** 917–923.

Kawasaki, H., Springett, G. M., Mochizuki, N., Toki, S., Nakaya, M., Matsuda, M., Housman, D. E., and Graybiel, A. M. (1998). A family of cAMP-binding proteins that directly activate Rap1. *Science* **282,** 2275–2279.

Keefe, R. S., Seidman, L. J., Christensen, B. K., Hamer, R. M., Sharma, T., Sitskoorn, M. M., Rock, S. L., Woolson, S., Tohen, M., Tollefson, G. D., Sanger, T. M., Lieberman, J. A., and HGDH Research Group. (2006). Long-term neurocognitive effects of olanzapine or low-dose haloperidol in first-episode psychosis. *Biol. Psychiatry* **59,** 97–105.

Kholmanskikh, S. S., Koeller, H. B., Wynshaw-Boris, A., Gomez, T., Letourneau, P. C., and Ross, M. E. (2006). Calcium-dependent interaction of Lis1 with IQGAP1 and Cdc42 promotes neuronal motility. *Nat. Neurosci.* **9,** 50–57.

Kimple, R. J., De Vries, L., Tronchere, H., Behe, C. I., Morris, R. A., Farquhar, M. G., and Siderovski, D. P. (2001). RGS12 and RGS14 GoLoco motifs are $G\alpha_i$ interaction sites with guanine nucleotide dissociation inhibitor activity. *J. Biol. Chem.* **276,** 29275–29281.

Klebe, G. (2000). Recent developments in structure-based drug design. *J. Mol. Med.* **78,** 269–281.

Kodama, M., Fujioka, T., and Duman, R. S. (2004). Chronic olanzapine or fluoxetine administration increases cell proliferation in hippocampus and prefrontal cortex of adult rat. *Biol. Psychiatry* **56,** 570–580.

Kolb, B., and Whishaw, I. Q. (1998). Brain plasticity and behavior. *Annu. Rev. Psychol.* **49,** 43–64.

Kovacevic, Z., and McGivan, J. D. (1983). Mitochondrial metabolism of glutamine and glutamate and its physiological significance. *Physiol. Rev.* **63,** 547–605.

Krishnan, K. S., Rikhy, R., Rao, S., Shivalkar, M., Mosko, M., Narayanan, R., Etter, P., Estes, P. S., and Ramaswami, M. (2001). Nucleoside diphosphate kinase, a source of GTP, is required for dynamin-dependent synaptic vesicle recycling. *Neuron* **30,** 197–210.

Laird, E. R., and Blake, J. F. (2004). Structure-based generation of viable leads from small combinatorial libraries. *Curr. Opin. Drug Discov. Devel.* **7,** 354–359.

Law, A. J., Hutchinson, L. J., Burnet, P. W., and Harrison, P. J. (2004). Antipsychotics increase microtubule-associated protein 2 mRNA but not spinophilin mRNA in rat hippocampus and cortex. *J. Neurosci. Res.* **76,** 376–382.

Lawlor, M. A., and Alessi, D. R. (2001). PKB/Akt: A key mediator of cell proliferation, survival and insulin responses? *J. Cell Sci.* **114,** 2903–2910.

Lawrence, M., Ho, D., Dash, R., and Sapolsky, R. (1995). Herpes simplex virus vectors overexpressing the glucose transporter gene protect against seizure-induced neuron loss. *Proc. Natl. Acad. Sci. USA* **92,** 7247–7251.

Lawrence, M., Sun, G., Kunis, T., Saydam, T., Dash, R., Ho, D., Sapolsky, R., and Steinberg, G. (1996). Overexpression of the glucose transporter gene with a herpes simplex viral vector protects striatal neurons against stroke. *J. Cerebr. Blood Flow Metab.* **16,** 181–185.

LeBlanc, A. C. (2005). The role of apoptotic pathways in Alzheimer's disease neurodegeneration and cell death. *Curr. Alzheimer Res.* **2,** 389–402.

Levi-Montalcini, R., and Angeletti, P. U. (1968). Nerve growth factor. *Physiol. Rev.* **48,** 534–569.

Lewis, D. A., and Levitt, P. (2002). Schizophrenia as a disorder of neurodevelopment. *Annu. Rev. Neurosci.* **25,** 409–432.

Lieberman, J. A., Stroup, T. S., McEvoy, J. P., Swartz, M. S., Rosenheck, R. A., and CATIE investigators. (2005a). Effectiveness of antipsychotic drugs in patients with chronic schizophrenia. *N. Engl. J. Med.* **353,** 1209–1223.

Lieberman, J. A., Tollefson, G. D., Charles, C., Zipursky, R., Sharma, T., Kahn, R. S., Keefe, R. S., Green, A. I., Gur, R. E., McEvoy, J., Perkins, D., Hamer, R. M., *et al.* (2005b). Antipsychotic drug effects on brain morphology in first-episode psychosis. *Arch. Gen. Psychiatry* **62,** 361–370.

Lo Nigro, C., Chong, S. S., Smith, A. C., Dobyns, W. B., Carrozzo, R., and Ledbetter, D. H. (1997). Point mutations and an intragenic deletion in *LIS1*, the lissencephaly causative gene in isolated lissencephaly sequence and Miller-Dieker syndrome. *Hum. Mol. Genet.* **6,** 157–164.

Lowenstein, J. M. (1972). Ammonia production in muscle and other tissues: The purine nucleotide cycle. *Physiol. Rev.* **52,** 382–414.

Lu, X.-H., and Dwyer, D. S. (2005). Second generation antipsychotic drugs, olanzapine, quetiapine, and clozapine enhance neurite outgrowth in PC12 cells via PI3K/Akt, ERK and pertussis toxin-sensitive pathways. *J. Mol. Neurosci.* **27,** 43–64.

Lu, X.-H., Bradley, R. J., and Dwyer, D. S. (2004). Olanzapine produces trophic effects *in vitro* and stimulates phosphorylation of Akt/PKB, ERK1/2, and the mitogen-activated protein kinase p38. *Brain Res.* **1011,** 58–68.

Maher, F., Vannucci, S. J., and Simpson, I. A. (1994). Glucose transporter proteins in brain. *FASEB J.* **8,** 1003–1011.

Maher, K. R., Harper, J. F., Macleay, A., and King, M. G. (1975). Peculiarities in the endocrine response to insulin stress in early infantile autism. *J. Nerv. Ment. Dis.* **161,** 180–184.

Mauer, I., Zierz, S., and Moller, H. (2001). Evidence for a mitochondrial oxidative phosphorylation defect in brains from patients with schizophrenia. *Schizophr. Res.* **48,** 125–136.

McGeer, E. G., and McGeer, P. L. (2005). Pharmacologic approaches to the treatment of amyotrophic lateral sclerosis. *BioDrugs* **19,** 31–37.

McGlashan, T. H., and Hoffman, R. E. (2000). Schizophrenia as a disorder of developmentally reduced synaptic connectivity. *Arch. Gen. Psychiatry* **57,** 637–648.

McKenna, M. C., Stevenson, J. H., Huang, X., and Hopkins, I. B. (2000). Differential distribution of the enzymes glutamate dehydrogenase and aspartate aminotransferase in cortical synaptic mitochondria contributes to metabolic compartmentation in cortical synaptic terminals. *Neurochem. Int.* **37,** 229–241.

Meininger, V., Bensimon, G., Bradley, W. R., Brooks, B., Douillet, P., Eisen, A. A., Lacomblez, L., Leigh, P. N., and Robberecht, W. (2004). Efficacy and safety of xaliproden in amyotrophic lateral sclerosis: Results of two phase III trials. *Amyotroph. Lateral Scler. Other Motor Neuron Disord.* **5,** 107–117.

Meltzer, H. Y., Li, Z., Kaneda, Y., and Ichikawa, J. (2003). Serotonin receptors: Their key role in drugs to treat schizophrenia. *Prog. Neuropsychopharmacol. Biol. Psychiatry* **27,** 1159–1172.

Middleton, F. A., Mirnics, K., Pierri, J. N., Lewis, D. A., and Levitt, P. (2002). Gene expression profiling reveals alterations of specific metabolic pathways in schizophrenia. *J. Neurosci.* **22,** 2718–2729.

Millar, J. K., Wilson-Annan, J. C., Anderson, S., Christie, S., Taylor, C. A., Semple, C. A., Devon, R. S., Clair, D. M., Muir, W. J., Blackwood, D. H., and Porteous, D. J. (2000). Disruption of two novel genes by a translocation co-segregating with schizophrenia. *Hum. Mol. Genet.* **9,** 1415–1423.

Millar, J. K., Pickard, B. S., Mackie, S., James, R., Christie, S., Buchanan, S. R., Malloy, M. P., Chubb, J. E., Huston, E., Baillie, G. S., Thomson, P. A., Hill, E. V., *et al.* (2005). DISC1 and PDE4B are interacting genetic factors in schizophrenia that regulate cAMP signaling. *Science* **310,** 1187–1191.

Mirnics, K., Middleton, F. A., Stanwood, G. D., Lewis, D. A., and Levitt, P. (2001). Disease-specific changes in regulator of G-protein signaling 4 (RGS4) expression in schizophrenia. *Mol. Psychiatry* **6,** 293–301.

Moberg, P. J., Agrin, R., Gur, R. E., Gur, R. C., Turetsky, B. I., and Doty, R. L. (1999). Olfactory dysfunction in schizophrenia: A qualitative and quantitative review. *Neuropsychopharmacology* **21,** 325–340.

Mohr, E., Nair, N. P., Sampson, M., Murtha, S., Belanger, G., Pappas, B., and Mendis, T. (1997). Treatment of Alzheimer's disease with sabeluzole: Functional and structural correlates. *Clin. Neuropharmacol.* **20,** 338–345.

Moises, H. W., Zoega, T., and Gottesman, I. I. (2002). The glial growth factors deficiency and synaptic stabilization hypothesis of schizophrenia. *BMC Psychiatry* **2,** 8–21.

Moley, K. H., and Mueckler, M. M. (2000). Glucose transport and apoptosis. *Apoptosis* **5,** 99–105.

Molina, V., Sanz, J., Benito, C., and Palomo, T. (2004). Direct association between orbitofrontal atrophy and the response of psychotic symptoms to olanzapine in schizophrenia. *Int. Clin. Psychopharmacol.* **19,** 221–228.

Moore, G. J., Bebchuk, J. M., Wilds, I. B., Chen, G., and Manji, H. K. (2000). Lithium-induced increase in human brain grey matter. *Lancet* **356,** 1241–1242.

Moreadith, R. W., and Lehninger, A. L. (1984). The pathways of glutamate and glutamine oxidation by tumor cell mitochondria. Role of mitochondrial $NAD(P)^+$-dependent malic enzyme. *J. Biol. Chem.* **259,** 6215–6221.

Morelli, A., Grasso, M., and Calissano, P. (1986). Effect of nerve growth factor on glucose utilization and nucleotide content of pheochromocytoma cells (clone PC12). *J. Neurochem.* **47,** 375–381.

Morris, A. J., Rudge, S. A., Mahlum, C. E., and Jenco, J. M. (1995). Regulation of phosphoinositide-3-kinase by G protein $\beta\gamma$ subunits in a rat osteosarcoma cell line. *Mol. Pharmacol.* **48,** 532–539.

Moxham, C. M., and Malbon, C. C. (1996). Insulin action impaired by deficiency of the G-protein subunit $G_{i\ alpha2}$. *Nature* **379,** 840–844.

Moxham, C. M., Hod, Y., and Malbon, C. C. (1993). Induction of $G_{\alpha i2}$-specific antisense RNA *in vivo* inhibits neonatal growth. *Science* **260,** 991–995.

Mueser, K. T., and McGurk, S. R. (2004). Schizophrenia. *Lancet* **363,** 2063–2072.

Mukherjee, S., Schnur, D. B., and Reddy, R. (1989). Family history of type 2 diabetes in schizophrenic patients. *Lancet* **1,** 495.

Neul, J. L., and Zoghbi, H. Y. (2004). Rett syndrome: A prototypical neurodevelopmental disorder. *Neuroscientist* **10,** 118–128.

Newcomer, J. W., Craft, S., Fucetola, R., Moldin, S. O., Selke, G., Paras, L., and Miller, R. (1999). Glucose-induced increase in memory performance in patients with schizophrenia. *Schizophr. Bull.* **25,** 321–335.

Newcomer, J. W., Haupt, D. W., Fucetola, R., Melson, A. K., Schweiger, J. A., Cooper, B. P., and Selke, G. (2002). Abnormalities in glucose regulation during antipsychotic treatment of schizophrenia. *Arch. Gen. Psychiatry* **59,** 337–345.

Niethammer, M., Smith, D. S., Ayala, R., Peng, J., Ko, J., Lee, M.-S., Morabito, M., and Tsai, L.-H. (2000). NUDEL is a novel Cdk5 substrate that associates with LIS1 and cytoplasmic dynein. *Neuron* **28,** 697–711.

Ninfali, P., Guidi, L., Aluigi, G., Biagotti, E., and Del Grande, P. (1999). High glucose-6-phosphate dehydrogenase activity contributes to the structural plasticity of periglomerular cells in the olfactory bulb of adult rats. *Brain Res.* **819,** 150–154.

Nozawa, Y., Sakai, N., Matsumoto, K., and Mizoue, K. (2002). A novel neuritogenic compound, NGA0187. *J. Antibiot.* **55,** 629–634.

Odagaki, Y., Nishi, N., and Koyama, T. (1998). Receptor-mediated and receptor-independent activation of G-proteins in rat brain membranes. *Life Sci.* **62,** 1537–1541.

Offermanns, S. (1999). New insights into the *in vivo* function of heterotrimeric G-proteins through gene deletion studies. *Naunyn Schmiedebergs Arch. Pharmacol.* **360,** 5–13.

Owen, O. E., Kalhan, S. C., and Hanson, R. W. (2002). The key role of anaplerosis and cataplerosis for citric acid cycle function. *J. Biol. Chem.* **277,** 30409–30412.

Palmen, S. J., van Engeland, H., Hof, P. R., and Schmitz, C. (2004). Neuropathological findings in autism. *Brain* **127,** 2572–2583.

Paul, J. W., Quach, T. T., Duchemin, A., Schrier, B. K., and DaVanzo, J. P. (1990). 1,1,3 Tricyano-2-amino-1-propene (Triap): A small molecule which mimics or potentiates nerve growth factor. *Brain Res. Dev. Brain Res.* **55,** 21–27.

Pellerin, L., Pellegri, G., Bittar, P. G., Charnay, Y., Bouras, C., Martin, J. L., Stella, N., and Magistretti, P. J. (1998). Evidence supporting the existence of an activity-dependent astrocyte-neuron lactate shuttle. *Dev. Neurosci.* **20,** 291–299.

Percy, A. K. (2002). Clinical trials and treatment prospects. *Ment. Retard. Dev. Disabil. Res. Rev.* **8,** 106–111.

Plas, D. R., and Thompson, C. B. (2005). Akt-dependent transformation: There is more to growth than just surviving. *Oncogene* **24,** 7435–7442.

Rabinovitz, M. (1992). The pleiotypic response to amino acid deprivation is the result of interactions between components of the glycolysis and protein synthesis pathways. *FEBS Lett.* **302,** 113–116.

Reiner, O. (2000). LIS1: Let's interact sometimes ... (Part 1). *Neuron* **28,** 633–636.

Reiner, O., Carrozzo, R., Shen, Y., Wehnert, M., Faustinella, F., Dobyns, W. B., Caskey, C. T., and Ledbetter, D. H. (1993). Isolation of a Miller-Dieker lissencephaly gene containing G protein β-subunit-like repeats. *Nature* **364,** 717–721.

Renshaw, P. F., Yurgelun-Todd, D. A., Wei, H., Charles, H. C., Tohen, M., Sharma, T., Zipursky, R. B., Kahn, R., Gur, R. E., Green, A. I., McEvoy, J. P., Perkins, D. O., *et al.* (2003). Olanzapine induced reductions in frontal lobe lactate levels correlate with treatment response in first episode psychosis. *Biol. Psychiatry* **53,** 67S.

Riese, U., Ziegler, E., and Hamburger, M. (2004). Militarinone A induces differentiation in PC12 cells via MAP and Akt kinase signal transduction pathways. *FEBS Lett.* **577,** 455–459.

Rosoklija, G., Toomayan, G., Ellis, S. P., Keilp, J., Mann, J. J., Latov, N., Hays, A. P., and Dwork, A. J. (2000). Structural abnormalities of subicular dendrites in subjects with schizophrenia and mood disorders: Preliminary findings. *Arch. Gen. Psychiatry* **57,** 349–356.

Ruigt, G. S., Makkink, W. K., and Konings, P. N. M. (1996). SR 57746A attenuates cytostatic drug-induced reduction of neurite outgrowth in co-cultures of rat dorsal root ganglia and Schwann cells. *Neurosci. Lett.* **203,** 9–12.

Rumsey, J. M., Duara, R., Grady, C., Rapoport, J. L., Margolin, R. A., Rapoport, S. I., and Cutler, N. R. (1985). Brain metabolism in autism. Resting cerebral glucose utilization rates as measured with positron emission tomography. *Arch. Gen. Psychiatry* **42,** 448–455.

Rutter, G. A., da Silva Xavier, G., and Leclerc, I. (2003). Roles of 5′-AMP-activated protein kinase (AMPK) in mammalian glucose homeostasis. *Biochem. J.* **375,** 1–16.

Sacks, W., Esser, A. H., Feitel, B., and Abbott, K. (1989). Acetazolamide and thiamine: An ancillary therapy for chronic mental illness. *Psychiatry Res.* **28,** 279–288.

Sakel, M. (1938). The methodical use of hypoglycemia in the treatment of psychoses. *Am. J. Psychiatry* **94,** 111–129.

Sanjo, N., Owada, K., Kobayashi, T., Mizusawa, H., Awaya, A., and Michikawa, M. (1998). A novel neurotrophic pyrimidine compound MS-818 enhances neurotrophic effects of basic fibroblast growth factor. *J. Neurosci. Res.* **54,** 604–612.

Saywell, V., Viola, A., Confort-Gouny, S., Le Fur, Y., Villard, L., and Cozzone, P. J. (2006). Brain magnetic resonance study of Mecp2 deletion effects on anatomy and metabolism. *Biochem. Biophys. Res. Commun.* **340,** 776–783.

Schifitto, G., Yiannoutsos, C., Simpson, D. M., Adornato, B. T., Singer, E. J., Hollander, H., Marra, C. M., Rubin, M., Cohen, B. A., Tucker, T., Koralnik, I. J., Katzenstein, D., *et al.* (2001). Long-term treatment with recombinant nerve growth factor for HIV-associated sensory neuropathy. *Neurology* **57,** 1313–1316.

Simpson, I. A., Chundu, K., Davies-Hill, T., Honer, W. G., and Davies, P. (1994). Decreased concentrations of GLUT1 and GLUT3 glucose transporters in the brains of patients with Alzheimer's disease. *Ann. Neurol.* **35,** 546–551.

Sofroniew, M. V., Howe, C. L., and Mobley, W. C. (2001). Nerve growth factor signaling, neuroprotection, and neural repair. *Annu. Rev. Neurosci.* **24,** 1217–1281.

Sokoloff, L. (1976). Circulation and energy metabolism in the brain. *In* "Basic Neurochemistry" (G. J. Siegel, R. W. Albers, R. Katzman, and B. W. Agranoff, Eds.), pp. 388–413. Little, Brown and Company, Boston.

Strange, P. G. (2001). Antipsychotic drugs: Importance of dopamine receptors for mechanisms of therapeutic actions and side effects. *Pharmacol. Rev.* **53,** 119–133.

Tailby, C., Wright, L. L., Metha, A. B., and Calford, M. B. (2005). Activity-dependent maintenance and growth of dendrites in adult cortex. *Proc. Natl. Acad. Sci. USA* **102,** 4631–4636.

Taylor, J. M., Jacob-Mosier, G. G., Lawton, R. G., Remmers, A. E., and Neubig, R. R. (1994). Binding of an α_2 adrenergic receptor third intracellular loop peptide to Gβ and the amino terminus of Gα. *J. Biol. Chem.* **269,** 27618–27624.

Thompson, P. M., Vidal, C., Giedd, J. N., Gochman, P., Blumenthal, J., Nicholson, R., Toga, A. W., and Rapoport, J. L. (2001). Mapping adolescent brain change reveals dynamic wave of accelerated gray matter loss in very early-onset schizophrenia. *Proc. Natl. Acad. Sci. USA* **98,** 11650–11655.

Tolosa, E. (2003). Advances in the pharmacological management of Parkinson disease. *J. Neural Transm. Suppl.* **64,** 65–78.

Tuszynski, M. H., Thal, L., Pay, M., Salmon, D. P., U. H. S., Bakay, R., Patel, P., Blesch, A., Vahlsing, H. L., Ho, G., Tong, G., Potkin, S. G., *et al.* (2005). A phase 1 clinical trial of nerve growth factor gene therapy for Alzheimer disease. *Nat. Med.* **11,** 551–555.

Ujike, H., and Nakata, K. (2003). Genetic study of schizophrenia: New targets for therapy. *Expert Rev. Neurother.* **3,** 41–50.

Van den Berghe, G., Bontemps, F., Vincent, M. F., and Van den Berghe, F. (1992). The purine nucleotide cycle and its molecular defects. *Prog. Neurobiol.* **39,** 547–561.

Vander Heiden, M. G., Plas, D. R., Rathmell, J. C., Fox, C. J., Harris, M. H., and Thompson, C. B. (2001). Growth factors can influence cell growth through effects on glucose metabolism. *Mol. Cell. Biol.* **21,** 5899–5912.

Vanhaesebroeck, B., Leevers, S. J., Ahmadi, K., Timms, J., Katso, R., Driscoll, P. C., Woscholski, R., Parker, P. J., and Waterfield, M. D. (2001). Synthesis and function of 3-phosphorylated inositol lipids. *Annu. Rev. Biochem.* **70,** 535–602.

Verdonk, M. L., and Hartshorn, M. J. (2004). Structure-guided fragment screening for lead discovery. *Curr. Opin. Drug Discov. Devel.* **7,** 404–410.

Volkmar, F. R., and Greenough, W. T. (1972). Rearing complexity affects branching of dendrites in the visual cortex of the rat. *Science* **176,** 1445–1447.

Wagle, A., Jivraj, S., Garlock, G. L., and Stapleton, S. R. (1998). Insulin regulation of glucose-6-phosphate dehydrogenase gene expression is rapamycin-sensitive and requires phosphatidylinositol 3-kinase. *J. Biol. Chem.* **273,** 14968–14974.

Waitzkin, L. (1966). A survey of unknown diabetics in a mental hospital. II. Men from age fifty. *Diabetes* **15,** 164–172.

Wakade, C. G., Mahadik, S. P., Waller, J. L., and Chiu, F. C. (2002). Atypical neuroleptics stimulate neurogenesis in adult rat brain. *J. Neurosci. Res.* **69,** 72–79.

Waki, A., Yano, R., Yoshimoto, M., Sadato, N., Yonekura, Y., and Fujibayashi, Y. (2001). Dynamic changes in glucose metabolism accompanying the expression of the neural phenotype after differentiation in PC12 cells. *Brain Res.* **894,** 88–94.

Wang, H. D., Dunnavant, F. D., Jarman, T., and Deutch, A. Y. (2004). Effects of antipsychotic drugs on neurogenesis in the forebrain of the adult rat. *Neuropsychopharmacology* **29,** 1230–1238.

Waters, C., Pyne, S., and Pyne, N. J. (2004). The role of G-protein coupled receptors and associated proteins in receptor tyrosine kinase signal transduction. *Semin. Cell Dev. Biol.* **15,** 309–323.

Wei, Z., Bai, O., Richardson, J. S., and Li, X.-M. (2003). Olanzapine protects PC12 cells from oxidative stress induced by hydrogen peroxide. *J. Neurosci. Res.* **73,** 364–368.

Weinberger, D. R. (1987). Implications of normal brain development for the pathogenesis of schizophrenia. *Arch. Gen. Psychiatry* **44,** 660–669.

Weinberger, D. R., Berman, K. F., and Zec, R. F. (1986). Physiologic dysfunction of dorsolateral prefrontal cortex in schizophrenia. I. Regional cerebral blood flow evidence. *Arch. Gen. Psychiatry* **43,** 114–124.

Wellmer, A., Misra, V. P., Sharief, M. K., Kopelman, P. G., and Anand, P. (2001). A double-blind placebo-controlled clinical trial of recombinant human brain-derived neurotrophic factor (rhBDNF) in diabetic polyneuropathy. *J. Peripher. Nerv. Syst.* **6,** 204–210.

Whiteman, E. L., Cho, H., and Birnbaum, M. J. (2002). Role of Akt/Protein kinase B in metabolism. *Trends Endocrinol. Metab.* **13,** 444–451.

Wolkin, A., Jaeger, J., Brodie, J. D., Wolf, A. P., Fowler, J., Rotrosen, J., Gomez-Mont, F., and Cancro, R. (1985). Persistence of cerebral metabolic abnormalities in chronic schizophrenia as determined by positron emission tomography. *Am. J. Psychiatry* **142,** 564–571.

Wu, E. H. T., and Wong, Y. H. (2005). Involvement of Gi/o proteins in nerve growth factor-stimulated phosphorylation and degradation of tuberin in PC-12 cells and cortical neurons. *Mol. Pharmacol.* **67,** 1195–1205.

Yoshikawa, H., Fueki, N., Suzuki, H., Sakuragawa, N., and Iio, M. (1992). Cerebral blood flow and oxygen metabolism in the Rett syndrome. *Brain Dev.* **14**(Suppl.), S69–S74.

Zhang, J., Zhang, J., Benovic, J. L., Sugai, M., Weyzker, R., Gout, I., and Rittenhouse, S. E. (1995). Sequestration of a G protein $\beta\gamma$ subunit or ADP-ribosylation of Rho can inhibit thrombin-induced activation of phosphoinositide 3-kinases. *J. Biol. Chem.* **270,** 6589–6594.

Zilbovicius, M., Garreau, B., Samson, Y., Remy, P., Barthelemy, C., Syrota, A., and Lelord, G. (1995). Delayed maturation of the frontal cortex in childhood autism. *Am. J. Psychiatry* **152,** 248–252.

Zoghbi, H. Y. (2003). Postnatal neurodevelopmental disorders: Meeting at the synapse? *Science* **302,** 826–830.

INDEX

A

AADC ES cell, 196
Aβ aggregation, 227
Aβ-amyloid, 14, 164
Aberrant neurite outgrowth, 255
Abortive regeneration, 58
Aβ-peptide, 226–228
ACC. *See* Anterior cingulate cortex
Acetylcholine, 33
Acetylcholinesterase, 277
  inhibitor, 3, 14
ACTH. *See* Adrenocorticotropin hormone
Acute thrombolysis, 8
Adaptive plasticity, 152
Adenosine kinase, 270
Adenylate kinase, 268, 270
Adenylyl cyclase, 70–71
Adrenocorticotropin hormone, 146
Adult
  born cell, 158
  CNS, 150
  hippocampus, 45
  mouse brain, 67
  neurogenesis, 2, 5, 12, 20, 152, 157–158
  neuron, morphology of, 46
  neuroplasticity, 158
  rat brain, 65
  rat hippocampus, 159
  stem cell, 5
  striatum, 158
Aex-3 promoter, 229
Agarose, 90
AICAR. *See* 5-Aminoimidazole-4-carboxamide ribonucleoside
Akinesia, 17
Akt, 121
  function, 277
  phosphorylation, 261
Alcohol toxicity, 253
Alpha-1 adrenergic receptor, 125
ALS, 19, 163. *See also* Amyotrophic lateral sclerosis
  treatment of, 251
Altman, Joseph, 2
Alzheimer's disease (AD), 1, 48, 129, 163–164, 182, 220, 248–249, 251, 274, 277
  brain, postmortem of, 227
  neurodegenerative pathology, 228
  tauopathy, 226–231
AMDA. *See* 5-Amino-5′-deoxyadenosine
Amino acid decarboxylase, 33, 38
5-Amino-5′-deoxyadenosine, 270
5-Aminoimidazole-4-carboxamide ribonucleoside, 270
Amnesia, 2
AMPA
  potentiator treatment, 201
  receptor, 180, 197, 202
  receptor-induced membrane depolarization, 199
  receptor-mediated transmission, 198
  receptor potentiator, 200–201, 203
AMPA-mediated Ca^{2+} flux, 198
Amphetamine, 125
  hyperactivity, 162
AMPK, 270, 276. *See* 5′-AMP kinase
5′-AMP kinase, 269
Amygdala, 150
Amyloid
  pathology, 228
  plaque formation, 15
  precursor protein, 164, 226
β-Amyloid peptide, 118–119
Amyotrophic lateral sclerosis, 1, 220, 236, 250
Anaerobic cell, 34
Anaerobic organism, 35
Antagonize, 255
Anterior cingulate cortex (ACC), 149–150
Anterior horn, 19
Anteroventral caudate, 148
Anti-apoptotic factor, 190

Anti-apoptotic protein, 8
Anticoagulant, 8
Anticonvulsant agent, 145
Antidepressant, 6, 109, 127, 129, 156–159, 163
  action, 156
  drug, 154
  effect, 157
  property, 145
  treatment, 159
Anti-inflammatory, 6
Antimitotic agent fluorodeoxyuridine, 81
Antioxidant, 32
Antiparkinsonian drug, 184
Antiplatelet agent, 8
Antipsychotic, 6, 108–109, 113, 119, 123–127, 129, 157, 159, 255–256, 261
Antipsychotic drug, 107–108, 112–114, 116, 118, 123–125, 129, 247, 250, 255, 265
Antipsychotic medication, 266, 274
Antipsychoticnaive, 111
Anxiety disorder, 48
Aphasia, 2
Apical ganglion, 37
Apoptosis, 163, 182
  oxidative stress-induced, 6, 9
Apoptosis pathway, 220
Apoptotic cell death, 156, 232
Apoptotic degeneration, 6
Apoptotic neurodegenerative change, 219
Apoptotic pathway, 230, 232
Apoptotic response, 232
APP, 15. *See also* Amyloid, precursor protein
Appetite, loss of, 192
Ara-C. *See* Cytosine arabinoside
ArgI. *See* Arginase I
ArgI level, 70
Arginase I, 70
Arginine 120, 63
Aripiprazole, 126
ASER neuron, 237
ASH neuronal dysfunction, 234
Astrocyte, 32, 45, 84, 150
Astrocytic protein, 80
ATX-2 protein, 235
Atypical antipsychotic, 125
Autism, 248–249, 276
Auxin, 31–32, 36
AVD interneruons, 220
AVG interneuron, 220
Axon, 41
  modeling of, 144
  myelination of, 77
  regeneration of myelinated, 76
Axonal
  degeneration, 231
  growth, 65
  outgrowth, 62
  regeneration, 57, 61–62, 66–68, 71, 72, 73, 77, 79, 84, 89
  remyelination, 84
  sprouting, 58, 67, 143

B

Basal ganglia, 192
Basso-Beattie-Bresnahan scale, 78
BBB, 71. *See* Basso-Beattie-Bresnahan scale
Bcl-2, 127
BDNF. *See* Brain-derived neurotrophic factor
BDNF-deficient mice, 194
BDNF-expressing fibroblast, 86
Betz cell, 19
Biarylpropylsulfonamide, 197, 199–201
Bipolar depression, 144, 154
Bipolar disorder (BPD), 143–145, 157–158, 164
  neuroplasticity, 145–150
  pathophysiology of, 145, 164–165
Blood flow, 40, 43
  interruption of, 8
Blood supply angiogenesis restore, 9
Blue-green algae, 34
Bone marrow cell, 189
BPD. *See* Bipolar disorder
Bradykinesia, 179
Brain
  adult, 1
  cell, 4
    replacement of, 4
  damage, 2
  disorder, 31
  glial scarring in, 60
  homeostasis, 31–32, 43, 47
  injury, 7, 20
  neurogenesis, 107, 116, 118
    regulation of, 108
  neuron, increased production of, 9
  pathology, 8
  rebuilding, 4
  regeneration of, 1

stem, 19, 38, 42
raphe nucleus, 76
vestibular nucleus, 76
trauma, 163
Brain deficit disorder, 249–250
Brain-derived neurotrophic factor (BDNF), 11, 18, 73, 76–77, 80, 86–87, 90, 108–112, 118, 127, 129, 152, 154–156, 163, 180, 190, 193, 195, 198, 250, 275
administration of, 76, 87
expression, 163
gene, 110
immunoreactivity, 112
level, 180, 200
mRNA, 200
production, 199
trophic effect of, 194
Brain-imaging technique, 264
BrdU. *See* Bromodeoxyuridine
BrdU-labeled cell, 116–118
BrdU monophosphate, 9
Breathing, 43
Bromodeoxyuridine (BrdU), 10, 12, 16–18, 115–116
cell, 18
immunopositive cell, 15
labeled cell, 15
positive cell, 11, 15, 17
Building brain
development, 3
evolution, 3

C

C3
enzyme, 69
neuroprotective action, 69
C. elegans Model
neurodegenerative disorder, 220–237
CA1, 14
Caenorhabditis elegans, 219–242
CAG trinucleotide, 233
Calbindin, 9
Calcineurin, 255
Calcitonin gene-related peptide, 82
cAMP-GEF. *See* cAMP regulated guanine nucleotide exchange factor
cAMP level, 72
elevation of, 79
cAMP regulated guanine nucleotide exchange factor, 268
cAMP response element-binding protein (CREB), 73, 118, 156, 162
activation of, 73
phosphorylation, 118
transcribes, 73
cAMP signaling, 71
Carbidopa combination, 17
Carbon dioxide chemoreceptor, 42
Cardiovascular diseases, 145, 148
Catecholamine, 33
Catecholaminergic axon, 86
Catecholaminergic marker, 76
Catechol-*O*-methyl transferase (COMT) inhibitor, 179
β-Catenin phosphorylation, 162
CBF, 148
CBP, 16
Cell
cycle regulation, 4
cytoskeleton, 36
death, 16, 35, 163–164
differentiation, 4–5, 31, 150, 258
division, 36
functional integration, 5
granule, 5
increased production, 9
loss, 3
migration, 5
periglomerular, 5
precursor, 4
production, 5
progenitor, 4
proliferation, 5, 11, 150, 152, 154, 163, 200, 248
inhibition of, 152
marker, 12
replacement, exogenous, 8
settling, 5
survival, 5
Cell-replacement therapy, 4, 7
Cell-Tracker dyes, 85
Cellular differentiation, 197, 260
Cellular mitosis, 44
Cellular plasticity, 144, 158–159, 165
Centralis superior nucleus pars medianus, 41
Central nervous system (CNS), 3, 7, 13, 57, 58, 66, 76–79, 83, 87, 89, 143, 159, 181
axon, 58, 76, 79, 86
regeneration of, 58–59, 75
function, 159

Central nervous system (CNS) (*cont.*)
  myelin, 62, 74, 77
  neuron, 81
  plasticity, 59
  repair, 59
Cerebellar granule, 64
Cerebellum, 41
Cerebral artery, 8
Cerebral cortex, 12, 16, 48, 60, 163
  ipsilateral, 10
Cerebral cortical, 14
  cultures, 11
Cerebral granule cells, 164
Cerebral injury, 2
Cerebral ischemia, 7, 163
Cerebrocortical neuron, 124
Cerebrocortical neurotoxicity, 124
CGN, 66. *See also* Cerebellar granule
CGN-induced cleavage, 69
CGRP. *See* Calcitonin gene-related peptide
CGRP-positive dorsal root axon, 82
Chaperone, 238
  proteins, 227
C3/4 hemisection lesion, 71
Chemorepulsive signaling, 7
Chemosensitive cell, 42
Chloroplast, 35
Chlorpromazine, 107–108, 118–119, 123–124, 255
Cholinergic axon, regeneration of, 75
Cholinergic neuron, 59, 110
Cholinergic neurotransmission, 229
Chondroitinase ABC, 61, 78
Chondroitin sulfate proteoglycan, 60
Choreoathetosis, 16
Chronic mood stabilizer, 156
Ciliary neurotrophic factor, 73
Cingulate, 43
Circadian rhythms, 41
Cisterna magna, 12
Claustrum, 148
Clinical medicine, 2
Clinical symptomatology, 184
Clot, 3
Clotting, 40
Clozapine, 36, 108, 112, 114, 116, 118–121, 125–127, 129, 257, 261, 265
CNTF, 80. *See also* Ciliary neurotrophic factor
Coerulospinal axon, 89
Cognition-enhancing effect, 125
Cognitive disease, 48
Cognitive dysfunction, 147
Cognitive function, 146
Cognitive impairment, 108
Collagen substratum, 75
Complex alkaloids, 35
COMT. *See* Catechol-O-methyl transferase
COMT inhibitor, 185–186
Corpus callosum, 64, 147
Cortical cell proliferation, 159
Cortical neuron, 80, 199
Cortical structure, degeneration of, 181
Corticospinal axon, 61, 78, 86, 87, 88–90
  regeneration of, 78
Corticospinal tract (CST), 19, 64
  axon, 82–83
    regeneration, 67
  injury, 67
  lesioning, 67
  regeneration, 68
Corticosteroid-induced psychosis, 126
Corticosteroids, adminstration of, 152
Corticothalamic connection, restoration of, 6
Corticothalamic neuron, 9
Corticotrophin-releasing hormone, 146
Cortisol, 146
CREB. *See* cAMP response element-binding protein
CRet ES cell, 196
CRH. *See* Corticotrophin-releasing hormone
Crystallographic analysis, 66
CSPG, 67, 69, 77. *See also* Chondroitin sulfate proteoglycan
  immunoreactivity, 77
  inhibitory activity, 61
  substrate, 61
CST, 67. *See also* Corticospinal tract
CX-516, 198
Cyclic AMP (cAMP), 63, 75, 93, 110, 270. *See also* Cyclic AMP
  elevation, 70, 78, 79
  intracellular level, 71
  up-regulation, 70
Cyst formation, 68
Cytoprotective effect, 107, 109, 119, 123–124
Cytosine arabinoside, 14
Cytoskeleton, 36
Cytotactin, 61

D

DAergic. *See* Dopaminergic
DAergic differentiation, 189–190
DAergic neuron, 189
  degeneration of, 190
Damaged neuron, 9
Dardarin. *See* LRRK2
DAT. *See* Dopamine transporter
DbcAMP, 70, 72, 73, 77–79, 87. *See also* Dibutyryl cAMP
DβHB. *See* D-β–hydroxybutyrate
D-β–hydroxybutyrate, 232
DCX, 9–10, 12–13, 18. *See also* Doublecortin
DCX-immunoreactive cell, 12
DCX-labeled cell, 18
Decarboxylase enzyme, 35
Defense withdrawal, 38
Deg phenotype, 221
Delusion, 108
Dementia, 14, 17, 31, 226
Dementing disease, 48
Demyelinating disease, 82
Dendrite, 41, 251
  outgrowth, 249–250
Dendritic remodeling, 143
Dentate gyral atrophy, 15
Dentate gyrus (DG), 15, 18, 116, 118, 150, 154, 158
  cell proliferation, 152
Depression, 31, 48, 129, 146
Depressive disorder, 48
Depressive symptoms, 146
DEX. *See* Dexamethasone
Dexamethasone, 146
DG-SGZ, 15
Diabetes mellitus, 145
Diacylglycerol (DAG), 160
Dibutyryl cAMP, 69, 270
Differentiation, 249
(2, 5-dimethoxy-4-bromophenyl)-2-aminopropane, 125
(2, 5-dimethoxy-4-iodophenyl)-2-aminopropane, 125
(2, 5-dimethoxy-4-methylphenyl)-2-aminopropane, 125
DISC1. *See* Disrupted in schizophrenia 1
Disorganization, 108
Disrupted in schizophrenia 1, 263, 275
Di(2-tolyl)guanidine, 124
Dizocilpine, 124, 126
DJ-1, 183
DLPFC, 149. *See also* Dorsolateral prefrontal cortex
D-lysergic acid diethylamide, 125
DMEM, 83. *See also* Dulbecco's modified Eagle's medium
DNA damage, 11
DOB. *See* 2, 5-dimethoxy-4-bromophenyl-2-aminopropane
DOI. *See* (2, 5-dimethoxy-4-iodophenyl)-2-aminopropane
DOM. *See* (2, 5-dimethoxy-4-methylphenyl)-2-aminopropane
Dopamine, 43, 185, 255
  activity, 186
  cell death, 182
    etiology of, 182
  depletion, 202
  neuron, 181–183
  nigrostriatal activity, 192
  receptor, 124, 186, 257–258, 264
    agonist, 17
Dopamine-depleted striatum, 18
Dopaminergic (DA), 18
  activity, 184, 192, 265
  cell, 193
    loss, 182
  damage, 187
  degeneration, 186
  depletion, 187
  marker, 189
  neurodegeneration, 182, 201, 231–233, 239
  neuron, 17, 19, 110, 181, 187, 191, 193, 195–196, 231, 241, 275
  neuronal death, 233
  neuronal loss, 183, 231
  neuronal perikarya, 191
  neuronal phenotype, 197
  neuron dysfunction, 233
  neurotransmission, 3, 180, 184
  nigrostriatal system, 191
  nigrostriatal tract, 183
Dopamine transporter (DAT)
  activity, 231
  ES cell, 196
Dorsal ACC, 150
Dorsal raphe nucleus, 48
Dorsal root axon, regeneration of, 82
Dorsal root entry zone, 82

Dorsal root ganglia, 60
Dorsolateral prefrontal cortex, 111, 147
Doublecortin, 5, 115
Doxycycline-dependent TET system, 190
DREZ. *See* Dorsal root entry zone
DRG, 66, 90. *See also* Dorsal root ganglia
  cell body, 73
  central axon, regeneration of, 71
  neurite, 83
  neuron, 66–68, 70–71, 73, 75, 81
Drugs, 232, 239, 241, 257–258, 264, 266, 274, 276, 278
  antidepressant, 156, 162
  antiparkinsonian, 184
  antipsychotic, 107–129, 247–248, 255–256, 258–263, 265
  HTS, 241, 242
  neuroprotective, 277
  neurorestorative, 186
  psychotropic, 144, 152, 154, 250
Drug abuse
  alcohol, 154
  morphine, 154
  nicotine, 154
  psychostimulant, 154
DTG, 124
Dulbecco's modified Eagle's medium, 257
Dyskinesia, 203
  L-dopa-induced, 192
Dystonia, 203

E

ECM. *See* Extracellular matrix
ECS, 155. *See also* Electroconvulsive seizures
  treatment, 159
Ectodomain crystal structure, 66
EGF, 89. *See also* Epidermal growth factor
EGF-dependent neurogenesis, 15
EGFR. *See* Epidermal growth factor receptor
  activation, 66
  kinase activity, 67
  phosphorylation, 67
EGF receptor-positive cell, 17
Egg-laying (Egl) defective phenotype, 221
Egl. *See* Egg laying
Egyptian mythology, 1
Electroconvulsive seizures, 154
Electroconvulsive therapy, 129
Electron microscopy, 230
Electrophysiology, 33
EM. *See* Electron microscopy
Embolic arterial occlusion, 8
Embryonic, 5
  cell, 232
  CNS tissue, 89
  hippocampus, 155
  neurosphere, 89
  spinal cord tissue, 71, 73, 87
  stem (ES) cell, 12, 89
ENCAM, 14
  immunoreactive cell, 10
  positive cell, 15
  positive neuronal progenitor, 10
Endocrine cell, 32
Endocrine system, 43
Endogenous cAMP level elevation, 70
Endogenous cell, 4
Endogenous neural precursor cell, 7
Endogenous neurogenesis, 17–18
Endoplasmic reticulum, 221
Endoproteolytic cleavage, 229
Endothelial cell, 32
Energy, loss of, 8
Entacapone, 185
Ependymal cell, 32
Ependymal layer, 41
EphA2 receptor, 68
Ephrin-B3, 64, 68
Epidermal growth factor receptor, 66
Epilepsy, 3, 7
EPO. *See* Erythropoietin
ER. *See* Endoplasmic reticulum
ERK, 121. *See also* Extracellular signal-regulated kinase
  pathway, 162–164
  phosphorylation, 121
ES. *See also* Embryonic stem
  cell, differentiation of, 190
Ethidium bromide, 81
E4 ubiquitin ligase, 233
Excitatory electrical influences, 33
Excitatory neurotransmission, 6
Excitotoxicity, 8, 182
Exogenous environmental toxins, 182
Exogenous myoinositol, 161
Exogenous NgR1, 65
External transcription spacer, 45
  Pet-1, 40
Extracellular Ig-like domain, 63

Extracellular senile, 14
Extracellular signal-regulated kinase (ERK), 71, 120
  signaling pathway, 156

F

FACE, 261. *See also* Fast-activated cell-based ELISA
Fasciculation, 19
Fast-activated cell-based ELISA, 261
FBS. *See* Fetal bovine serum
FDG-PET, 148
Feeding, 40
Fenpyroximate, 232
Ferromagnetic particles, 12
Fetal bovine serum, 257
Fetal cortical cell, 194
FGF-2, 11, 17, 129. *See also* Fibroblast growth factor-2
  infusion, 18
FGF-8, 17. *See also* Fibroblast growth factor-8
Fibroblast growth factor-2, 89, 118
Fibroblast growth factor-8, 189
Fibroblast proliferation, 75
Fibronectin, 81
Flatworm, 36
FLIPR. *See* Fluorescent imaging plate reader
Fluorescent imaging plate reader, 197
Fluoxetine, 129
Fluphenazine, 119, 255–256, 265
Focal cerebral ischemia, 9, 12, 14
Focal necrosis, 8
Food intake, 38
Forebrain, 41, 45
Forskolin, 70
Frontotemporal dementia with parkinsonism chromosome 17 type (FTDP - 17), 226
  mutant lines, 230
  neurodegeneration in, 227
Functional neuron generation, 2

G

GABA, 33, 89, 155
GABAergic interneuron, 155
GABAergic medium, 184
GABAergic motoneuron, 230
GABAergic neurogenesis, 156
GABAergic neurons, 110, 149, 163, 230
GABAergic phenotype, 155–156
GAG. *See* Glycosaminoglycan
Gα_3-induced neurodegeneration, 226
Gait disorder, 2
Ganglia of Annelids, 38
GAP. *See* GTPase activating protein
GAP-43, 71, 82
GCL, 9, 14
GDNF, 191, 193, 196, 203. *See also* Glial cell line-derived neurotrophic factor
  administration, 192
  expression, 192
Gene, 19
GFAP, 80, 83, 158
GFAP-positive astrocytes, 159
GFP. *See* Green fluorescent protein
Gi/Go activation, 71
Gi/Go protein, 71
Glia, 66
Glial activity, 43
Glial cells, 32, 41, 44–45, 80, 144, 180, 193–194
Glial cell-derived neurotrophic factor, 190
Glial cell line-derived neurotrophic factor, 80
Glial protein, 46
Glial scar, 60, 63, 78, 83
  formation, 88
Glial scarring, 71
Gliogenesis, 9
Global ischemia, 11
Globus pallidus, 185
Glucocorticoids, 6, 45, 146, 152, 155–156
  effects, 146
  levels, 152
  receptor, 146
  secretion of, 147
Glucose-6-phosphate dehydrogenase, 260
GLUT, 260, 277. *See also* Glucose transporter
  expression, 253
  protein, 252, 265
Glutamate, 33
Glutamate dehydrogenase, 266
Glutamate-glutamine cycle, 276
Glutamate receptor antagonist, 3
Glutamine, 253, 271
Glutathione peroxidase, 119
Glycine, 33
Glycogen synthase kinase 3, 161
Glycogen synthase kinase 3β, 259
Glycolytic pathway, 265
Glycosaminoglycan, 61

Glycosylphosphatidylinositol, 64
GPCR. *See* G protein-coupled receptor
GPI. *See* Glycosylphosphatidylinositol
G protein-coupled receptor, 263
GR. *See* Glucocorticoid receptor
Granule cell, apoptosis of, 154
Granule neuron, 45
Greek mythology, 1
Green fluorescent protein, 230
Griffonia simplicifolia, 35
Growth axon, 57
Growth cone, 58
Growth factor, 6
GRs, 152
GSK-3, 162. *See also* Glycogen synthase kinase 3
GSK3β. *See* Glycogen synthase kinase 3β
GTPase activating protein, 276
GTPase protein, 69

H

Hallucination, 108
Haloperidol, 107–111, 114–116, 118, 124, 155, 255
  administration, 112
HB-EGF. *See* Heparinbinding EGF-like growth factor
HCM. *See* Hypoxia-conditioned medium
HD, 16. *See also* Huntington's disease
Heme oxygenase-2 HO-2, 127
Hemiparesis, 2
Heparinbinding EGF-like growth factor, 6
Hepatic toxicity, 185
Hereditary polyglutamine disorder, 3
Hexamethylpropyleneamine oxime, 148
High-throughput screening, 241, 272
Hippocampal formation, 111
Hippocampal GABAergic interneurons, 156
Hippocampal level, 127, 129
Hippocampal neurogenesis, 118, 127, 152, 154–155, 157–158
Hippocampal neuron, 199
  cultures, 162
Hippocampal progenitor cell proliferation, 152
Hippocampus, 43, 110–112, 125, 147, 149–150
  of rat, 112
Histamine, 33
Histones, 45
HMPAO. *See* Hexamethylpropyleneamine oxime
HO-2, 129
Hoechst-labeled OEC, 83
Homeostasis, 41
  loss of, 47
  regulator of, 38, 43–44, 47
Hormone secretion, 40
Horseradish peroxidase, 59
HPA. *See* Hypothalamic pituitary adrenal; Hypothalamic pituitary-adrenocortical
HPA hyperactivity, 146
HRP. *See* Horseradish peroxidase
5-HT, 35–36, 42–43, 47. *See also* 5-hydroxytryptamine
  axon, 32
  brainstem system, 43
  dysfunction, 48
  neuron, 40, 43–45
  raphe neuron, 43
  receptor, 36, 38
  synthesis, 32
    inhibition of, 154
5-HT1A receptor, 45–46
  inhibition, 47
Htn. *See* Human huntingtin
5-HTP. *See* 5-hydroxytryptophan
HTS. *See also* High-throughput screening
  methodology, 242
  use of, 241, 242
HTT, mutant form of, 16
5-HTT, 40, 48. *See also* 5-HT transporter
  protein, 47
5-HTT-IR axon, 47
Hu, 9
Human adult adrenomedullary, 17
Human disease gene, 220
Human endothelial cell, 45
Human ES cell, 19
  transplantation of, 189
Human fetal mesencephalic tissue, 17
Human huntingtin, 234
Human neuroblastoma SH-SY5Y cell, 162
Human neurodegenerative disease, 219
Human peripheral nerve, 75
Human tau isoform, 229
Human tissue, regeneration of, 1
Huntington's disease, 1, 163–164, 220, 249
Hydra, 36
6-Hydroxydopamine (6-OHDA), 180
  exposure, 231
Hydroxylase enzyme, 35

5-Hydroxytryptamine, 36
5-Hydroxytryptophan, 31
Hyperglycemia, 265
Hyperphosphorylated tau protein, 14, 226
Hypoglycemia, 159
Hypothalamic pituitary adrenal (HPA) axis, 146
Hypothalamic-pituitary-adrenocortical, 126
Hypothalamus, 41, 110
Hypoxia, 11, 220
increased BrdU, 11
stimulated proliferation, 10

I

IBM. *See* Inclusion body myositis
Icv. *See* Intracerebroventricular
Icv administration, 192
IG. *See* Immunoglobulin
IGF-I. *See* Insulin-like growth factor-I
Ig-like domain, 63
βIII-tubulin, 5, 9–10, 13, 16–18, 89
IL1 sensory neurons, 220
Imipramine, 232
Immature neuronal marker, 11
Immunoelectron microscopy, 227
Immunoglobulin, 63
Impaired judgment, 145
IN-1, 61
antibody, 61
Inclusion body myositis, 228
Indole alkaloid, 32, 35
Inhibition, 70, 162
Inhibitor, 17
Inhibitory electrical influences, 33
Injury-induced neurogenesis, 20
Inositol-monophosphatase, 160
Inositol-1, 4, 5-trisphosphate, 160
Institutional care, 2
Insula, 150
Insulin-like growth factor-I, 6
Interleukin-6, administration of, 91
Interleukin-10, 78
Interneuron, 38
Intracellular adaptation, 8
Intracellular calcium overload, 8
Intracellular cAMP, 77
Intracellular neurofibrillary tangle, 14
Intracellular organelle, 35
Intracerebroventricular (icv) infusion, 191
Intracytoplasmic inclusion, 16
Intranuclear inclusion, 16
Intravascular suture occlusion, 8
Ionic homeostasis, 8
Ischemia, 11, 20, 220, 241, 253
sublethal forebrain, 7
Ischemia-induced neurogenesis, 11, 14
Ischemia-induced neuromigration, 13
Ischemia-induced neuroproliferation, 13
Ischemic brain lesion, 13
Ischemic cell, 8
Ischemic cerebral cortex, 12
Ischemic heart disease, 48
Ischemic penumbra, 12
Ischemic striatum, 12

K

Kainic acid, 125
Keratin sulfate proteoglycan NG2, 60
Ketanserine, 36
α-Ketoglutarate, 266, 271
α-Ketoglutarate dehydrogenase, 277
Kinase signaling pathway, 248
Korsakoff's psychosis, 265
Korsakoff's syndrome, 2

L

Lactate, 253
LacZ positive cell, 19
Lamina propria OEC, 83
Laminin, 45
Lateral cortical stream, 13
Lateral ventricles, 150
LB. *See also* Lewy body
disease, 182
pathology, 182
L-3, 4-dihydroxyphenylalanine, 184
L-dopa, 186, 201. *See also* Levodopa
L4 DRG, 87
Left dorsal anterolateral, 148
Leucine-rich repeat, 63
Levodopa, 17, 179. *See also* L-3, 4-dihydroxyphenylalanine
Lewy body, 48
absence of, 231
disease, 181
LIM homeodomain, 40
LINGO-1, 66
receptor complex, 69

Lisuride, 184
Liver growth, 2
LLRK2, 233
Long-term potentiation, 152, 197
Loss of appetite, 192
Loss of depression, 192
Loss of hallucinations, 192
Loss of nausea, 192
Loss of tingling, 192
Low-affinity neurotrophin receptor, 65
Lower motor neuron, 19
Loxapine, 124
LP-OEC morphology, 84
LRR, 65. *See also* Leucine-rich repeat
LRRK2. *See also* Leucine-rich repeat kinase-2
  gene, 183
  mutation, 183
LSD. *See* D-lysergic acid diethylamide; Lysergic acid diethylamid
LTP. *See* Long-term potentiation
Lumbar spinal cord, 19
LY404187, 201–202
LY503430, 198, 201–202

M

MAG, 62–66, 68–70, 73. *See also* Myelin-associated glycoprotein
  binding, 65
  deficient, 66
  inhibition, 64, 67, 69–70
  transducer for, 66
MAG-deficient mice, 67
Magnetic resonance imaging, 146
Magnetic resonance spectroscopy, 148–149
MANF. *See* Mesencephalic astrocyte-derived neurotrophic factor
Mania, 145–146
MAO, 17. *See also* Monoamine oxidase
MAO-B inhibitor, 186
  selegiline, 184
MAP-2, 9–10
MAP, phosphorylation of, 45
MAPK
  inhibitor, 199
  pathway, 199
Maprotiline, 129
MAP-specific inhibition, 45
MAPT. *See* Microtubuleassociated protein tau
MAPT-encoding, 183
Marrow stromal cell, 87
Mast cell, 37–38
Maturation, 35
Maturational pattern, 41
Maturation of plant cell, 36
Mature neuronal marker, 10
MCAO, 10, 12–14. *See also* Middle cerebral artery occlusion
MCAO-induced neurogenesis, 10
Mec. *See* Mechanosensation phenotype
Mechanosensation phenotype, 230
MeCP2. *See* Methyl-CpG-binding protein
Mec-7 promoter, 229
Medulla oblongata, 60
Medullary raphe neuron, 42
Melatonin, 31–32
Membrane-spanning domain, 229
Memory, loss of, 226
Meninges, 83
Mental behavior, 38
Mental disorder, 43
Mental health, 47
Mesencephalic astrocyte-derived neurotrophic factor, 193
Mesencephalic dopaminergic neuron, 110
Mesencephalic primary neuron, 191
Mesencephalic stem cell, 18
Mesencephalon tissue, 193
Methyl-CpG-binding protein, 275
Methylprednisolone, 76–77
Microtubule-destabilizing protein, 13
Microtubule-stabilizing protein, 13
Middle cerebral artery occlusion, 10
Migraine, 3
  headache, 42
Migrational pattern, 41
Mineralocorticoid receptor, 146
Mitochondrial complex I inhibitor, 183
Mitogen-activated protein kinase kinase (MAPKK) pathway, 193
Mitosis, 31
Mitotic cleavage, 10
Monoamine inhibitor, 185
Monoamine oxidase (MAO), 252
Monoamine oxidase (MAO)-B, 182
Monoamine oxidase (MAO) inhibitor, 179
Monoterpenoid indole, 35
Monozygotic twins, 147
Mood stabilizers, 145, 155, 157
  treatment, 161

Motor cortex, 19
Motor neuron, 37
  disease, 8, 19
Motor neuron-targeted expression, 232
Mouse model, 17
MPP^+. *See* N-Methyl-4-phenylpyridinium ion
MPTP, 17, 182, 187
  administration, 189
  induced neurotoxicity, 201
  model, 195
  toxicity, 185
MPTP-induced nigral cell death, 182
MPTP-mediated cell death, 191
MPTP-treated monkey, 191
MPTP-treated mouse, 191
MR. *See* Mineralocorticoid receptor
MRI. *See* Magnetic resonance imaging
mRNA, 45
MRS. *See* Magnetic resonance spectroscopy
MSC. *See* Marrow stromal cell
Multicellular organism, 35–36
Multiple sclerosis, 19
Multipotent neural stem cell, 150
Murine cell culture, 240
Musashi1- immunopositive cell, 10
Muscarinic anticholinergics, 17
Muscle atrophy, 19
Muscular rigidity, 179
Mutant superoxide dismutase, 19
Mutant tau sequences, 229
Mutant tau transgenic lines, 229
Myelin, inhibition, 73
Myelin-associated glycoprotein, 63
Myelin-associated inhibitor, 61, 64
Myelinated axon, 76, 78, 79, 80
Myelinating oligodendrocytes, 158
Myelinating phenotype, 81
Myelin-based inhibitor, 64, 67, 69
Myelin inhibition, 64, 69–70
Myelin inhibitor, 64–65, 69, 73
Myelin signal transduction, 69
Myoinositol, 160
Myristoylated alanine-rich C kinase substrate (MARCKS), 160

N

NAA. *See* N-acetyl aspartate
NAA level, 149
N-acetyl aspartate, 148
NADH, 34
NADPH, 34
  neurotransmitter synthesis, 252
N-cadherin, 73
N-CAM, 80. *See also* Neural cell adhesion molecule
NCR-1, 237
NDPK. *See* Nucleoside diphosphate kinase
NE, 42
Necrosis, 182
Necrotic cell death, 8, 220, 241
Necrotic neurodegeneration, 232
  genetic modifiers, 238
Necrotic neurodegenerative change, 219
Necrotic pathway, 220
Nematode, 38
Neocortex, 48
Neocortical neurogenesis, 158
NEP1-40, 65
Nerve fiber, 80
Nerve growth factor (NGF), 73, 109, 250
  BDNF, 162
Nerve regeneration, 57
Nervous system development, 70
NeuN, 9–10, 12, 17
NeuN-positive cell, 11
Neural cell, 32
  adhesion molecule, 73
  lineages, 159
Neural precursor cell, 89
Neural progenitor cell, 73, 89, 90
Neural stem cell, 10, 90, 116, 155, 157
Neuregulin, 255
Neurite extension, 143
Neurite outgrowth, 68–70, 73, 249–252, 255, 257, 260, 275, 278
  promotion of, 272
Neuroblast, maturation of, 156
Neuroblastoma cell, 196
Neurodegeneration, 2, 124, 164, 182, 186–187, 230, 234, 240–241, 242
  chronic, 3
  modifier genes, 238
  molecular inhibitors, 239
  phenotype, 230
  progression of, 185
  torsion proteins, 239
Neurodegenerative disease, 7, 182, 194, 220, 242, 249, 274
  acute, 3
  chronic, 3

Neurodegenerative disorder, 16–17, 180, 219, 226, 236
Neurodegenerative phenotype, 221
Neurodevelopmental disorder, 248, 254, 274
Neurodifferentiation, 4
Neuroenhancement, 247, 268
Neuroenhancing agent, 250
Neuroenhancing drug, 277
Neurofibrillary tangles, 226
Neurogenesis, 6, 9, 33, 89, 114, 129, 143, 154–155, 201, 249
 adult, 1, 4
 endogenous, 1
 impaired, 15
 induction of, 15
 injury-induced, 6–7, 9
 ischemia-induced, 10
 stimulation, 2, 6, 11
 in stroke, 8
Neurogenesis-promoting effect, 11
Neurogenesis response, 20
Neurohumoral agent, 115
Neuroleptic drug, 107, 115, 123
Neuroleptic effect, 119
Neurological disease, 1
Neurological disorder, 3, 48, 164
Neuromigration, 13
Neuron(s), 45, 63, 66, 150
 birth of, 2
 corticothalamic, 6
 degeneration, 203, 231
 endogenous, 4
 exogenous sources, 4
 loss of, 14, 219
 morphology of, 41
 newborn, 4
 number, 4
 production, 152
 size, 4
Neuronal, 20
 activity, 42
 cell, 59, 160, 162, 193, 232, 251, 266, 270
  cultures, 162
 cytoskeleton, 45
 dendrites, 46
 differentiation, 4, 18, 45, 155–156, 191, 251, 276
  antigen NeuroD, 5
 dysfunction, 229–231, 240
 function, 247, 250
 lineage marker protein, 19
 loss, 231
 marker, 9, 18
  protein, 14
 migration, 4, 249
 precursor, 5, 19, 45
 progenitor, 189–190
  cell, 189
 proliferation, 1, 4, 275
 protein, 12
 settling, 4
 stem cell, 190
 survival, 45, 249
 system, 32, 43
Neuropathology, 239
Neuropeptide, 33, 44
Neuropil, loss of, 14
Neuroplasticity, 31–32, 43, 47, 143–146, 160, 164–165
 in mammal, 36
Neuroplastic pathway, 201
Neuroplastic processes, 143
Neuroproliferative capacity of brain, 1
Neuroproliferative SGZ, 14
Neuroproliferative zone, 6, 10, 19
Neuroprotection, 125, 252, 272
 outgrowth, 251–252
Neuroprotective action, 203
Neuroprotective agent, 186, 251
Neuroprotective drug, 186, 277
Neuroprotective effect, 107, 127, 195, 260
Neuroprotective properties, 129, 163, 184
Neuroprotective therapy, 186–187
Neuropsychological dysfunction, 146
Neuroreceptor, 108
Neurorestorative agent, 186, 197
Neurorestorative approach, 181
Neurorestorative drug, 186
Neurorestorative treatment, 190
Neurosurgical technique, 186
Neurotoxicity, 163, 195, 239
Neurotoxin-induced toxicity, 197
Neurotoxin (6-OHDA), 202
Neurotransmission, 15, 44
Neurotransmitter, 33, 36, 40, 276
 receptor, 45
 release, 144
 serotonin, 32
 synthesis, 44
Neurotrophic action, 43, 201, 203
Neurotrophic activity, 191

Neurotrophic agent, 187, 197
Neurotrophic approach, 180–181
Neurotrophic effect, 107, 109
Neurotrophic factors, 80, 86, 152, 156, 180, 190, 196, 250
Neurotrophic pathway, 201
Neurotrophin, 70, 76–77, 90, 108, 193
  receptor, 64
Neurotrophin-3, 76
Neurotrophin-expressing fibroblast, 71
Neurovascular position, 41
Neurovegetative symptoms, 144
New born cell, 155
Newborn neuron, 11–12
NFT. *See* Neurofibrillary tangle
NFT pathology, 231
NG2, 158
NGF, 77, 80, 86, 110, 251, 257–258, 260, 276. *See also* Nerve growth factor
  cell, 121
  level, 109
  receptor, 113
NGF-responsive axon, 76, 85
NG2-positive cells, 159
NG2-positive progenitors, 159
NgR1, 64–67. *See also* Nogo receptor
NgR2, 65
Ngr1-deficient mice, 67
Ngr1 gene, 67
Nicotinic acetylcholine receptor, 221
Niemann-Pick type C1 disease, 236
Nigral cell, 17
Nigral dopaminergic neuron, 194
Nigrostriatal degeneration, 17, 203
Nigrostriatal dopaminergic neuron, 194
Nigrostriatal dopaminergic system, 184
Nigrostriatal lesion, 203
Nigrostriatal neuron, 196
Nigrostriatal pathway, 180–181, 186, 193
Nigrostriatal tract, 17
Nisoxetine, 232
NMDA, 125, 152. *See also* N-methyl-D-aspartate
  antagonist, 3, 6
  receptor, 124
NMDA-type glutamate receptor, 14
NMDA-type glutamate receptor antagonist riluzole, 19
N-methyl-D-aspartate, 124
N-methyl-4-phenylpyridinium ion, 119, 182
Nociceptive dorsal root axon, 59
Nogo, 62, 67
Nogo-66, 62, 64–68
Nogo-A, 62, 68–69
  elimination, 67
Nogo-A-24, 64
Nogo-A/B knockout mice, 68
Nogo-A knockout, 68
Nogo-B, 67, 68
  transcript, 68
  up-regulation, 68
Nogo-C, 62
Nogo receptor, 64
Nonimmunogenic polymer, 90
Nonischemic hemisphere, 13
Nonneurogenic zones, 156
Nonneuronal cell, 42
Nonneuronal system, 32
Noradrenergic axon, 76, 83
Noradrenergic locus coeruleus, 181
Noradrenergic neuron, 59, 110
Norepinephrine, 42
Novelty suppressed feeding task, 157
NPC. *See* Niemann-Pick type C1
NPC1
  disorder, 237
  function, 237
  gene, 236
NR4A2, 196
NSC. *See* Neural stem cell
NSF. *See* Novelty suppressed feeding task
NT, 110, 198. *See also* Neurotrophin
NT-3, 76–77, 86–87, 109
  administration of, 87
NTs, 109
Nuclear hormone receptor, 197
Nucleoside diphosphate kinase, 268
Nucleus raphe magnus, 41
Nucleus raphe pallidus, 41
Nucleus raphe ventricularis, 41
Nuronal differentiation, 1
Nuronal migration, 1
Nurr1, 196
Nurr1-RXR heterodimer activator, 197

O

OB, 12. *See also* Olfactory bulb
Obesity, 145
Obsessive-compulsive disorder, 48, 129
Occipital cortex, 110–111

OEC, 73, 78, 79–86. *See also* Olfactory ensheathing cell
myelination by, 84
of tracking, 85
transplantation, 79–86
OEC-transplanted animal, 83
OHDA, 186
6-OHDA, 18, 191, 195–196, 200–202. *See also* 6-hydroxydopamine
lesion, 194
model, 187, 203
neurotoxicity, 232, 239
toxicity, 197
6-OHDA-mediated cell death, 191
6-OHDA-positive cell, 18
Okadaic acid, 125–126
Olanzapine, 108, 110, 112, 114, 116, 119–121, 125–127, 129, 159, 248, 250, 255, 257–258, 260–261, 273–274, 276
Olfactory axon, 80–82
Olfactory bulb, 80, 118, 150, 158
Olfactory ensheathing cell, 73
Olfactory epithelium, 79
Olfactory lamina propria, 83
Olfactory nerve, 80
Olfactory neurogenesis, 80
Olfactory neuron precursor, differentiation of, 80
Olfactory system, 85
Oligodendrocyte, 61, 64, 83, 90, 150, 159
myelin glycoprotein, 64
Olivopontocerebellar atrophy, 181
OLP. *See* Olfactory lamina propria
OLP transplantation, 84
OMgp, 64–68. *See also* Oligodendrocyte myelin glycoprotein
OMgp binding, 65
Optic nerve, 58, 60, 61
crush, 68
lesioning, 67
regeneration, 66
Orbital PFC, 147
Ornithine decarboxylase
inhibition, 70
Osmolarity, 40
Oxidative phosphorylation, 252, 271

P

pAkt. *See* Phosphorylated Akt
Panic disorder, 42, 48
Pan-neuronal expression, 232
Para-chloroamphetamine PCPA, 46
Paraquat, 17
Parkin, 16, 183, 232
Parkinsonian degeneration, 188
Parkinsonian symptom, 19, 181
Parkinsonian toxin MPTP, 18
Parkinsonism, 18
Parkinson's disease, 1, 48, 163–164, 179, 220, 249, 251
Pathogenic mutation
DJ-1, 183
LRRK2, 183
PINK-1, 183
PRKN, 183
SNCA, 183
Pathological phosphoepitope, 231
PC12 cell, 121
PCNA, 9–10, 16. *See also* Proliferating cell nuclear antigen
PCP, 125. *See also* Phencyclidine
PCPA. *See* Para-chloroamphetamine
PD, 180. *See also* Parkinson's disease
depression in, 181
diagnosis of, 180
etiology of, 180, 195
neurodegeneration in, 182
neurodegenerative disorder, 231
pathology, 181–182
patient, Postmortem study, 186
symptom, 179
symptom of, 180, 184
treatment, 184, 186, 191
PDE. *See* Phosphodiesters
PDE4. *See also* Phosphodiesterase 4
inhibition, 70
inhibitor, 69
PDGFβ. *See* Platelet-derived growth factor-β
PD-like syndrome, 182
PD-related gene, 233
Pentose phosphate pathway, 252
Pergolide, 184
Periinfarction cortex, 10
Perinatal rat, 80
Peripheral axon, 58
Pertussis toxin, 71
PET, 147. *See also* Positron emission tomography
PFC. *See* Left dorsal anterolateral
PFK. *See* Phosphofructokinase
Phencyclidine, 124

Phenylalanine, 33, 35
Phosphatidylinositol 3-kinase, 120, 259
Phosphodiesterase 4 inhibition, 70
Phosphodiesterase inhibitor, 6
Phosphodiesters, 149
Phosphoepitopes, 230
Phosphofructokinase, 268
Phosphoinositide phosphate (PIP-2), 160
Phosphoinositol, 149
Phospholipase C (PLC), 160
Phosphor-monoesters, 149
Phosphorylated Akt, 261
PI. *See* Phosphoinositol
Pick's disease, 48
PI3K. *See* Phosphatidylinositol 3-kinase
PI3K-Akt signaling pathway, 263, 277
PINK-1, 233. *See also* PTEN-induced putative kinase-1
Pirenzepine, 257
P13K. *See* Phosphatidylinositol 3-kinase
PKA, 70
  inhibitor, 268
P3K-AKT/mitogen-activated protein kinase kinase (MAPKK) pathway, 193
PKB. *See* Protein kinase B
PKC, 45. *See also* Protein kinase C
  phosphorylation, 45
Plaque, 14
Plasma cortisol, 41
Pluripotent precursors, 144
PME. *See* Phosphor-monoesters
PNS, 44, 82
PNS–CNS, 82
$P75^{NTR}$, 65–71
Polyamine, 73
  synthesis, 70
Polyglutamine aggregation, genetic modifiers, 238
Polyglutamine disorders, 233
Polyglutamine peptides, 233
Polyglutamine (polyQ)-mediated neurotoxicity, 234
Polyhydroxybutyrate, 90
Polylactic acid, 90
PolyQ. *See also* Polyglutamine
  aggregation, 239
  cytotoxicity, 236, 239
  disorder, 236
  modifiers, 238
  protein, 234
  toxicity, 234–235, 239
  transgenic animal, 235
Positron emission tomography, 186
Postmortem brain analyses, 149–150
Postnatal hippocampus, 159
Postsynaptic cell, 33
Postsynaptic neuron, 31
Postsynaptic neuronal firing, 43
Postural instability, 17
PPP, 253. *See also* Pentose phosphate pathway
Pramipexole, 184–185, 251
Prefrontal cortex, 43, 111–112, 129, 150
Prelimbic cortex, 116
Presenilin 1, 226
Presenilin function, 228–229
Primary neuronal culture, 199
Priming, 70
Progenitor cell, 17, 114, 156
  proliferation of, 155
Progenitor proliferation, 156
Programmed cell death, 4, 8, 65, 109, 220
Proliferating cell, 150, 159
Promethazine, 119
Prophylactic effects, 145
Propriospinal neuron, 77
Proteasome activity, 183
Proteasome function, 228
Proteasome inhibitor, 187
Protein
  kinase B, 248, 258
  kinase C (PKC), 69, 160
  tyrosine kinase, 199
Protein 43, 71
Proteoglycan, 61, 158
PS. *See* Presenilin
PS1, 15
PSA-NCAM, 115. *See also* Polysialylated neural cell adhesion molecule
PSI. *See* Presenilin 1
Psychiatric illness, 126
Psychomotor behavior, 144
Psychosis, 16
Psychosocial function, 146
Psychosocial stress, 145
Psychostimulants, 161
Psychotropic drug, 143–144, 152, 154, 250
PTEN-induced kinase 1, 16
PTEN-induced putative kinase-1, 183
PTX. *See* Pertussis toxin
Putative reticular function, 41

Pyridaben, 232
Pyruvate, 253
Pyruvate dehydrogenase, 265
Pyruvate kinase, 268

Q

Quadriplegic victim, 2
Quetiapine, 108, 114, 118–121, 125, 127, 129, 257, 260, 264

R

R118. *See* Arginine 120
RAGE, 45. *See also* Receptor for advanced glycation endproduct
Raphe neuron, 43
Raphespinal axon, 83, 89
Rat substantia nigra, 17
Reactive oxygen species, 119
Receptor for advanced glycation endproduct, 45
Regeneration axon, 57
Regional cerebral blood flow, 148
Reserpine, 107
Respiration, 40
Resting tremor, 179
Restoring brain function, 3
Reticulospinal axon, 77
Retinal ganglion cell, 162
Retraction bulb, 58
Retroviral gene trapping, 67
Rett syndrome, 249, 275
Rho activation, 69
Rho-associated kinase, 60
Rhodopsin, 34
Rho GDP, 69
Rho GTPase, 61
Rho inactivation, 69
Ribulose bisphosphate, 34
Right anterior cingulate cortex, 149
Right insula, 148
Rigidity, 17
Riluzole, 3, 251
Rimcazole, 124
Risperidone, 108–109, 111–112, 114, 116, 118–119, 125, 127, 257–258
RMS, 12. *See also* Rostral migratory stream
ROCK, 69. *See* Rho-associated kinase
Rodent embryonic stem (ES) cell, 189
Rolipram, 70–71, 78–79
Ropinirole, 184–185
ROS. *See* Reactive oxygen species
Rotenone, 17, 183, 187, 232
  toxicity, 240
RTK. *See* Tyrosine kinase
Rubisco, 34
RuBP. *See* Monoamine oxidase
RuBP carboxylase, 34
Rubrospinal axon, regeneration of, 76, 86
RXR, 197

S

Sabeluzole, 251
SAR. *See* Structure-activity relationship
SAR analysis, 273
SBMA. *See* Spinal and bulbar muscular atrophy
SCA. *See* Spinocerebellar ataxias
SCF. *See* Stem-cell factor
Schizophrenia, 107, 123–124, 126, 129, 248–250
  etiology, 254–256
  pathology, 254–256
  treatment, 254–256
Schwann cell, 73, 75–79, 82–86, 90
  channel, implantation of, 77
  graft, 77
  migration of, 85
  myelination, 82
  transplant, 71, 74–75, 77, 79, 84, 89
Schwann cell-like OEC, 82
Sciatic nerve graft, 60
Sea star's arms, regrowth of, 2
α-Secretase, 69
γ-Secretase, 226, 228
Seizures, 253
Sensorimotor cortex, 82
Sensory neurons, 38
Septo-hippocampal pathway, 75
Serotonergic, 37
Serotonergic axon, 78, 84–85
  regeneration of, 78
Serotonergic fiber, 71
Serotonergic (5-hydroxytryptamine, 5-HT) nucleus, 181
Serotonergic marker, 75
Serotonergic neuron, 32, 40–42, 46, 110
Serotonergic system, 37
Serotonin, 31–33, 35, 43, 125, 155, 255
  level, 43, 47–48

neuron, 36, 38, 43
receptor, 257–258, 264
synthesis inhibitor, 46
Serotoninergic neuron, 42
Serotonin 5-HT, 33
transmission, 154
Serotonin-producing cell, 37
Sex hormone, 6
SGZ, 9. *See also* Subgranular zone
Shh, 197. *See also* Sonic hedgehog
Sialic acid, 65
SIDS. *See* Sudden infant death syndrome
Single nucleotide polymorphism, 260
Single-photon emission computed tomography, 186
Skin pigment cell, 34
SNc, 182, 195. *See also* Substantia nigra pars compacta
SNCA mutation, 183
SNP. *See* Single nucleotide polymorphism
Social behavior, 38
Social phobia, 48
SOD. *See* Superoxide dismutase
SOD1. *See* Superoxide dismutase 1 gene
SOD1-mediated toxicity, 236
SOD1 mutation, 19, 236
Somatic disease, 48
Sonic hedgehog, 189
SPECT, 147. *See also* Single-photon emission computed tomography
Spinal cord, 19, 32, 41, 58–60
axon, greater myelination of, 78
cervical segment, 76
contusion, 78, 79, 84, 89
demyelinated, 82
glial scarring in, 60
injury, 2, 19, 60, 66–68, 71, 74–75, 79–80, 84, 87
cure for, 91
transplantation, 86
treatment of, 90
interneuron, 76
lesioning, 67
neurogenesis, 19
neuron, 82
regeneration, 75
sacral segment, 77
trauma, 85–87
Spinal muscular atrophy, 233
Spinocerebellar ataxias, 233
Spongiform encephalopathies, 48
SSRI. *See* Serotonin-specific reuptake inhibitor
SSRI fluoxetine, 159
Statin, 6
Stem cell
adult, 5
factor, 6
hematopoietic, 87
Steroid
addition, 68
administration, 68
Stigmatellin, 232
STN. *See* Subthalamic nucleus
Striatal neurogenesis, 18, 194
Striatal-pallidal excitotoxic injections of ibotenic acid, 195
Striatal TH, 202
immunoreactivity, 202
Striatum, 48
Stroke, 1–2, 8, 220, 241
aftermath of, 3
survivor of, 2
Stromal cell, 73, 189
Structure-activity relationship, 261
Subcutaneous FGF-2, 16
Subgranular zone, 114, 150
Substantia nigra, 201–202
Substantia nigra pars compacta, 180
Subthalamic nucleus, 185
Subventricular zone, 114, 150
Sulpiride, 110
Superoxide dismutase, 119
Superoxide dismutase 1 gene, 236
Supraleminiscal nucleus, 41
Supraspinal regeneration, 76
SVZ, 10, 12–15, 18, 118, 154, 158. *See also* Subventricular zone
neurogenesis, 12
neuronal progenitor, 10
Symptomatic agent, 186
Synaptic activity, 268
Synaptic plasticity, 144, 149–150
Synaptic proteins, 149
Synaptic strength, 143
Synaptic transmission, 197
Synaptogenesis, 162
Synaptophysin, 149
Synaptophysin staining, 15
α-Synuclein, 16–17, 194–196, 233 181

α-Synuclein, mutation in
DJ-1, 231
LRRK2, 231
parkin, 231
PINK, 231
α-Synuclein-induced toxicity, 232
α-Synuclein neurotoxicity, 241
α-Synucleinopathy, 232
α-Synuclein Pathology, 232

T

TAJ, 67
Tau-induced Mec phenotype, 230–231
Tau kinase genes
cdk-5, 230
gsk-3, 230
jkk-1, 230
mek-1, 230
Tau protein function, 227
Taurosodeoxycholic acid, 232
Tauroursodeoxycholic acid, 240
TCA cycle, 266
Telenzepine, 257
Tello's peripheral nerve, 59
Temperature, 43
Temporal cortex, 148
Tenascin, 60
TH, 17. *See also* Tyrosine hydroxylase
in ES cell, 196
immunoreactivity, 202
Thalamus, 41
TH and DAT. *See* Dopamine transporter
Thiamine, 250
deficiency, 265
Thienobenzodiazepine, 273
Thioflavin S, 227
Thioridazine, 124
TH-positive cell, 18
Thrombolytic agent, 2
Thrombotic arterial occlusion, 8
Thymidine, 2, 5, 9
marker, 115
Tissue regeneration
annelids, 2
flatworms, 2
hydra, 2
newts, 2
salamanders, 2
seastars, 2
sponges, 2
Tolcapone, 185
TOR1A gene, 239
Toxic protein aggregation, 233
Transected axon, 58
Transgenic mice, 15–16, 19–20
Transgenic model, 18
Transmembrane domains, 228
Transthyretin peptides, 227
Trauma, 220, 241
Tremor, 17
Trifluoperazine, 118–119, 124
Trk. *See* Tyrosine kinase
TrkB. *See* Tyrosine kinase receptor B
TrkB-Fc fusion protein, 11
TrkB mRNA-containing neuron, 194
Trophic factor stimulation, 4
Trophotrophic system, 42
TROY, 66
TROY/TAJ-deficient mice, 67
Tryptophan, 31, 33–36
biosynthetic enzyme, 35
hydroxylase, 33, 35, 38, 40
synthesis, 35
TSC2, 274. *See also* Tuberous sclerosis complex protein
Tuberin, 276
Tuberous sclerosis complex protein, 259
β-Tubulin, 115
TUC-4, 14
TUDCA. *See* Tauroursodeoxycholic acid
Tumor necrosis factor receptor, 65
Tyrosine, 33, 35
Tyrosine hydroxylase (TH)-expressing cell, 17
Tyrosine hydroxylase (TH)-immunoreactive labeling, 112
Tyrosine hydroxylase (TH)-positive neuron, 189
Tyrosine kinases (Trk), 109, 264
neurotrophins, 162
receptor, 70
receptor B, 193
Tyrosine phosphorylation, 66

U

UBD. *See* Ubiquitin-like domain
Ubiquitin, 17
Ubiquitin-like domain, 232
Ubiquitin-proteasomal activity, 182
Unc. *See* Uncoordinated locomotion

Uncoordinated locomotion, 221
Unc phenotype, 232, 238
Unicellular organism, 43
Unified Parkinson's Diease Rating Score, 185
Unilateral focal ischemia, 12
Unique parkin domain, 232
Unmyelinated axon, 76, 80
UPD, 232. *See also* Unique parkin domain
UPDRS, 191–192, 203. *See also* Unified Parkinson's Diease Rating Score
Upper motor neuron, 19

V

Valproic acid, 145, 149
Vascular, system, 43
Vascular degeneration, 148
Vascular endothelial growth factor, 6
VEGF, 11. *See also* Vascular endothelial growth factor
Venlafaxine, 129
Ventral anterior cingulate gyrus, 150
Ventral mesencephalic (VM) tissue, 188
Ventral migratory mass, 13
Ventral PFC, 147
Ventral striatum, 150
Vestibulospinal axon, 77
Vimentin, 80
VM. *See* Ventral mesencephalic
VMAT2 ES cell, 196
VM tissue, 188–189
  transplantation, 188
VPA, 153, 155, 160–162, 164

W

Wallerian degeneration, 73
WGA. *See* Wheat germ agglutinase
Wheat germ agglutinase -HRP, 78
White matter hyperintensities, 147
WMH. *See* White matter hyperintensities
Wnt
  glycoproteins, 162
  pathway, 161
  signaling, 162
  signaling pathway, 162
Wound healing, 2, 68

X

Xaliproden, 251
XXX cell, 237
Xylosyltransferase-1, 61

Y

Yellow fluorescent protein, 234
YFP. *See* Yellow fluorescent protein

Z

Ziprasidone, 125–126, 257–258
Zotepine, 126

CONTENTS OF RECENT VOLUMES

Volume 37

Section I: Selectionist Ideas and Neurobiology

Selectionist and Instructionist Ideas in Neuroscience
Olaf Sporns

Population Thinking and Neuronal Selection: Metaphors or Concepts?
Ernst Mayr

Selection and the Origin of Information
Manfred Eigen

Section II: Development and Neuronal Populations

Morphoregulatory Molecules and Selectional Dynamics during Development
Kathryn L. Crossin

Exploration and Selection in the Early Acquisition of Skill
Esther Thelen and Daniela Corbetta

Population Activity in the Control of Movement
Apostolos P. Georgopoulos

Section III: Functional Segregation and Integration in the Brain

Reentry and the Problem of Cortical Integration
Giulio Tononi

Coherence as an Organizing Principle of Cortical Functions
Wolf Singerl

Temporal Mechanisms in Perception
Ernst Pöppel

Section IV: Memory and Models

Selection versus Instruction: Use of Computer Models to Compare Brain Theories
George N. Reeke, Jr.

Memory and Forgetting: Long-Term and Gradual Changes in Memory Storage
Larry R. Squire

Implicit Knowledge: New Perspectives on Unconscious Processes
Daniel L. Schacter

Section V: Psychophysics, Psychoanalysis, and Neuropsychology

Phantom Limbs, Neglect Syndromes, Repressed Memories, and Freudian Psychology
V. S. Ramachandran

Neural Darwinism and a Conceptual Crisis in Psychoanalysis
Arnold H. Modell

A New Vision of the Mind
Oliver Sacks

INDEX

Volume 38

Regulation of $GABA_A$ Receptor Function and Gene Expression in the Central Nervous System
A. Leslie Morrow

Genetics and the Organization of the Basal Ganglia
Robert Hitzemann, Yeang Olan, Stephen Kanes, Katherine Dains, and Barbara Hitzemann

Structure and Pharmacology of Vertebrate $GABA_A$ Receptor Subtypes
Paul J. Whiting, Ruth M. McKernan, and Keith A. Wafford

Neurotransmitter Transporters: Molecular Biology, Function, and Regulation
Beth Borowsky and Beth J. Hoffman

Presynaptic Excitability
Meyer B. Jackson

Monoamine Neurotransmitters in Invertebrates and Vertebrates: An Examination of the Diverse Enzymatic Pathways Utilized to Synthesize and Inactivate Biogenic Amines
B. D. Sloley and A. V. Juorio

Neurotransmitter Systems in Schizophrenia
Gavin P. Reynolds

Physiology of Bergmann Glial Cells
Thomas Müller and Helmut Kettenmann

INDEX

Volume 39

Modulation of Amino Acid-Gated Ion Channels by Protein Phosphorylation
Stephen J. Moss and Trevor G. Smart

Use-Dependent Regulation of $GABA_A$ Receptors
Eugene M. Barnes, Jr.

Synaptic Transmission and Modulation in the Neostriatum
David M. Lovinger and Elizabeth Tyler

The Cytoskeleton and Neurotransmitter Receptors
Valerie J. Whatley and R. Adron Harris

Endogenous Opioid Regulation of Hippocampal Function
Michele L. Simmons and Charles Chavkin

Molecular Neurobiology of the Cannabinoid Receptor
Mary E. Abood and Billy R. Martin

Genetic Models in the Study of Anesthetic Drug Action
Victoria J. Simpson and Thomas E. Johnson

Neurochemical Bases of Locomotion and Ethanol Stimulant Effects
Tamara J. Phillips and Elaine H. Shen

Effects of Ethanol on Ion Channels
Fulton T. Crews, A. Leslie Morrow, Hugh Criswell, and George Breese

INDEX

Volume 40

Mechanisms of Nerve Cell Death: Apoptosis or Necrosis after Cerebral Ischemia
R. M. E. Chalmers-Redman, A. D. Fraser, W. Y. H. Ju, J. Wadia, N. A. Tatton, and W. G. Tatton

Changes in Ionic Fluxes during Cerebral Ischemia
Tibor Kristian and Bo K. Siesjo

Techniques for Examining Neuroprotective Drugs *in Vitro*
A. Richard Green and Alan J. Cross

Techniques for Examining Neuroprotective Drugs *in Vivo*
Mark P. Goldberg, Uta Strasser, and Laura L. Dugan

Calcium Antagonists: Their Role in Neuroprotection
A. Jacqueline Hunter

Sodium and Potassium Channel Modulators: Their Role in Neuroprotection
Tihomir P. Obrenovich

NMDA Antagonists: Their Role in Neuroprotection
Danial L. Small

Development of the NMDA Ion-Channel Blocker, Aptiganel Hydrochloride, as a Neuroprotective Agent for Acute CNS Injury
Robert N. McBurney

The Pharmacology of AMPA Antagonists and Their Role in Neuroprotection
Rammy Gill and David Lodge

GABA and Neuroprotection
Patrick D. Lyden

Adenosine and Neuroprotection
Bertil B. Fredholm

Interleukins and Cerebral Ischemia
Nancy J. Rothwell, Sarah A. Loddick, and Paul Stroemer

Nitrone-Based Free Radical Traps as Neuroprotective Agents in Cerebral Ischemia and Other Pathologies
Kenneth Hensley, John M. Carney, Charles A. Stewart, Tahera Tabatabaie, Quentin Pye, and Robert A. Floyd

Neurotoxic and Neuroprotective Roles of Nitric Oxide in Cerebral Ischemia
Turgay Dalkara and Michael A. Moskowitz

A Review of Earlier Clinical Studies on Neuroprotective Agents and Current Approaches
Nils-Gunnar Wahlgren

INDEX

Volume 41

Section I: Historical Overview

Rediscovery of an Early Concept
Jeremy D. Schmahmann

Section II: Anatomic Substrates

The Cerebrocerebellar System
Jeremy D. Schmahmann and Deepak N. Pandya

Cerebellar Output Channels
Frank A. Middleton and Peter L. Strick

Cerebellar-Hypothalamic Axis: Basic Circuits and Clinical Observations
Duane E. Haines, Espen Dietrichs, Gregory A. Mihailoff, and E. Frank McDonald

Section III. Physiological Observations

Amelioration of Aggression: Response to Selective Cerebellar Lesions in the Rhesus Monkey
Aaron J. Berman

Autonomic and Vasomotor Regulation
Donald J. Reis and Eugene V. Golanov

Associative Learning
Richard F. Thompson, Shaowen Bao, Lu Chen, Benjamin D. Cipriano, Jeffrey S. Grethe, Jeansok J. Kim, Judith K. Thompson, Jo Anne Tracy, Martha S. Weninger, and David J. Krupa

Visuospatial Abilities
Robert Lalonde

Spatial Event Processing
Marco Molinari, Laura Petrosini, and Liliana G. Grammaldo

Section IV: Functional Neuroimaging Studies

Linguistic Processing
Julie A. Fiez and Marcus E. Raichle

Sensory and Cognitive Functions
Lawrence M. Parsons and Peter T. Fox

Skill Learning
Julien Doyon

Section V: Clinical and Neuropsychological Observations

Executive Function and Motor Skill Learning
Mark Hallett and Jordon Grafman

Verbal Fluency and Agrammatism
Marco Molinari, Maria G. Leggio, and Maria C. Silveri

Classical Conditioning
Diana S. Woodruff-Pak

Early Infantile Autism
Margaret L. Bauman, Pauline A. Filipek, and Thomas L. Kemper

Olivopontocerebellar Atrophy and Friedreich's Ataxia: Neuropsychological Consequences of Bilateral versus Unilateral Cerebellar Lesions
Thérèse Botez-Marquard and Mihai I. Botez

Posterior Fossa Syndrome
Ian F. Pollack

Cerebellar Cognitive Affective Syndrome
Jeremy D. Schmahmann and Janet C. Sherman

Inherited Cerebellar Diseases
Claus W. Wallesch and Claudius Bartels

Neuropsychological Abnormalities in Cerebellar Syndromes—Fact or Fiction?
Irene Daum and Hermann Ackermann

Section VI: Theoretical Considerations

Cerebellar Microcomplexes
Masao Ito

Control of Sensory Data Acquisition
James M. Bower

Neural Representations of Moving Systems
Michael Paulin

How Fibers Subserve Computing Capabilities: Similarities between Brains and Machines
Henrietta C. Leiner and Alan L. Leiner

Cerebellar Timing Systems
Richard Ivry

Attention Coordination and Anticipatory Control
Natacha A. Akshoomoff, Eric Courchesne, and Jeanne Townsend

Context-Response Linkage
W. Thomas Thach

Duality of Cerebellar Motor and Cognitive Functions
James R. Bloedel and Vlastislav Bracha

Section VII: Future Directions

Therapeutic and Research Implications
Jeremy D. Schmahmann

Volume 42

Alzheimer Disease
Mark A. Smith

Neurobiology of Stroke
W. Dalton Dietrich

Free Radicals, Calcium, and the Synaptic Plasticity-Cell Death Continuum: Emerging Roles of the Trascription Factor NFκB
Mark P. Mattson

AP-I Transcription Factors: Short- and Long-Term Modulators of Gene Expression in the Brain
Keith Pennypacker

Ion Channels in Epilepsy
Istvan Mody

Posttranslational Regulation of Ionotropic Glutamate Receptors and Synaptic Plasticity
Xiaoning Bi, Steve Standley, and Michel Baudry

Heritable Mutations in the Glycine, $GABA_A$, and Nicotinic Acetylcholine Receptors Provide New Insights into the Ligand-Gated Ion Channel Receptor Superfamily
Behnaz Vafa and Peter R. Schofield

INDEX

Volume 43

Early Development of the *Drosophila* Neuromuscular Junction: A Model for Studying Neuronal Networks in Development
Akira Chiba

Development of Larval Body Wall Muscles
Michael Bate, Matthias Landgraf, and Mar Ruiz Gmez Bate

Development of Electrical Properties and Synaptic Transmission at the Embryonic Neuromuscular Junction
Kendal S. Broadie

Ultrastructural Correlates of Neuromuscular Junction Development
Mary B. Rheuben, Motojiro Yoshihara, and Yoshiaki Kidokoro

Assembly and Maturation of the *Drosophila* Larval Neuromuscular Junction
L. Sian Gramates and Vivian Budnik

Second Messenger Systems Underlying Plasticity at the Neuromuscular Junction
Frances Hannan and Yi Zhong

Mechanisms of Neurotransmitter Release
J. Troy Littleton, Leo Pallanck, and Barry Ganetzky

Vesicle Recycling at the *Drosophila* Neuromuscular Junction
Daniel T. Stimson and Mani Ramaswami

Ionic Currents in Larval Muscles of *Drosophila*
Satpal Singh and Chun-Fang Wu

Development of the Adult Neuromuscular System
Joyce J. Fernandes and Haig Keshishian

Controlling the Motor Neuron
James R. Trimarchi, Ping Jin, and Rodney K. Murphey

Volume 44

Human Ego-Motion Perception
A. V. van den Berg

Optic Flow and Eye Movements
M. Lappe and K.-P. Hoffman

The Role of MST Neurons during Ocular Tracking in 3D Space
K. Kawano, U. Inoue, A. Takemura, Y. Kodaka, and F. A. Miles

Visual Navigation in Flying Insects
M. V. Srinivasan and S.-W. Zhang

Neuronal Matched Filters for Optic Flow Processing in Flying Insects
H. G. Krapp

A Common Frame of Reference for the Analysis of Optic Flow and Vestibular Information
B. J. Frost and D. R. W. Wylie

Optic Flow and the Visual Guidance of Locomotion in the Cat
H. Sherk and G. A. Fowler

Stages of Self-Motion Processing in Primate Posterior Parietal Cortex
F. Bremmer, J.-R. Duhamel, S. B. Hamed, and W. Graf

Optic Flow Analysis for Self-Movement Perception
C. J. Duffy

Neural Mechanisms for Self-Motion Perception in Area MST
R. A. Andersen, K. V. Shenoy, J. A. Crowell, and D. C. Bradley

Computational Mechanisms for Optic Flow Analysis in Primate Cortex
M. Lappe

Human Cortical Areas Underlying the Perception of Optic Flow: Brain Imaging Studies
M. W. Greenlee

What Neurological Patients Tell Us about the Use of Optic Flow
L. M. Vaina and S. K. Rushton

INDEX

Volume 45

Mechanisms of Brain Plasticity: From Normal Brain Function to Pathology
Philip. A. Schwartzkroin

Brain Development and Generation of Brain Pathologies
Gregory L. Holmes and Bridget McCabe

Maturation of Channels and Receptors: Consequences for Excitability
David F. Owens and Arnold R. Kriegstein

Neuronal Activity and the Establishment of Normal and Epileptic Circuits during Brain Development
John W. Swann, Karen L. Smith, and Chong L. Lee

The Effects of Seizures of the Hippocampus of the Immature Brain
Ellen F. Sperber and Solomon L. Moshe

Abnormal Development and Catastrophic Epilepsies: The Clinical Picture and Relation to Neuroimaging
Harry T. Chugani and Diane C. Chugani

Cortical Reorganization and Seizure Generation in Dysplastic Cortex
G. Avanzini, R. Preafico, S. Franceschetti, G. Sancini, G. Battaglia, and V. Scaioli

Rasmussen's Syndrome with Particular Reference to Cerebral Plasticity: A Tribute to Frank Morrell
Fredrick Andermann and Yuonne Hart

Structural Reorganization of Hippocampal Networks Caused by Seizure Activity
Daniel H. Lowenstein

Epilepsy-Associated Plasticity in gamma-Amniobutyric Acid Receptor Expression, Function and Inhibitory Synaptic Properties
Douglas A. Coulter

Synaptic Plasticity and Secondary Epileptogenesis
Timothy J. Teyler, Steven L. Morgan, Rebecca N. Russell, and Brian L. Woodside

Synaptic Plasticity in Epileptogenesis: Cellular Mechanisms Underlying Long-Lasting Synaptic Modifications that Require New Gene Expression
Oswald Steward, Christopher S. Wallace, and Paul F. Worley

Cellular Correlates of Behavior
Emma R. Wood, Paul A. Dudchenko, and Howard Eichenbaum

Mechanisms of Neuronal Conditioning
David A. T. King, David J. Krupa, Michael R. Foy, and Richard F. Thompson

Plasticity in the Aging Central Nervous System
C. A. Barnes

Secondary Epileptogenesis, Kindling, and Intractable Epilepsy: A Reappraisal from the Perspective of Neuronal Plasticity
Thomas P. Sutula

Kindling and the Mirror Focus
Dan C. McIntyre and Michael O. Poulter

Partial Kindling and Behavioral Pathologies
Robert E. Adamec

The Mirror Focus and Secondary Epileptogenesis
B. J. Wilder

Hippocampal Lesions in Epilepsy: A Historical Review
Robert Naquet

Clinical Evidence for Secondary Epileptogensis
Hans O. Luders

Epilepsy as a Progressive (or Nonprogressive "Benign") Disorder
John A. Wada

Pathophysiological Aspects of Landau-Kleffner Syndrome: From the Active Epileptic Phase to Recovery
Marie-Noelle Metz-Lutz, Pierre Maquet, Annd De Saint Martin, Gabrielle Rudolf, Norma Wioland, Edouard Hirsch, and Chriatian Marescaux

Local Pathways of Seizure Propagation in Neocortex
Barry W. Connors, David J. Pinto, and Albert E. Telefeian

Multiple Subpial Transection: A Clinical Assessment
C. E. Polkey

The Legacy of Frank Morrell
Jerome Engel, Jr.

Volume 46

Neurosteroids: Beginning of the Story
Etienne E. Baulieu, P. Robel, and M. Schumacher

Biosynthesis of Neurosteroids and Regulation of Their Synthesis
Synthia H. Mellon and Hubert Vaudry

Neurosteroid 7-Hydroxylation Products in the Brain
Robert Morfin and Luboslav Stárka

Neurosteroid Analysis
Ahmed A. Alomary, Robert L. Fitzgerald, and Robert H. Purdy

Role of the Peripheral-Type Benzodiazepine Receptor in Adrenal and Brain Steroidogenesis
Rachel C. Brown and Vassilios Papadopoulos

Formation and Effects of Neuroactive Steroids in the Central and Peripheral Nervous System
Roberto Cosimo Melcangi, Valerio Magnaghi, Mariarita Galbiati, and Luciano Martini

Neurosteroid Modulation of Recombinant and Synaptic $GABA_A$ Receptors
Jeremy J. Lambert, Sarah C. Harney, Delia Belelli, and John A. Peters

$GABA_A$-Receptor Plasticity during Long-Term Exposure to and Withdrawal from Progesterone
Giovanni Biggio, Paolo Follesa, Enrico Sanna, Robert H. Purdy, and Alessandra Concas

Stress and Neuroactive Steroids
Maria Luisa Barbaccia, Mariangela Serra, Robert H. Purdy, and Giovanni Biggio

Neurosteroids in Learning and Memory Processes
Monique Vallée, Willy Mayo, George F. Koob, and Michel Le Moal

Neurosteroids and Behavior
Sharon R. Engel and Kathleen A. Grant

Ethanol and Neurosteroid Interactions in the Brain
A. Leslie Morrow, Margaret J. VanDoren, Rebekah Fleming, and Shannon Penland

Preclinical Development of Neurosteroids as Neuroprotective Agents for the Treatment of Neurodegenerative Diseases
Paul A. Lapchak and Dalia M. Araujo

Clinical Implications of Circulating Neurosteroids
Andrea R. Genazzani, Patrizia Monteleone, Massimo Stomati, Francesca Bernardi, Luigi Cobellis, Elena Casarosa, Michele Luisi, Stefano Luisi, and Felice Petraglia

Neuroactive Steroids and Central Nervous System Disorders
Mingde Wang, Torbjörn Bäckström, Inger Sundström, Göran Wahlström, Tommy Olsson, Di Zhu, Inga-Maj Johansson, Inger Björn, and Marie Bixo

Neuroactive Steroids in Neuropsychopharmacology
Rainer Rupprecht and Florian Holsboer

Current Perspectives on the Role of Neurosteroids in PMS and Depression
Lisa D. Griffin, Susan C. Conrad, and Synthia H. Mellon

INDEX

Volume 47

Introduction: Studying Gene Expression in Neural Tissues by *in Situ* Hybridization
W. Wisden and B. J. Morris

Part I: *In Situ* Hybridization with Radiolabelled Oligonucleotides

In Situ Hybridization with Oligonucleotide Probes
Wl. Wisden and B. J. Morris

Cryostat Sectioning of Brains
Victoria Revilla and Alison Jones

Processing Rodent Embryonic and Early Postnatal Tissue for *in Situ* Hybridization with Radiolabelled Oligonucleotides
David J. Laurie, Petra C. U. Schrotz, Hannah Monyer, and Ulla Amtmann

Processing of Retinal Tissue for *in Situ* Hybridization
Frank Müller

Processing the Spinal Cord for *in Situ* Hybridization with Radiolabelled Oligonucleotides
A. Berthele and T. R. Tölle

Processing Human Brain Tissue for *in Situ* Hybridization with Radiolabelled Oligonucleotides
Louise F. B. Nicholson

In Situ Hybridization of Astrocytes and Neurons Cultured *in Vitro*
L. A. Arizza-McNaughton, C. De Felipe, and S. P. Hunt

In Situ Hybridization on Organotypic Slice Cultures
A. Gerfin-Moser and H. Monyer

Quantitative Analysis of *in Situ* Hybridization Histochemistry
Andrew L. Gundlach and Ross D. O'Shea

Part II: Nonradioactive *in Situ* hybridization

Nonradioactive *in Situ* Hybridization Using Alkaline Phosphatase-Labelled Oligonucleotides
S. J. Augood, E. M. McGowan, B. R. Finsen, B. Heppelmann, and P. C. Emson

Combining Nonradioactive *in Situ* Hybridization with Immunohistological and Anatomical Techniques
Petra Wahle

Nonradioactive *in Situ* Hybridization: Simplified Procedures for Use in Whole Mounts of Mouse and Chick Embryos
Linda Ariza-McNaughton and Robb Krumlauf

INDEX

Volume 48

Assembly and Intracellular Trafficking of $GABA_A$ Receptors Eugene
Barnes

Subcellular Localization and Regulation of $GABA_A$ Receptors and Associated Proteins Bernhard Lüscher and Jean-Marc Fritschy D_1 Dopamine Receptors
Richard Mailman

Molecular Modeling of Ligand-Gated Ion Channels: Progress and Challenges
Ed Bertaccini and James R. Trudel

Alzheimer's Disease: Its Diagnosis and Pathogenesis
Jillian J. Kril and Glenda M. Halliday

DNA Arrays and Functional Genomics in Neurobiology
Christelle Thibault, Long Wang, Li Zhang, and Michael F. Miles

INDEX

Volume 49

What Is West Syndrome?
Olivier Dulac, Christine Soufflet, Catherine Chiron, and Anna Kaminski

The Relationship between encephalopathy and Abnormal Neuronal Activity in the Developing Brain
Frances E. Jensen

Hypotheses from Functional Neuroimaging Studies
Csaba Juhász, Harry T. Chugani, Ouo Muzik, and Diane C. Chugani

Infantile Spasms: Unique Sydrome or General Age-Dependent Manifestation of a Diffuse Encephalopathy?
M. A. Koehn and M. Duchowny

Histopathology of Brain Tissue from Patients with Infantile Spasms
Harry V. Vinters

Generators of Ictal and Interictal Electroencephalograms Associated with Infantile Spasms: Intracellular Studies of Cortical and Thalamic Neurons
M. Steriade and I. Timofeev

Cortical and Subcortical Generators of Normal and Abnormal Rhythmicity
David A. McCormick

Role of Subcortical Structures in the Pathogenesis of Infantile Spasms: What Are Possible Subcortical Mediators?
F. A. Lado and S. L. Moshé

What Must We Know to Develop Better Therapies?
Jean Aicardi

The Treatment of Infantile Spasms: An Evidence-Based Approach
Mark Mackay, Shelly Weiss, and O. Carter Snead III

ACTH Treatment of Infantile Spasms: Mechanisms of Its Effects in Modulation of Neuronal Excitability
K. L. Brunson, S. Avishai-Eliner, and T. Z. Baram

Neurosteroids and Infantile Spasms: The Deoxycorticosterone Hypothesis
Michael A. Rogawski and Doodipala S. Reddy

Are there Specific Anatomical and/or Transmitter Systems (Cortical or Subcortical) That Should Be Targeted?
Phillip C. Jobe

Medical versus Surgical Treatment: Which Treatment When
W. Donald Shields

Developmental Outcome with and without Successful Intervention
Rochelle Caplan, Prabha Siddarth, Gary Mathern, Harry Vinters, Susan Curtiss, Jennifer Levitt, Robert Asarnow, and W. Donald Shields

Infantile Spasms versus Myoclonus: Is There a Connection?
Michael R. Pranzatelli

Tuberous Sclerosis as an Underlying Basis for Infantile Spasm
Raymond S. Yeung

Brain Malformation, Epilepsy, and Infantile Spasms
M. Elizabeth Ross

Brain Maturational Aspects Relevant to Pathophysiology of Infantile Spasms
G. Auanzini, F. Panzica, and S. Franceschetti

Gene Expression Analysis as a Strategy to Understand the Molecular Pathogenesis of Infantile Spasms
Peter B. Crino

Infantile Spasms: Criteria for an Animal Model
Carl E. Stafstrom and Gregory L. Holmes

INDEX

Volume 50

Part I: Primary Mechanisms

How Does Glucose Generate Oxidative Stress In Peripheral Nerve?
Irina G. Obrosova

Glycation in Diabetic Neuropathy: Characteristics, Consequences, Causes, and Therapeutic Options
Paul J. Thornalley

Part II: Secondary Changes

Protein Kinase C Changes in Diabetes: Is the Concept Relevant to Neuropathy?
Joseph Eichberg

Are Mitogen-Activated Protein Kinases Glucose Transducers for Diabetic Neuropathies?
Tertia D. Purves and David R. Tomlinson

Neurofilaments in Diabetic Neuropathy
Paul Fernyhough and Robert E. Schmidt

Apoptosis in Diabetic Neuropathy
Aviva Tolkovsky

Nerve and Ganglion Blood Flow in Diabetes: An Appraisal
Douglas W. Zochodne

Part III: Manifestations

Potential Mechanisms of Neuropathic Pain in Diabetes
Nigel A. Calcutt

Electrophysiologic Measures of Diabetic Neuropathy: Mechanism and Meaning
Joseph C. Arezzo and Elena Zotova

Neuropathology and Pathogenesis of Diabetic Autonomic Neuropathy
Robert E. Schmidt

Role of the Schwann Cell in Diabetic Neuropathy
Luke Eckersley

Part IV: Potential Treatment

Polyol Pathway and Diabetic Peripheral Neuropathy
Peter J. Oates

Nerve Growth Factor for the Treatment of Diabetic Neuropathy: What Went Wrong, What Went Right, and What Does the Future Hold?
Stuart C. Apfel

Angiotensin-Converting Enzyme Inhibitors: Are there Credible Mechanisms for Beneficial Effects in Diabetic Neuropathy?
Rayaz A. Malik and David R. Tomlinson

Clinical Trials for Drugs Against Diabetic Neuropathy: Can We Combine Scientific Needs With Clinical Practicalities?
Dan Ziegler and Dieter Luft

INDEX

Volume 50

Energy Metabolism in the Brain
Leif Hertz and Gerald A. Dienel

The Cerebral Glucose-Fatty Acid Cycle: Evolutionary Roots, Regulation, and (Patho) physiological Importance
Kurt Heininger

Expression, Regulation, and Functional Role of Glucose Transporters (GLUTs) in Brain
Donard S. Dwyer, Susan J. Vannucci, and Ian A. Simpson

Insulin-Like Growth Factor-1 Promotes Neuronal Glucose Utilization During Brain Development and Repair Processes
Carolyn A. Bondy and Clara M. Cheng

CNS Sensing and Regulation of Peripheral Glucose Levels
Barry E. Levin, Ambrose A. Dunn-Meynell, and Vanessa H. Routh

Glucose Transporter Protein Syndromes
Darryl C. De Vivo, Dong Wang, Juan M. Pascual, and Yuan Yuan Ho

Glucose, Stress, and Hippocampal Neuronal Vulnerability
Lawrence P. Reagan

Glucose/Mitochondria in Neurological Conditions
John P. Blass

Energy Utilization in the Ischemic/Reperfused Brain
John W. Phillis and Michael H. O'Regan

Diabetes Mellitus and the Central Nervous System
Anthony L. McCall

Diabetes, the Brain, and Behavior: Is There a Biological Mechanism Underlying the Association between Diabetes and Depression?
A. M. Jacobson, J. A. Samson, K. Weinger, and C. M. Ryan

Schizophrenia and Diabetes
David C. Henderson and Elissa R. Ettinger

Psychoactive Drugs Affect Glucose Transport and the Regulation of Glucose Metabolism
Donard S. Dwyer, Timothy D. Ardizzone, and Ronald J. Bradley

INDEX

Volume 52

Neuroimmune Relationships in Perspective
Frank Hucklebridge and Angela Clow

Sympathetic Nervous System Interaction with the Immune System
Virginia M. Sanders and Adam P. Kohm

Mechanisms by Which Cytokines Signal the Brain
Adrian J. Dunn

Neuropeptides: Modulators of Immune Responses in Health and Disease
David S. Jessop

Brain–Immune Interactions in Sleep
Lisa Marshall and Jan Born

Neuroendocrinology of Autoimmunity
Michael Harbuz

Systemic Stress-Induced Th2 Shift and Its Clinical Implications
Ibia J. Elenkov

Neural Control of Salivary S-IgA Secretion
Gordon B. Proctor and Guy H. Carpenter

Stress and Secretory Immunity
Jos A. Bosch, Christopher Ring, Eco J. C. de Geus, Enno C. I. Veerman, and Arie V. Nieuw Amerongen

Cytokines and Depression
Angela Clow

Immunity and Schizophrenia: Autoimmunity, Cytokines, and Immune Responses
Fiona Gaughran

Cerebral Lateralization and the Immune System
Pierre J. Neveu

Behavioral Conditioning of the Immune System
Frank Hucklebridge

Psychological and Neuroendocrine Correlates of Disease Progression
Julie M. Turner-Cobb

The Role of Psychological Intervention in Modulating Aspects of Immune Function in Relation to Health and Well-Being
J. H. Gruzelier

INDEX

Volume 53

Section I: Mitochondrial Structure and Function

Mitochondrial DNA Structure and Function
Carlos T. Moraes, Sarika Srivastava, Ilias Kirkinezos, Jose Oca-Cossio, Corina van Waveren, Markus Woischnick, and Francisca Diaz

Oxidative Phosphorylation: Structure, Function, and Intermediary Metabolism
Simon J. R. Heales, Matthew E. Gegg, and John B. Clark

Import of Mitochondrial Proteins
Matthias F. Bauer, Sabine Hofmann, and Walter Neupert

Section II: Primary Respiratory Chain Disorders

Mitochondrial Disorders of the Nervous System: Clinical, Biochemical, and Molecular Genetic Features
Dominic Thyagarajan and Edward Byrne

Section III: Secondary Respiratory Chain Disorders

Friedreich's Ataxia
J. M. Cooper and J. L. Bradley

Wilson Disease
C. A. Davie and A. H. V. Schapira

Hereditary Spastic Paraplegia
Christopher J. McDermott and Pamela J. Shaw

Cytochrome c Oxidase Deficiency
Giacomo P. Comi, Sandra Strazzer, Sara Galbiati, and Nereo Bresolin

Section IV: Toxin Induced Mitochondrial Dysfunction

Toxin-Induced Mitochondrial Dysfunction
Susan E. Browne and M. Flint Beal

Section V: Neurodegenerative Disorders

Parkinson's Disease
L. V. P. Korlipara and A. H. V. Schapira

Huntington's Disease: The Mystery Unfolds?
Åsa Petersén and Patrik Brundin

Mitochondria in Alzheimer's Disease
Russell H. Swerdlow and Stephen J. Kish

Contributions of Mitochondrial Alterations, Resulting from Bad Genes and a Hostile Environment, to the Pathogenesis of Alzheimer's Disease
Mark P. Mattson

Mitochondria and Amyotrophic Lateral Sclerosis
Richard W. Orrell and Anthony H. V. Schapira

Section VI: Models of Mitochondrial Disease

Models of Mitochondrial Disease
Danae Liolitsa and Michael G. Hanna

Section VII: Defects of β Oxidation Including Carnitine Deficiency

Defects of β Oxidation Including Carnitine Deficiency
K. Bartlett and M. Pourfarzam

Section VIII: Mitochondrial Involvement in Aging

The Mitochondrial Theory of Aging: Involvement of Mitochondrial DNA Damage and Repair
Nadja C. de Souza-Pinto and Vilhelm A. Bohr

INDEX

Volume 54

Unique General Anesthetic Binding Sites Within Distinct Conformational States of the Nicotinic Acetylcholine Receptor
Hugo R. Ariaas, William, R. Kem, James R. Truddell, and Michael P. Blanton

Signaling Molecules and Receptor Transduction Cascades That Regulate NMDA Receptor-Mediated Synaptic Transmission
Suhas. A. Kotecha and John F. MacDonald

Behavioral Measures of Alcohol Self-Administration and Intake Control: Rodent Models
Herman H. Samson and Cristine L. Czachowski

Dopaminergic Mouse Mutants: Investigating the Roles of the Different Dopamine Receptor Subtypes and the Dopamine Transporter
Shirlee Tan, Bettina Hermann, and Emiliana Borrelli

Drosophila melanogaster, A Genetic Model System for Alcohol Research
Douglas J. Guarnieri and Ulrike Heberlein

INDEX

Volume 55

Section I: Virsu Vectors For Use in the Nervous System

Non-Neurotropic Adenovirus: a Vector for Gene Transfer to the Brain and Gene Therapy of Neurological Disorders
P. R. Lowenstein, D. Suwelack, J. Hu, X. Yuan, M. Jimenez-Dalmaroni, S. Goverdhama, and M.G. Castro

Adeno-Associated Virus Vectors
E. Lehtonen and L. Tenenbaum

Problems in the Use of Herpes Simplex Virus as a Vector
L. T. Feldman

Lentiviral Vectors
J. Jakobsson, C. Ericson, N. Rosenquist, and C. Lundberg

Retroviral Vectors for Gene Delivery to Neural Precursor Cells
K. Kageyama, H. Hirata, and J. Hatakeyama

Section II: Gene Therapy with Virus Vectors for Specific Disease of the Nervous System

The Principles of Molecular Therapies for Glioblastoma
G. Karpati and J. Nalbatonglu

Oncolytic Herpes Simplex Virus
J. C. C. Hu and R. S. Coffin

Recombinant Retrovirus Vectors for Treatment of Brain Tumors
N. G. Rainov and C. M. Kramm

Adeno-Associated Viral Vectors for Parkinson's Disease
I. Muramatsu, L. Wang, K. Ikeguchi, K-i Fujimoto, T. Okada, H. Mizukami, Y. Hanazono, A. Kume, I. Nakano, and K. Ozawa

HSV Vectors for Parkinson's Disease
D. S. Latchman

Gene Therapy for Stroke
K. Abe and W. R. Zhang

Gene Therapy for Mucopolysaccharidosis
A. Bosch and J. M. Heard

INDEX

Volume 56

Behavioral Mechanisms and the Neurobiology of Conditioned Sexual Responding
Mark Krause

NMDA Receptors in Alcoholism
Paula L. Hoffman

Processing and Representation of Species-Specific Communication Calls in the Auditory System of Bats
George D. Pollak, Achim Klug, and Eric E. Bauer

Central Nervous System Control of Micturition
Gert Holstege and Leonora J. Mouton

The Structure and Physiology of the Rat Auditory System: An Overview
Manuel Malmierca

Neurobiology of Cat and Human Sexual Behavior
Gert Holstege and J. R. Georgiadis

INDEX

Volume 57

Cumulative Subject Index of Volumes 1–25

Volume 58

Cumulative Subject Index of Volumes 26–50

Volume 59

Loss of Spines and Neuropil
Liesl B. Jones

Schizophrenia as a Disorder of Neuroplasticity
Robert E. McCullumsmith, Sarah M. Clinton, and James H. Meador-Woodruff

The Synaptic Pathology of Schizophrenia: Is Aberrant Neurodevelopment and Plasticity to Blame?
Sharon L. Eastwood

Neurochemical Basis for an Epigenetic Vision of Synaptic Organization
E. Costa, D. R. Grayson, M. Veldic, and A. Guidotti

Muscarinic Receptors in Schizophrenia: Is There a Role for Synaptic Plasticity?
Thomas J. Raedler

Serotonin and Brain Development
Monsheel S. K. Sodhi and Elaine Sanders-Bush

Presynaptic Proteins and Schizophrenia
William G. Honer and Clint E. Young

Mitogen-Activated Protein Kinase Signaling
Svetlana V. Kyosseva

Postsynaptic Density Scaffolding Proteins at Excitatory Synapse and Disorders of Synaptic Plasticity: Implications for Human Behavior Pathologies
Andrea de Bartolomeis and Germano Fiore

Prostaglandin-Mediated Signaling in Schizophrenia
S. Smesny

Mitochondria, Synaptic Plasticity, and Schizophrenia
Dorit Ben-Shachar and Daphna Laifenfeld

Membrane Phospholipids and Cytokine Interaction in Schizophrenia
Jeffrey K. Yao and Daniel P. van Kammen

Neurotensin, Schizophrenia, and Antipsychotic Drug Action
Becky Kinkead and Charles B. Nemeroff

Schizophrenia, Vitamin D, and Brain Development
Alan Mackay-Sim, François Féron, Darryl Eyles, Thomas Burne, and John McGrath

Possible Contributions of Myelin and Oligodendrocyte Dysfunction to Schizophrenia
Daniel G. Stewart and Kenneth L. Davis

Brain-Derived Neurotrophic Factor and the Plasticity of the Mesolimbic Dopamine Pathway
Oliver Guillin, Nathalie Griffon, Jorge Diaz, Bernard Le Foll, Erwan Bezard, Christian Gross, Chris Lammers, Holger Stark, Patrick Carroll, Jean-Charles Schwartz, and Pierre Sokoloff

S100B in Schizophrenic Psychosis
Matthias Rothermundt, Gerald Ponath, and Volker Arolt

Oct-6 Transcription Factor
Maria Ilia

NMDA Receptor Function, Neuroplasticity, and the Pathophysiology of Schizophrenia
Joseph T. Coyle and Guochuan Tsai

INDEX

Volume 60

Microarray Platforms: Introduction and Application to Neurobiology
Stanislav L. Karsten, Lili C. Kudo, and Daniel H. Geschwind

Experimental Design and Low-Level Analysis of Microarray Data
B. M. Bolstad, F. Collin, K. M. Simpson, R. A. Irizarry, and T. P. Speed

Brain Gene Expression: Genomics and Genetics
Elissa J. Chesler and Robert W. Williams

DNA Microarrays and Animal Models of Learning and Memory
Sebastiano Cavallaro

Microarray Analysis of Human Nervous System Gene Expression in Neurological Disease
Steven A. Greenberg

DNA Microarray Analysis of Postmortem Brain Tissue
Károly Mirnics, Pat Levitt, and David A. Lewis

INDEX

Volume 61

Section I: High-Throughput Technologies

Biomarker Discovery Using Molecular Profiling Approaches
Stephen J. Walker and Arron Xu

Proteomic Analysis of Mitochondrial Proteins
Mary F. Lopez, Simon Melov, Felicity Johnson, Nicole Nagulko, Eva Golenko, Scott Kuzdzal, Suzanne Ackloo, and Alvydas Mikulskis

Section II: Proteomic Applications

NMDA Receptors, Neural Pathways, and Protein Interaction Databases
Holger Husi

Dopamine Transporter Network and Pathways
Rajani Maiya and R. Dayne Mayfield

Proteomic Approaches in Drug Discovery and Development
Holly D. Soares, Stephen A. Williams,

Peter J. Snyder, Feng Gao, Tom Stiger, Christian Rohlff, Athula Herath, Trey Sunderland, Karen Putnam, and W. Frost White

Section III: Informatics

Proteomic Informatics
Steven Russell, William Old, Katheryn Resing, and Lawrence Hunter

Section IV: Changes in the Proteome by Disease

Proteomics Analysis in Alzheimer's Disease: New Insights into Mechanisms of Neurodegeneration
D. Allan Butterfield and Debra Boyd-Kimball

Proteomics and Alcoholism
Frank A. Witzmann and Wendy N. Strother

Proteomics Studies of Traumatic Brain Injury
Kevin K. W. Wang, Andrew Ottens, William Haskins, Ming Cheng Liu, Firas Kobeissy, Nancy Denslow, SuShing Chen, and Ronald L. Hayes

Influence of Huntington's Disease on the Human and Mouse Proteome
Claus Zabel and Joachim Klose

Section V: Overview of the Neuroproteome

Proteomics—Application to the Brain
Katrin Marcus, Oliver Schmidt, Heike Schaefer, Michael Hamacher, AndrÅ van Hall, and Helmut E. Meyer

INDEX

Volume 62

$GABA_A$ Receptor Structure–Function Studies: A Reexamination in Light of New Acetylcholine Receptor Structures
Myles H. Akabas

Dopamine Mechanisms and Cocaine Reward
Aiko Ikegami and Christine L. Duvauchelle

Proteolytic Dysfunction in Neurodegenerative Disorders
Kevin St. P. McNaught

Neuroimaging Studies in Bipolar Children and Adolescents
Rene L. Olvera, David C. Glahn, Sheila C. Caetano, Steven R. Pliszka, and Jair C. Soares

Chemosensory G-Protein-Coupled Receptor Signaling in the Brain
Geoffrey E. Woodard

Disturbances of Emotion Regulation after Focal Brain Lesions
Antoine Bechara

The Use of *Caenorhabditis elegans* in Molecular Neuropharmacology
Jill C. Bettinger, Lucinda Carnell, Andrew G. Davies, and Steven L. McIntire

INDEX

Volume 63

Mapping Neuroreceptors at work: On the Definition and Interpretation of Binding Potentials after 20 years of Progress
Albert Gjedde, Dean F. Wong, Pedro Rosa-Neto, and Paul Cumming

Mitochondriál Dysfunction in Bipolar Disorder: From ^{31}P-Magnetic Resonance Spectroscopic Findings to Their Molecular Mechanisms
Tadafumi Kato

Large-Scale Microarray Studies of Gene Expression in Multiple Regions of the Brain in Schizophrenia and Alzeimer's Disease
Pavel L. Katsel, Kenneth L. Davis, and Vahram Haroutunian

Regulation of Serotonin 2C Receptor PRE-mRNA Editing By Serotonin
Claudia Schmauss

The Dopamine Hypothesis of Drug Addiction: Hypodopaminergic State
Miriam Melis, Saturnino Spiga, and Marco Diana

Human and Animal Spongiform Encephalopathies are Autoimmune Diseases: A Novel Theory and Its supporting Evidence
Bao Ting Zhu

Adenosine and Brain Function
Bertil B. Fredholm, Jiang-Fan Chen, Rodrigo A. Cunha, Per Svenningsson, and Jean-Marie Vaugeois

INDEX

Volume 64

Section I. The Cholinergic System
John Smythies

Section II. The Dopamine System
John Symythies

Section III. The Norepinephrine System
John Smythies

Section IV. The Adrenaline System
John Smythies

Section V. Serotonin System
John Smythies

INDEX

Volume 65

Insulin Resistance: Causes and Consequences
Zachary T. Bloomgarden

Antidepressant-Induced Manic Conversion: A Developmentally Informed Synthesis of the Literature
Christine J. Lim, James F. Leckman, Christopher Young, and Andrés Martin

Sites of Alcohol and Volatile Anesthetic Action on Glycine Receptors
Ingrid A. Lobo and R. Adron Harris

Role of the Orbitofrontal Cortex in Reinforcement Processing and Inhibitory Control: Evidence from Functional Magnetic Resonance Imaging Studies in Healthy Human Subjects
Rebecca Elliott and Bill Deakin

Common Substrates of Dysphoria in Stimulant Drug Abuse and Primary Depression: Therapeutic Targets
Kate Baicy, Carrie E. Bearden, John Monterosso, Arthur L. Brody, Andrew J. Isaacson, and Edythe D. London

The Role of cAMP Response Element–Binding Proteins in Mediating Stress-Induced Vulnerability to Drug Abuse
Arati Sadalge Kreibich and Julie A. Blendy

G-Protein–Coupled Receptor Deorphanizations
Yumiko Saito and Olivier Civelli

Mechanistic Connections Between Glucose/Lipid Disturbances and Weight Gain Induced by Antipsychotic Drugs
Donard S. Dwyer, Dallas Donohoe, Xiao-Hong Lu, and Eric J. Aamodt

Serotonin Firing Activity as a Marker for Mood Disorders: Lessons from Knockout Mice
Gabriella Gobbi

INDEX

Volume 66

Brain Atlases of Normal and Diseased Populations
Arthur W. Toga and Paul M. Thompson

Neuroimaging Databases as a Resource for Scientific Discovery
John Darrell Van Horn, John Wolfe, Autumn Agnoli, Jeffrey Woodward, Michael Schmitt, James Dobson, Sarene Schumacher, and Bennet Vance

Modeling Brain Responses
Karl J. Friston, William Penny, and Olivier David

Voxel-Based Morphometric Analysis Using Shape Transformations
Christos Davatzikos

The Cutting Edge of fMRI and High-Field fMRI
Dae-Shik Kim

Quantification of White Matter Using Diffusion-Tensor Imaging
Hae-Jeong Park

Perfusion fMRI for Functional Neuroimaging
Geoffrey K. Aguirre, John A. Detre, and Jiongjiong Wang

Functional Near-Infrared Spectroscopy: Potential and Limitations in Neuroimaging Studies
Yoko Hoshi

Neural Modeling and Functional Brain Imaging: The Interplay Between the Data-Fitting and Simulation Approaches
Barry Horwitz and Michael F. Glabus

Combined EEG and fMRI Studies of Human Brain Function
V. Menon and S. Crottaz-Herbette

INDEX

Volume 67

Distinguishing Neural Substrates of Heterogeneity Among Anxiety Disorders
Jack B. Nitschke and Wendy Heller

Neuroimaging in Dementia
K. P. Ebmeier, C. Donaghey, and N. J. Dougall

Prefrontal and Anterior Cingulate Contributions to Volition in Depression
Jack B. Nitschke and Kristen L. Mackiewicz

Functional Imaging Research in Schizophrenia
H. Tost, G. Ende, M. Ruf, F. A. Henn, and A. Meyer-Lindenberg

Neuroimaging in Functional Somatic Syndromes
Patrick B. Wood

Neuroimaging in Multiple Sclerosis
Alireza Minagar, Eduardo Gonzalez-Toledo, James Pinkston, and Stephen L. Jaffe

Stroke
Roger E. Kelley and Eduardo Gonzalez-Toledo

Functional MRI in Pediatric Neurobehavioral Disorders
Michael Seyffert and F. Xavier Castellanos

Structural MRI and Brain Development
Paul M. Thompson, Elizabeth R. Sowell, Nitin Gogtay, Jay N. Giedd, Christine N. Vidal, Kiralee M. Hayashi, Alex Leow, Rob Nicolson, Judith L. Rapoport, and Arthur W. Toga

Neuroimaging and Human Genetics
Georg Winterer, Ahmad R. Hariri, David Goldman, and Daniel R. Weinberger

Neuroreceptor Imaging in Psychiatry: Theory and Applications
W. Gordon Frankle, Mark Slifstein, Peter S. Talbot, and Marc Laruelle

INDEX

Volume 68

Fetal Magnetoencephalography: Viewing the Developing Brain *In Utero*
Hubert Preissl, Curtis L. Lowery, and Hari Eswaran

Magnetoencephalography in Studies of Infants and Children
Minna Huotilainen

Let's Talk Together: Memory Traces Revealed by Cooperative Activation in the Cerebral Cortex
Jochen Kaiser, Susanne Leiberg, and Werner Lutzenberger

Human Communication Investigated With Magnetoencephalography: Speech, Music, and Gestures
Thomas R. Knösche, Burkhard Maess, Akinori Nakamura, and Angela D. Friederici

Combining Magnetoencephalography and Functional Magnetic Resonance Imaging
Klaus Mathiak and Andreas J. Fallgatter

Beamformer Analysis of MEG Data
Arjan Hillebrand and Gareth R. Barnes

Functional Connectivity Analysis in Magnetoencephalography
Alfons Schnitzler and Joachim Gross

Human Visual Processing as Revealed by Magnetoencephalographys
Yoshiki Kaneoke, Shoko Watanabe, and Ryusuke Kakigi

A Review of Clinical Applications of Magnetoencephalography
Andrew C. Papanicolaou, Eduardo M. Castillo, Rebecca Billingsley-Marshall, Ekaterina Pataraia, and Panagiotis G. Simos

INDEX

Volume 69

Nematode Neurons: Anatomy and Anatomical Methods in Caenorhabditis elegans
David H. Hall, Robyn Lints, and Zeynep Altun

Investigations of Learning and Memory in Caenorhabditis elegans
Andrew C. Giles, Jacqueline K. Rose, and Catharine H. Rankin

Neural Specification and Differentiation
Eric Aamodt and Stephanie Aamodt

Sexual Behavior of the Caenorhabditis elegans Male
Scott W. Emmons

The Motor Circuit
Stephen E. Von Stetina, Millet Treinin, and David M. Miller III

Mechanosensation in Caenorhabditis elegans
Robert O'Hagan and Martin Chalfie

Volume 70

Spectral Processing by the Peripheral Auditory System Facts and Models
Enrique A. Lopez-Poveda

Basic Psychophysics of Human Spectral Processing
Brian C. J. Moore

Across-Channel Spectral Processing
John H. Grose, Joseph W. Hall III, and Emily Buss

Speech and Music Have Different Requirements for Spectral Resolution
Robert V. Shannon

Non-Linearities and the Representation of Auditory Spectra
Eric D. Young, Jane J. Yu, and Lina A. J. Reiss

Spectral Processing in the Inferior Colliculus
Kevin A. Davis

Neural Mechanisms for Spectral Analysis in the Auditory Midbrain, Thalamus, and Cortex
Monty A. Escab and Heather L. Read

Spectral Processing in the Auditory Cortex
Mitchell L. Sutter

Processing of Dynamic Spectral Properties of Sounds
Adrian Rees and Manuel S. Malmierca

Representations of Spectral Coding in the Human Brain
Deborah A. Hall, PhD

Spectral Processing and Sound Source Determination
Donal G. Sinex

Spectral Information in Sound Localization
Simon Carlile, Russell Martin, and Ken McAnally

Plasticity of Spectral Processing
Dexter R. F. Irvine and Beverly A. Wright

Spectral Processing In Cochlear Implants
Colette M. McKay

INDEX

Volume 71

Autism: Neuropathology, Alterations of the GABAergic System, and Animal Models
Christoph Schmitz, Imke A. J. van Kooten, Patrick R. Hof, Herman van Engeland, Paul H. Patterson, and Harry W. M. Steinbusch

The Role of GABA in the Early Neuronal Development
Marta Jelitai and Emı´lia Madarasz

GABAergic Signaling in the Developing Cerebellum
Chitoshi Takayama

Insights into GABA Functions in the Developing Cerebellum
Mo´nica L. Fiszman

Role of GABA in the Mechanism of the Onset of Puberty in Non-Human Primates
Ei Terasawa

Rett Syndrome: A Rosetta Stone for Understanding the Molecular Pathogenesis of Autism
Janine M. LaSalle, Amber Hogart, and Karen N. Thatcher

GABAergic Cerebellar System in Autism: A Neuropathological and Developmental Perspective
Gene J. Blatt

Reelin Glycoprotein in Autism and Schizophrenia
S. Hossein Fatemi

Is There A Connection Between Autism, Prader-Willi Syndrome, Catatonia, and GABA?
Dirk M. Dhossche, Yaru Song, and Yiming Liu

Alcohol, GABA Receptors, and Neurodevelopmental Disorders
Ujjwal K. Rout

Effects of Secretin on Extracellular GABA and Other Amino Acid Concentrations in the Rat Hippocampus
Hans-Willi Clement, Alexander Pschibul, and Eberhard Schulz

Predicted Role of Secretin and Oxytocin in the Treatment of Behavioral and Developmental Disorders: Implications for Autism
Martha G. Welch and David A. Ruggiero

Immunological Findings in Autism
Hari Har Parshad Cohly and Asit Panja

Correlates of Psychomotor Symptoms in Autism
Laura Stoppelbein, Sara Sytsma-Jordan, and Leilani Greening

GABRB3 Gene Deficient Mice: A Potential Model of Autism Spectrum Disorder
Timothy M. DeLorey

The Reeler Mouse: Anatomy of a Mutant
Gabriella D'Arcangelo

Shared Chromosomal Susceptibility Regions Between Autism and Other Mental Disorders
Yvon C. Chagnon index

INDEX

Volume 72

Classification Matters for Catatonia and Autism in Children
Klaus-Jürgen Neumärker

A Systematic Examination of Catatonia-Like Clinical Pictures in Autism Spectrum Disorders
Lorna Wing and Amitta Shah

Catatonia in Individuals with Autism Spectrum Disorders in Adolescence and Early Adulthood: A Long-Term Prospective Study
Masataka Ohta, Yukiko Kano, and Yoko Nagai

Are Autistic and Catatonic Regression Related? A Few Working Hypotheses Involving GABA, Purkinje Cell Survival, Neurogenesis, and ECT
Dirk Marcel Dhossche and Ujjwal Rout

Psychomotor Development and Psychopathology in Childhood
Dirk M. J. De Raeymaecker

The Importance of Catatonia and Stereotypies in Autistic Spectrum Disorders
Laura Stoppelbein, Leilani Greening, and Angelina Kakooza

Prader–Willi Syndrome: Atypical Psychoses and Motor Dysfunctions
Willem M. A. Verhoeven and Siegfried Tuinier

Towards a Valid Nosography and Psychopathology of Catatonia in Children and Adolescents
David Cohen

Is There a Common Neuronal Basis for Autism and Catatonia?
Dirk Marcel Dhossche, Brendan T. Carroll, and Tressa D. Carroll

Shared Susceptibility Region on Chromosome 15 Between Autism and Catatonia
Yvon C. Chagnon

Current Trends in Behavioral Interventions for Children with Autism
Dorothy Scattone and Kimberly R. Knight

Case Reports with a Child Psychiatric Exploration of Catatonia, Autism, and Delirium
Jan N. M. Schieveld

ECT and the Youth: Catatonia in Context
Frank K. M. Zaw

Catatonia in Autistic Spectrum Disorders: A Medical Treatment Algorithm
Max Fink, Michael A. Taylor, and Neera Ghaziuddin

Psychological Approaches to Chronic Catatonia-Like Deterioration in Autism Spectrum Disorders
Amitta Shah and Lorna Wing

Section V: Blueprints
Blueprints for the Assessment, Treatment, and Future Study of Catatonia in Autism Spectrum Disorders
Dirk Marcel, Dhossche, Amitta Shah, and Lorna Wing

INDEX

Volume 73

Chromosome 22 Deletion Syndrome and Schizophrenia
Nigel M. Williams, Michael C. O'Donovan, and Michael J. Owen

Characterization of Proteome of Human Cerebrospinal Fluid
Jing Xu, Jinzhi Chen, Elaine R. Peskind, Jinghua Jin, Jimmy Eng, Catherine Pan, Thomas J. Montine, David R. Goodlett, and Jing Zhang

Hormonal Pathways Regulating Intermale and Interfemale Aggression
Neal G. Simon, Qianxing Mo, Shan Hu, Carrie Garippa, and Shi-Fang Lu

Neuronal GAP Junctions: Expression, Function, and Implications for Behavior
Clinton B. McCracken and David C. S. Roberts

Effects of Genes and Stress on the Neurobiology of Depression
J. John Mann and Dianne Currier

Quantitative Imaging with the Micropet Small-Animal Pet Tomograph
Paul Vaska, Daniel J. Rubins, David L. Alexoff, and Wynne K. Schiffer

Understanding Myelination through Studying its Evolution
Rüdiger Schweigreiter, Betty I. Roots, Christine Bandtlow, and Robert M. Gould

INDEX

Volume 74

Evolutionary Neurobiology and Art
C. U. M. Smith

Section I: Visual Aspects

Perceptual Portraits
Nicholas Wade

The Neuropsychology of Visual Art: Conferring Capacity
Anjan Chatterjee

Vision, Illusions, and Reality
Christopher Kennard

Localization in the Visual Brain
George K. York

Section II: Episodic Disorders

Neurology, Synaesthesia, and Painting
Amy Ione

Fainting in Classical Art
Philip Smith

Migraine Art in the Internet: A Study of 450 Contemporary Artists
Klaus Podoll

Sarah Raphael's Migraine with Aura as Inspiration for the Foray of Her Work into Abstraction
Klaus Podoll and Debbie Ayles

The Visual Art of Contemporary Artists with Epilepsy
Steven C. Schachter

Section III: Brain Damage

Creativity in Painting and Style in Brain-Damaged Artists
Julien Bogousslavsky

Artistic Changes in Alzheimer's Disease
Sebastian J. Crutch and Martin N. Rossor

Section IV: Cerebrovascular Disease

Stroke in Painters
H. Bäzner and M. Hennerici

Visuospatial Neglect in Lovis Corinth's Self-Portraits
Olaf Blanke

Art, Constructional Apraxia, and the Brain
Louis Caplan

Section V: Genetic Diseases

Neurogenetics in Art
Alan E. H. Emery

A Naïve Artist of St Ives
F. Clifford Rose

Van Gogh's Madness
F. Clifford Rose

Absinthe, The Nervous System and Painting
Tiina Rekand

Section VI: Neurologists as Artists

Sir Charles Bell, KGH, FRS, FRSE (1774–1842)
Christopher Gardner-Thorpe

Section VII: Miscellaneous

Peg Leg Frieda
Espen Dietrichs

The Deafness of Goya (1746–1828)
F. Clifford Rose

INDEX

Volume 75

Introduction on the Use of the *Drosophila* Embryonic/Larval Neuromuscular Junction as a Model System to Study Synapse Development and Function, and a Brief Summary of Pathfinding and Target Recognition
Catalina Ruiz-Cañada and Vivian Budnik

Development and Structure of Motoneurons
Matthias Landgraf and Stefan Thor

The Development of the *Drosophila* Larval Body Wall Muscles
Karen Beckett and Mary K. Baylies

Organization of the Efferent System and Structure of Neuromuscular Junctions in *Drosophila*
Andreas Prokop

Development of Motoneuron Electrical Properties and Motor Output
Richard A. Baines

Transmitter Release at the Neuromuscular Junction
Thomas L. Schwarz

Vesicle Trafficking and Recycling at the Neuromuscular Junction: Two Pathways for Endocytosis
Yoshiaki Kidokoro

Glutamate Receptors at the *Drosophila* Neuromuscular Junction
Aaron DiAntonio

Scaffolding Proteins at the *Drosophila* Neuromuscular Junction
Bulent Ataman, Vivian Budnik, and Ulrich Thomas

Synaptic Cytoskeleton at the Neuromuscular Junction
Catalina Ruiz-Cañada and Vivian Budnik

Plasticity and Second Messengers During Synapse Development
Leslie C. Griffith and Vivian Budnik

Retrograde Signaling that Regulates Synaptic Development and Function at the *Drosophila* Neuromuscular Junction
Guillermo Marqués and Bing Zhang

Activity-Dependent Regulation of Transcription During Development of Synapses
Subhabrata Sanyal and Mani Ramaswami

Experience-Dependent Potentiation of Larval Neuromuscular Synapses
Christoph M. Schuster

Selected Methods for the Anatomical Study of *Drosophila* Embryonic and Larval Neuromuscular Junctions
Vivian Budnik, Michael Gorczyca, and Andreas Prokop

INDEX

Volume 76

Section I: Physiological Correlates of Freud's Theories

The ID, the Ego, and the Temporal Lobe
Shirley M. Ferguson and Mark Rayport

ID, Ego, and Temporal Lobe Revisited
Shirley M. Ferguson and Mark Rayport

Section II: Stereotaxic Studies

Olfactory Gustatory Responses Evoked by Electrical Stimulation of Amygdalar Region in Man Are Qualitatively Modifiable by Interview Content: Case Report and Review
Mark Rayport, Sepehr Sani, and Shirley M. Ferguson

Section III: Controversy in Definition of Behavioral Disturbance

Pathogenesis of Psychosis in Epilepsy. The "Seesaw" Theory: Myth or Reality?
Shirley M. Ferguson and Mark Rayport

Section IV: Outcome of Temporal Lobectomy

Memory Function After Temporal Lobectomy for Seizure Control: A Comparative Neuropsy chiatric and Neuropsychological Study
Shirley M. Ferguson, A. John McSweeny, and Mark Rayport

Life After Surgery for Temporolimbic Seizures
Shirley M. Ferguson, Mark Rayport, and Carolyn A. Schell

Appendix I
Mark Rayport

Appendix II: Conceptual Foundations of Studies of Patients Undergoing Temporal Lobe Surgery for Seizure Control
Mark Rayport

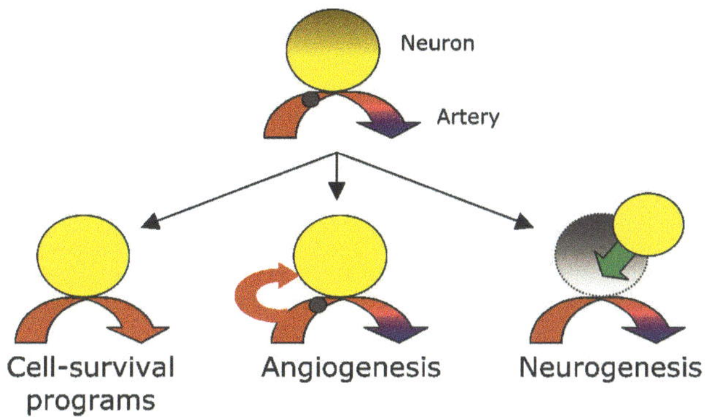

GREENBERG AND JIN, CHAPTER 1, FIG. 1. (See Legend in Text.)

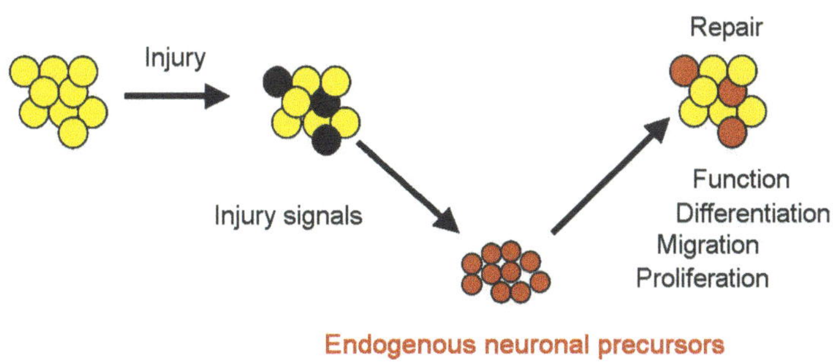

GREENBERG AND JIN, CHAPTER 1, FIG. 2. (See Legend in Text.)

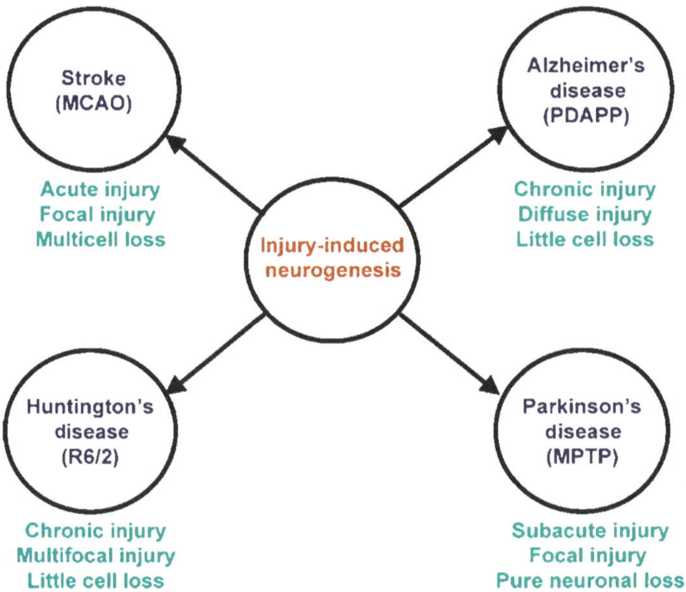

GREENBERG AND JIN, CHAPTER 1, FIG. 3. (See Legend in Text.)

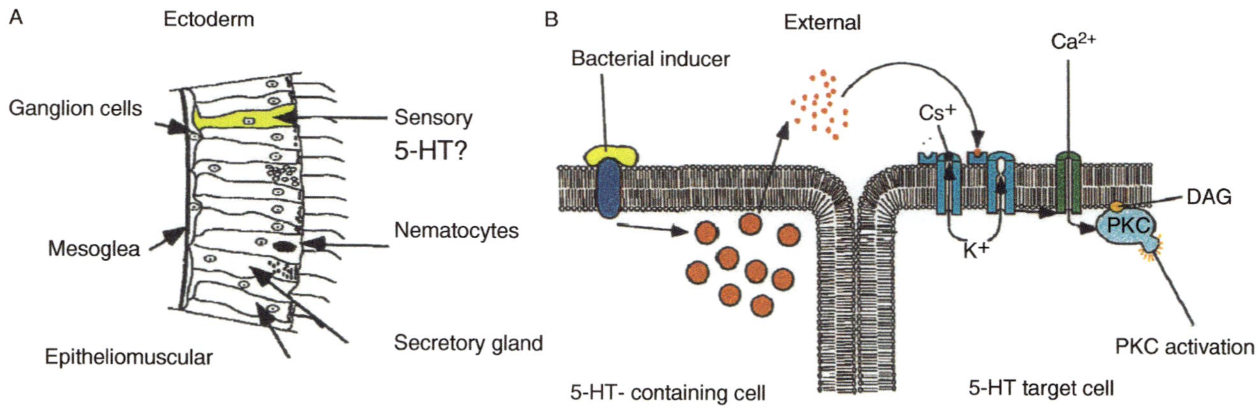

AZMITIA, CHAPTER 2, FIG. 3. (See Legend in Text.)

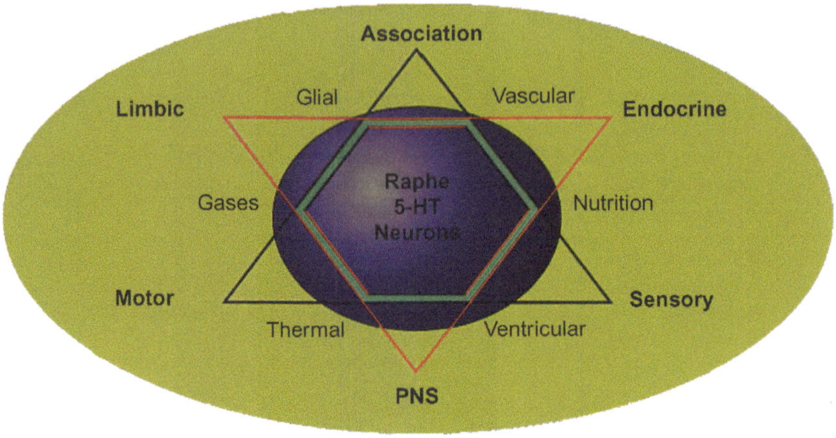

AZMITIA, CHAPTER 2, FIG. 6. (See Legend in Text.)

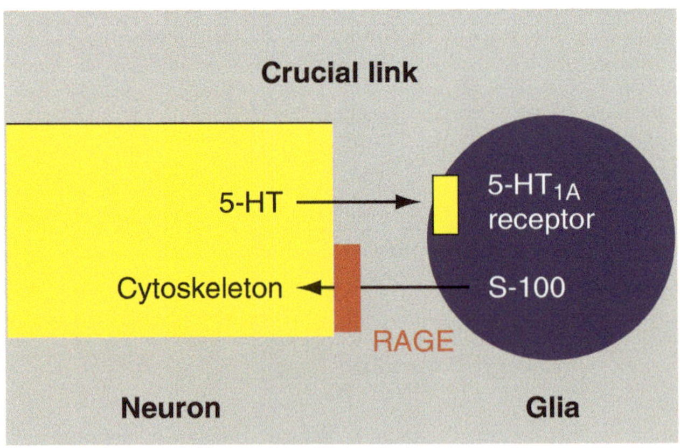

AZMITIA, CHAPTER 2, FIG. 7. (See Legend in Text.)

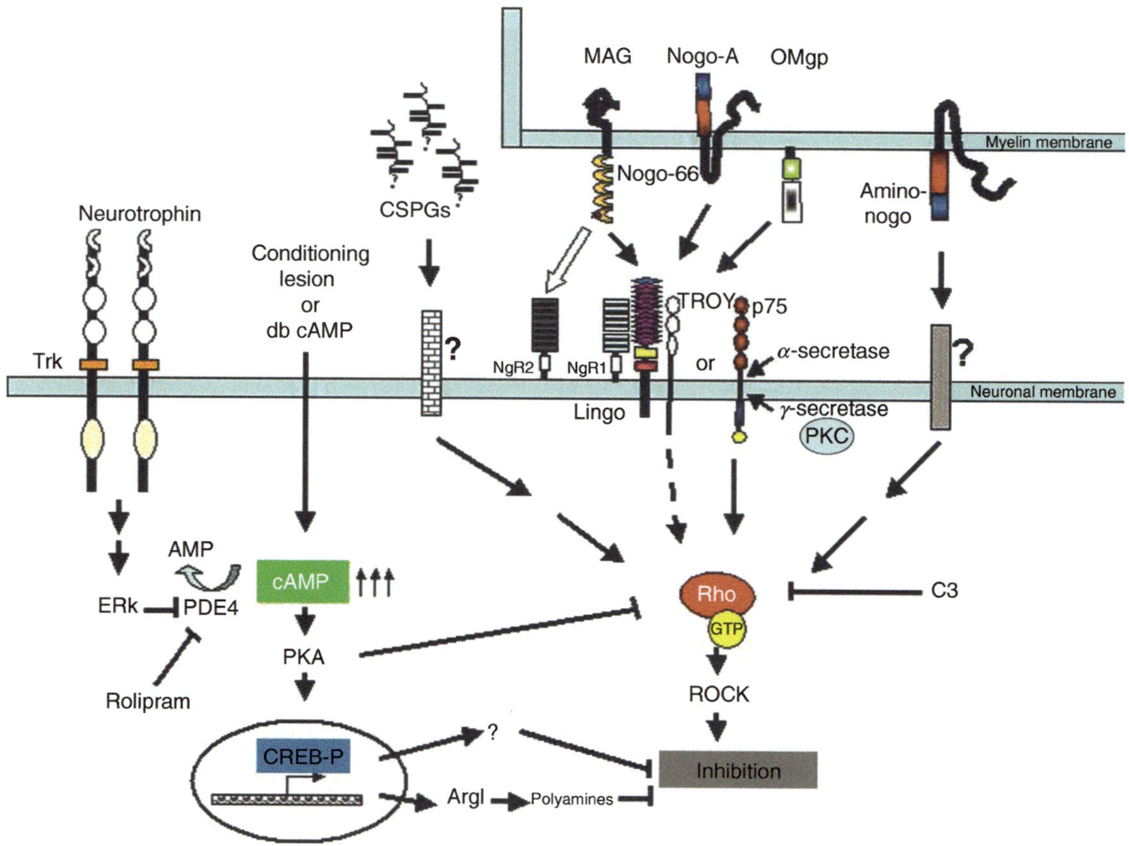

HANNILA *ET AL.*, CHAPTER 3, FIG. 2. (See Legend in Text.)

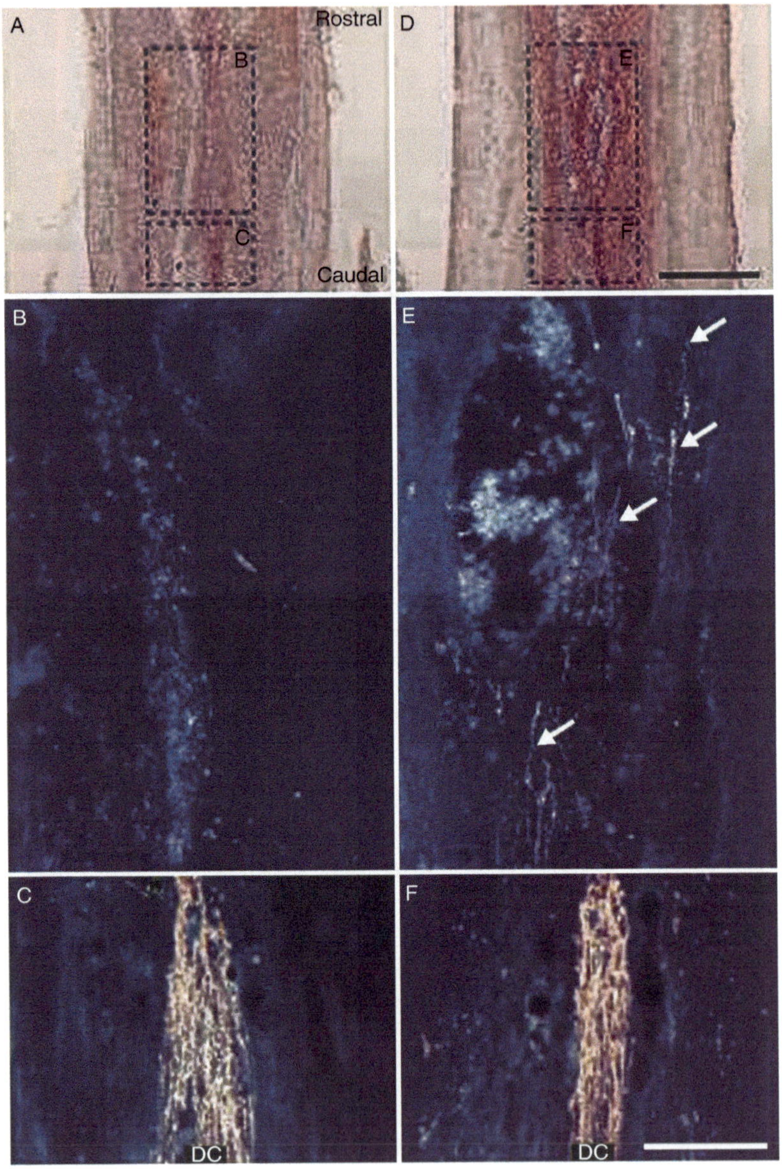

HANNILA *ET AL.*, CHAPTER 3, FIG. 3. (See Legend in Text.)

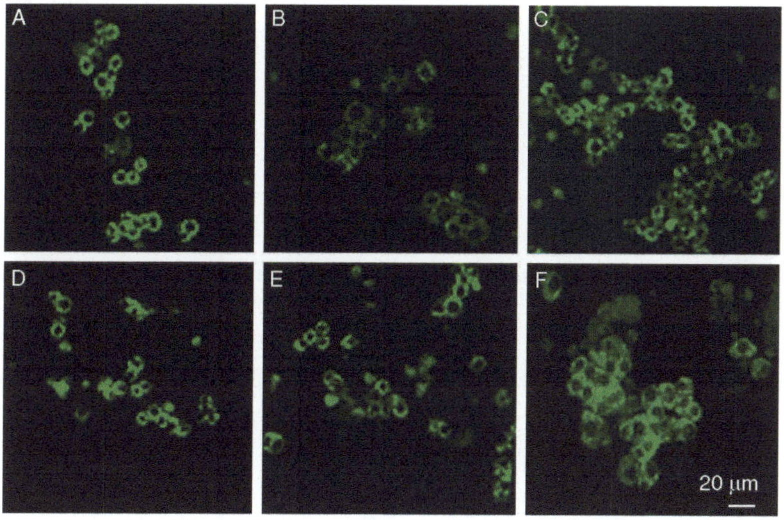

LI AND XU, CHAPTER 4, FIG. 6. (See Legend in Text.)

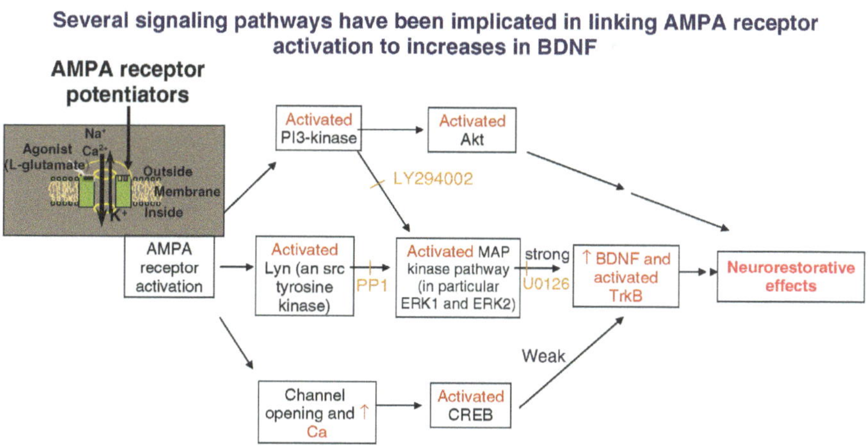

O'NEILL *ET AL.*, CHAPTER 6, FIG. 2. (See Legend in Text.)

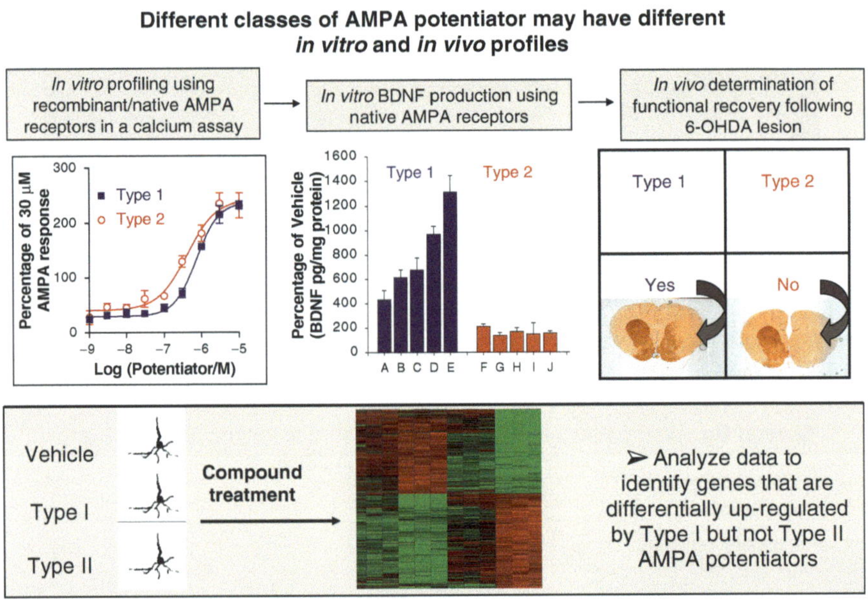

O'Neill *et al.*, Chapter 6, Fig. 3. (See Legend in Text.)

Kolibri 360 GmbH
Ferdinand-Jühlke-Straße 7
99095 Erfurt, Deutschland
produktsicherheit@kolibri360.de